AAPS Advances in the Pharmaceutical Sciences Series

The AAPS Advances in the Pharmaceutical Sciences Series, published in partnership with the American Association of Pharmaceutical Scientists, is designed to deliver well written volumes authored by opinion leaders and authorities from around the globe, addressing innovations in drug research and development, and best practice for scientists and industry professionals in the pharma and biotech industries. For more details and to see a list of titles in the Series please visit http://www.springer.com/series/8825

More information about this series at http://www.springer.com/series/8825

Amy Rosenberg • Barthélemy Demeule
Editors

Biobetters

Protein Engineering to Approach the Curative

Springer

Editors
Amy Rosenberg
Division of Biotechnology Review
and Research III
Office of Biotechnology Products
Center for Drug Evaluation and Research
Food and Drug Administration
Silver Spring, MD, USA

Barthélemy Demeule
Genentech, Inc.
South San Francisco, CA, USA

ISSN 2210-7371 ISSN 2210-738X (electronic)
AAPS Advances in the Pharmaceutical Sciences Series
ISBN 978-1-4939-2542-1 ISBN 978-1-4939-2543-8 (eBook)
DOI 10.1007/978-1-4939-2543-8

Library of Congress Control Number: 2015939911

Springer New York Heidelberg Dordrecht London

Printed on acid-free paper

Springer Science+Business Media LLC New York is part of Springer Science+Business Media (www.springer.com)

Preface

> ...and what never frees us from the cost of knowledge, which is to act on what we know again and again
>
> Marge Piercy, American poet

Curative or preventive therapies for human and animal diseases would provide optimal benefit, not only to individual patients in terms of quality and quantity of life but as well to the healthcare system in minimizing the need for chronic and often very expensive treatments and supportive measures. Preventive therapies in the form of vaccines have been termed our "most effective public health measures,[1]" yet their development provides limited economic incentive. Other preventive therapies may be restricted to small subsets of patients who have a relatively high probability of manifesting disease, such as prophylactic mastectomy/oophorectomy in patients with known high-risk mutations for breast and ovarian cancers (e.g., BRCA mutation positive patients). While these measures are critically important to the patients, the savings to the healthcare system are much more modest. In contrast to therapeutic proteins such as monoclonal antibodies that address peripheral mediators of disease (e.g., TNF inhibitors), curative therapies, aimed at the underlying pathophysiology of the disease (e.g., RA), remain elusive despite renewed emphasis.[2] These unfortunate circumstances highlight the need for optimal, not just effective, therapeutics to bridge the gap between an effective but expensive and chronically administered therapeutic and a curative therapy for diverse clinical entities. Thus, while the approval of a new therapeutic protein for an unmet clinical need is always an important advance, once approved, there appears to be limited impetus to improve the clinical performance of the initial product even though the means and ways to do so may be known and feasible. The advent of "biosimilars" and their economic impact (captured in the chapter by Berndt et al.) will hopefully change this landscape and offer strong economic incentive for development of biobetters.

---

[1] Bulletin of the World Health Organization, Volume 86 Number 2, February 2008, 81–160

[2] http://www.ncats.nih.gov/funding-and-notices/can/can.html

This book was conceived to address the bridge to curative therapies, to improve upon the gains of current therapeutics by enhancing their efficacy and safety pending the development of curative therapies. We have focused on two types of therapeutic proteins as providing illustrations for important concepts pertaining to biobetters: monoclonal antibodies, the most rapidly growing class of therapeutic proteins, and therapeutic enzymes for lysosomal storage diseases, therapeutics which are frequently the sole treatments for rare and often rapidly fatal "inborn errors of metabolism."

The means to enhance pharmacokinetics (PK), to prevent degradation in the in vivo environment, to minimize immunogenicity, and to enhance product efficacy without incurring novel safety issues are common aspirations for both types of therapeutic proteins. While enhancements in PK by technologies such as pegylation are important improvements, especially for maximizing dosing interval, and thus important for patient quality of life, the gains to be accrued from more highly effective therapies, both to the patient and to the healthcare system, are the key focus of this book. Thus, the chapters regarding the means to enhance efficacy, such as improved targeting to critical target tissues (e.g., penetrating the blood–brain barrier in the absence of inflammation; targeting muscle with its low expression of critical receptors for therapeutic enzymes), minimizing immunogenicity via protein engineering or tolerance induction regimens, optimizing affinity of mAbs for receptors or improving effector function via engineering of CDR and Fc regions, respectively, or via employment of novel scaffolds, are of key importance.

Of course, great caution is warranted in such undertakings, as it may be possible to be too "biobetter." For example, sustained activity of some therapeutic protein hormones may have unintended outcomes if downstream mediators which are key factors in induction/proliferation of malignancies are induced and sustained at high levels rather than having a transient exposure profile. For enzyme deficiency disorders, sustained prolonged activity of the therapeutic enzyme leading to efficient and near complete substrate depletion has the potential to cause serious problems. For example, in the setting of Gaucher Disease, an overly efficient enzyme replacement therapy (ERT) could cause rapid conversion of glucocerebrosides to ceramides, high levels of which may cause cellular apoptosis. Similarly, in phenylketonuria (PKU), there is concern that an overly active phenylalanine lyase could drop phenylalanine to such low levels that protein production may potentially be limited in younger children and those with metabolic stress. As to monoclonal antibodies and their derivatives, the design of biobetters should take into account the fact that higher affinities do not necessarily translate to clinical benefit and that new constructs targeting multiple pathways (e.g., bispecifics) should be carefully evaluated for their potential to generate unintended adverse effects. Use of preclinical animal models of diseases as well as carefully conducted clinical studies to test more highly active biobetters should mitigate against the specter of "too biobetter."

We would like to conclude with a reminder of addressing great efforts from a great President facing a monumental task, that of President John F. Kennedy in considering a program to put a human on the moon. In considering the magnitude of the effort, he said, "We choose to go to the moon. We choose to go to the moon in this decade and do the other things, not because they are easy, but because they are hard, because that goal will serve to organize and measure the best of our energies and skills, because that challenge is one that we are willing to accept, one we are unwilling to postpone, and one which we intend to win…." Unlike putting a human on the moon, we already have the ways and means and knowledge to optimize our therapeutic protein products. So let us too proceed to utilize the best of our energies, skills, and knowledge to improve therapeutic proteins to optimize clinical outcomes for suffering patients.

Silver Spring, MD, USA — Amy Rosenberg
South San Francisco, CA, USA — Barthélemy Demeule

Contents

Part I Therapeutic Enzymes

Targeting Glucocerebrosidase to Macrophages for Effective Treatment of Patients with Gaucher Disease: Setting the Paradigm of a "Fit for Purpose" Approach to Enzyme Replacement Therapy 3
Roscoe O. Brady

Challenges of Enzyme Replacement Therapy: Poor Tissue Distribution in Lysosomal Diseases Using Pompe Disease as a Model 9
Priya S. Kishnani

Muscle Targeting 23
Nancy M. Dahms

Blood–Brain Barrier Targeting of Therapeutic Lysosomal Enzymes 41
William M. Pardridge

Novel Methods for Addressing Immunogenicity of Therapeutic Enzymes 63
Leslie P. Cousens, Leonard Moise, and Anne S. De Groot

Part II Monoclonal Antibodies: Degradation Mechanisms and Potential Improvements

Structure of Monoclonal Antibodies 81
Balakrishnan S. Moorthy, Bo Xie, Ehab M. Moussa, Lavanya K. Iyer, Saradha Chandrasekhar, Jainik P. Panchal, and Elizabeth M. Topp

Prediction of Aggregation In Vivo by Studies of Therapeutic Proteins in Human Plasma 91
Tudor Arvinte, Emilie Poirier, and Caroline Palais

Effect of Hydrolytic Degradation on the In Vivo Properties of Monoclonal Antibodies ... 105
Balakrishnan S. Moorthy, Bo Xie, Ehab M. Moussa, Lavanya K. Iyer, Saradha Chandrasekhar, Jainik P. Panchal, and Elizabeth M. Topp

Oxidation of Proteins in the In Vivo Environment: What We Know; What We Need to Study and Potential Mitigation Strategies ... 137
Christian Schöneich

Molecular Assessment of Monoclonal Antibody-Based Therapeutics Enabling Lead Selection for Clinical Development ... 153
Vikas K. Sharma and Robert F. Kelley

Part III New Platforms

Perspectives on Engineering Biobetter Therapeutic Proteins with Greater Stability in Inflammatory Environments ... 183
V. Ashutosh Rao

Antibody-Like Molecules Designed for Superior Targeting and Pharmacokinetics ... 203
Alexey A. Lugovskoy and Melissa L. Geddie

Alternative Protein Scaffolds as Novel Biotherapeutics ... 221
Michaela Gebauer and Arne Skerra

Current Strategies for Pharmacokinetic Optimization ... 269
Uli Binder and Arne Skerra

Part IV Economic and Regulatory Considerations

Biosimilar and Biobetter Scenarios for the US and Europe: What Should We Expect? ... 315
Ernst R. Berndt and Mark R. Trusheim

Building Biobetters: The Regulatory Landscape ... 361
Emanuela Lacana, Lynne Yao, Anne Pariser, Amy Rosenberg, and Janet Woodcock

Index ... 371

Contributors

Tudor Arvinte Therapeomic Inc., Basel, Switzerland

Department of Pharmaceutical Sciences, University of Geneva–University of Lausanne, Geneva, Switzerland

Ernst R. Berndt Alfred P. Sloan School of Management, Massachusetts Institute of Technology, Cambridge, MA, USA

Uli Binder XL-protein GmbH, Freising, Germany

Roscoe O. Brady National Institute of Neurological Disorders and Stroke, National Institutes of Health, Bethesda, MD, USA

Saradha Chandrasekhar Department of Industrial and Physical Pharmacy, Purdue University, West Lafayette, IN, USA

Leslie P. Cousens EpiVax Inc., Providence, RI, USA

Nancy M. Dahms Department of Biochemistry, Medical College of Wisconsin, Milwaukee, WI, USA

Michaela Gebauer XL-protein GmbH, Freising-Weihenstephan, Germany

Melissa L. Geddie Merrimack Pharmaceuticals Inc., Cambridge, MA, USA

Anne S. De Groot EpiVax Inc., Providence, RI, USA

Institute of Immunology and Informatics, University of Rhode Island, Kingston, RI, USA

Lavanya K. Iyer Department of Industrial and Physical Pharmacy, Purdue University, West Lafayette, IN, USA

Robert F. Kelley Pharmaceutical Development, Genentech, Inc., South San Francisco, CA, USA

Priya S. Kishnani Department of Pediatrics—Medical Genetics, Duke University Medical Center, Durham, NC, USA

Emanuela Lacana Office of Biotechnology Products, Center for Drug Evaluation and Research, Food and Drug Administration, Silver Spring, MD, USA

Alexey A. Lugovskoy Merrimack Pharmaceuticals, Inc., Cambridge, MA, USA

Leonard Moise EpiVax Inc., Providence, RI, USA

Institute of Immunology and Informatics, University of Rhode Island, Kingston, RI, USA

Balakrishnan S. Moorthy Department of Industrial and Physical Pharmacy, Purdue University, West Lafayette, IN, USA

Ehab M. Moussa Department of Industrial and Physical Pharmacy, Purdue University, West Lafayette, IN, USA

Caroline Palais Therapeomic Inc., Basel, Switzerland

Jainik P. Panchal Department of Industrial and Physical Pharmacy, Purdue University, West Lafayette, IN, USA

William M. Pardridge ArmaGen Technologies Inc., Calabasas, CA, USA

Anne Pariser Office of Translational Sciences, Center for Drug Evaluation and Research, Food and Drug Administration, Silver Spring, MD, USA

Emilie Poirier Therapeomic Inc., Basel, Switzerland

V. Ashutosh Rao Division of Biotechnology Review and Research III, Office of Biotechnology Products, Center for Drug Evaluation and Research, Food and Drug Administration, Silver Spring, MD, USA

Amy Rosenberg Division of Biotechnology Review and Research III, Office of Biotechnology Products, Center for Drug Evaluation and Research, Food and Drug Administration, Silver Spring, MD, USA

Christian Schöneich Department of Pharmaceutical Chemistry, The University of Kansas, Lawrence, KS, USA

Vikas K. Sharma Pharmaceutical Development, Genentech, Inc., South San Francisco, CA, USA

Arne Skerra Lehrstuhl für Biologische Chemie, Technische Universität München, Freising-Weihenstephan, Germany

Elizabeth M. Topp Department of Industrial and Physical Pharmacy, Purdue University, West Lafayette, IN, USA

Mark R. Trusheim Alfred P. Sloan School of Management, Massachusetts Institute of Technology, Cambridge, MA, USA

Janet Woodcock Center for Drug Evaluation and Research, Food and Drug Administration, Silver Spring, MD, USA

Bo Xie Department of Industrial and Physical Pharmacy, Purdue University, West Lafayette, IN, USA

Lynne Yao Office of New Drugs, Center for Drug Evaluation and Research, Food and Drug Administration, Silver Spring, MD, USA

About the Editors

Dr. Amy Rosenberg received her M.D. from Albert Einstein College of Medicine and trained in internal medicine and infectious diseases. She was a postdoctoral fellow in Al Singer's Laboratory in the NCI before coming to CBER, FDA. She is the director of the Division of Biotechnology Review and Research III, Office of Biotechnology Products in CDER, FDA which regulates diverse protein therapeutics, including monoclonal antibodies, fusion proteins, enzyme replacement therapies, hematologic and somatic cell growth factors, and immunomodulatory agents. Her particular interest is in immune tolerance induction in diverse clinical settings including autoimmunity, therapeutic protein immunogenicity, and transplantation.

Dr. Barthélemy Demeule obtained his Ph.D. at the University of Geneva, Switzerland, where he started his investigations on the physicochemical stability of biopharmaceuticals. After a postdoctoral work at Genentech, Inc. focused on the effect of the in vivo environment on antibody–antigen interactions, he stayed in the company where he held positions of increasing responsibilities. He currently leads a group of scientists responsible for the pharmaceutical development of monoclonal antibodies in the last phases of clinical development. He also serves on the editorial board of the *European Journal of Pharmaceutics and Biopharmaceutics*.

Part I
Therapeutic Enzymes

Targeting Glucocerebrosidase to Macrophages for Effective Treatment of Patients with Gaucher Disease: Setting the Paradigm of a "Fit for Purpose" Approach to Enzyme Replacement Therapy

Roscoe O. Brady

Gaucher disease is one of the most prevalent hereditary metabolic storage disorders of humans. A patient with an enlarged spleen was described by the French medical student Phillipe C. E. Gaucher who thought she had a splenic neoplasm (Gaucher 1882). Brill (1901) suggested that patients with such a presentation represented a familial disorder. It was reported that the spleen of these patients contained a hyaline-like material (Marchand 1907). Lieb (1924) believed that the accumulating material in the spleen was galactocerebroside. However, the optical rotation of the sugar released by acid hydrolysis was inconsistent with this assumption. Aghion (1934) demonstrated that glucocerebroside was the substance that accumulated (Fig. 1a). The kinetics of the formation of glucocerebroside was found to be normal in patients with Gaucher disease (Trams and Brady 1960). It was postulated that the metabolic defect in these patients was of a catabolic nature. Several years later, an enzyme was discovered in mammalian organs that catalyzed the hydrolytic cleavage of glucose from glucocerebroside (Brady et al. 1965a) (Fig. 1b). Reduced activity of this enzyme was shown to be the cause of Gaucher disease (Brady et al. 1965b, 1966). The possibility of overcoming the insufficient glucocerebrosidase by enzyme replacement was proposed (Brady 1966).

I wished to obtain a human source of glucocerebrosidase if it were possible. One evening it occurred to me that the placenta might be useful in this regard. The next day I homogenized some fresh placental tissue and found that it did indeed contain glucocerebrosidase. My colleagues and I were able to obtain small amounts of comparatively pure glucocerebrosidase from this tissue (Pentchev et al. 1973). When we injected it into two patients with Gaucher disease, we found a significant decrease in the quantity of glucocerebroside that had accumulated in the liver (Brady et al. 1974). Moreover, there was marked decrease of the elevated glucocerebroside that

R.O. Brady, M.D. (✉)
National Institute of Neurological Disorders and Stroke, National Institutes of Health, Building 10 Room 3D03, Bethesda, MD 20892-1260, USA
e-mail: bradyr@ninds.nih.gov

A. Rosenberg, B. Demeule (eds.), *Biobetters*, AAPS Advances in the Pharmaceutical Sciences Series 19, DOI 10.1007/978-1-4939-2543-8_1

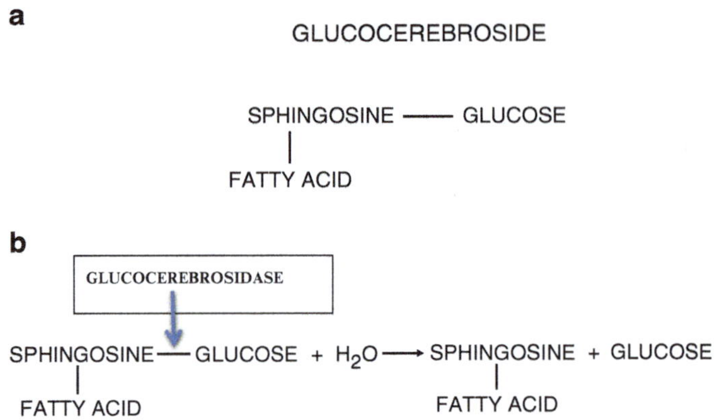

Fig. 1 (**a**) Accumulating material in Gaucher Disease. (**b**) The catabolism of glucocerebroside is initiated by the enzyme glucocerebrosidase

was associated with red blood cells in the circulation (Brady et al. 1974). Of particular interest was the lengthy period of time following the injection of glucocerebrosidase before glucocerebroside associated with red blood cells rose toward pre-injection levels (Pentchev et al. 1975).

We were quite surprised when next patient we treated with glucocerebrosidase showed an insignificant clearance of glucocerebroside. We discovered that she had accumulated 24 times the quantity of glucocerebroside in her liver than the first recipient and 11 times more than the second. We realized we would have to improve our purification procedure in order to obtain sufficient quantities of glucocerebrosidase to treat such patients. We achieved this goal by developing a technique to isolate the enzyme based on the incorporation of two hydrophobic column chromatography steps in the purification procedure (Furbish et al. 1977). We were quite startled with the findings when we injected enzyme purified in this fashion into seven patients with Gaucher disease. Three of the patients had significant reductions of glucocerebrosidase but four showed no change at all. This was not caused by any lack of catalytic activity of the preparation. Glucocerebroside specifically accumulates in macrophages (Kupffer cells) in the liver. We suspected that we probably were not *targeting* the glucocerebrosidase to macrophages that are involved in biodegrading sphingolipids arising from rapidly turning over cells such as white and red blood cells and blood platelets. We felt that the inability to deliver glucocerebrosidase to macrophages was caused by the requirement to treat the placental extract with butanol in order to remove lipids that prevented the binding of glucocerebrosidase to the hydrophobic columns. Among the lipids that were extracted by this method was phosphatidylserine that had two specific effects on glucocerebrosidase. It was shown by Dale et al. (1976) and Choy (1984) that it markedly stimulated the activity of this enzyme. Moreover, Schroit et al. (1984)

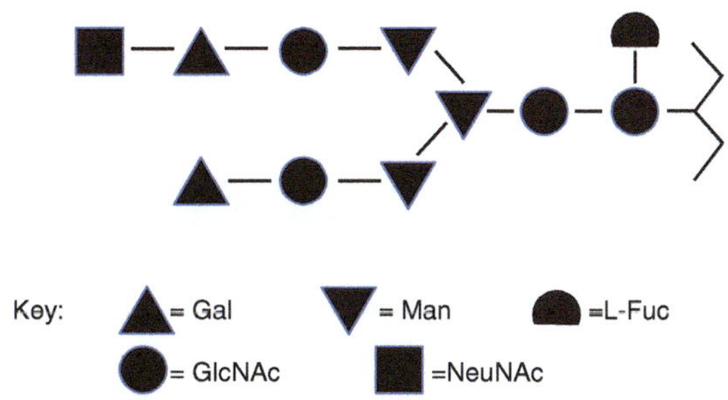

Abbreviations: Gal = galactose; Man = mannose; L-Fuc = fucose; GlcNAc = N-acetylglucosamine; NeuNAc = N-acetylneuraminic acid

Fig. 2 Carbohydrate unit of native glucocerebrosidase

discovered that phosphatidylserine was specifically recognized by macrophages. We tried to re-lipidate glucocerebrosidase that had been purified by hydrophobic column chromatography with phosphatidylserine but achieved only modest success in increasing its delivery to macrophages in which glucocerebroside accumulates.

Thus, my associates and I embarked on a different approach to target glucocerebrosidase to macrophages. It was known that macrophages have a lectin (carbohydrate binding protein) on their surface that has a high affinity for mannose-terminal oligosaccharides (Stahl et al. 1978). Glucocerebrosidase is a glycoprotein with four oligosaccharide side chains, three of which have a complex array of sugars terminating with *N*-acetylneuraminic acid that shields three underlying mannose residues (Takasaki et al. 1984) (Fig. 2). A series of investigations was undertaken to determine whether altering the oligosaccharide chains of glucocerebrosidase to expose such mannose residues would affect the cellular uptake of the enzyme. We sequentially removed the three external moieties of the oligosaccharide chains with exoglycosidases (Fig. 3) producing the mannose-terminal glycoform of glucocerebrosidase (Fig. 4) (Furbish et al. 1978, 1981; Steer et al 1978; Brady and Furbish 1982). We discovered that mannose-terminated glucocerebrosidase was taken up by macrophages 50 times more effectively than native placental glucocerebrosidase. We began to administer glucocerebrosidase modified in this fashion intravenously to patients with Gaucher disease. The first trial consisted of seven adults and one child with Gaucher disease. Only the child showed evidence of benefit (Barton et al 1990). We realized that we should have carried out a dose–response study before such a trial. We therefore undertook that investigation and found that a consistent reduction of accumulated glucocerebroside was obtained by administering 60 IU of mannose-terminal glucocerebrosidase per kg of body weight. An investigation of this amount of enzyme administered to 12 patients with Gaucher disease revealed

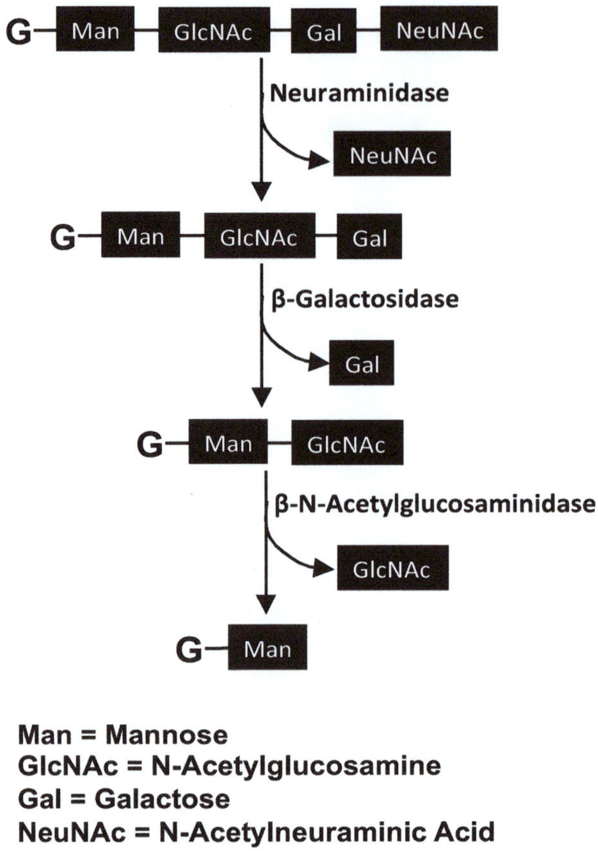

Fig. 3 Enzymatic modification of glucocerebrosidase (G)

Fig. 4 Carbohydrate unit of mannose terminated glucocerebrosidase

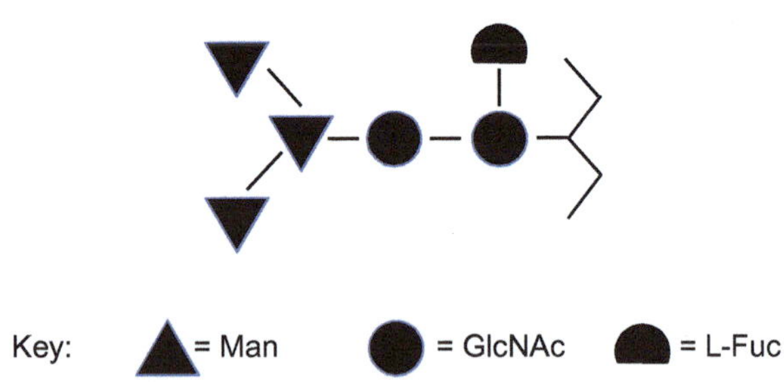

highly beneficial responses in all recipients. There was a reduction of the size of the enlarged liver and spleen, an increase in blood platelets, an increase in hemoglobin and improvement of the skeleton in all recipients (Barton et al. 1991). Based on these findings, enzyme replacement therapy was approved by the U.S. Food and Drug Administration (FDA) for patients with Gaucher disease on April 5, 1991. It was quickly realized that the collection and processing of sufficient placentas to treat all of the patients with Gaucher disease who required this therapy would be extremely difficult, if not impossible. The Genzyme Corporation decided to produce the enzyme by recombinant technology in Chinese hamster ovary cells in large bioreactors. The oligosaccharide side chains of glucocerebrosidase obtained in this process also required modificationin the same manner as the placental glucocerebrosidase. Recombinant glucocerebrosidase was approved for the treatment of patients with Gaucher disease by the U.S. FDA in 1994. It was shown to be as effective as oligosaccharide-modified placental glucocerebrosidase (Grabowski et al. 1995).

More than 6,000 patients with Gaucher disease throughout the world are now being successfully treated with macrophage-targeted glucocerebrosidase. Other hereditary metabolic disorders affect different target tissues bearing different lectins and thus, will require related procedures to deliver a therapeutic protein to treat those conditions effectively. Alternatively, genetic regulation of the activity of the glycotransfereases involved in creating the glycoforms of required enzymes may evolve as a useful strategy to obtain effective products.

References

Aghion A (1934) La maladie de Gaucher dans l'enfance. Thèse, Paris

Barton NW, Furbish FS, Murray GJ, Garfield M, Brady RO (1990) Therapeutic response to intravenous infusions of glucocerebrosidase in a patient with Gaucher disease. Proc Natl Acad Sci U S A 87:1913–1916

Barton NW, Brady RO, Dambrosia JM, DiBisceglie AM, Doppelt SH, Hill SC, Mankin HJ, Murray GJ, Parker RI, Argoff CE, Grewal RP, Yu K-T (1991) Replacement therapy for inherited enzyme deficiency—macrophage-targeted glucocerebrosidase for Gaucher's disease. N Engl J Med 324:1464–1470

Brady RO (1966) The sphingolipidoses. N Engl J Med 275:312–318

Brady RO, Furbish FS (1982) Enzyme replacement therapy: specific targeting of exogenous enzymes to storage cells. In: Martonosi AN (ed) Membranes and transport, vol 2. Plenum, New York, pp 587–592

Brady RO, Kanfer J, Shapiro D (1965a) The metabolism of glucocerebrosides. I. Purification and properties of a glucocerebroside-cleaving enzyme from spleen tissue. J Biol Chem 240:39–42

Brady RO, Kanfer JN, Shapiro D (1965b) The metabolism of glucocerebrosides. II. Evidence of an enzymatic deficiency in Gaucher's disease. Biochem Biophys Res Commun 18:221–225

Brady RO, Kanfer JN, Bradley RM, Shapiro D (1966) Demonstration of a deficiency of glucocerebroside-cleaving enzyme in Gaucher's disease. J Clin Invest 45:1112–1115

Brady RO, Pentchev PG, Gal AE, Hibbet SR, Dekaban AS (1974) Replacement therapy for inherited enzyme deficiency: use of purified glucocerebrosidase in Gaucher's disease. N Engl J Med 291:989–993

Brill NE (1901) Primary splenomegaly with a report of three cases occurring in one family. Am J Med Sci 121:377–392

Choy FYM (1984) Gaucher disease: the effects of phosphaticylserine on glucocerebrosidase from normal and Gaucher fibroblasts. Hum Genet 67:432–436

Dale GL, Villacorte DG, Beutler E (1976) Solubilization of glucocerebrosidase from human placenta and demonstration of a phospholipid requirement for its catalytic activity. Biochem Biophys Res Commun 71:1048–1053

Furbish FS, Blair HE, Shiloach J, Pentchev PG, Brady RO (1977) Enzyme replacement therapy in Gaucher's disease: large-scale purification of glucocerebrosidase suitable for human administration. Proc Natl Acad Sci U S A 74:3560–3563

Furbish FS, Steer CJ, Barranger JA, Jones EA, Brady RO (1978) The uptake of native and desialylated glucocerebrosidase by rat hepatocytges and Kupffer cells. Biochem Biophys Res Commun 81:1047–1053

Furbish FS, Steer CJ, Krett NL, Barranger JA (1981) Uptake and distribution of placental glucocerebrosidase in rat hepatic cells and effects of sequential deglycosylation. Biochim Biophys Acta 673:425–434

Gaucher PCE (1882) De l'épithélioma primitif de la rate. Thèse de Paris

Grabowski GA, Barton NW, Pastores G, Dambrosia JM, Banerjee TK, McKee MA, Parker C, Schiffmann R, Hill SC, Brady RO (1995) Enzyme therapy in Gaucher disease type 1: comparative efficacy of mannose-terminated glucocerebrosidase from natural and recombinant sources. Ann Intern Med 122:33–39

Lieb H (1924) Cerebrosidespeicherung bei Splenomegalie Typus Gaucher. Ztschr Physiol Chem 140:305–313

Marchand F (1907) Über Sogennante idiopathische Splenomegalie (Typus Gaucher). Munchen med Wchnschr 54:1102–1103

Pentchev PG, Brady RO, Hibbert SR, Gal AE, Shapiro D (1973) Isolation and characterization of glucocerebrosidase from human placental tissue. J Biol Chem 248:5256–5261

Pentchev PG, Brady RO, Gal AE, Hibbert SR (1975) Replacement therapy for inherited enyzme deficiency. Sustained clearance of accumulated glucocerebroside in Gaucher's disease following infusion of purified glucocerebrosidase. J Mol Med 1:73–78

Schroit AJ, Tanaka Y, Madsen J, Fidler IJ (1984) The recognition of red blood cells by macrophages: role of phosphatidylserine and possible implications of membrane phospholipid asymmetry. Biol Cell 51:227–238

Stahl PD, Rodman JS, Miller MJ, Schlesinger PH (1978) Evidence for receptor-mediated binding of glycoproteins, glycoconjugates, and lysosomal glycosidases by alveolar macrophages. Proc Natl Acad Sci U S A 75:1399–1403

Steer CJ, Furbish FS, Barranger JA, Brady RO, Jones EA (1978) The uptake of agalacto-glucocerebrosidase by rat hepatocytes and Kupffer cells. FEBS Lett 91:202–205

Takasaki S, Murray GJ, Furbish F, Brady RO, Barranger JA, Kobata A (1984) Structure of N-asparagine-linked oligosaccharide units of human placental beta-glucocerebrosidase. J Biol Chem 259:10112–10117

Trams EG, Brady RO (1960) Cerebroside synthesis in Gaucher's disease. J Clin Invest 39: 1546–1550

Challenges of Enzyme Replacement Therapy: Poor Tissue Distribution in Lysosomal Diseases Using Pompe Disease as a Model

Priya S. Kishnani

After over two decades of concept studies, animal model studies, and safety trials, the FDA approved the first human enzyme replacement therapy (ERT) for Gaucher disease type I, a lysosomal storage disease (LSD), in 1991 (Barton et al. 1991). The therapeutic enzyme effectively lowered buildup of glycosylceramide in the liver, spleen, and bone marrow, among other tissues, resulting in notable clinical improvements including reduced organomegaly, and improvements in hematologic, and skeletal parameters. This landmark achievement marked not only the progress towards reaching a life-saving treatment for Gaucher disease, one of the most common LSDs, but also planted the seeds of hope that ERT could be utilized for the other LSDs. To date, seven LSDs are being treated with ERT including Mucopolysaccharidosis I (MPS I), MPS II, MPS IV, MPS VI, Gaucher disease, Fabry disease, and Pompe disease. For Gaucher disease there are three different ERTs currently approved and two oral medications that reduce substrate accumulation. For Fabry disease there are two approved ERTs. Several other disease-specific ERTs are currently in development (Table 1).

ERT has revolutionized treatment for patients with LSDs, dramatically improving lifespan, increasing overall quality of life, and diminishing the extent of organ involvement. Despite this progress, certain challenges have been identified in patients receiving ERT due to a number of factors including the following: minimal or no enzyme delivery to all necessary target sites; delay in diagnosis enabling substrate buildup with often irreversible consequences prior to the start of ERT; inability of the therapeutic enzyme to reach certain sanctuary sites, including the central nervous

P.S. Kishnani, M.D. (✉)
Department of Pediatrics—Medical Genetics, Duke University Medical Center,
DUMC 103856, 595 Lasalle Street, GSRB 1, 4th Floor, Room 4010,
Durham, NC 27710, USA
e-mail: priya.kishnani@duke.edu

A. Rosenberg, B. Demeule (eds.), *Biobetters*, AAPS Advances
in the Pharmaceutical Sciences Series 19, DOI 10.1007/978-1-4939-2543-8_2

Table 1 Overview of therapies either FDA approved or in commercial development for LSDs

Lysosomal storage disease	Deficient enzyme	Current therapy trade name (recombinant enzyme)	FDA/EMA approval or clinical trial status	Clinical improvements	Limitations of current therapy (persistent issues)
Gaucher disease, Type I	Acid beta-glucosidase	ERT	FDA 1991; EMA 1997	Reduction of organomegaly, hematological improvements, improvement in BMD, reduction of bone pain, improved quality of life (Anderson et al. 2014)	Osteopenia, pulmonary infiltrates, mesenteric lymphadenopathy, Gaucheromas (Gaucher-cell pseudo tumors), Parkinson's-like symptoms, pulmonary hypertension, no correction of CNS disease (Poll and Vom Dahl 2009)
		Cerezyme (imiglucerase)			
		Zavesca (miglustat)	EMA 2002; FDA 2003		
		VPRIV (velaglucerase alfa)	FDA 2010; EMA 2010		
		Elelyso (taliglucerase alfa)	FDA 2012		
		SRT	FDA 2014		
		Cerdelga (eliglustat)			
Fabry disease, classic, late-onset	Alpha-galactosidase A	ERT	EMA 2001; FDA 2003	Improvements in neuropathic pain, improved nephropathy, improved cardiomyopathy	Cardiac fibrosis, arrhythmia risk, stroke risk, kidney failure, white matter changes in brain (Pieroni et al. 2013), increase in antienzyme antibodies
		Fabrazyme (agalsidase beta)			
		Replagal (agalsidase alfa)	EMA 2001		
Mucopolysaccharidosis I, Hurler-Sheie spectrum	Alpha-L-iduronidase	ERT	FDA 2003; EMA 2003	Stabilization or improvement of joint range of motion, sleep apnea, and left ventricular hypertrophy (Clarke et al. 2009)	Neurological, cardiac, ophthalmologic (corneal clouding) and skeletal complications remain (Clarke et al. 2009); antienzyme antibodies
		Aldurazyme (laronidase)			
		Hematopoietic stem cell transplant			

Lysosomal storage disease	Deficient enzyme	Current therapy trade name (recombinant enzyme)	FDA/EMA approval or clinical trial status	Clinical improvements	Limitations of current therapy (persistent issues)
Mucopolysaccharidosis VI	*N*-Acetylgalactosamine-4-sulfatase	ERT Naglazyme (galsulfase)	FDA 2005; EMA 2006	Improvements in gait, respiratory ability and general quality of life (Valayannopoulos and Wijburg 2011)	Opthalmologic and skeletal complications remain
Mucopolysaccharidosis II, severe and attenuated	Iduronate-2-sulfatase	ERT Elaprase (idursulfase)	FDA 2006; EMA 2007	Improvements in gait, decreased liver and spleen volumes, reduced urinary GAG levels (Muenzer et al. 2006)	Neurological and skeletal complications remain (Muenzer et al. 2006)
Pompe disease	Acid alpha-glucosidase	ERT	FDA 2006	Improved motor function, respiratory function and cardiac function	Musculoskeletal damage and progression despite ERT, anterior horn cell disease and bulbar involvement
Infantile		Myozyme (aglucosidase alfa)			
Late-onset		Lumizyme (aglucosidase alfa)	FDA 2010		Cardiac arrhythmias, fibrosis, antienzyme antibodies
Mucopolysaccharidosis IV, type A	*N*-Acetylgalactosamine-6-sulfatase	ERT Vimizin (elosulfase alfa)	FDA 2014	Decrease in urinary keratin sulfate, increase in walk distance and improvement in stair climbing overall	Skeletal complications remain, significant morbidity

(continued)

Table 1 (continued)

Lysosomal storage disease	Deficient enzyme	Current therapy trade name (recombinant enzyme)	FDA/EMA approval or clinical trial status	Clinical improvements	Limitations of current therapy (persistent issues)
Lysosomal acid lipase deficiency, CESD and Wolman disease	Lysosomal acid lipase	ERT Sebilipase alfa	For CESD, phase 1 completed in 2012; For Wolman disease, phase 2/3	Successful completion of Phase 3 tral	Disease progression remains
Alpha-mannosidosis	Alpha-mannosidase	ERT Lamazym (alpha-mannosidase)	Phase 3	Urinary and cerebrospinal fluid oligosaccharides decreased, motor function improved	Low permeability of the blood–brain barrier
Niemann Pick, type B	Acid sphingomyelinase	ERT Acid sphingomyelinase	Phase 2 trial	Drug well tolerated at lower doses, but at doses of 0.6 and 1.0 mg/kg the cytokine and bilirubin levels were elevated	Progression of lung disease
Mucopolysaccharidosis VII	Beta-glucuronidase	ERT UX003	Phase 1/2 enrolled in 2013	Improvements in walking capabilities, respiratory ability, and general quality of life (Valayannopoulos and Wijburg 2011)	Neurological and skeletal complications remain
Metachromatic leukodystrophy and Krabbe disease	Arylsulfatase A and galactocerebroside β-galactosidase, respectively	ERT HSCT	N/A	Stalls disease progression	Limitation of ERT is that intravenously delivered enzymes generally do not cross the blood–brain barrier
Niemann Pick, type A	Acid sphingomyelinase	ERT	N/A	N/A	Severe neurovisceral form, poor prognosis and limited survival

system (CNS), bone, and cartilage (Hollak and Wijburg 2014); and immune responses to the exogenous enzyme abrogating its effectiveness, especially in patients who are cross-reactive immunologic material (CRIM) negative.

These therapeutic deficiencies are evidenced in varying ways in the different LSDs. In Gaucher disease, patients receiving ERT can still experience osteopenia, pulmonary infiltrates, mesenteric lymphadenopathy, Gaucheromas (Gaucher-cell pseudo tumors), seizures, Parkinson-like symptoms, and pulmonary hypertension (Bennett and Mohan 2013; Poll and Vom Dahl 2009). An increased incidence of multiple myeloma is also present, which is believed to be caused by elevated levels of IL-6 causing clonal expansion of B cells (Rosenbloom et al. 2005). ERT, unable to cross the blood–brain barrier, fails to halt neurologic symptoms in the neuronopathic subtype of Gaucher disease (Schiffmann et al. 2008), MPS I, and MPS II (Muenzer 2014). In Fabry disease, an LSD caused by a deficiency of the enzyme alpha-galactosidase A resulting in glycosphingolipid buildup in vascular endothelium, ERT with agalsidase alfa reduces and/or stabilizes symptoms of neuropathic pain, nephropathy, and cardiomyopathy (Eng et al. 2001). Nonetheless, even with ERT, patients with Fabry disease continue to experience complications such as renal failure, strokes, arrhythmias, proteinuria, chronic neuropathic pain, and myocardial fibrosis. Likely contributing to therapeutic failure is inefficient delivery to certain target sites, and development of antibodies with neutralizing activity to ERT in male patients with Fabry disease (Linthorst et al. 2004). Finally, myocardial fibrosis, a common finding in older patients with Fabry disease, may be present prior to treatment initiation and creates a hurdle, sometimes a barrier, to the efficacy of exogenous enzyme (Weidemann et al. 2013, 2014).

The mucopolysaccharidoses, another subgroup of LSDs, have also encountered both successes and shortcomings through ERT (Muenzer 2014). In patients with MPS I, II, and IV, complications of the heart, skeletal system, lungs, and gastrointestinal tract (organomegaly, hernias) visibly improve with ERT (Noh and Lee 2014). At the same time, several symptoms tend to persist with ERT, including cardiac valve disease (stagnation of mitral/aortic valve stenosis; progressive aortic valve regurgitation), skeletal/joint disease (dysostosis multiplex), and airway disease (Muenzer 2014) likely attributable to failure of ERT to penetrate such tissues sufficiently. As in neuronopathic Gaucher disease, the CNS is an elusive treatment area highlighted in MPS and neurological progression persists despite ERT (Noh and Lee 2014).

In infantile Pompe disease (IPD), ERT with alglucosidase alfa has changed the natural history of the disease. Improvements in cardiac and motor function have been observed in long-term survivors of IPD. However a new emerging phenotype is evident due to the longer term survival of these patients which includes proximal and distal myopathy, sensorineural hearing loss, risk for arrhythmias, hypernasal speech, dysphagia (with risk for aspiration), ptosis, and osteopenia (Jones et al. 2010; Nicolino et al. 2009; Yanovitch et al. 2010; Prater et al. 2012) again portraying lack of effective penetration/activity of ERT in critical target tissues. A study of long-term ERT in IPD patients indicates that skeletal muscle damage persists in patients despite ERT, including those started within the first month of life (albeit to a lesser extent than in patients who start ERT at an older age) (Prater et al. 2013). There is also involvement of the anterior horn cells, and other CNS manifestations including delayed processing speed that is noted in long term IPD survivors (Spiridigliozzi et al. 2013). Residual deficits in late onset Pompe disease treated with ERT include respiratory insufficiency and difficulty walking and/or climbing (Kobayashi et al. 2010; Strothotte et al. 2010).

Such persistent debilitating complications seen in LSDs treated with ERT exemplify how ERT, although a monumental step towards improving the quality of life for patients, is not optimal and certainly not curative.

Present efforts towards bridging the gap between currently available therapies and a cure for LSDs are best viewed through the lens of our collective experience with Pompe disease. About 9 years after ERT was established for Gaucher disease, clinical trials for ERT in Pompe disease unfurled with very promising results. The intravenous administration of alglucosidase alfa (GAA) for Pompe disease blazed the trail as proof of concept for ERT use in neuromuscular disorders. Pompe disease, a true disease spectrum, presents broadly as an infantile and adult onset form (Kishnani et al. 2012). Across the continuum of Pompe disease, the enzyme GAA is partially or completely deficient, leading to the accumulation of glycogen in many tissues, especially the cardiac, skeletal, and smooth muscles. Since alglucosidase alfa's approval by the FDA in 2006, many benefits, challenges, and underlying issues have been highlighted. ERT dramatically transformed the prognosis for Pompe disease: the infantile onset form is no longer fatal within the first year of life and the adult onset form improves or stabilizes instead of worsening in many of the patients treated. With the adult onset patients living longer and the infantile onset patients surviving past 1 year, the natural history of the disease spectrum is becoming better understood and increasingly useful to parallel with other GSDs and LSDs. Although progress with ERT is notable in Pompe disease, long-term issues stemming from delayed therapy initiation and/or poor ERT delivery to certain sanctuary sites are becoming evident. While enhanced newborn screening programs, which aim to target neonates and infants prior to symptom onset and enable early treatment initiation (Liao et al. 2014), have reduced delayed administration of ERT, residual motor deficits, which can cause hypotonia and fatigue, are still noted in IPD (Chien et al. 2013).

Pompe disease serves as a useful treatment model for several reasons, including the applicability of ERT across the whole disease continuum spanning from infantile (severe) to adult (less severe) onset and the multi-systemic nature of Pompe disease. The efficacy of ERT can be tested on patients of all ages and its effects on different disease manifestations, *i.e.,* cardiac, skeletal muscle involvement and respiratory involvement in infantile and late onset disease, can be compared. The extent of musculoskeletal involvement is also monitored across the disease spectrum, particularly in relation to the effect of ERT on its progression and patient's age at start of therapy. The infantile presentation, in particular, provides insight into the long-term effects of treatment, as well as on the importance of early treatment. Pompe disease especially serves as a beneficial model due to the rapidity of disease progression: IPD left untreated proves fatal by 1 year of age and the clinical effects of ERT and of immune responses to ERT are readily apparent. The availability of a mouse model for Pompe disease is another strong point in this disease. Animal testing enables better understanding of cellular mechanisms of Pompe disease as well as the effects of potential treatments on said mechanisms.

Although ERT in Pompe disease has proven to be very successful in many regards, several limitations exist: the efficiency of ERT distribution and uptake in the extensive muscular system, inability of ERT to cross the blood–brain barrier, pathological preconditions, defective cellular machinery, and immune responses to ERT. The biggest challenge remains the enormity of the target organ, muscle. ERT is required in high doses, 30–100 times greater than in other LSDs, to attempt to saturate muscle, which makes up 40 % of body mass (Desnick 2004). The heterogeneity of the muscular system and the muscle fiber type may also contribute to the variable response of different tissues. It has been shown that cardiac muscle responds much better to ERT than skeletal muscle, one theory being that because skeletal muscle has both an endothelial barrier and endomysium, an exogenously introduced enzyme may be deterred. Another theory for suboptimal skeletal muscle response lies in the density of cation-independent mannose 6-phosphate (M6P) receptors essential for the binding and uptake of ERT into the muscle and trafficking to lysosomes. Cardiac muscle tends to have a much higher density of M6P receptors than does skeletal muscle, corresponding with the level of response to ERT (Raben et al. 2003; Winkel et al. 2004; Zhu et al. 2004). M6P receptors recognize ERT molecules via their expression of M6P residues, which target them for transport to the lysosome and are therefore key in delivering exogenous enzyme to the lysosome for glycogen degradation.

Biobetters, in which the ERT is engineered to express high levels of bis-mannose 6-P residues, could substantially improve uptake of ERT into skeletal muscle cells, as could enhancing expression of M6P receptors on skeletal muscle cells. Similar to the inhibitory effects of the endomysium and endothelial linings surrounding skeletal muscle tissue, ERT also cannot cross the blood–brain barrier to break down glycogen accumulation in the CNS. This section of the book includes a chapter discussing the means to address CNS penetration of therapeutic proteins.

A number of pathological factors in Pompe disease also govern response to ERT. The degree of muscular and lysosomal damage present at the time of ERT administration affects outcome: patients with little damage fare better (Kishnani et al. 2007). This finding highlights the importance of accurate and early diagnostic techniques, i.e., newborn screening, which in turn would lead to an early treatment (Chien et al. 2011; Burton 2012; Shigeto et al. 2011). The muscle fiber type also contributes to outcome. GAA enzyme breaks down less glycogen in "fast twitch" type II muscle fibers, whereas "slow twitch" type I muscle fibers tend to undergo significant glycogen clearance by ERT. In one study, muscle biopsies of eight infantile onset patients who were on ERT showed varied reduction of glycogen accumulation (Thurberg et al. 2006). Those patients who had lower glycogen storage, milder cellular damage, more type I muscle fibers, and earlier initiation of ERT typically exhibited better clinical outcome.

Defective autophagy, contributing to disease pathology, is being noted in patients treated with ERT. Lysosomal storage of substrate has been found to impair proper autophagic function, which, in both Pompe disease patients and in GAA knock out mice, correlated to muscle weakness, buildup of dysfunctional mitochondria, and muscle atrophy (Shea and Raben 2009; Raben et al. 2012; Nascimbeni et al. 2012).

Although impaired autophagy has historically been more visible in adult cases, infantile cases reveal autophagic buildup after as early as 6 months of ERT (Raben et al. 2010). Raben et al. also noted greater autophagosome buildup in human type II muscle fibers in comparison to the slow twitch type I muscle fibers, leading the group to believe that defective autophagy may lead to enzymatic buildup within the autophagosomes thus rendering ERT ineffective (Raben et al. 2007). This may clarify why ERT appears to improve functionality in certain muscles based on the majority fiber type with type II fibers tending to respond poorly to ERT (Raben et al. 2005).

The development of inhibitory antibodies in response to ERT poses a challenge to successful sustained enzyme uptake and activity. Patients deemed CRIM positive have some residual natural enzyme protein production that tolerizes their immune systems to some extent, whereas those who are CRIM negative lack natural enzyme protein and therefore tend to mount vigorous immune responses to ERT upon treatment (de Vries et al. 2010; Kishnani et al. 2010). Of note, however, a subset of CRIM positive patients also develop high-sustained antibody titers (Banugaria et al. 2011). The basis for lack of immune tolerance in such patients, and the ability to predict which CRIM positive patients will mount such responses is of paramount interest. Immune tolerance induction (ITI) has been successfully used to abrogate or prevent development of such antibody responses (Messinger et al. 2012; Lacaná et al. 2012; Banugaria et al. 2013a, b)

In addition for the need to improve targeting of ERT via increased expression of M6P residues and receptors, dose is also an issue. The recommended ERT dose of 20 mg/kg every 2 weeks administered to IPD patients seems insufficient for long-term treatment and is unable to keep pace with the requirements of a developing child. Cases of clinical plateau and even decline have been documented, and as a result some patients are now receiving up to 40 mg/kg of ERT once a week or once every 2 weeks (Prater et al. 2012, 2013). One improvement correlated with an increased dose is improvement in ptosis (Yanovitch et al. 2010). Other improvements are also being noted such as symptoms of urinary incontinence, reduced fatigue, and improvements in activities of daily living (PSK personal experience, manuscript in preparation)

Many efforts to address the myriad of challenges in treating LSDs in addition to or in place of ERT are currently under development. Immunomodulation, adjunctive therapies, and reduction of defective autophagy supplement current ERT whereas small molecule chaperone therapy, second generation (biobetter) ERT, and gene therapy aim to supplant current methods.

'ERT carried the torch a long way, enabling patients with once-fatal LSDs (such as classic IPD) to live, and others (such as those with late-onset Pompe disease) to live with an improved quality of life.' Much has been learned in the context of ERT: the shortcomings and adverse effects of current therapies such as immune response and fibrosis in Pompe disease and Fabry disease call for a therapy that bridges the current improvements to definitive, curative treatments.

ERTs currently available are clearly life-saving. However, they have several limitations as outlined above. Questions remain: does the perfect ERT exist, or could a fine adjustment of timing, dosage, and/or combination therapy be the missing link?

The concept of a biobetter, an ERT that would be highly efficient in reducing substrate burden in all tissues, not just a subset of affected tissues and one that does not elicit an immune response, would be a clear advance in the field and a boon to patients. One could envision an ERT that is safe and highly efficacious with the ability to target all organs affected by disease as a next research milestone in the field.

There are several theoretical methods to achieve these goals, which may become a reality in the near future. Due to the high demand for enzyme production, glycosylation engineering can be utilized to produce not only a greater quantity of enzyme, but enzyme with much better expression of product attributes critical for efficacy such as enhanced levels of M6P moieties.

Inadequate delivery to tissues including the CNS is another problem which could be altered by engineering ERT with other receptor tags or moieties to allow penetration of such tissues (Grubb et al. 2008, Pardridge et al. this issue).

Reducing or eliminating the immune response elicited by ERT has been shown to improve or restore therapeutic efficacy. Protocols to most efficiently induce immune tolerance to ERT should be pursued. Research considerations for immunomodulation include identification of at-risk patients prior to ERT, improving administration of immune modulation simultaneously with ERT, and development of protocols to efficiently abrogate an established immune response.

Given the complexity of diseases such as Pompe disease, with numerous manifestations that require adjunctive therapy, it is likely that ERT alone will not be sufficient to optimally treat this disease and other LSDs. Indeed, the defective autophagy in LSDs, characterized by massive autophagic buildup in muscle fibers, muscle degeneration, the presence and increasing burden of dysfunctional mitochondria, has a compounding effect. Therapies to mitigate defective autophagy would reduce substrate burden, enabling ERTs to produce results in a faster manner.

The use of small molecules to treat LSDs is yet another possible therapeutic option. The stabilization of misfolded proteins may be the cause of enzyme deficiency, where such exist (i.e., in CRIM positive patients), that can be enhanced by the use of pharmacological chaperones - specific, low molecular weight, hydrophobic ligands that bind to and rescue specific misfolded or mistargeted proteins thereby increasing protein function at the target site and improving clinical benefit. However, although chaperones aid in the proper formation of the target enzyme, some of these molecules bind to the active site of the enzyme and thereby act competitively with the substrate upon which the enzyme is meant to act. If these issues can be addressed, chaperone therapy may be able to help patients.

Gene therapy, via administration of AAV vectors encoding genes for defective enzymes may also be able to correct lysosomal storage issues in LSDs. Again, it is likely that the most effective approach to treatment would involve combining these treatment approaches, which may address the majority or all of the challenges.

Thus, it is apparent that much critical work remains to be done in order to optimize therapy for these rare, but devastating diseases.

References

Anderson LJ, Henley W, Wyatt KM, Nikolaou V, Waldek S, Hughes DA, Pastores GM, Logan S (2014) Long-term effectiveness of enzyme replacement therapy in children with Gaucher disease: results from the NCS-LSD cohort study. J Inherit Metab Dis 37(6):961–968

Banugaria SG, Prater SN, Ng YK, Kobori JA, Finkel RS, Ladda RL, Chen YT, Rosenberg AS, Kishnani PS (2011) The impact of antibodies on clinical outcomes in diseases treated with therapeutic protein: lessons learned from infantile Pompe disease. Genet Med 13(8):729–736

Banugaria SG, Prater SN, McGann JK, Feldman JD, Tannenbaum JA, Bailey C, Gera R, Conway RL, Viskochil D, Kobori JA, Rosenberg AS, Kishnani PS (2013a) Bortezomib in the rapid reduction of high sustained antibody titers in disorders treated with therapeutic protein: lessons learned from Pompe disease. Genet Med 15(2):123–131

Banugaria SG, Prater SN, Patel TT, Dearmey SM, Milleson C, Sheets KB, Bali DS, Rehder CW, Raiman JA, Wang RA, Labarthe F, Charrow J, Harmatz P, Chakraborty P, Rosenberg AS, Kishnani PS (2013b) Algorithm for the early diagnosis and treatment of patients with cross reactive immunologic material-negative classic infantile pompe disease: a step towards improving the efficacy of ERT. PLoS One 8(6):e67052

Barton NW, Brady RO, Damrosia JM et al (1991) Replacement therapy for inherited enzyme deficiency-macrophage-targeted glucocerebrosidase for Gaucher's disease. N Engl J Med 324:1464–1470

Bennett LL, Mohan D (2013) Gaucher disease and its treatment options. Ann Pharmacother 47(9):1182–1193. doi:10.1177/1060028013500469

Burton BK (2012) Newborn screening for Pompe disease: an update, 2011. Am J Med Genet C Semin Med Genet 160C(1):8–12

Chien YH, Lee NC, Huang HJ, Thurberg BL, Tsai FJ, Hwu WL (2011) Later-onset Pompe disease: early detection and early treatment initiation enabled by newborn screening. J Pediatr 158(6):1023–1027.e1

Chien YH, Hwu WL, Lee NC (2013) Pompe disease: early diagnosis and early treatment make a difference. Pediatr Neonatol 54(4):219–227, Epub 2013 Apr 28

Clarke L, Wraith JE, Beck M et al (2009) Long-term efficacy and safety of laronidase in the treatment of mucopolysaccharidosis I. Pediatrics 123:229–240

de Vries JM, van der Beek NA, Kroos MA, Ozkan L, van Doorn PA, Richards SM, Sung CC, Brugma JD, Zandbergen AA, van der Ploeg AT, Reuser AJ (2010) High antibody titer in an adult with Pompe disease affects treatment with alglucosidase alfa. Mol Genet Metab 101(4):338–345, Epub 2010 Aug 14

Desnick RJ (2004) Enzyme replacement and enhancement therapies for lysosomal diseases. J Inherit Metab Dis 27(3):385–410

Eng CM, Guffon N, Wilcox WR, Germain DP, Lee P, Waldek S, Caplan L, Linthorst GE, Desnick RJ, International Collaborative Fabry Disease Study Group (2001) Safety and efficacy of recombinant human α-galactosidase A replacement therapy in Fabry's disease. N Engl J Med 345(1):9–16

Grubb JH, Vogler C, Levy B, Galvin N, Tan Y, Sly WS (2008) Chemically modified β-glucuronidase crosses blood–brain barrier and clears neuronal storage in murine mucopolysaccharidosis VII. Proc Natl Acad Sci 105(7):2616–2621

Hollak CE, Wijburg FA (2014) Treatment of lysosomal storage disorders: successes and challenges. J Inherit Metab Dis 37(4):587-598

Jones HN, Muller CW, Lin M, Banugaria SG, Case LE, Li JS, O'Grady G, Heller JH, Kishnani PS (2010) Oropharyngeal dysphagia in infants and children with infantile Pompe disease. Dysphagia 25(4):277–283

Kishnani PS, Corzo D, Nicolino M, Byrne B, Mandel H, Hwu WL, Leslie N, Levine J, Spencer C, McDonald M, Li J, Dumontier J, Halberthal M, Chien YH, Hopkin R, Vijayaraghavan S, Gruskin D, Bartholomew D, van der Ploeg A, Clancy JP, Parini R, Morin G, Beck M, De la Gastine GS, Jokic M, Thurberg B, Richards S, Bali D, Davison M, Worden MA, Chen YT,

Wraith JE (2007) Recombinant human acid [alpha]-glucosidase: major clinical benefits in infantile onset Pompe disease. Neurology 68:99–109

Kishnani PS, Goldenberg PC, DeArmey SL, Heller J, Benjamin D, Young S, Bali D, Smith SA, Li JS, Mandel H, Koeberl D, Rosenberg A, Chen YT (2010) Cross-reactive immunologic material status affects treatment outcomes in Pompe disease infants. Mol Genet Metab 99(1):26–33

Kishnani PS, Beckemeyer AA, Mendelsohn NJ (2012) The new era of Pompe disease: advances in the detection, understanding of the phenotypic spectrum, pathophysiology, and management. Am J Med Genet C Semin Med Genet 160(10):1–7

Kobayashi H, Shimada Y, Ikegami M, Kawai T, Sakurai K, Urashima T, Ijima M, Fujiwara M, Kaneshiro E, Ohashi T, Eto Y, Ishigaki K, Osawa M, Kyosen SO, Ida H (2010) Prognostic factors for the late onset Pompe disease with enzyme replacement therapy: from our experience of 4 cases including an autopsy case. Mol Genet Metab 100(1):14–19

Lacaná E, Yao LP, Pariser AR, Rosenberg AS (2012) The role of immune tolerance induction in restoration of the efficacy of ERT in Pompe disease. Am J Med Genet C Semin Med Genet 160C(1):30–39

Liao HC, Chiang CC, Niu DM, Wang CH, Kao SM, Tsai FJ, Huang YH, Liu HC, Huang CK, Gao HJ, Yang CF, Chan MJ, Lin WD, Chen YJ (2014) Detecting multiple lysosomal storage diseases by tandem mass spectrometry – a national newborn screening program in Taiwan. Clin Chim Acta 431:80–86, Epub 2014 Feb 7

Linthorst GE, Hollak CE, Donker-Koopman WE, Strijland A, Aerts JM (2004) Enzyme therapy for Fabry disease: neutralizing antibodies toward agalsidase alpha and beta. Kidney Int 66(4):1589–1595

Messinger YH, Mendelsohn NJ, Rhead W, Dimmock D, Hershkovitz E, Champion M, Jones SA, Olson R, White A, Wells C, Bali D, Case LE, Young SP, Rosenberg AS, Kishnani PS (2012) Successful immune tolerance induction to enzyme replacement therapy in CRIM-negative infantile Pompe disease. Genet Med 14(1):135–142

Muenzer J (2014) Early initiation of enzyme replacement therapy for the mucopolysaccharidoses. Mol Genet Metab 111(2):63-72

Muenzer J, Wraith JE, Beck M et al (2006) A phase II/III clinical study of enzyme replacement therapy with idursulfase in mucopolysaccharidosis II (Hunter syndrome). Genet Med 8:465–473

Nascimbeni AC, Fanin M, Masiero E, Angelini C, Sandri M (2012) Impaired autophagy contributes to muscle atrophy in glycogen storage disease type II patients. Autophagy 8(11):1697–1700. doi:10.4161/auto.21691, Epub 2012 Aug 31

Nicolino M, Byrne B, Wraith JE, Leslie N, Mandel H, Freyer DR, Arnold GL, Pivnick EK, Ottinger CJ, Robinson PH, Loo JC, Smitka M, Jardine P, Tatò L, Chabrol B, McCandless S, Kimura S, Mehta L, Bali D, Skrinar A, Morgan C, Rangachari L, Corzo D, Kishnani PS (2009) Clinical outcomes after long-term treatment with alglucosidase alfa in infants and children with advanced Pompe disease. Genet Med 11(3):210–219

Noh H, Lee JI (2014) Current and potential therapeutic strategies for mucopolysaccharidoses. J Clin Pharm Ther 39(3):215–224. doi:10.1111/jcpt.12136, Epub 2014 Feb 25

Pieroni M, Camporeale A, Della Bona R, Sabini A, Cosmi D, Magnolfi A, Bolognese L (2013) Progression of Fabry cardiomyopathy despite enzyme replacement therapy. Circulation 128(15):1687–1688

Poll LW, Vom Dahl S (2009) Image of the month. Hepatic Gaucheroma mimicking focal nodular hyperplasia. Hepatology 50(3):985–986

Prater SN, Banugaria SG, DeArmey SM, Botha EG, Stege EM, Case LE, Jones HN, Phornphutkul C, Wang RY, Young SP, Kishnani PS (2012) The emerging phenotype of long-term survivors with infantile Pompe disease. Genet Med 14(9):800–810

Prater SN, Patel TT, Buckley AF, Mandel H, Vlodavski E, Banugaria S, Feeney EJ, Raben N, Kishnani PS (2013) Skeletal muscle pathology of infantile Pompe disease during long-term enzyme replacement therapy. Orphanet J Rare Dis 8:90

Raben N, Danon M, Gilbert AL, Dwivedi S, Collins B, Thurberg BL, Mattaliano RJ, Nagaraju K, Plotz PH (2003) Enzyme replacement therapy in the mouse model of Pompe disease. Mol Genet Metab 80(1–2):159–169

Raben N, Fukuda T, Gilbert AL et al (2005) Replacing acid alpha-glucosidase in Pompe disease: recombinant and transgenic enzymes are equipotent, but neither completely clears glycogen from type II muscle fibers. Mol Ther 11:48–56

Raben N, Takikita S, Pittis MG, Bembi B, Marie SK, Roberts A, Page L, Kishnani PS, Schoser BG, Chien YH, Ralston E, Nagaraju K, Plotz PH (2007) Deconstructing Pompe disease by analyzing single muscle fibers: to see a world in a grain of sand…. Autophagy 3(6):546–552, Epub 2007 Jun 15

Raben N, Ralston E, Chien YH, Baum R, Schreiner C, Hwu WL, Zaal KJ, Plotz PH (2010) Differences in the predominance of lysosomal and autophagic pathologies between infants and adults with Pompe disease: implications for therapy. Mol Genet Metab 8:324–331

Raben N, Wong A, Ralston E, Myerowitz R (2012) Autophagy and mitochondria in Pompe disease: nothing is so new as what has long been forgotten. Am J Med Genet C Semin Med Genet 160C(1):13–21

Rosenbloom BE, Weinreb NJ, Zimran A, Kacena KA, Charrow J, Ward E (2005) Gaucher disease and cancer incidence: a study from the Gaucher Registry. Blood 105(12):4569–4572

Schiffmann R, Fitzgibbon EJ, Harris C, DeVile C, Davies EH, Abel L, van Schaik IN, Benko W, Timmons M, Ries M, Vellodi A (2008) Randomized, controlled trial of miglustat in Gaucher's disease type 3. Ann Neurol 64(5):514–522

Shea L, Raben N (2009) Autophagy in skeletal muscle: implications for Pompe disease. Int J Clin Pharmacol Ther 47(Suppl 1):S42–S47

Shigeto S, Katafuchi T, Okada Y, Nakamura K, Endo F, Okuyama T, Takeuchi H, Kroos MA, Verheijen FW, Reuser AJ, Okumiya T (2011) Improved assay for differential diagnosis between Pompe disease and acid α-glucosidase pseudodeficiency on dried blood spots. Mol Genet Metab 103(1):12–17, Epub 2011 Jan 22

Spiridigliozzi GA, Heller JH, Kishnani PS, Van der Ploeg AT, Ebbink BJ, Aarsen FK, van Gelder CM, Van den Hout JMP (2013) Cognitive outcome of patients with classic infantile Pompe disease receiving enzyme therapy. Neurology 80(12):1173–1173

Strothotte S, Strigl-Pill N, Grunert B, Kornblum C, Eger K, Wessig C, Deschauer M, Breunig F, Glocker FX, Vielhaber S, Brejova A, Hilz M, Reiners K, Müller-Felber W, Mengel E, Spranger M, Schoser B (2010) Enzyme replacement therapy with alglucosidase alfa in 44 patients with late-onset glycogen storage disease type 2: 12-month results of an observational clinical trial. J Neurol 257(1):91–97

Thurberg BL, Lynch Maloney C, Vaccaro C, Afonso K, Tsai AC, Bossen E, Kishnani PS, O'Callaghan M (2006) Characterization of pre- and post-treatment pathology after enzyme replacement therapy for Pompe disease. Lab Invest 8:1208–1220

Valayannopoulos V, Wijburg FA (2011) Therapy for the mucopolysaccharidoses. Rheumatology (Oxford) 50(Suppl 5):v49–v59

Weidemann F, Niemann M, Störk S, Breunig F, Beer M, Sommer C, Herrmann S, Ertl G, Wanner C (2013) Long-term outcome of enzyme-replacement therapy in advanced Fabry disease: evidence for disease progression towards serious complications. J Intern Med 274(4):331–341, Epub 2013 May 6

Weidemann F, Krämer J, Duning T, Lenders M, Canaan-Kühl S, Krebs A, González HG, Sommer C, Uçeyler N, Niemann M, Störk S, Schelleckes M, Reiermann S, Stypmann J, Brand SM, Wanner C, Brand E (2014) Patients with Fabry disease after enzyme replacement therapy dose reduction versus treatment switch. J Am Soc Nephrol 25(4):837–849

Winkel LP, Van den Hout JM, Kamphoven JH, Disseldorp JA, Remmerswaal M, Arts WF, Loonen MC, Vulto AG, Van Doorn PA, De Jong G, Hop W, Smit GP, Shapira SK, Boer MA, van Diggelen OP, Reuser AJ, Van der Ploeg AT (2004) Enzyme replacement therapy in late-onset Pompe's disease: a three-year follow-up. Ann Neurol 55(4):495–502

Yanovitch TL, Casey R, Banugaria SG, Kishnani PS (2010) Improvement of bilateral ptosis on higher dose enzyme replacement therapy in Pompe disease. J Neuroophthalmol 30(2): 165–166

Zhu Y, Li X, Kyazike J, Zhou Q, Thurberg BL, Raben N, Mattaliano RJ, Cheng SH (2004) Conjugation of mannose 6-phosphate-containing oligosaccharides to acid alpha-glucosidase improves the clearance of glycogen in pompe mice. J Biol Chem 279(48):50336–50341, Epub 2004 Sep 21

Muscle Targeting

Nancy M. Dahms

Introduction

Pompe disease (OMIM #232300), also known as glycogen storage disorder type II (GSD II) or acid maltase deficiency, is an autosomal recessive disorder caused by a deficiency of the lysosomal enzyme acid α-glucosidase (GAA; acid maltase, E.C.3.2.1.20) (Cori 1952; Hers 1963; Hirschhorn and Reuser 2001). GAA is solely responsible for hydrolyzing the α-1,4 and α-1,6 linkages in branched glycogen molecules within the acidic environment of lysosomes (Brown et al. 1970). Because this is a systemic disorder, the lack of GAA activity results in the aberrant accumulation of glycogen in numerous cell types, with some cell types being more severely impacted than others. Notably, striated (cardiac and skeletal) muscle and liver are more dependent on GAA activity as evidenced by the massive glycogen storage observed in these tissues (Hirschhorn and Reuser 2001; Malicdan et al. 2008; Shea and Raben 2009). However, the clinical manifestations of Pompe disease are most prominent in cardiac and skeletal muscle. Patients exhibit a wide range of phenotypes, with the onset and clinical course of Pompe disease correlating with the level of residual enzyme activity. The most severely affected are infants with a complete loss of enzyme function. This classical infantile form exhibits rapid progression, and typically death occurs prior to 2 years of age due to cardiomyopathy and cardiorespiratory insufficiency (van den Hout et al. 2003; Kishnani et al. 2006). In contrast, patients with residual enzyme function can survive into adulthood, and individuals with very mild forms of the disease can live to 60 or 70 years of age. In this adult or late-onset form of the disease, patients experience progressive skeletal

This work was supported by National Institutes of Health grant R01DK042667

N.M. Dahms (✉)
Department of Biochemistry, Medical College of Wisconsin,
8701 Watertown Plank Road, Milwaukee, WI 53226, USA
e-mail: ndahms@mcw.edu

A. Rosenberg, B. Demeule (eds.), *Biobetters*, AAPS Advances
in the Pharmaceutical Sciences Series 19, DOI 10.1007/978-1-4939-2543-8_3

muscle weakness, generally without cardiac involvement, and eventually succumb to respiratory failure as a result of diaphragmatic weakness (Winkel et al. 2005; Wokke et al. 2008). The clinical heterogeneity observed in this monogenic disease can be explained, in part, by the diversity of mutations (over 460 sequence variants of the GAA gene have been reported (see database at www.pompecenter.nl)) that results in the variable level of GAA activity. The clinical phenotype is further complicated by the fact that most Pompe patients are compound heterozygotes expressing two different mutant alleles (Kroos et al. 2012a, b).

The pathogenesis of Pompe disease is complex, and it is not fully understood why cardiac and muscle cells are more severely impacted by glycogen accumulation than other cell types. A general picture of the disease process includes: (1) an increase in the number and size of glycogen-filled lysosomes that mechanically disrupt contractile units, leading to a reduced myofibrillar transmission force, (2) rupture of swollen lysosomes, with the release of their contents (i.e., hydrolytic enzymes and the massive accumulation of glycogen) into the cytoplasm causing muscle damage (Hesselink et al. 2003; Thurberg et al. 2006), and (3) impaired autophagy and mitochondrial dysfunction which contribute to muscle damage (Raben et al. 2012). Currently, the only FDA-approved treatment for Pompe disease is enzyme replacement therapy (ERT) involving bi-weekly intravenous infusion of recombinant human GAA (rhGAA) (Kishnani et al. 2007; Byrne et al. 2011). Clinical studies have shown that ERT increased the ventilator-free survival of infant-onset Pompe patients (Kishnani et al. 2006, 2007; Amalfitano et al. 2001). Although administration of rhGAA reduced hypertrophic cardiomyopathy in these patients, it was less effective in treating the cellular pathology of skeletal muscle. These results and others (Nicolino et al. 2009) are consistent with data from a number of studies in the Pompe mouse knockout model which show that administration of rhGAA results in clearance of stored glycogen from liver and cardiac muscle, with only modest effects observed in skeletal muscle (Raben et al. 1998, 2003; Zhu et al. 2005). Respiratory failure is the leading cause of morbidity in adult-onset Pompe patients. Given that the progressive respiratory and proximal muscle weakness cause most adult-onset Pompe patients to require ambulatory and ventilator support (Hagemans et al. 2005a, b; Oba-Shinjo et al. 2009), there is a great need to develop new and improved treatments aimed at reducing the aberrant glycogen load in skeletal muscle.

Delivery of GAA to Lysosomes

Lysosomes are acidified organelles which break down macromolecules that are delivered to them by endocytic, phagocytic, and autophagic pathways. The activity of lysosomes is dependent upon the acquisition of a diverse set of over 60 soluble acid hydrolases that are capable of degrading a heterogeneous population of proteins, lipids, and glycans (Holtzman 1989). These hydrolytic enzymes are glycoproteins that are synthesized in the endoplasmic reticulum. As they travel through the secretory pathway, the acid hydrolases' *N*-glycans undergo selective modification and

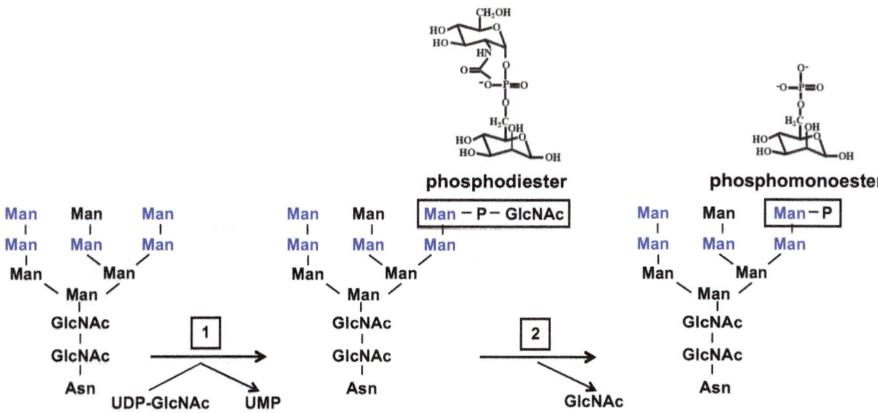

Fig. 1 Synthesis of the M6P tag on *N*-linked glycans. In the *cis* Golgi, GlcNAc-1-phosphotransferase specifically recognizes newly synthesized acid hydrolases and transfers GlcNAc-1-phosphate from UDP-GlcNAc to selected mannose residues (*blue*) on the acid hydrolases' *N*-glycans to form the phosphodiester Man-P-GlcNAc (step 1). In the *trans* Golgi network, the uncovering enzyme removes the GlcNAc to generate the phosphomonoester M6P (step 2)

acquire a specific tag, mannose 6-phosphate (M6P), that marks them for delivery to lysosomes by specific receptors (Kornfeld and Sly 2001; Braulke and Bonifacino 2009) (Fig. 1). In the *cis* Golgi, UDP-*N*-acetylglucosamine:lysosomal enzyme *N*-acetylglucosamine-1-phosphotransferase (GlcNAc-1-phosphotransferase; EC 2.7.8.17) distinguishes the ~60 acid hydrolases from among the myriad of proteins traveling through the secretory pathway and adds GlcNAc-1-phosphate to the C-6 hydroxyl group of selected mannose residues to form a phosphodiester, M6P-GlcNAc (Hasilik et al. 1980; Varki and Kornfeld 1980; Bao et al. 1996; Kudo et al. 2005). In the *trans* Golgi network (TGN), A second enzyme, *N*-acetylglucosamine-1-phosphodiester α-*N*-acetylglucosaminidase (uncovering enzyme; EC 3.1.4.45), removes the GlcNAc residue to reveal a phosphomonoester, M6P, that serves as a high affinity ligand for M6P receptors (MPRs) (Waheed et al. 1981; Varki et al. 1983; Kornfeld et al. 1999; Rohrer and Kornfeld 2001; Do et al. 2002).

MPRs, which are expressed on most cell types, are defined by their unique ability to bind phosphomannosyl residues. There are two MPRs, the ~46 kDa cation-dependent MPR (CD-MPR) and the ~300 kDa cation-independent MPR (CI-MPR), both of which are dimeric type I membrane glycoproteins (Fig. 2). CD-MPR is encoded by the *M6PR* gene and has an ~160-residue extracytoplasmic region containing a single carbohydrate-binding domain, termed the MRH domain (Mannose 6-phosphate Receptor Homology). In contrast, CI-MPR is encoded by the *IGF2R* gene and contains a large, ~2,300-residue extracytoplasmic region with 15 contiguous MRH domains, three of which bind carbohydrate (domains 3, 5 and 9) (reviewed in Kim et al. 2009; Brown et al. 2009). MPRs travel continuously between the TGN, endosomes, and the plasma membrane during their lifetime (reviewed in Braulke and Bonifacino 2009). MPRs carry newly synthesized

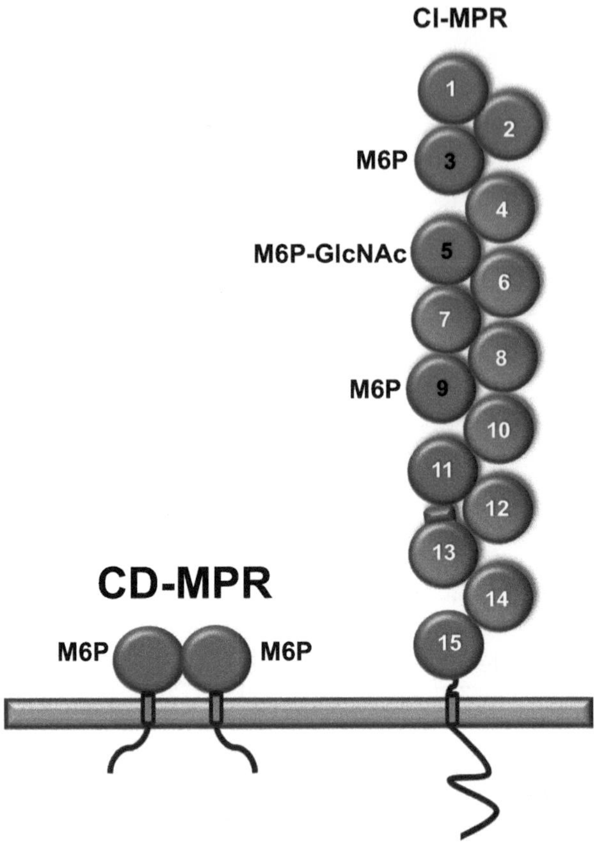

Fig. 2 Schematic diagram of the MPR proteins. The 46 kDa CD-MPR and 300 kDa CI-MPR are type I transmembrane glycoproteins that contain a single hydrophobic membrane-spanning domain. The CD-MPR is a homodimer with ~70-residue cytoplasmic region and a single extracytoplasmic MRH domain that binds the phosphomonoester M6P. The CI-MPR has an ~165-residue cytoplasmic region and a large extracellular region comprised of 15 contiguous MRH domains. MRH domains 3 and 9 bind M6P whereas MRH domain 5 of the CI-MPR interacts preferentially with the phosphodiester (M6P-GlcNAc). For simplicity, only the single polypeptide of the homodimeric CI-MPR is shown

M6P-tagged acid hydrolases from the TGN to late endosomal compartments where the acidic environment of the late endosome causes MPRs to release their cargo. The hydrolytic enzymes are packaged into lysosomes whereas the receptors recycle back to the TGN to repeat the process or move to the cell surface where the CI-MPR, but not the CD-MPR, can internalize exogenous ligands (Tong and Kornfeld 1989; Hoflack and Kornfeld 1985; Distler et al. 1987; Watanabe et al. 1990; Hoflack et al. 1987). Thus, the cell surface internalization of exogenously supplied ligands, as occurs in ERT, is dependent upon receptor-mediated endocytosis carried out by the CI-MPR.

The *GAA* gene encodes a 952-residue precursor protein of 110 kDa. Like the other soluble acid hydrolases, GAA is a glycoprotein that acquires the M6P tag on its *N*-glycans. GAA is a heavily glycosylated protein: it contains seven *N*-glycosylation sites, all of which are glycosylated (Hermans et al. 1993). Following its MPR-mediated delivery to lysosomes, GAA undergoes proteolytic processing to form the 76 and 70 kDa catalytically active, mature forms of the enzyme found within lysosomes (Wisselaar et al. 1993; Moreland et al. 2005).

ERT and rhGAA Efficacy

The rationale behind ERT for Pompe disease is based on providing cells with a functional GAA enzyme. The excessive accumulation of glycogen in cardiac and skeletal muscle of Pompe patients is caused by insufficient amounts of catalytically active GAA in the lysosome as a result of a genetic mutation of the *GAA* gene (Hirschhorn and Reuser 2001). The ability to deliver adequate levels of functional GAA to the lysosome by ERT is key for the degradation of the accumulated lysosomal glycogen. The goal of reversing the underlying cellular pathology (e.g., aberrant glycogen accumulation) is to significantly improve the clinical outcomes for Pompe patients.

Large-scale production of recombinant human GAA (rhGAA) was obtained in transgenic rabbit milk (Bijvoet et al. 1999) and Chinese hamster ovary (CHO) cells (Van Hove et al. 1996). Although administration of rhGAA from both origins was effective in clearing glycogen from cardiac muscle and liver of Pompe mice, ERT was much less efficient in removing aberrant glycogen stores from skeletal muscle (Kishnani et al. 2006, 2007; Amalfitano et al. 2001; Raben et al. 2005; Hawes et al. 2007). Currently, ERT using rhGAA (alglucosidase alfa, Genzyme Corporation, Inc.) produced in CHO cells is the only FDA-approved treatment for Pompe disease. The approved dose of rhGAA is 20–40 times higher than that used for other lysosomal storage diseases (20 mg/kg biweekly dose versus between 0.5 and 1 mg/kg for the lysosomal storage diseases Gaucher, Fabry and mucopolysaccharidosis I). Injection of this relatively high dose of exogenous rhGAA protein every 2 weeks is associated with induction of a sustained immune response to the recombinant enzyme, especially in cross-reactive immunological material (CRIM)-negative patients (e.g., patients unable to produce native enzyme protein due to deleterious mutations) (Kishnani et al. 2010; Amalfitano et al. 2001).

Several factors likely contribute to this relatively high dose, which has limited efficacy in reducing glycogen accumulation in skeletal muscle (Nicolino et al. 2009; Raben et al. 1998, 2003; Zhu et al. 2005). These include the properties of the targeted tissue itself in which there is a relatively low blood flow to skeletal muscle tissue compared to liver, heart and other tissues, and the expression level of CI-MPR, which is low in skeletal muscle cells (Wenk et al. 1991; Funk et al. 1992). In addition, rhGAA has only about 0.9–1.2 mol M6P per mole of enzyme and only a small fraction of rhGAA glycans contain two M6P moieties (McVie-Wylie et al. 2008;

Yang et al. 1998). Because glycans with a single M6P bind CI-MPR exhibit a lower affinity than glycans containing two phosphomannosyl residues (Tong et al. 1989; Tong and Kornfeld 1989; Varki and Kornfeld 1983; Bohnsack et al. 2009), the existing rhGAA preparations used for ERT are not efficiently targeted to the lysosome via receptor-mediated endocytosis due to their low affinity interaction with the CI-MPR on the cell surface.

In recent years, several strategies described below have been employed to address the limitations of the existing ERT for Pompe disease, with the ultimate goal of reducing glycogen accumulation in skeletal muscle of Pompe patients and thereby eliminating the need for ambulatory and ventilator support.

Glycoengineering of rhGAA

Because the existing rhGAA preparations produced in CHO cells have a low content of M6P, several strategies have been used to increase the number of phosphomannosyl groups on rhGAA, including enzymatic modification of rhGAA, chemical conjugation of synthetic glycans onto rhGAA, and production of rhGAA in engineered yeast strains followed by enzymatic modification of the glycans. These studies tested the hypothesis that increased M6P content on rhGAA will increase its affinity of interaction with CI-MPRs, and result in higher amounts of intravenously administered rhGAA being internalized by CI-MPRs and delivered to lysosomes.

Stepwise Enzymatic Modification of rhGAA

The *N*-glycans of acid hydrolases undergo a two-step modification in the Golgi to acquire the M6P tag: (1) GlcNAc-1-phosphotransferase recognizes a lysine-containing, conformation-dependent motif unique to soluble acid hydrolases and adds GlcNAc-1-phosphate to selected mannose residues to form a phosphodiester, M6P-GlcNAc (Hasilik et al. 1980; Varki and Kornfeld 1980; Reitman and Kornfeld 1981a, b; Bao et al. 1996; Cuozzo et al. 1998; Kudo et al. 2005), and (2) uncovering enzyme removes the GlcNAc residue to generate the phosphomonoester, M6P. The cloning and expression of recombinant GlcNAc-1-phosphotransferase (Kudo and Canfield 2006) and the purification of the uncovering enzyme (Kornfeld et al. 1998; Do et al. 2002) allowed for the generation of a form of rhGAA with a higher M6P content (referred to as HP-GAA) using the enzymes that in vivo normally modify GAA's *N*-glycans. rhGAA was produced in CHO-K1 cells that were cultured in the presence of the α-mannosidase inhibitor kifunensine, and the purified rhGAA containing high mannose-type glycans was incubated sequentially in vitro with recombinant GlcNAc-1-phosphotransferase and purified uncovering enzyme (Chavez et al. 2007). Although this HP-GAA preparation had increased M6P content compared to rhGAA produced in non-kifunensine-treated CHO cells (~3.5 mol/mol for

HP-GAA versus ~1.3 mol/mol for CHO-GAA), a higher level of glycans containing two M6P, and bound CI-MPR in vitro with higher affinity than CHO-GAA, testing of this hyper-M6P variant revealed that it was no more effective than the CHO-GAA at clearing accumulated glycogen from the tissues of Pompe mice (McVie-Wylie et al. 2008). The authors hypothesized that this was due to HP-GAA presenting additional mannose residues that led to the nonproductive sequestration of the infused enzyme by mannose receptors on endothelial cells and/or fibroblasts, rather than GAA being internalized via CI-MPRs on the desired skeletal muscle cells. Although three different doses (i.e., 20, 60 and 100 mg/kg) were used in these studies (McVie-Wylie et al. 2008), the use of a broader range of doses in their protocol may have revealed differences in the efficacy of the different rhGAA preparations.

Chemical Conjugation of Synthetic Glycans onto rhGAA

Synthetic glycans of the high mannose-type bearing M6P have been conjugated to the *N*-glycans of rhGAA produced in CHO cells using several different chemistries. Key to this approach is that the catalytic activity and stability of GAA must remain intact following the coupling chemistry and conjugation (Zhou et al. 2011), and with no significant negative impact on the enzyme's pharmacokinetics. In addition, the method must be suitable for large-scale manufacture if it is to be used for ERT. Cheng and co-workers synthesized a bi-antennary glycan containing six mannose residues, with the two terminal mannose residues being phosphorylated and in an α1,2-linkage to the penultimate mannose. The design of the glycan was based on the higher affinity of the CI-MPR for α1,2-linked terminal M6P residues compared to α1,3- and α1,6-linked M6P-containing glycans (Distler et al. 1991), and CI-MPR's higher affinity for glycans bearing two M6P residues rather than one (Tong et al. 1989; Tong and Kornfeld 1989; Varki and Kornfeld 1983; Bohnsack et al. 2009). In addition, the design of this bi-antennary glycan lacking free terminal mannose residues, in contrast to HP-GAA, was chosen to reduce the potential interaction of the glycan with mannose receptors present on other cell types, such as endothelial cells. The synthetic glycan was conjugated to periodate-oxidized sialic acid residues of rhGAA using either carbonyl-coupled hydrazone chemistry (neo-rhGAA) (Zhu et al. 2004, 2005) or carbonyl-coupled oxime chemistry (oxime-neo-rhGAA) (Zhu et al. 2009). Both approaches resulted in the modified rhGAA exhibiting a higher affinity for the CI-MPR than the unmodified enzyme. Importantly, in comparison to the unmodified rhGAA, neo-rhGAA and oxime-neo-rhGAA displayed an increase in cellular uptake both in vitro and in vivo, and an increase in glycogen clearance from muscle (heart, diaphragm, quadriceps, and triceps) of the Pompe mouse. Together, these studies show that the cellular uptake and delivery of rhGAA to lysosomes can be improved by providing high-affinity ligands for the CI-MPR.

Zhou et al. has extended these studies by synthesizing a series of glycans with mono-, di-, tri-, tetra-, or hexamannoses containing one or two phosphates

(Zhou et al. 2013). The synthesized glycans were attached to an α or β aminooxy linker, then conjugated to oxidized terminal sialic acid residues of rhGAA. The results showed that the β-linked hexamannose with two phosphates was superior to the α-linked hexamannose with two phosphates with respect to receptor-mediated cellular uptake and glycogen clearance in the Pompe mouse. Importantly, a two to fourfold improvement in the efficacy of glycogen clearance in quadriceps was observed for the β-linked hexamannose with two phosphates compared to rhGAA. The authors suggest that the β-linked hexamannose may mimic the naturally occurring *N*-glycans that are attached to the core GlcNAc via a β linkage.

Chen and co-workers have recently generated synthetic glycans containing M6P phosphomonoesters and/or M6P-GlcNAc phosphodiesters (Liu and Chen 2011; Liu et al. 2013). Because the CI-MPR contains three carbohydrate-binding domains, with MRH domain 3 and MRH domain 9 specific for M6P and MRH domain 5 specific for M6P-GlcNAc (Song et al. 2009; Bohnsack et al. 2009) (Fig. 2), the aim here is to generate high affinity ligands that interact with the CI-MPR in a multivalent fashion. To date, these synthetic glycans have not yet been evaluated for binding to the CI-MPR. Additional studies are needed to determine whether these glycans will prove useful for enhancing CI-MPR-mediated uptake of rhGAA by muscle cells.

Designer Yeast Strains Plus Glycosidase Treatment

Rather than producing rhGAA in mammalian cells, Tiels et al. took another approach using yeast strains *Yarrowia lipolytica* and *Pichia pastoris* that have been "humanized" to express human-type Man8/Man9-containing glycans rather than the hypermannosylated glycans typically found in yeast (Tiels et al. 2012). Yeast synthesize M6P containing *N*-glycans on their secreted and cell wall-associated mannoproteins as phosphodiesters that are capped with a mannose (Ballou 1975), rather than with a GlcNAc residue as found on mammalian acid hydrolases. It should be noted that yeast do not express MPRs and do not use phosphomannosyl residues to target their acid hydrolases to lysosomes (vacuoles). However, this *N*-glycan biosynthetic pathway of yeast provides a template for generating M6P-containing recombinant acid hydrolases. The challenge in the field had been identifying a method to efficiently remove the capping mannose residue, thereby revealing the M6P phosphomonoester. Tiel et al. recently solved this problem by identifying and solving the crystal structure of an enzyme from the bacterium *Cellulosimicrobium cellulans* that has this uncapping activity (Tiels et al. 2012). rhGAA purified from an optimized M6P-Man-modifying *Y. lipolytica* strain (yrhGAA) was incubated with the newly discovered uncapping enzyme (CcGh92_5) followed by α-mannosidase. α-mannosidase treatment was performed to remove terminal, non-phosphorylated mannose residues, thereby reducing the overall mannose content as a means to decrease the likelihood of mannose receptor-mediated clearance of the infused recombinant enzyme. yrhGAA was found to have a 15-fold higher M6P content than alglucosidase alfa (Myozyme®) and showed improved therapeutic efficacy compared to Myozyme®

with respect to CI-MPR-mediated uptake by Pompe patient cells and glycogen clearance from heart, but not quadriceps muscle, in the GAA-knock out mouse model of Pompe disease. However, only a single dose (i.e., 20 mg/kg once a week for 4 weeks) was evaluated in these studies (Tiels et al. 2012). Additional dosing regimens and glycoengineering may be warranted in this system to obtain glycogen clearance from skeletal muscle.

Generation of rhGAA with a Peptide-Based Tag (IGF-2)

ERT for the FDA approved treatments of six different lysosomal storage diseases is dependent upon glycan recognition of the recombinant enzyme, with CI-MPR being the receptor targeted in five out of the six lysosomal storage diseases (Brady 2006). The large-scale production of glycoproteins such as acid hydrolases that rely on specific glycosylation for their biological activity (e.g., M6P-dependent lysosomal targeting) is challenging due to the heterogeneity of glycans and the influence culture conditions can have on glycosylation. A glycosylation-independent strategy has been developed to target acid hydrolases to the lysosome that takes advantage of the multifunctional nature of the CI-MPR: in addition to its three carbohydrate-binding MRH domains, the CI-MPR binds the small (~7.5 kDa) polypeptide insulin-like growth factor 2 (IGF-2) with high affinity ($K_d = \sim 1–2$ nM) via its MRH domain 11 (Brown et al. 2008). This novel peptide-based strategy was developed by LeBowitz et al. and involved the generation of a cDNA construct encoding a chimeric protein in which a portion of IGF-2 was fused to the acid hydrolase β-glucuronidase (hGUS) (LeBowitz et al. 2004), which is the enzyme deficient in the lysosomal storage disease mucopolysaccharidosis VII (MPS VII). The resulting chimeric protein was efficiently internalized into cultured MPS VII fibroblasts by binding to CI-MPR's IGF-2 binding site. Importantly, in vivo studies showed that the IGF-2-tagged hGUS was more effective than untagged hGUS at delivering active enzyme to a wide range of cell types (LeBowitz et al. 2004).

In a recent report, this approach was applied to Pompe disease in which the IGF-2 tag was fused to the N-terminus of rhGAA to create a chimeric enzyme (BMN 701) that binds to the CI-MPR with high affinity (Maga et al. 2013). BMN 701 is delivered more effectively to lysosomes of L6 myoblasts in vitro and infusion of BMN 701 into Pompe mice greatly increases clearance of accumulated glycogen in skeletal muscle tissue compared to equivalent doses of rhGAA. BMN 701 is more potent than untagged rhGAA in muscle tissue glycogen clearance: in most muscle tissues examined, BMN 701 doses of 5 mg/kg led to greater clearance of glycogen than untagged rhGAA doses of 20 mg/kg. Furthermore, BMN 701 appears to be most effective in clearing glycogen from heart, soleus, tibialis anterior, extensor digitorum longus, gastrocnemius, and quadriceps. Unlike rhGAA produced in CHO cells in which only a small fraction (<5 %) of its glycans bear M6P (Tiels et al. 2012), every BMN 701 enzyme contains the IGF-2 tag. This contributes to the finding that BMN 701 is internalized and delivered to the lysosome 26-fold more efficiently than rhGAA, enabling BMN 701 to be about fivefold more effective than rhGAA in its ability to clear glycogen from skeletal muscle of Pompe mice (Maga et al. 2013).

This approach has also been shown to be successful in cell-based assays for another lysosomal storage disorder, mucopolysaccharidosis type IIIB (MPS IIIB or Sanfilippo B syndrome). Fusion of IGF-2 to α-*N*-acetylglucosaminidase (rhNAGLU-IGF2) resulted in enhanced cellular uptake and reduced cellular storage of glycosaminoglycans when compared to the untagged enzyme (Kan et al. 2013). Taken together, these studies demonstrate that the glycosylation-independent IGF-2 tag is a promising alternate method for generating high affinity ligands for the CI-MPR that are suitable for ERT.

Increase CI-MPR Expression on Muscle Cells

The expression level of CI-MPR is relatively low in skeletal muscle cells (Wenk et al. 1991; Funk et al. 1992; Koeberl et al. 2011), which supports the hypothesis that CI-MPR is limiting for ERT in Pompe disease. Consistent with this view is the observation that increasing M6P content on rhGAA or adding a different, high affinity tag specific for the CI-MPR (i.e., IGF-2) improves the efficacy of ERT in Pompe disease (see "Chemical Conjugation of Synthetic Glycans onto rhGAA", "Designer Yeast Strains Plus Glycosidase Treatment", and " Generation of rhGAA with a Peptide-Based Tag (IGF-2)") (Zhu et al. 2005, 2009; McVie-Wylie et al. 2008; Maga et al. 2013). The critical role of the CI-MPR is further supported by a comparison of a transgenic mouse deficient in GAA alone and a double knockout mouse generated by crossing the GAA knockout mouse with a muscle-specific conditional CI-MPR knockout mouse (Koeberl et al. 2011). The results showed that the double knockout mouse exhibits a markedly lower level of correction of aberrant glycogen stores compared to the GAA knockout mouse following administration of the same therapeutic dose of GAA (Koeberl et al. 2011). In addition, fibroblasts from Pompe patients exhibit impaired CI-MPR trafficking and internalization of rhGAA is inhibited (Cardone et al. 2008). The impaired internalization of rhGAA can be explained, in part, by the observed decrease in CI-MPR levels on the cell surface and the reduced recycling of the CI-MPR between the plasma membrane and endosomal compartments (Cardone et al. 2008). Because the existing FDA approved ERT targets the CI-MPR for the treatment of Pompe disease, one strategy to deliver more exogenous enzyme to muscle is to increase CI-MPR protein levels and/or CI-MPR surface expression in muscle cells.

Previous studies by Matsumoto et al. showed that oral administration of clenbuterol, a β_2-agonist known to induce hypertrophy of skeletal muscles such as soleus and masseter muscles, increased the mRNA expression level of CI-MPR in the masseter muscle of rats (Matsumoto et al. 2006). Studies by Koeberl and co-workers have extended these studies and evaluated the effect of clenbuterol treatment on Pompe mice. Western blot analysis demonstrated administration of clenbuterol significantly increased CI-MPR protein levels in heart and quadriceps, and efficacy of rhGAA was improved in these animals as assessed by glycogen content in selected muscle tissue (Koeberl et al. 2011). Subsequent studies indicate that clenbuterol has both CI-MPR-dependent and CI-MPR-independent effects on the efficacy of rhGAA treatment in Pompe mice (Farah et al. 2014). Although preliminary, as this pilot study did not include placebo controls, β_2-agonist treatment of adult Pompe patients with albuterol showed that CI-MPR protein expression was increased in three of the patients as assessed by Western blot analysis of quadriceps muscle biopsy, and improvements were noted in the 6-min walk test for all seven subjects (Koeberl et al. 2014). Clearly more studies are needed to determine whether albuterol treatment will be a useful combination therapy to improve the efficacy of rhGAA treatment in Pompe patients.

Nanocarriers

Polymer nanocarriers have been used for the delivery of acid hydrolases in lysosomal storage diseases. Instead of CI-MPR, intercellular adhesion molecule-1 (ICAM-1), a transmembrane protein involved in inflammation and overexpressed on many cell types under pathological conditions, including lysosomal storage diseases (Marlin and Springer 1987), was the cell surface receptor targeted for the uptake of an exogenously delivered acid hydrolase in Fabry disease (Hsu et al. 2011; Muro et al. 2006). These findings have been expanded upon to include rhGAA. Hsu et al. have shown that rhGAA coupled to polymer nanocarriers coated with an antibody specific to ICAM-1 can be internalized by cells and reduce aberrant glycogen accumulation. In these studies, endothelial cells (human umbilical endothelial cells, HUVEC) were treated with turanose, a competitive inhibitor of GAA that induces intracellular glycogen accumulation, to generate a cellular model of Pompe disease. ICAM-1-specific nanocarriers exhibited a threefold enhancement of glycogen degradation compared to nontargeted rhGAA in this cellular model of Pompe disease (Hsu et al. 2012). Importantly, ICAM-1-specific nanocarriers administered to mice enhanced delivery of GAA in vivo over nontargeted enzyme to every tissue examined, including quadriceps and gastrocnemius muscles. Taken together, these results are promising and indicate that ICAM-1 may serve as an alternate strategy to target enzymes to a number of tissues, including muscle.

Polymer nanocarriers have also been coated with anti-CI-MPR antibodies, and to enhance the internalization of polymer (polystyrene) carriers, the surface of the polymer carriers was functionalized with sphingomyelinase (Ansar et al. 2013). The rationale behind this approach is that the exogenous sphingomyelinase will hydrolyze the phospholipid sphingomyelin on the cell membrane into ceramide. The resulting increased ceramide levels will alter the biophysical properties of membranes, thereby facilitating the formation of engulfment structures to enhance the intracellular transport of nanocarriers. These sphingomyelinase modified nanocarriers coated with anti-CI-MPR displayed improved delivery to lung and kidneys of C57/BL6 mice compared to non-functionalized nanocarriers (Ansar et al. 2013). However, additional studies are required to determine if these anti-CI-MPR coated nanocarriers can be efficiently delivered to skeletal muscle.

Perspectives

The glycobiology surrounding the targeted delivery of acid hydrolases to the lysosome is well understood and has been instrumental in the development of FDA approved ERT for six lysosomal storage diseases. Although the production of synthetic glycans containing up to six mannose residues and two phosphates has been achieved, the in vivo production of M6P-containing glycans that are homogeneous remains a challenge. Increasing the M6P content on rhGAA has been shown to improve the efficacy of ERT in Pompe disease (Zhu et al. 2005, 2009; McVie-Wylie et al. 2008). In order to improve the design of high affinity acid hydrolases that will interact with the receptor in a multivalent fashion, the spatial orientation of the CI-MPR's three carbohydrate binding sites (six in the dimeric structure) is needed. To date, the three-dimensional structures of eight (MRH domains 1–3, domain 5, domains 11–14) out of the 15 MRH domains of the CI-MPR have been determined (Brown et al. 2002, 2008; Uson et al. 2003; Olson et al. 2004a, b, 2010). However, the structure of the entire extracytoplasmic region of the receptor is needed to determine how the three carbohydrate binding sites are oriented spatially with respect to each other. This information will be critical for the design of multivalent ligands in which M6P and/or M6P-GlcNAc residues are properly spaced in order to align and engage with the corresponding carbohydrate binding sites of the CI-MPR.

Clinical trials are ongoing to determine whether the glycosylation-independent IGF-2 tagged version of rhGAA (BMN 701) (Maga et al. 2013) exhibits enhanced delivery to skeletal muscle and other tissues in Pompe patients. Currently, the IGF-2 tag has been engineered successfully into three different acid hydrolases without significant loss of catalytic activity (LeBowitz et al. 2004; Maga et al. 2013; Kan et al. 2013). Thus, this promising strategy may be applicable to a wide range of proteins whose delivery is needed to numerous tissues that express the CI-MPR.

Given the relatively low level of expression of the CI-MPR in skeletal muscle, other receptors that are more abundantly expressed in this tissue should be evaluated. Along these lines, rhGAA coupled nanocarriers coated with an antibody to another receptor, ICAM-1, showed promising results in its ability to deliver rhGAA to mouse quadriceps and gastrocnemius muscles (Hsu et al. 2012). In addition to β_2-agonists, are there other ways to upregulate CI-MPR expression? Studies by Urayama et al. showed that epinephrine treatment of MPS VII mice enhanced the delivery of the missing lysosomal enzyme, β-glucuronidase across the blood brain barrier in a M6P-dependent fashion. The authors hypothesized that the adrenergic effects of epinephrine may modulate the cellular distribution of the CI-MPR (Urayama et al. 2007). Additional studies are needed to understand the molecular mechanism underlying this effect and whether epinephrine may modulate the efficacy of ERT in other tissues.

It is likely that a combination of approaches may be needed to efficiently eliminate the glycogen burden in skeletal muscle of Pompe patients. However, the pathology of Pompe disease is complex and other factors are likely important contributors to muscle damage. For example, accumulation of autophagic debris is found in skeletal muscle fibers and recent studies have shown that overexpression of transcription factors (TFEB and TFE3) that regulate autophagy and autophagosome exocytosis facilitates the clearance of cellular debris in Pompe disease (Spampanato et al. 2013; Martina et al. 2014). These intriguing findings indicate there remain exciting avenues to explore in the treatment of Pompe disease.

References

Amalfitano A, Bengur AR, Morse RP, Majure JM, Case LE, Veerling DL, Mackey J, Kishnani P, Smith W, McVie-Wylie A, Sullivan JA, Hoganson GE, Phillips JA 3rd, Schaefer GB, Charrow J, Ware RE, Bossen EH, Chen YT (2001) Recombinant human acid alpha-glucosidase enzyme therapy for infantile glycogen storage disease type II: results of a phase I/II clinical trial. Genet Med 3:132–138

Ansar M, Serrano D, Papademetriou I, Bhowmick TK, Muro S (2013) Biological functionalization of drug delivery carriers to bypass size restrictions of receptor-mediated endocytosis independently from receptor targeting. ACS Nano 7:10597–10611

Ballou DL (1975) Genetic control of yeast mannan structure: mapping genes mnn2 and mnn4 in *Saccharomyces cerevisiae*. J Bacteriol 123:616–619

Bao M, Booth JL, Elmendorf BJ, Canfield WM (1996) Bovine UDP-N-acetylglucosamine:lysosomal-enzyme N-acetylglucosamine-1-phosphotransferase. I. Purification and subunit structure. J Biol Chem 271:31437–31445

Bijvoet AG, Van Hirtum H, Kroos MA, van de Kamp EH, Schoneveld O, Visser P, Brakenhoff JP, Weggeman M, Van Corven EJ, van der Ploeg AT, Reuser AJ (1999) Human acid alpha-glucosidase from rabbit milk has therapeutic effect in mice with glycogen storage disease type II. Hum Mol Genet 8:2145–2153

Bohnsack RN, Song X, Olson LJ, Kudo M, Gotschall RR, Canfield WM, Cummings RD, Smith DF, Dahms NM (2009) Cation-independent mannose 6-phosphate receptor: a composite of distinct phosphomannosyl binding sites. J Biol Chem 284:35215–35226

Brady RO (2006) Enzyme replacement for lysosomal diseases. Annu Rev Med 57:283–296

Braulke T, Bonifacino JS (2009) Sorting of lysosomal proteins. Biochim Biophys Acta 1793: 605–614

Brown BI, Brown DH, Jeffrey PL (1970) Simultaneous absence of alpha-1,4-glucosidase and alpha-1,6-glucosidase activities (pH 4) in tissues of children with type II glycogen storage disease. Biochemistry 9:1423–1428

Brown J, Esnouf RM, Jones MA, Linnell J, Harlos K, Hassan AB, Jones EY (2002) Structure of a functional IGF2R fragment determined from the anomalous scattering of sulfur. EMBO J 21:1054–1062

Brown J, Delaine C, Zaccheo OJ, Siebold C, Gilbert RJ, Van Boxel G, Denley A, Wallace JC, Hassan AB, Forbes BE, Jones EY (2008) Structure and functional analysis of the IGF-II/IGF2R interaction. EMBO J 27:265–276

Brown J, Jones EY, Forbes BE (2009) Keeping IGF-II under control: lessons from the IGF-II-IGF2R crystal structure. Trends Biochem Sci 34:612–619

Byrne BJ, Falk DJ, Pacak CA, Nayak S, Herzog RW, Elder ME, Collins SW, Conlon TJ, Clement N, Cleaver BD, Cloutier DA, Porvasnik SL, Islam S, Elmallah MK, Martin A, Smith BK, Fuller DD, Lawson LA, Mah CS (2011) Pompe disease gene therapy. Hum Mol Genet 20:R61–R68

Cardone M, Porto C, Tarallo A, Vicinanza M, Rossi B, Polishchuk E, Donaudy F, Andria G, DE Matteis MA, Parenti G (2008) Abnormal mannose-6-phosphate receptor trafficking impairs recombinant alpha-glucosidase uptake in Pompe disease fibroblasts. Pathogenetics 1:6

Chavez CA, Bohnsack RN, Kudo M, Gotschall RR, Canfield WM, Dahms NM (2007) Domain 5 of the cation-independent mannose 6-phosphate receptor preferentially binds phosphodiesters (mannose 6-phosphate N-acetylglucosamine ester). Biochemistry 46:12604–12617

Cori GT (1952) Glycogen structure and enzyme deficiencies in glycogen storage disease. Harvey Lect 48:145–171

Cuozzo JW, Tao K, Cygler M, Mort JS, Sahagian GG (1998) Lysine-based structure responsible for selective mannose phosphorylation of cathepsin D and cathepsin L defines a common structural motif for lysosomal enzyme targeting. J Biol Chem 273:21067–21076

Distler JJ, Patel R, Jourdian GW (1987) Immobilization and assay of low-molecular-weight phosphomannosyl receptor in multiwell plates. Anal Biochem 166:65–71

Distler JJ, Guo JF, Jourdian GW, Srivastava OP, Hindsgaul O (1991) The binding specificity of high and low molecular weight phosphomannosyl receptors from bovine testes. Inhibition studies with chemically synthesized 6-O-phosphorylated oligomannosides. J Biol Chem 266:21687–21692

Do H, Lee WS, Ghosh P, Hollowell T, Canfield W, Kornfeld S (2002) Human mannose 6-phosphate-uncovering enzyme is synthesized as a proenzyme that is activated by the endoprotease furin. J Biol Chem 277:29737–29744

Farah BL, Madden L, Li S, Nance S, Bird A, Bursac N, Yen PM, Young SP, Koeberl DD (2014) Adjunctive beta2-agonist treatment reduces glycogen independently of receptor-mediated acid alpha-glucosidase uptake in the limb muscles of mice with Pompe disease. FASEB J 28(5):2272–2280

Funk B, Kessler U, Eisenmenger W, Hansmann A, Kolb HJ, Kiess W (1992) Expression of the insulin-like growth factor-II/mannose-6-phosphate receptor in multiple human tissues during fetal life and early infancy. J Clin Endocrinol Metab 75:424–431

Hagemans ML, Winkel LP, Hop WC, Reuser AJ, Van Doorn PA, van der Ploeg AT (2005a) Disease severity in children and adults with Pompe disease related to age and disease duration. Neurology 64:2139–2141

Hagemans ML, Winkel LP, Van Doorn PA, Hop WJ, Loonen MC, Reuser AJ, van der Ploeg AT (2005b) Clinical manifestation and natural course of late-onset Pompe's disease in 54 Dutch patients. Brain 128:671–677

Hasilik A, Klein U, Waheed A, Strecker G, von Figura K (1980) Phosphorylated oligosaccharides in lysosomal enzymes: identification of alpha-N-acetylglucosamine(1)phospho(6)mannose diester groups. Proc Natl Acad Sci U S A 77:7074–7078

Hawes ML, Kennedy W, O'Callaghan MW, Thurberg BL (2007) Differential muscular glycogen clearance after enzyme replacement therapy in a mouse model of Pompe disease. Mol Genet Metab 91:343–351

Hermans MM, Wisselaar HA, Kroos MA, Oostra BA, Reuser AJ (1993) Human lysosomal alpha-glucosidase: functional characterization of the glycosylation sites. Biochem J 289:681–686

Hers HG (1963) Alpha-glucosidase deficiency in generalized glycogen storage disease (Pompe's disease). Biochem J 86:11–16

Hesselink RP, Wagenmakers AJ, Drost MR, van der Vusse GJ (2003) Lysosomal dysfunction in muscle with special reference to glycogen storage disease type II. Biochim Biophys Acta 1637:164–170

Hirschhorn R, Reuser AJJ (2001) Glycogen storage disease type II: acid alpha-glucosidase (acid maltase) deficiency. In: Scriver CR, Beaudet al, Sly WS, Valle D (eds) The metabolic and molecular bases of inherited disease, 8th edn. McGraw-Hill, New York

Hoflack B, Kornfeld S (1985) Lysosomal enzyme binding to mouse P388D1 macrophage membranes lacking the 215-kDa mannose 6-phosphate receptor: evidence for the existence of a second mannose 6-phosphate receptor. Proc Natl Acad Sci U S A 82:4428–4432

Hoflack B, Fujimoto K, Kornfeld S (1987) The interaction of phosphorylated oligosaccharides and lysosomal enzymes with bovine liver cation-dependent mannose 6-phosphate receptor. J Biol Chem 262:123–129

Holtzman E (1989) Lysosomes. Plenum, New York

Hsu J, Serrano D, Bhowmick T, Kumar K, Shen Y, Kuo YC, Garnacho C, Muro S (2011) Enhanced endothelial delivery and biochemical effects of alpha-galactosidase by ICAM-1-targeted nanocarriers for Fabry disease. J Control Release 149:323–331

Hsu J, Northrup L, Bhowmick T, Muro S (2012) Enhanced delivery of alpha-glucosidase for Pompe disease by ICAM-1-targeted nanocarriers: comparative performance of a strategy for three distinct lysosomal storage disorders. Nanomedicine 8:731–739

Kan SH, Troitskaya LA, Sinow CS, Haitz K, Todd AK, di Stefano A, Le SQ, Dickson PI, Tippin BL (2013) Insulin-like growth factor II peptide fusion enables uptake and lysosomal delivery of alpha-N-acetylglucosaminidase to mucopolysaccharidosis type IIIB fibroblasts. Biochem J 458(2):281–289

Kim JJ, Olson LJ, Dahms NM (2009) Carbohydrate recognition by the mannose-6-phosphate receptors. Curr Opin Struct Biol 19:534–542

Kishnani PS, Nicolino M, Voit T, Rogers RC, Tsai AC, Waterson J, Herman GE, Amalfitano A, Thurberg BL, Richards S, Davison M, Corzo D, Chen YT (2006) Chinese hamster ovary cell-derived recombinant human acid alpha-glucosidase in infantile-onset Pompe disease. J Pediatr 149:89–97

Kishnani PS, Corzo D, Nicolino M, Byrne B, Mandel H, Hwu WL, Leslie N, Levine J, Spencer C, McDonald M, Li J, Dumontier J, Halberthal M, Chien YH, Hopkin R, Vijayaraghavan S, Gruskin D, Bartholomew D, van der Ploeg A, Clancy JP, Parini R, Morin G, Beck M, de la Gastine GS, Jokic M, Thurberg B, Richards S, Bali D, Davison M, Worden MA, Chen YT, Wraith JE (2007) Recombinant human acid [alpha]-glucosidase: major clinical benefits in infantile-onset Pompe disease. Neurology 68:99–109

Kishnani PS, Goldenberg PC, Dearmey SL, Heller J, Benjamin D, Young S, Bali D, Smith SA, Li JS, Mandel H, Koeberl D, Rosenberg A, Chen YT (2010) Cross-reactive immunologic material status affects treatment outcomes in Pompe disease infants. Mol Genet Metab 99:26–33

Koeberl DD, Luo X, Sun B, McVie-Wylie A, Dai J, Li S, Banugaria SG, Chen YT, Bali DS (2011) Enhanced efficacy of enzyme replacement therapy in Pompe disease through mannose-6-phosphate receptor expression in skeletal muscle. Mol Genet Metab 103:107–112

Koeberl DD, Austin S, Case LE, Smith EC, Buckley AF, Young SP, Bali D, Kishnani PS (2014) Adjunctive albuterol enhances the response to enzyme replacement therapy in late-onset Pompe disease. FASEB J 28(5):2171–2176

Kornfeld S, Sly WS (2001) I cell disease and pseudo-Hurler polydystrophy: disorders of lysosomal enzyme phosphorylation and localization. In: Scriver CR, Beaudet al, Sly WS, Valle D (eds) Metabolic and molecular bases of inherited diseases, 8th edn. McGraw Hill, New York

Kornfeld R, Bao M, Brewer K, Noll C, Canfield WM (1998) Purification and multimeric structure of bovine N-acetylglucosamine-1- phosphodiester alpha-N-acetylglucosaminidase. J Biol Chem 273:23203–23210

Kornfeld R, Bao M, Brewer K, Noll C, Canfield W (1999) Molecular cloning and functional expression of two splice forms of human N-acetylglucosamine-1-phosphodiester alpha-N-acetylglucosaminidase. J Biol Chem 274:32778–32785

Kroos M, Hoogeveen-Westerveld M, Michelakakis H, Pomponio R, van der Ploeg A, Halley D, Reuser A, Consortium GAAD (2012a) Update of the pompe disease mutation database with 60 novel GAA sequence variants and additional studies on the functional effect of 34 previously reported variants. Hum Mutat 33:1161–5

Kroos M, Hoogeveen-Westerveld M, van der Ploeg A, Reuser AJ (2012b) The genotype–phenotype correlation in Pompe disease. Am J Med Genet C Semin Med Genet 160C:59–68

Kudo M, Canfield WM (2006) Structural requirements for efficient processing and activation of recombinant human UDP-N-acetylglucosamine:lysosomal-enzyme-N-acetylglucosamine-1-phosphotransferase. J Biol Chem 281:11761–11768

Kudo M, Bao M, D'Souza A, Ying F, Pan H, Roe BA, Canfield WM (2005) The alpha- and beta-subunits of the human UDP-N-acetylglucosamine:lysosomal enzyme phosphotransferase are encoded by a single cDNA. J Biol Chem 280:36141–36149

LeBowitz JH, Grubb JH, Maga JA, Schmiel DH, Vogler C, Sly WS (2004) Glycosylation-independent targeting enhances enzyme delivery to lysosomes and decreases storage in mucopolysaccharidosis type VII mice. Proc Natl Acad Sci U S A 101:3083–3088

Liu Y, Chen G (2011) Chemical synthesis of N-linked glycans carrying both mannose-6-phosphate and GlcNAc-mannose-6-phosphate motifs. J Org Chem 76:8682–8689

Liu Y, Marshall J, Li Q, Edwards N, Chen G (2013) Synthesis of novel bivalent mimetic ligands for mannose-6-phosphate receptors. Bioorg Med Chem Lett 23:2328–2331

Maga JA, Zhou J, Kambampati R, Peng S, Wang X, Bohnsack RN, Thomm A, Golata S, Tom P, Dahms NM, Byrne BJ, LeBowitz JH (2013) Glycosylation-independent lysosomal targeting of acid alpha-glucosidase enhances muscle glycogen clearance in pompe mice. J Biol Chem 288:1428–1438

Malicdan MC, Noguchi S, Nonaka I, Saftig P, Nishino I (2008) Lysosomal myopathies: an excessive build-up in autophagosomes is too much to handle. Neuromuscul Disord 18:521–529

Marlin SD, Springer TA (1987) Purified intercellular adhesion molecule-1 (ICAM-1) is a ligand for lymphocyte function-associated antigen 1 (LFA-1). Cell 51:813–819

Martina JA, Diab HI, Lishu L, Jeong AL, Patange S, Raben N, Puertollano R (2014) The nutrient-responsive transcription factor TFE3 promotes autophagy, lysosomal biogenesis, and clearance of cellular debris. Sci Signal 7:ra9

Matsumoto T, Akutsu S, Wakana N, Morito M, Shimada A, Yamane A (2006) The expressions of insulin-like growth factors, their receptors, and binding proteins are related to the mechanism regulating masseter muscle mass in the rat. Arch Oral Biol 51:603–611

McVie-Wylie AJ, Lee KL, Qiu H, Jin X, Do H, Gotschall R, Thurberg BL, Rogers C, Raben N, O'Callaghan M, Canfield W, Andrews L, McPherson JM, Mattaliano RJ (2008) Biochemical and pharmacological characterization of different recombinant acid alpha-glucosidase preparations evaluated for the treatment of Pompe disease. Mol Genet Metab 94:448–455

Moreland RJ, Jin X, Zhang XK, Decker RW, Albee KL, Lee KL, Cauthron RD, Brewer K, Edmunds T, Canfield WM (2005) Lysosomal acid alpha-glucosidase consists of four different peptides processed from a single chain precursor. J Biol Chem 280:6780–6791

Muro S, Schuchman EH, Muzykantov VR (2006) Lysosomal enzyme delivery by ICAM-1-targeted nanocarriers bypassing glycosylation- and clathrin-dependent endocytosis. Mol Ther 13:135–141

Nicolino M, Byrne B, Wraith JE, Leslie N, Mandel H, Freyer DR, Arnold GL, Pivnick EK, Ottinger CJ, Robinson PH, Loo JC, Smitka M, Jardine P, Tato L, Chabrol B, McCandless S, Kimura S, Mehta L, Bali D, Skrinar A, Morgan C, Rangachari L, Corzo D, Kishnani PS (2009) Clinical outcomes after long-term treatment with alglucosidase alfa in infants and children with advanced Pompe disease. Genet Med 11:210–219

Oba-Shinjo SM, Da Silva R, Andrade FG, Palmer RE, Pomponio RJ, Ciociola KM, Carvalho MS, Gutierrez PS, Porta G, Marrone CD, Munoz V, Grzesiuk AK, Llerena JC Jr, Berditchevsky CR, Sobreira C, Horovitz D, Hatem TP, Frota ER, Pecchini R, Kouyoumdjian JA, Werneck L, Amado VM, Camelo JS Jr, Mattaliano RJ, Marie SK (2009) Pompe disease in a Brazilian series: clinical and molecular analyses with identification of nine new mutations. J Neurol 256:1881–1890

Olson LJ, Dahms NM, Kim JJ (2004a) The N-terminal carbohydrate recognition site of the cation-independent mannose 6-phosphate receptor. J Biol Chem 279:34000–34009

Olson LJ, Yammani RD, Dahms NM, Kim JJ (2004b) Structure of uPAR, plasminogen, and sugar-binding sites of the 300 kDa mannose 6-phosphate receptor. EMBO J 23:2019–2028

Olson LJ, Peterson FC, Castonguay A, Bohnsack RN, Kudo M, Gotschall RR, Canfield WM, Volkman BF, Dahms NM (2010) Structural basis for recognition of phosphodiester-containing lysosomal enzymes by the cation-independent mannose 6-phosphate receptor. Proc Natl Acad Sci U S A 107:12493–12498

Raben N, Nagaraju K, Lee E, Kessler P, Byrne B, Lee L, Lamarca M, King C, Ward J, Sauer B, Plotz P (1998) Targeted disruption of the acid alpha-glucosidase gene in mice causes an illness with critical features of both infantile and adult human glycogen storage disease type II. J Biol Chem 273:19086–19092

Raben N, Danon M, Gilbert AL, Dwivedi S, Collins B, Thurberg BL, Mattaliano RJ, Nagaraju K, Plotz PH (2003) Enzyme replacement therapy in the mouse model of Pompe disease. Mol Genet Metab 80:159–169

Raben N, Fukuda T, Gilbert AL, de Jong D, Thurberg BL, Mattaliano RJ, Meikle P, Hopwood JJ, Nagashima K, Nagaraju K, Plotz PH (2005) Replacing acid alpha-glucosidase in Pompe disease: recombinant and transgenic enzymes are equipotent, but neither completely clears glycogen from type II muscle fibers. Mol Ther 11:48–56

Raben N, Wong A, Ralston E, Myerowitz R (2012) Autophagy and mitochondria in Pompe disease: nothing is so new as what has long been forgotten. Am J Med Genet C Semin Med Genet 160C:13–21

Reitman ML, Kornfeld S (1981a) Lysosomal enzyme targeting. N-Acetylglucosaminylphosphotransferase selectively phosphorylates native lysosomal enzymes. J Biol Chem 256:11977–11980

Reitman ML, Kornfeld S (1981b) UDP-N-acetylglucosamine:glycoprotein N-acetylglucosamine-1-phosphotransferase. Proposed enzyme for the phosphorylation of the high mannose oligosaccharide units of lysosomal enzymes. J Biol Chem 256:4275–4281

Rohrer J, Kornfeld R (2001) Lysosomal hydrolase mannose 6-phosphate uncovering enzyme resides in the trans-Golgi network. Mol Biol Cell 12:1623–1631

Shea L, Raben N (2009) Autophagy in skeletal muscle: implications for Pompe disease. Int J Clin Pharmacol Ther 47(Suppl 1):S42–S47

Song X, Lasanajak Y, Olson LJ, Boonen M, Dahms NM, Kornfeld S, Cummings RD, Smith DF (2009) Glycan microarray analysis of P-type lectins reveals distinct phosphomannose glycan recognition. J Biol Chem 284:35201–35214

Spampanato C, Feeney E, Li L, Cardone M, Lim JA, Annunziata F, Zare H, Polishchuk R, Puertollano R, Parenti G, Ballabio A, Raben N (2013) Transcription factor EB (TFEB) is a new therapeutic target for Pompe disease. EMBO Mol Med 5:691–706

Thurberg BL, Lynch Maloney C, Vaccaro C, Afonso K, Tsai AC, Bossen E, Kishnani PS, O'Callaghan M (2006) Characterization of pre- and post-treatment pathology after enzyme replacement therapy for Pompe disease. Lab Invest 86:1208–1220

Tiels P, Baranova E, Piens K, de Visscher C, Pynaert G, Nerinckx W, Stout J, Fudalej F, Hulpiau P, Tannler S, Geysens S, Van Hecke A, Valevska A, Vervecken W, Remaut H, Callewaert N (2012) A bacterial glycosidase enables mannose-6-phosphate modification and improved cellular uptake of yeast-produced recombinant human lysosomal enzymes. Nat Biotechnol 30:1225–1231

Tong PY, Kornfeld S (1989) Ligand interactions of the cation-dependent mannose 6-phosphate receptor. Comparison with the cation-independent mannose 6-phosphate receptor. J Biol Chem 264:7970–7975

Tong PY, Gregory W, Kornfeld S (1989) Ligand interactions of the cation-independent mannose 6-phosphate receptor. The stoichiometry of mannose 6-phosphate binding. J Biol Chem 264:7962–7969

Urayama A, Grubb JH, Banks WA, Sly WS (2007) Epinephrine enhances lysosomal enzyme delivery across the blood brain barrier by up-regulation of the mannose 6-phosphate receptor. Proc Natl Acad Sci U S A 104:12873–12878

Uson I, Schmidt B, von Bulow R, Grimme S, von Figura K, Dauter M, Rajashankar KR, Dauter Z, Sheldrick GM (2003) Locating the anomalous scatterer substructures in halide and sulfur phasing. Acta Crystallogr D Biol Crystallogr 59:57–66

van den Hout HM, Hop W, Van Diggelen OP, Smeitink JA, Smit GP, Poll-The BT, Bakker HD, Loonen MC, DE Klerk JB, Reuser AJ, van der Ploeg AT (2003) The natural course of infantile Pompe's disease: 20 original cases compared with 133 cases from the literature. Pediatrics 112:332–340

Van Hove JL, Yang HW, Wu JY, Brady RO, Chen YT (1996) High-level production of recombinant human lysosomal acid alpha-glucosidase in Chinese hamster ovary cells which targets to heart muscle and corrects glycogen accumulation in fibroblasts from patients with Pompe disease. Proc Natl Acad Sci U S A 93:65–70

Varki A, Kornfeld S (1980) Structural studies of phosphorylated high mannose-type oligosaccharides. J Biol Chem 255:10847–10858

Varki A, Kornfeld S (1983) The spectrum of anionic oligosaccharides released by endo-beta-N-acetylglucosaminidase H from glycoproteins. Structural studies and interactions with the phosphomannosyl receptor. J Biol Chem 258:2808–2818

Varki A, Sherman W, Kornfeld S (1983) Demonstration of the enzymatic mechanisms of alpha-N-acetyl-D-glucosamine-1-phosphodiester N-acetylglucosaminidase (formerly called alpha-N-acetylglucosaminylphosphodiesterase) and lysosomal alpha-N-acetylglucosaminidase. Arch Biochem Biophysics 222:145–149

Waheed A, Hasilik A, von Figura K (1981) Processing of the phosphorylated recognition marker in lysosomal enzymes. Characterization and partial purification of a microsomal alpha-N-acetylglucosaminyl phosphodiesterase. J Biol Chem 256:5717–5721

Watanabe H, Grubb JH, Sly WS (1990) The overexpressed human 46-kDa mannose 6-phosphate receptor mediates endocytosis and sorting of beta-glucuronidase. Proc Natl Acad Sci U S A 87:8036–8040

Wenk J, Hille A, von Figura K (1991) Quantitation of Mr 46000 and Mr 300000 mannose 6-phosphate receptors in human cells and tissues. Biochem Int 23:723–731

Winkel LP, Hagemans ML, Van Doorn PA, Loonen MC, Hop WJ, Reuser AJ, van der Ploeg AT (2005) The natural course of non-classic Pompe's disease: a review of 225 published cases. J Neurol 252:875–884

Wisselaar HA, Kroos MA, Hermans MM, Van Beeumen J, Reuser AJ (1993) Structural and functional changes of lysosomal acid alpha-glucosidase during intracellular transport and maturation. J Biol Chem 268:2223–2231

Wokke JH, Escolar DM, Pestronk A, Jaffe KM, Carter GT, van den Berg LH, Florence JM, Mayhew J, Skrinar A, Corzo D, Laforet P (2008) Clinical features of late-onset Pompe disease: a prospective cohort study. Muscle Nerve 38:1236–1245

Yang HW, Kikuchi T, Hagiwara Y, Mizutani M, Chen YT, Van Hove JL (1998) Recombinant human acid alpha-glucosidase corrects acid alpha- glucosidase-deficient human fibroblasts, quail fibroblasts, and quail myoblasts. Pediatr Res 43:374–380

Zhou Q, Stefano JE, Harrahy J, Finn P, Avila L, Kyazike J, Wei R, Van Patten SM, Gotschall R, Zheng X, Zhu Y, Edmunds T, Pan CQ (2011) Strategies for Neoglycan conjugation to human acid alpha-glucosidase. Bioconjug Chem 22:741–751

Zhou Q, Avila LZ, Konowicz PA, Harrahy J, Finn P, Kim J, Reardon MR, Kyazike J, Brunyak E, Zheng X, Patten SM, Miller RJ, pan CQ (2013) Glycan structure determinants for cation-independent mannose 6-phosphate receptor binding and cellular uptake of a recombinant protein. Bioconjug Chem 24:2025–2035

Zhu Y, Li X, Kyazike J, Zhou Q, Thurberg BL, Raben N, Mattaliano RJ, Cheng SH (2004) Conjugation of mannose 6-phosphate-containing oligosaccharides to acid alpha-glucosidase improves the clearance of glycogen in pompe mice. J Biol Chem 279:50336–50341

Zhu Y, Li X, McVie-Wylie A, Jiang C, Thurberg BL, Raben N, Mattaliano RJ, Cheng SH (2005) Carbohydrate-remodelled acid alpha-glucosidase with higher affinity for the cation-independent mannose 6-phosphate receptor demonstrates improved delivery to muscles of Pompe mice. Biochem J 389:619–628

Zhu Y, Jiang JL, Gumlaw NK, Zhang J, Bercury SD, Ziegler RJ, Lee K, Kudo M, Canfield WM, Edmunds T, Jiang C, Mattaliano RJ, Cheng SH (2009) Glycoengineered acid alpha-glucosidase with improved efficacy at correcting the metabolic aberrations and motor function deficits in a mouse model of Pompe disease. Mol Ther 17:954–963

Blood–Brain Barrier Targeting of Therapeutic Lysosomal Enzymes

William M. Pardridge

Abbreviations

ADA	Anti-drug antibody
ASA	Arylsulfatase A
ASM	Acid sphingomyelinase
AUC	Area under the plasma concentration curve
BBB	Blood–brain barrier
BDNF	Brain derived neurotrophic factor
CHO	Chinese hamster ovary
CI MPR	Cation independent mannose 6-phosphate receptor
C_{max}	Maximal plasma concentration
CMT	Carrier-mediated transport
CNS	Central nervous system
CSF	Cerebrospinal fluid
cTfRMAb	Chimeric MAb against mouse TfR
E	Enzyme
ERT	Enzyme replacement therapy
GAG	Glycosoaminoglycan
HC	Heavy chain
HIR	Human insulin receptor
HIRMAb	MAb against HIR
HPLC	High performance liquid chromatography
ID	Injection dose
IDS	Iduronate 2-sulfatase
IDUA	Iduronidase

W.M. Pardridge, M.D. (✉)
ArmaGen Technologies, Inc., 26679 Agoura Road, Ste. 100, Calabasas, CA 91302, USA
e-mail: wpardridge@armagen.com

A. Rosenberg, B. Demeule (eds.), *Biobetters*, AAPS Advances
in the Pharmaceutical Sciences Series 19, DOI 10.1007/978-1-4939-2543-8_4

IR	Insulin receptor
IT	Intrathecal
IV	Intravenous
LAMP1	Lysosomal associated membrane protein type 1
LC	Light chain
M6P	Mannose 6-phosphate
MAb	Monoclonal antibody
MPS	Mucopolysaccharidosis
MTH	Molecular Trojan horse
NAb	Neutralizing antibody
OD	Optical density
PK	Pharmacokinetics
RMT	Receptor-mediated transport
SFM	Serum free medium
Tf	Transferrin
TfR	Transferrin receptor
TH	Trojan horse
VD	Organ volume of distribution
V_{ss}	Systemic volume of distribution

Introduction

There are over 50 lysosomal storage disorders (LSDs), and approximately 75 % of these conditions affect the central nervous system (CNS) (Cheng and Smith 2003). A partial list of LSDs that affect the CNS is shown in Table 1. The principal therapy for LSDs is Enzyme Replacement Therapy or ERT, where the patient is administered the recombinant enzyme by intravenous (IV) infusion (Brady and Schiffmann 2004). This has shown to be an effective therapy for the extra-CNS manifestations of many LSDs. However, the recombinant enzymes are large molecule drugs that do not cross the blood–brain barrier (BBB), and do not penetrate the brain from blood (Al Sawaf et al. 2008). Therefore, ERT is not effective for LSDs that affect the CNS, and this represents a premier challenge in current therapy for such disorders (Wraith et al. 2008).

Pharmaceutical companies that developed recombinant lysosomal enzymes for ERT did so without a parallel development of a BBB drug targeting technology. Consequently, the only approach available for enzyme delivery to brain was the intra-thecal (IT) route, whereby recombinant enzyme is invasively injected into the cerebrospinal fluid (CSF) compartment (Kakkis et al. 2004).

Intra-thecal Enzyme Delivery to the Brain

IT injection of enzyme, or any other drug, into the CSF compartment is not an effective brain delivery strategy when the goal is drug delivery to the parenchyma of human brain (Pardridge 2012). The IT approach is suitable when the brain disorder

Table 1 Inborn Errors of Metabolism: Candidates for CNS Enzyme Replacement Therapy

Group	Disease	Enzyme	Genbank
MPS	MPS-I (Hurler)	IDUA	NM_000203
	MPS-II (Hunter)	IDS	NM_000202
	MPS-III (Sanfillipo type A)	SGSH	NM_000199
	MPS-IIIB (Sanfillipo type B)	NAGLU	NM_000263
	MPS-IV (Morquio type A)	GALNS	NM_000512
	MPS-IV (Morquio type B)	GLB1	NM_000404
	MPS-VI (Maroteaux-Lamy)	ARSB	NM_000046
	MPS-VII (Sly)	GUSB	NM_000181
GSD	GSD-II (Pompe)	GAA	NM_000152
Sphingolipidoses	Gaucher	GBA	NM_000157
	Fabry	GLA	NM_000169
	Tay Sachs	HEXA	NM_000520
	Niemann-Pick type A	ASM	NM_000543
	Krabbe	GALC	NM_000153
	GM1-gangliosidosis	GLB1	NM_000404
	MLD	ARSA	NM_000487
	Farber	ASAH1	NM_177924
Leukodystrophy	Canavan	ASPA	NM_000049
NCL	Type 1	PPT1	NM_000310
	Type 2	TPP1	NM_000391

MPS mucopolysaccharidosis, *GSD* glycogen storage disease, *NCL* neuronal ceroid lipofuscinoses, *MPS* mucopolysaccharidosis, *GSD* glycogen storage disease, *GM* gangliosidosis, *MLD* metachromatic leukodystrophy, *IDUA* iduronidase, *IDS* iduronate 2-sulfatase, *SGSH* *N*-sulfoglucosamine sulfohydrolase, *NAGLU* alpha-*N*-acetylglucosaminidase, *GALNS* *N*-acetyl-galatosamine-6-sulfatase, *GLB1* beta galactosidase, *ARSA* arylsulfatase A, *ARSB* arylsulfatase B, *GUSB* beta glucuronisase, *GAA* acid alpha-glucosidase, *GBA* beta acid-glucosidase, *GLA* alpha-galactosidase A, *HEXA* hexosaminidase A, *ASM* acid shingomyelinase, *GALC* galactosylceramidase, *ASA* arysulfatase A, *ASAH1* acid ceramidase, *ASPA* aspartoacylase, *PPT1* palmitoyl-protein thioesterase type 1, *TPP1* tripeptidyl amino peptidase type 1

affects the surface of the brain that is adjacent to the CSF flow tracts. However, drug or enzyme does not effectively penetrate into brain parenchyma from the CSF surface following IT administration. This is because CSF exits the brain rapidly via bulk flow (convection), whereas drug diffuses from CSF to brain parenchyma slowly via diffusion (Pardridge 2012).

The rapid rate of CSF exit from brain to blood is a product of the anatomical properties of CSF formation and elimination within the brain. The CSF is produced in the choroid plexus of the ventricles, moves over the surface of the brain, and is absorbed into the venous circulation across the arachnoid villi into the superior sagittal sinus (Davson 1969). In the human brain, there is about 100–140 mL of CSF, and this entire volume is turned over completely every 4–5 h, or 4–5 times per day (Cutler et al. 1968). In the mouse brain, there is about 40 uL of CSF, and the entire volume is turned over every 2 h, or about 12 times per day (Rudick et al. 1982). Thus, the differential rates of convection of CSF (rapid) and diffusion of drug (slow) create the paradox that drug injected into the CSF distributes easily to blood and poorly to brain beyond the ependymal surface. An IT injection of an enzyme is, therefore, equivalent to a slow intravenous infusion.

Transport of drug from the CSF to brain tissue is limited by diffusion, which decreases with the square of the distance. This is illustrated by the poor distribution of a neurotrophin into brain parenchyma, as shown by autoradiography of rat brain prepared 20 h after a single injection of [^{125}I]-brain derived neurotrophic factor (BDNF) into a lateral ventricle (Yan et al. 1994). The BDNF distributed only to the ependymal surface of brain ipsilateral to the ventricular injection. This limited distribution into brain from the CSF is observed even for small molecules. The concentration of water soluble small molecules in brain tissue at increasing distances from the CSF surface was measured in the primate following the intra-cerebroventricular injection of the drug (Blasberg et al. 1975), and revealed a logarithmically decreased drug concentration in brain removed from the CSF surface. An approximate tenfold decrease in drug concentration of small molecule is found with each mm of distance removed from the brain surface (Blasberg et al. 1975). In the case of a lipid soluble small molecule, which can diffuse into brain cells, there was a tenfold decrease in drug concentration with each 500 μm of distance into the brain (Fung et al. 1996). The logarithmic decrease in drug concentration in brain is even steeper for a large molecule drug such as a lysosomal enzyme, which has a lower diffusion coefficient as compared to a small molecule. In order to achieve a therapeutic drug level in brain at a 5 mm distance from the CSF surface, it may be necessary to inject a dose of drug that generates at least a 5-log increase in CSF drug concentration. This approach may cover drug distribution in the mouse brain, where all parts of brain are within 3–5 mm of the CSF surface. However, review of an atlas of the human brain shows the distance between critical internal brain structures such as the striatum and the CSF surface is up to 50 mm, which precludes significant drug distribution into brain parenchyma from the CSF compartment (Pardridge 2012). This limited distribution of a lysosomal enzyme in the primate brain was shown for acid sphingomyelinase (ASM), the enzyme mutated in Nieman-Pick Type A (Table 1). A large amount of ASM, 960 mg, was infused into the CSF of Rhesus monkeys for 4 h (Ziegler et al. 2011). However, the distribution of the ASM to primate brain was limited to the ependymal surface of the brain. Moreover, there may be consequences to such very high concentrations of drug that localize near the CSF surface of the brain. For example, leptomeningeal changes with axonal sprouting and astrogliosis at the ependymal surface of the brain were observed in rats and monkeys following IT administration of neurotrophins (Yamada et al. 1991; Day-Lollini et al. 1997).

Blood–Brain Barrier Molecular Trojan Horse Technology

In contrast to the limitations imposed on intrathecal drug delivery by diffusion (Pardridge 2012), diffusion plays no role in the trans-vascular route of drug delivery to the brain. Since the distance between capillaries in brain is about 40 μm and even a large molecule enzyme therapeutic diffuses 40 μm within a second, once enzyme is targeted across the BBB from blood, there is instantaneous delivery of the enzyme to virtually every cell within the brain. However, the trans-vascular route to brain is not available for enzyme drugs, unless a BBB drug delivery technology is used to enable enzyme transfer across the BBB. BBB delivery of protein therapeutics has been attempted with BBB disruption, nanoparticles, or molecular Trojan horse technology. BBB disruption, such as with focused ultrasound following the intravenous administration of microbubbles has been developed (Burgess and Hynynen 2013). This approach delivers drug to a focal area of brain, whereas lysosomal enzyme therapeutics must be delivered to all parts of the brain and spinal cord. BBB disruption leads to the brain uptake of plasma proteins, which are neurotoxic and can lead to chronic neuropathologic changes in the brain (Salahuddin et al. 1988). Nanoparticles are large structures that do not cross the BBB in the absence of a mediated delivery system. Pegylated immunonanoparticles have been produced that target an endogenous receptor on the BBB such as the transferrin receptor (Olivier et al. 2002). Molecular Trojan horses are peptides or peptidomimetic monoclonal antibodies (MAb) that target an endogenous BBB receptor transporter such as the insulin receptor or the transferrin receptor (TfR). The molecular Trojan horse (MTH) technology for BBB transport will now be discussed in the context of lysosomal enzyme delivery to brain.

The BBB MTH technology was developed following the discovery that certain peptides or proteins in the blood penetrate the BBB via receptor-mediated transport (RMT) (Pardridge 2012). The BBB insulin receptor mediates the BBB transport of insulin (Duffy and Pardridge 1987), and the BBB TfR mediates the BBB transport of transferrin (Tf) (Skarlatos et al. 1995). The RMT peptide transporters at the BBB operate in parallel with the carrier-mediated transport (CMT) systems that transport small molecules such as glucose or amino acids into the CNS (Pardridge 2012). Knowledge of the molecular properties of the BBB CMT and RMT systems allows the drug developer to re-engineer small molecule drugs, and large molecule drugs, respectively, so that the pharmaceutical penetrates the BBB via the targeted endogenous transport system. For example, dopamine does not cross the BBB. However, L-DOPA, which is the precursor to dopamine, is a large neutral amino acid, and therefore crosses the BBB via the endogenous large neutral amino acid transporter type 1 or LAT1 (Pardridge 2012).

Insulin or Tf could be used as a BBB MTH. However, insulin is not a preferred MTH, because insulin administration would cause hypoglycemia. Tf is not a preferred MTH, because the concentration of endogenous Tf in the plasma is so high that the Tf binding site on the BBB TfR is >99 % saturated by endogenous Tf. However, peptidomimetic monoclonal antibodies (MAb) can be used as a BBB MTH (Pardridge et al. 1991, 1995). The MAb binds an exofacial epitope on the BBB receptor, which is spatially removed from the binding site of the endogenous ligand, and the MAb undergoes RMT across the BBB. The most potent MTH for the human BBB is a MAb against the human insulin receptor (HIR), and both chimeric and humanized forms of this antibody have been developed, which are designated the HIRMAb (Boado et al. 2007). The HIRMAb cross reacts with the insulin receptor in Old World primates, such as the Rhesus monkey, but does not cross react with the insulin receptor in New World primates or lower animals such as rodents (Zhou et al. 2012a). There is no known MAb against the rodent insulin receptor that could be used as a BBB MTH. Therefore, drug delivery across the mouse BBB is achieved with a genetically engineered chimeric MAb against the mouse TfR, designated the cTfRMAb (Boado et al. 2009a). Following the identification of a MAb with properties of high receptor-mediated BBB transport, it is then possible to genetically engineer IgG fusion proteins comprised of the transporting MAb and a protein therapeutic that alone does not cross the BBB (Pardridge and Boado 2012).

BBB Delivery of Trojan Horse-Enzyme Fusion Proteins

The delivery of a therapeutic enzyme from blood to brain requires transport across two membranes in series, the BBB and the brain cell membrane, as depicted in Fig. 1. The advantage of targeting the BBB HIR or TfR is that these receptors are expressed on the plasma membranes of brain cells as well as the BBB. Therefore, a lysosomal enzyme can be delivered from blood to the intracellular compartment of brain cells, following the re-engineering of the enzyme as a Trojan horse fusion protein (Fig. 1). Whereas the MAb domain of the MAb-enzyme fusion protein triggers transport across the BBB and the brain cell membrane, the enzyme domain then triggers triage to the lysosomal compartment (Fig. 1). Enzyme triage to the

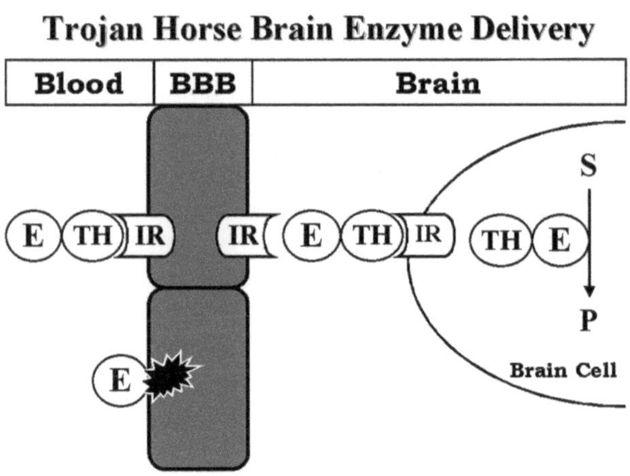

Fig. 1 The lysosomal enzyme (E) alone does not cross the blood–brain barrier (BBB). However, re-engineering of the enzyme as a Trojan horse (TH) fusion protein enables penetration through the BBB. The Trojan horse is an engineered MAb to the human insulin receptor (IR), which triggers transport across both the BBB, and the brain cell membrane, on the endogenous insulin receptor. The MAb-enzyme fusion protein is then triaged to the lysosomal compartment of brain cells, where accumulated lysosomal products (S) are degraded to low molecular weight products (P). From Boado et al. (2008)

lysosome is mediated by a M6P-dependent pathway, as well as a M6P-independent pathway (Braulke and Bonifacino 2009). The M6P-dependent pathway is mediated by the cation independent M6P receptor (CI MPR), which binds to the mannose 6-phosphorylated glycan via a divalent binding reaction that requires the presence of dual M6P moieties on the enzyme for high affinity (Tong and Kornfeld 1989). Entry of the IgG-enzyme fusion protein into the lysosomal compartment has been demonstrated for MAb-enzyme fusion proteins of IDUA, IDS, or ASA using confocal microscopy (Boado et al. 2008, 2013a; Lu et al. 2011).

The structure of the MAb-enzyme fusion protein is shown in Fig. 2. The lysosomal enzyme, without its signal peptide, is fused to the carboxyl terminus of the heavy chain of the HIRMAb, for BBB delivery in humans or Rhesus monkeys, or to the carboxyl terminal of the cTfRMAb for BBB delivery in the mouse. The construct shown in Fig. 2 places the lysosomal enzyme in a dimeric configuration, which replicates the dimeric structure native to many lysosomal enzymes (Ruth et al. 2000). The amino terminus of the heavy chain and the light chain of the MAb is free of steric hindrance from the enzyme that is fused at the carboxyl terminus. This is advantageous, since the complementarity determining regions of the MAb, which bind the target antigen, are located near the amino terminus of the IgG. When engineering an IgG-enzyme fusion protein, such as that shown in Fig. 2, it is essential that the functionality of both the MAb and the enzyme be retained following

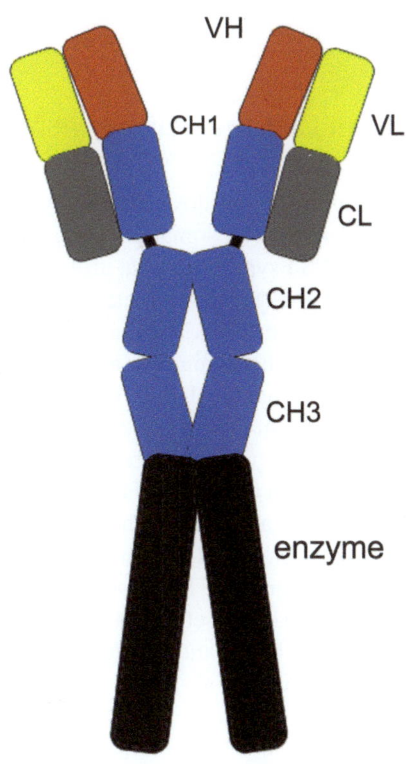

Fig. 2 The IgG-enzyme fusion protein is formed by fusion of the amino terminus of the mature enzyme to the carboxyl terminus of the CH3 region of the heavy chain of the targeting MAb. The fusion protein is a bi-functional molecule: the fusion protein binds the targeted receptor, such as the insulin receptor or transferrin receptor at the BBB, to mediate transport into the brain, and expresses enzyme activity in the brain following transport across the BBB

expression of the IgG-enzyme fusion protein. That is, the MAb-enzyme fusion protein should bind the target BBB receptor with the same high affinity as the native MAb. In addition, both the lysosomal enzyme activity, and the lysosome targeting activity, of the MAb-enzyme fusion protein should be comparable to the native recombinant enzyme. The IgG domain of the fusion protein may be engineered to either include or exclude the Fc region of the IgG. The Fc region is intact in the IgG-enzyme fusion protein shown in Fig. 2. The advantage of inclusion of the Fc region is that the fusion protein may be purified by protein A affinity chromatography, which simplifies the downstream processing in the fusion protein manufacturing. Another advantage of Fc domain retention is that the fusion protein is a ligand for the BBB Fc receptor (FcR), which binds the CH2–CH3 region of the IgG domain. The principal FcR expressed at the BBB is the neonatal FcR or FcRn (Schlachetzki, et al. 2002). However, the BBB FcR only mediates the efflux of IgG in the brain to blood direction, but does not allow for IgG transport from blood to brain (Zhang and Pardridge 2001). The FcR binding properties of the MAb-enzyme fusion protein do not contribute to enhanced brain uptake from blood, but do allow for a route of egress from brain back to blood. However, the rate of efflux of the IgG from brain is slow (Wu et al. 1997), because the surface area of the neuron is so much larger than the surface area of the brain capillary endothelium. If the goal is to engineer a MAb-enzyme fusion protein that does not bind to the FcR, this can be accomplished by site directed mutagenesis of certain amino acid residues in the constant region of the heavy chain.

Brain Enzyme Delivery in Mucopolysaccharidosis Type I

Mucopolysaccharidosis (MPS) Type I, MPSI, is caused by mutations in the gene encoding the lysosomal enzyme, iduronidase (IDUA) (Table 1). Patients with distinct MPS I syndromes that affect the CNS (Hurler syndrome or Hurler-Scheie syndrome) are currently treated with weekly IV infusions of recombinant IDUA. However, the IDUA does not penetrate the BBB, and thus, the ERT of MPSI does not treat the brain manifestations of disease. To produce a BBB penetrating form of IDUA, the human IDUA enzyme was fused to the HIRMAb, and the resulting fusion protein is designated HIRMAb-IDUA (Boado et al. 2008). The HIRMAb-IDUA fusion protein can be tested in Rhesus monkeys, but not in the mouse, because the HIRMAb domain of the fusion protein does not bind to the mouse insulin receptor (Zhou et al. 2012a). The reactivity of the HIRMAb for the dog insulin receptor has not been tested. To enable treatment of the brain in mouse models of neural disease, a surrogate MTH was developed that targets the mouse TfR, designated the cTfRMAb (Boado et al. 2009a). Subsequently, the mouse IDUA enzyme was fused to the cTfRMAb, and this fusion protein is designated the cTfRMAb-IDUA fusion protein (Boado et al. 2011), as discussed below.

HIRMAb-IDUA Fusion Protein for Humans and Monkeys

The gene encoding the 627 amino acid human IDUA enzyme was fused to the 3′-terminus of the gene encoding the heavy chain (HC) of the HIRMAb, which produced a new fusion gene encoding the fusion protein of the HIRMAb HC and IDUA (Boado et al. 2008). The HIRMAb-IDUA fusion protein was expressed in stably transfected Chinese hamster ovary (CHO) cells grown in serum free medium (Boado et al. 2009b), and the drug substance was purified principally with a protein A affinity column (see below for additional manufacturing details). The critical attributes of the protein, consisting of BBB and brain cell targeting (HIRMAb domain) and lysosomal enzymatic activity (IDUA domain) were retained in the fusion protein, as demonstrated with multiple assays. The IDUA enzyme activity of the HIRMAb-IDUA fusion protein was comparable to that of recombinant IDUA; and the fusion protein was taken up by Hurler fibroblasts and triaged to the lysosomal compartment, as demonstrated by confocal microscopy (Fig. 3a–c). Treatment of Hurler fibroblasts with the HIRMAb-IDUA fusion protein caused a reduction in the cell content of glycosoaminoglycans (GAGs) (Boado et al. 2008).

The affinity of the HIRMAb-IDUA fusion protein for the HIR was comparable to that of the native HIRMAb (Boado et al. 2008). This property predicted that the HIRMAb-IDUA fusion protein should rapidly penetrate the BBB in the Rhesus monkey at a rate comparable to that of the native HIRMAb. Indeed this was the case as film autoradiography of the primate brain removed 2 h after IV injection showed a global distribution of the HIRMAb-IDUA fusion protein in the brain (Fig. 3d), with higher uptake in gray matter as compared to white matter, owing to the higher vascular density in gray matter. The brain scan in Fig. 3d is comparable to a 2-deoxyglucose scan, and is produced by rapid transport of the fusion protein across the BBB followed by penetration of the parenchyma of brain. The transport of the fusion protein through the BBB was confirmed with the capillary depletion method (Boado et al. 2008). The capillary depletion method separates the vascular tissue in brain from the post-vascular compartment based on density centrifugation (Triguero et al. 1990). Based on measurements of the specific activity of brain capillary-specific enzymes, such as γ-glutamyl transpeptidase or alkaline phosphatase, the post-vascular supernatant is >95 % depleted of brain vasculature. The penetration of HIRMAb fusion proteins into brain parenchyma has also been demonstrated with emulsion autoradiography of primate brain (Boado et al. 2013a). The HIRMAb-IDUA fusion protein also distributes broadly to peripheral organs in the primate (Boado et al. 2008). The uptake by peripheral organs is also discussed below in the context of the reduction in GAG content in peripheral organs in the Hurler mouse.

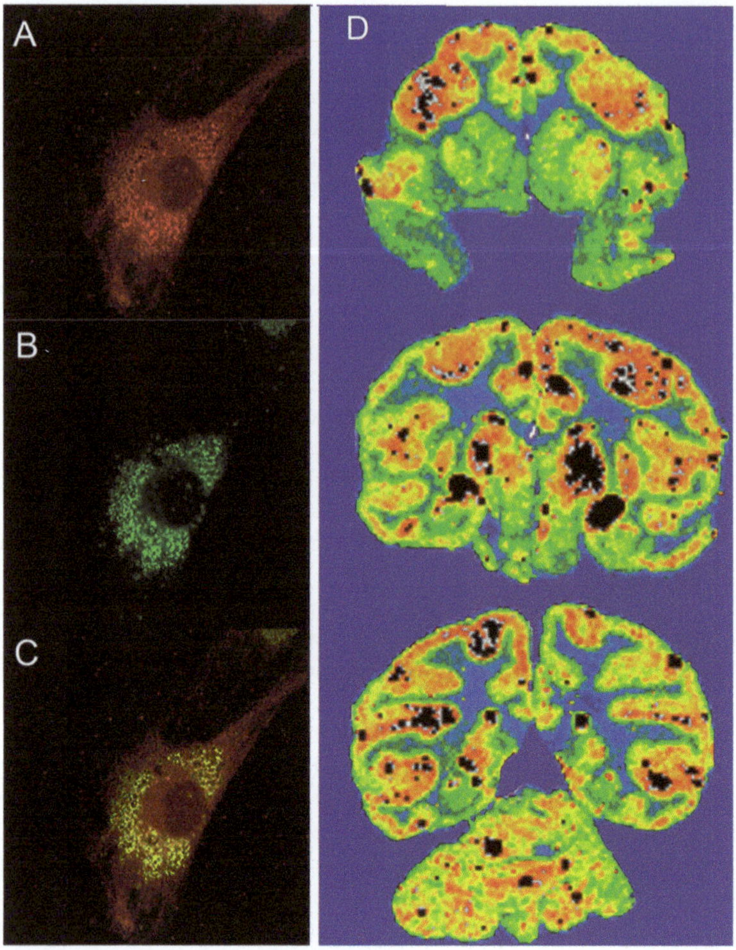

Fig. 3 (**a**, **b**, **c**) Hurler fibroblasts were incubated with HIRMAb-IDUA fusion protein for 24 h and then fixed and immune stained for confocal microscopy. The fixed cells were stained with a rabbit polyclonal antibody to human IDUA (panel a: *red channel*), and a mouse monoclonal antibody to human lysosomal associated membrane protein (LAMP)-1 (panel b: *green channel*). The overlap image in panel c shows sequestration of the HIRMAb-IDUA fusion protein within lysosomes. (**d**) Film autoradiography of Rhesus monkey brain removed 2 h after an intravenous administration of [^{125}I]-HIRMAb-IDUA fusion protein. Coronal sections through the forebrain (*top panel*), midbrain (*middle panel*), and hindbrain/cerebellum (*bottom panel*) are shown. From Boado et al. (2008)

Pharmacokinetics of HIRMAb-IDUA Fusion Protein

The plasma IDUA enzyme activity in Rhesus monkeys was determined following a 30 min IV infusion of 3, 9, or 30 mg/kg of the HIRMAb-IDUA fusion protein, and the enzyme activity data are plotted in Fig. 4. There was no change in either the PK parameters or in the rate of removal of the IDUA enzyme activity from plasma at the start (week 1) or end (week 25) of the chronic treatment study, indicating that the

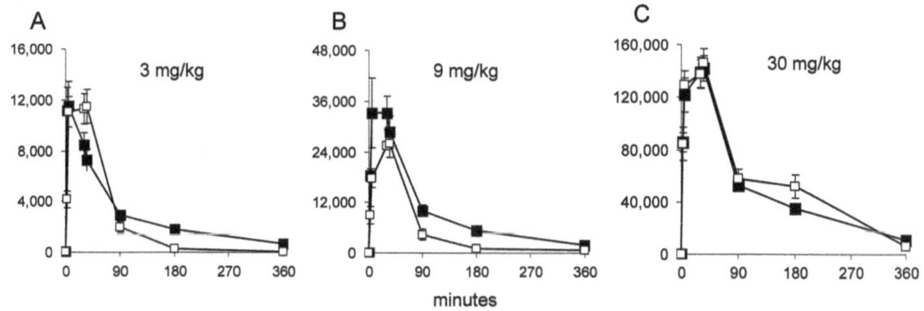

Fig. 4 Plasma IDUA enzyme activity (nmol/h/mL) in the Rhesus monkey between 5 and 360 min after a 30 min infusion of the HIRMAb-IDUA fusion protein in the low dose (3 mg/kg), mid dose (9 mg/kg) and high dose (30 mg/kg) treatment groups. Data are mean ± SE at each time point for combined sexes. There were no significant differences between male and female enzyme activity. Plasma profiles determined at the start of the study (week 1) are shown by *solid squares* (■), and plasma profiles determined at the end of the study (week 25) are shown by the *open squares* (□). From Boado et al. (2013c)

Table 2 Pharmacokinetic parameters of clearance of IDUA enzyme activity from plasma following a 30 min IV infusion in Rhesus monkeys

	HIRMAb-IDUA infusion dose		
Parameter	3 mg/kg	9 mg/kg	30 mg/kg
C_{max} (kilounits/mL)	11.56 ± 1.90	33.19 ± 8.24	141.92 ± 9.39
AUC (kilounits · min/mL)	1,059 ± 145	3,461 ± 474	18,171 ± 1,896
V_{ss} (mL/kg)	294 ± 62	243 ± 51	182 ± 27
CL (mL/kg/min)	2.25 ± 0.31	2.06 ± 0.28	1.31 ± 0.13
$T_{1/2}$ (min)	89.8 ± 16.0	81.6 ± 13.2	95.9 ± 11.9
Body weight (kg)	3.33 ± 0.39	3.36 ± 0.38	3.31 ± 0.85
Injected dose (kilounits)	7,955 ± 954	24,043 ± 2,706	79,003 ± 8,353

Parameters determined from data in Fig. 4 for the start of study infusion. From Boado et al. (2013c)

anti-drug antibodies (ADA) that formed during the course of the 6 months of treatment (vide infra) had no effect either on the plasma IDUA enzyme activity or on the rate of removal of the fusion protein from blood (Boado et al. 2013c). The pharmacokinetic (PK) parameters of the fusion protein removal from plasma are given in Table 2. The plasma AUC is proportional to the injection dose (ID), which indicates the fusion protein clearance from plasma follows a linear PK profile. Finally, the HIRMAb-IDUA fusion protein remained intact in plasma, which was also demonstrated by several orthogonal studies (Boado et al. 2009b).

Immune Responses to HIRMAb-IDUA Fusion Protein

Recombinant human proteins, such as the HIRMAb-IDUA fusion protein, can be immunogenic in primates, and the immune response in primates, usually directed to xenogeneic determinants, may thus not be predictive of the immune response in

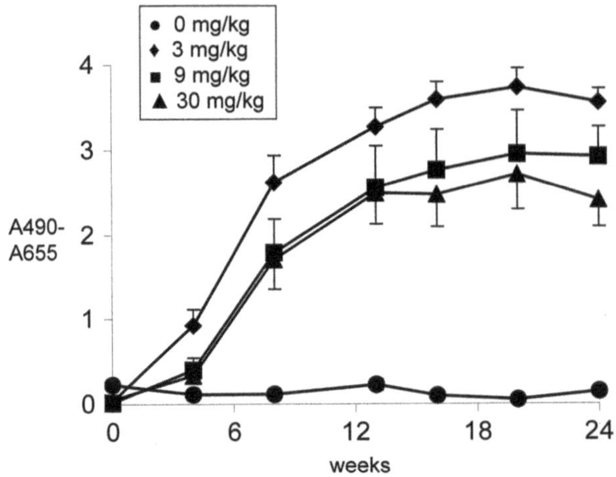

Fig. 5 Time course of ADA formation against the HIRMAb-IDUA fusion protein following chronic treatment in the Rhesus monkey. Data are mean ± SE (n = 8–12 monkeys per time point) for combined sexes. There were no differences between male and female absorbance readings at any time point. A490-A655 is the difference between the absorbance at 490 and 655 nm. The A655, which accounts for light scattering, was <0.05. All plasma samples were diluted 1:50 in phosphate buffered saline. From Boado et al. (2013c)

humans (Weinberg et al. 2005; Ponce et al. 2009). The sequence of the full length Rhesus monkey IDUA is not known. Potential immunogenic domains of the HIRMAb-IDUA fusion protein include the antibody variable (V)-region, the antibody constant (C)-region, and the IDUA domain. Chronic weekly IV infusion of Rhesus monkeys with the HIRMAb-IDUA fusion protein over a period of 6 months resulted in the formation of anti-drug antibodies (ADAs) as detected in plasma by a sandwich ELISA assay where the HIRMAb-IDUA fusion protein is used as the capture reagent, and biotinylated HIRMAb-IDUA fusion protein is used as the detector reagent (Fig. 5) (Boado et al. 2009b, 2013c). The ADA titer was quantitated as optical density (OD)/uL plasma. The ADA titer reached a plateau by 21 weeks, and then began to decrease. The ADA titer was not measured after 26 weeks, because of study termination. The ADAs were primarily directed against the HIRMAb V-region and the IDUA, and, to a lesser extent, to the HIRMAb constant-region (Boado et al. 2013c).

ADAs directed against the V-region of the HIRMAb could either block or enhance antibody binding to the insulin receptor. However, such anti-idiotypic antibodies should alter the clearance of the fusion protein from plasma, which is mediated by the endogenous insulin receptor. The ADAs formed against the HIRMAb-IDUA fusion protein neither altered the plasma clearance of the fusion protein, as demonstrated by the comparable rate of clearance of plasma IDUA enzyme activity at the start (week 1) and end (week 25) of the study (Fig. 4), nor displayed neutralizing activity of the IDUA enzyme. However, though not altering PK, it is possible for ADAs to alter the cellular targeting of their protein therapeutic ligands by directing the fusion protein into FcR bearing cells, thus limiting delivery to the target cells of the disease. The ADAs formed against the HIRMAb-IDUA fusion protein were observed in all primates, and some monkeys demonstrated hypersensitivity reactions during the course of the 26-week treatment (Boado et al. 2013c); such adverse reactions have been reported in chronic dosing of primates with human recombinant proteins (Ponce et al. 2009; Chapman et al. 2012).

cTfRMAb-IDUA Fusion Protein for the Mouse

A mouse model of Hurler's disease was generated with an IDUA knockout line of transgenic mice (Ohmi et al. 2003). This Hurler mouse model could not be treated with the HIRMAb-IDUA fusion protein, because the HIRMAb domain does not recognize the insulin receptor in the mouse (Zhou et al. 2012a). Therefore, the cTfRMAb-IDUA fusion protein was engineered, expressed in host cells, and purified by protein G affinity chromatography (Boado et al. 2011). The cTfRMAb-IDUA fusion protein demonstrated the following properties critical for in vivo activity:

- High affinity binding to the mouse TfR with a KI of 0.67 ± 0.11 nM using a radio-receptor assay and mouse fibroblasts as the source of the mouse TfR
- IDUA enzyme specific activity of 776 ± 79 units/ug protein, where 1 unit = 1 nmol/h, based on a fluorometric assay using 4-methylumbelliferyl α-L-iduronide as substrate.
- Treatment of mouse fibroblasts with a single 2-hour exposure to the cTfRMAb-IDUA fusion protein caused a 20-fold increase in intracellular IDUA enzyme activity that decayed with a half-time of 67 h

Supporting in vitro assessments of critical attributes for treatment of MPSI, the in vivo activity of the fusion protein was demonstrated in the Hurler mouse model, revealing the following important measures of efficacy:

- Elevation of IDUA enzyme activity in plasma, brain, and peripheral organs following the IV injection of 1 mg/kg of the cTfRMAb-IDUA fusion protein in Hurler mice
- Reduction in GAG content: the GAG content in peripheral organs was measured with an alcian blue dye binding method; GAG content was reduced from 24 % in kidney to >95 % in liver in aged (6 months old) Hurler mice treated with twice-weekly IV injections of the cTfRMAb-IDUA fusion protein at a dose of 1 mg/kg for 8 consecutive weeks. The alcian blue method measures total GAGs in the tissue, and total GAGs are not elevated in the brain of the Hurler mouse (Chung et al. 2007). Therefore, the number of multivacuolated cells in brain was measured with microscopy (Boado et al. 2011).

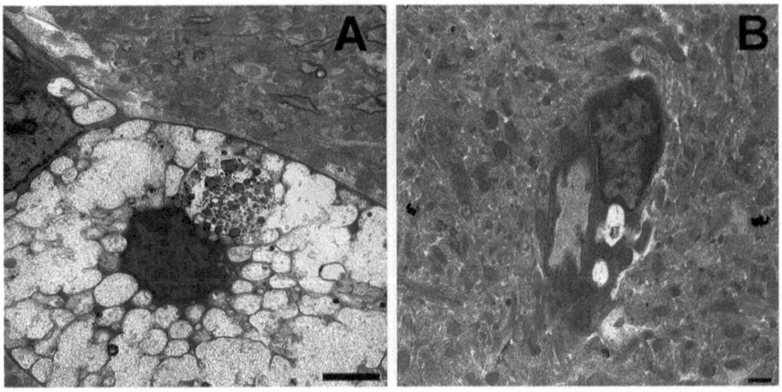

Fig. 6 Electron microscopy of MPSI mouse brain treated with saline (**a**) or the cTfRMAb-IDUA fusion protein (**b**). Large perivascular lysosomal inclusion bodies seen in the brain of the saline treated mouse (**a**) are diminished in the fusion protein treated mouse (**b**). Magnification bars in panels a and b are 2 and 0.5 μm, respectively. The cells shown are representative of formal quantitative counting of hundreds of cells in *o*-toluidine blue-stained sections of mouse brain. From Boado et al. (2011)

- Reduction in the number of multivacuolated brain cells. The number of multivacuolated brain cells (reflecting lysosomal inclusion bodies) was quantified by light microscopy as the number of inclusions per 100 brain cell nucleoli (Boado et al. 2011). The number of lysosomal inclusion bodies was reduced 73 % by chronic treatment with the cTfRMAb-IDUA fusion protein in brain of the MPS I mice following 8 weeks of treatment with the cTfRMAb-IDUA fusion protein. An electron micrograph of the lysosomal inclusion bodies in the brain of Hurler mice treated with saline or the cTfRMAb-IDUA fusion protein is shown in Fig. 6. These largest multi-vesicular bodies were often perivascular as shown in Fig. 6a.

Chronic treatment of the Hurler mice with the cTfRMAb-IDUA fusion protein caused a measureable elevation of ADA in plasma in some mice, but the titer (OD/uL) was low (Boado et al. 2011). The ADA titer produced in mice treated chronically with the cTfRMAb-IDUA fusion protein (Boado et al. 2011) was tenfold lower than the ADA titer produced in Rhesus monkeys treated chronically with the HIRMAb-IDUA fusion protein (Boado et al. 2013c). These findings are in agreement with the exaggerated ADA response to human proteins in primates (Chapman et al. 2012). The effect of the ADAs in the Hurler mice on fusion protein distribution was not investigated. However, in a parallel study, it was shown that the ADAs formed against a cTfRMAb fusion protein after 12 weeks of chronic treatment have no effect on fusion protein distribution to peripheral organs or on BBB transport (Zhou et al. 2011).

In conclusion, the Hurler mouse treatment study demonstrated that the IgG-IDUA fusion protein is a bio-better, relative to IDUA alone, in that chronic treatment with the cTfRMAb-IDUA fusion protein caused a reduction in lysosomal inclusion bodies not only in peripheral organs, but within the CNS.

Brain Enzyme Delivery in Mucopolysaccharidosis Type II

MPS Type II, MPSII, also called Hunter's syndrome, is caused by mutations in the gene encoding the lysosomal enzyme, iduronate 2-sulfatase (IDS) (Table 1). Patients with MPSII are currently treated with weekly IV infusions of recombinant IDS. However, as with other ERTs, the IDS does not penetrate the BBB, and, therefore, ERT does not treat the brain manifestations of MPS II. To produce a BBB penetrating form of IDS, the human enzyme was fused to the chimeric HIRMAb, producing the HIRMAb-IDS fusion protein (Lu et al. 2010, 2011). Again, to utilize mouse models of disease, the human IDS enzyme was also fused to the cTfRMAb, and this fusion protein is designated the cTfRMAb-IDS fusion protein (Zhou et al. 2012b).

HIRMAb-IDS Fusion Protein for Humans and Monkeys

The IDS enzyme activity of the HIRMAb-IDS fusion protein was comparable to recombinant IDS. The fusion protein was taken up by Hunter fibroblasts and triaged to the lysosomal compartment, as demonstrated by confocal microscopy (Lu et al. 2011). Treatment of Hunter fibroblasts with the HIRMAb-IDS fusion protein caused a dose-dependent increase in intracellular IDS enzyme activity (Fig. 7) which persisted for several days. Following a 2-hour exposure of the cells to the HIRMAb-IDS fusion protein, intracellular IDS enzyme activity remained high for at least 96 h and declined with a half-time of 72 h (Table 3).

The affinity of the HIRMAb-IDS fusion protein for the HIR was comparable to that of the native HIRMAb (Lu et al. 2011), which predicted a high rate of uptake by the Rhesus monkey brain in vivo. This was confirmed with a pharmacokinetics/brain uptake study in the Rhesus monkey using [^{3}H]-HIRMAb-IDS fusion protein (Lu et al. 2011). The brain uptake of the HIRMAb-IDS fusion protein, 0.8 % ID/100 g brain, is high compared to the brain uptake of a large molecule that does not cross the BBB in the Rhesus monkey. The brain uptake in the Rhesus monkey of GDNF, which has a comparable plasma pharmacokinetic profile as the HIRMAb-IDS fusion protein, at 2 h after IV injection, is 0.033 % ID/100 g brain (Boado and Pardridge 2009). Therefore, the brain uptake of the HIRMAb-IDS fusion protein is >20-fold greater than the uptake expected for a molecule that is confined to the plasma compartment of brain in the Rhesus monkey. Delivery of the HIRMAb-IDS fusion protein across the BBB was verified by the capillary depletion method (Fig. 8a). The brain volume of distribution (VD) of the fusion protein is high compared to proteins that are confined to the brain plasma volume, such as an isotype IgG or GDNF (Fig. 8a).

The uptake by brain and peripheral organs of the HIRMAb-IDS fusion protein was compared with the uptake of recombinant human IDS (idursulfase) in the Rhesus monkey (Boado et al. 2013b). Both the IDS and the HIRMAb-IDS fusion

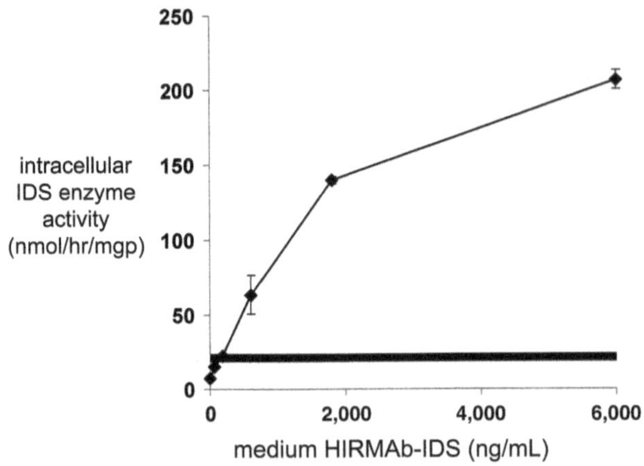

Fig. 7 Intracellular IDS enzyme activity is increased in Hunter fibroblasts in proportion to the concentration of medium HIRMAb-IDS fusion protein. Data are mean ± SE (n = 3 dishes/point). The *horizontal bar* is the IDS enzyme activity in healthy human fibroblasts (17 ± 2 units/mg protein). From Lu et al. (2011)

Table 3 Time-response study of intracellular IDS enzyme activity in Hunter fibroblasts after a 2-hour exposure to the HIRMAb-IDS fusion protein

Time (h)	Intracellular IDS activity (nmol/h/mgp)
0	<5
2	140 ± 11
24	105 ± 7
48	77 ± 4
72	76 ± 3
96	56 ± 6

Mean ± SE (n = 4 plates/time point). Cells were exposed to the HIRMAb-IDS fusion protein (6 ug/mL) in the medium for 2 h, washed extensively to remove the HIRMAb-IDS fusion protein, and incubated up to 96 h in fresh medium without HIRMAb-IDS fusion protein. From Lu et al. (2011)

protein distribute equally to non-brain organs (Table 4). However, the brain uptake of the HIRMAb-IDS fusion protein is 35-fold higher than the brain uptake of IDS (Table 4). When the brain uptake of IDS is expressed as a brain volume of distribution (VD), it is shown that the brain VD of IDS is equal to the brain blood volume. That is, the small distribution of IDS to brain that is detected simply represents IDS sequestration within the blood volume of brain without any *trans*-BBB transfer of the IDS (Boado et al. 2013b). The differential distribution to brain of recombinant IDS vs the HIRMAb-IDS fusion protein is shown by the 2 h brain scan of the uptake of the two proteins (Fig. 9). In contrast to the absence of any brain penetration by the IDS, there is global distribution in brain of the HIRMAb-IDS fusion protein (Boado et al. 2013b). Emulsion autoradiography complements the film autoradiography, and shows the HIRMAb-enzyme fusion protein distributes beyond the BBB to cells of the parenchyma of brain (Boado et al. 2013a).

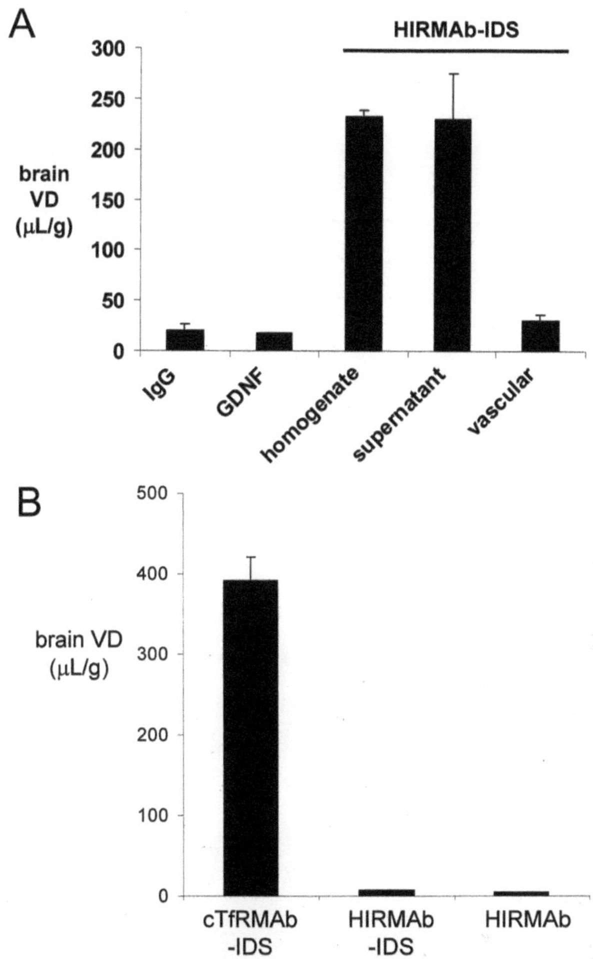

Fig. 8 (**a**) The brain homogenate volume of distribution (VD) is shown for three proteins: (1) a non-specific IgG, which is a marker of the brain blood volume, (2) GDNF, which is a molecule that does not cross the BBB (Boado and Pardridge 2009), and (3) the HIRMAb-IDS fusion protein (Lu et al. 2011). The brain VD for the HIRMAb-IDS fusion protein is shown for the post-vascular supernatant and the vascular pellet in primate brain. Comparison of the VD of the HIRMAb-IDS fusion protein in the post-vascular supernatant and the vascular pellet shows that 90 % of the HIRMAb-IDS fusion protein taken up by the Rhesus monkey brain has moved through the vascular barrier and penetrated brain parenchyma. (**b**) Brain VD is shown for the HIRMAb, the HIRMAb-IDS fusion protein, or the cTfRMAb-IDS fusion protein in the mouse at 60 min after an IV injection. The absence of transport of the HIRMAb-IDS fusion protein, or the HIRMAb alone, into the brain of the mouse is due to the lack of reactivity of the HIRMAb domain with the insulin receptor of the mouse. Data are mean ± SE (n = 4 mice). From Zhou et al. (2012a)

Table 4 Organ uptake of IDS compared to the organ uptake of the HIRMAb-IDS fusion protein in the Rhesus monkey

	Organ uptake (%ID/100 g)	
Organ	[^{125}I]-IDS	[^{125}I]-HIRMAb-IDS
Brain-gray matter	0.030 ± 0.004	1.04 ± 0.07
Brain-white matter	0.026 ± 0.007	0.99 ± 0.19
Heart	1.1 ± 0.2	1.3 ± 0.1
Liver	27.8 ± 0.7	33.2 ± 4.5
Spleen	12.4 ± 0.2	19.2 ± 0.2
Lung	3.4 ± 0.2	3.0 ± 0.1
Skeletal muscle	0.20 ± 0.05	0.23 ± 0.05
Fat	0.27 ± 0.03	0.25 ± 0.02

Data are mean ± SE of triplicates samples removed at 2 h after a single IV injection. The recombinant human IDS or the HIRMAb-IDS fusion protein was radiolabeled with the [^{125}I]-Bolton-Hunter reagent. From Boado et al. (2013b)

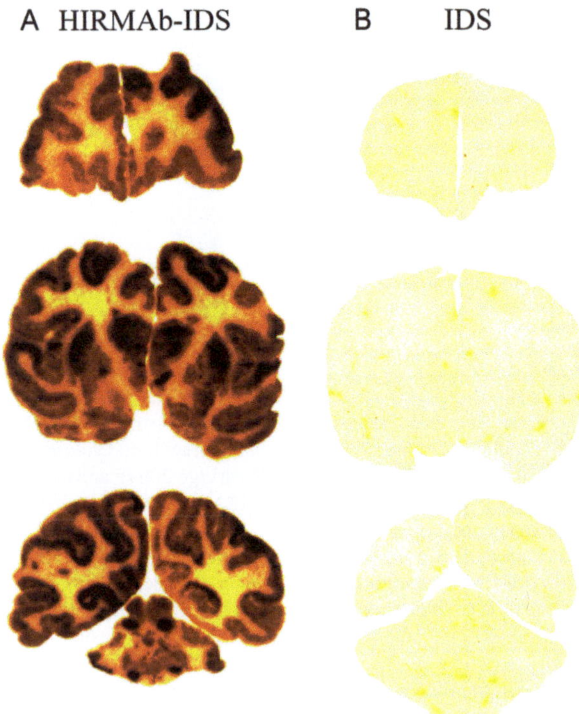

Fig. 9 Film autoradiogram of 20 μm sections of rhesus monkey brain removed 120 min after IV injection of the HIRMAb-IDS fusion protein (**a**) or IDS (**b**). The forebrain section is on the *top*, the midbrain section is in the *middle*, and the hindbrain section with cerebellum is on the *bottom*. Scans produced after labeling of the HIRMAb-IDS fusion protein or IDS with [^{125}I]-Bolton-Hunter. X-ray films exposed for 7 days. From Boado et al. (2013b)

cTfRMAb-IDS Fusion Protein for the Mouse

The affinity of the cTfRMAb-IDS fusion protein for the mouse TfR is equal to the affinity of the cTfRMAb (Zhou et al. 2012b). Fusion of the IDS protein to the carboxyl terminus of the heavy chain of the cTfRMAb had no effect on binding to the mouse TfR. The high affinity of the cTfRMAb-IDS fusion protein for the mouse TfR predicted a high rate of brain uptake in the mouse, which was verified with in vivo studies: brain uptake in the mouse was high for the cTfRMAb-IDS fusion protein, and low for the HIRMAb-IDS fusion protein (Zhou et al. 2012a, b). The brain volume of distribution (VD) measurement of the cTfRMAb-IDS fusion protein was high, 392 ± 29 uL/g (Fig. 8b), indicating rapid penetration of the brain parenchyma from blood. In contrast, the uptake of the IDS enzyme alone by the brain was 100-fold lower than that of the cTfRMAb-IDS fusion protein (Zhou et al. 2012b), owing to lack of BBB transport of the native enzyme.

Manufacturing of MAb-Enzyme Fusion Proteins

The manufacturing of IgG-enzyme fusion proteins is simpler than the manufacturing of the enzyme alone, because a protein A capture column can be used in the downstream processing of the IgG-enzyme fusion protein. Thus, fusion proteins are usually ~99 % purified with a single protein A column run. Subsequent cation exchange and anion exchange columns remove residual traces of DNA and endotoxin (Boado et al. 2009b, 2013c). The final drug product is purified and characterized with respect to quality (appearance, pH, osmolality), strength (A280), identity (human IgG and enzyme Western blotting), purity (reducing and non-reducing SDS-PAGE, size exclusion HPLC, anion exchange HPLC), potency (HIR binding affinity, enzyme specific activity), and both process and product related impurities including protein A, CHO host cell protein, CHO host cell DNA, endotoxin, bioburden, and sub-visible particles. Protein characterization studies include amino acid analysis, peptide mapping, mass spectrometry, monosaccharide, sialic acid, and mannose 6-phosphate content. The proteins are N-glycosylated based on enzyme deglycosylation studies, and N-glycan analysis. Product stability is also a critical parameter, and the HIRMAb-enzyme fusion proteins are stable at 4 °C for 2 years, based on purity (electrophoresis, size exclusion chromatography), identity (IgG and enzyme Western blotting, isoelectric focusing), potency (HIR binding affinity and enzyme specific activity), and sterility.

Conclusions

There are several advantages to the re-engineering of lysosomal enzymes as IgG-enzyme fusion proteins, as compared to the native enzyme. First, the presence of the IgG domain allows for use of a protein A affinity capture column in the product purification, which simplifies the downstream processing of the manufacturing process. Second, the IgG domain may target a specific receptor, such as the IR or TfR, to enable receptor-mediated delivery of the fusion protein into target organs including the CNS. Third, antibodies against the mannose 6-phosphate (M6P) moiety of the enzyme may block target organ uptake via the M6P receptor, but such antibodies have no effect in organ targeting via the insulin or transferrin receptors (Zhou et al. 2011; Boado et al. 2013c). Anti-M6P antibodies could theoretically alter intracellular triage of the bound enzyme to the lysosome via the lysosomal CI MPR, but such M6P bound antibodies might be restricted from ingress into cells not bearing FcR and thus have limited access to the lysosomal membrane within the intracellular compartment of target cells. Fourth, the constant region of the IgG domain contains certain amino acid sequences, called Tregitopes, which induce T cell immune tolerance (De Groot et al. 2008). The ability to treat the CNS with the IgG-enzyme fusion protein warrants consideration of the use of bio-better therapeutic enzymes that are re-engineered as IgG-enzyme fusion proteins.

References

Al Sawaf S, Mayatepek E, Hoffmann B (2008) Neurological findings in Hunter disease: pathology and possible therapeutic effects reviewed. J Inherit Metab Dis 31:473–480

Blasberg RG, Patlak C, Fenstermacher JD (1975) Intrathecal chemotherapy: brain tissue profiles after ventriculocisternal perfusion. J Pharmacol Exp Ther 195:73–83

Boado RJ, Pardridge WM (2009) Comparison of blood–brain barrier transport of GDNF and an IgG-GDNF fusion protein in the Rhesus monkey. Drug Metab Dispos 37:2299–2304

Boado RJ, Zhang Y, Pardridge WM (2007) Humanization of anti-human insulin receptor antibody for drug targeting across the human blood–brain barrier. Biotechnol Bioeng 96:381–391

Boado RJ, Zhang Y, Xia CF, Wang Y, Pardridge WM (2008) Genetic engineering of a lysosomal enzyme fusion protein for targeted delivery across the human blood–brain barrier. Biotechnol Bioeng 99:475–484

Boado RJ, Zhang Y, Wang Y, Pardridge WM (2009a) Engineering and expression of a chimeric transferrin receptor monoclonal antibody for blood–brain barrier delivery in the mouse. Biotechnol Bioeng 102:1251–1258

Boado RJ, Hui EK, Lu JZ, Pardridge WM (2009b) AGT-181: expression in CHO cells and pharmacokinetics, safety, and plasma iduronidase enzyme activity in Rhesus monkeys. J Biotechnol 144:135–141

Boado RJ, Hui EK, Lu JZ, Zhou QH, Pardridge WM (2011) Reversal of lysosomal storage in brain of adult MPS-I mice with intravenous Trojan horse-iduronidase fusion protein. Mol Pharm 8:1342–1350

Boado RJ, Lu JZ, Hui WK-W, Sumbria RK, Pardridge WM (2013a) Pharmacokinetics and brain uptake in the Rhesus monkey of a fusion protein of arylsulfatase A and a monoclonal antibody against the human insulin receptor. Biotechnol Bioeng 110:1456–1465

Boado RJ, Hui EK-W, Lu JZ, Sumbria RK, Pardridge WM (2013b) Blood–brain barrier molecular Trojan horse enables brain imaging of radioiodinated recombinant protein in the Rhesus monkey. Bioconjug Chem 24:1741–1749

Boado RJ, Hui EK, Lu JZ, Pardridge WM (2013c) IgG-enzyme fusion protein: pharmacokinetics and anti-drug antibody response in Rhesus monkeys. Bioconjug Chem 24:97–104

Brady RO, Schiffmann R (2004) Enzyme-replacement therapy for metabolic storage disorders. Lancet Neurol 3:752–756

Braulke T, Bonifacino JS (2009) Sorting of lysosomal proteins. Biochim Biophys Acta 1793: 605–614

Burgess A, Hynynen K (2013) Noninvasive and targeted drug delivery to the brain using focused ultrasound. ACS Chem Neurosci 4:519–526
Chapman KL, Andrews L, Bajramovic JJ, Baldrick P, Black LE et al (2012) The design of chronic toxicology studies of monoclonal antibodies: implications for the reduction in use of non-human primates. Regul Toxicol Pharmacol 62:347–354
Cheng SH, Smith AE (2003) Gene therapy progress and prospects: gene therapy of lysosomal storage disorders. Gene Ther 10:1275–1281
Chung S, Ma X, Liu Y, Lee D, Tittiger M, Ponder KP (2007) Effect of neonatal administration of a retroviral vector expressing alpha-L-iduronidase upon lysosomal storage in brain and other organs in mucopolysaccharidosis I mice. Mol Genet Metab 90:181–192
Cutler RW, Page L, Galicich J, Watters GV (1968) Formation and absorption of cerebrospinal fluid in man. Brain 91:707–720
Davson H (1969) The cerebrospinal fluid. Handb Neurochem 2:23–48
Day-Lollini PA, Stewart GR, Taylor MJ, Johnson RM, Chellman GJ (1997) Hyperplastic changes within the leptomeninges of the rat and monkey in response to chronic intracerebroventricular infusion of nerve growth factor. Exp Neurol 145:24–37
De Groot AS, Moise L, McMurry JA, Wambre E, Van Overtvelt L et al (2008) Activation of natural regulatory T cells by IgG Fc-derived peptide "Tregitopes". Blood 112:3303–3311
Duffy KR, Pardridge WM (1987) Blood–brain barrier transcytosis of insulin in developing rabbits. Brain Res 420:32–38
Fung LK, Shin M, Tyler B, Brem H, Saltzman WM (1996) Chemotherapeutic drugs released from polymers: distribution of 1,3-bis(2-chloroethyl)-1-nitrosourea in the rat brain. Pharm Res 13:671–682
Kakkis E, McEntee M, Vogler C, Le S, Levy B et al (2004) Intrathecal enzyme replacement therapy reduces lysosomal storage in the brain and meninges of the canine model of MPS I. Mol Genet Metab 83:163–174
Lu JZ, Hui EK, Boado RJ, Pardridge WM (2010) Genetic engineering of a bifunctional IgG fusion protein with iduronate-2-sulfatase. Bioconjug Chem 21:151–156
Lu JZ, Boado RJ, Hui EK, Zhou QH, Pardridge WM (2011) Expression in CHO cells and pharmacokinetics and brain uptake in the Rhesus monkey of an IgG-iduronate-2-sulfatase fusion protein. Biotechnol Bioeng 108:1954–1964
Ohmi K, Greenberg DS, Rajavel KS, Ryazantsev S, Li HH, Neufeld EF (2003) Activated microglia in cortex of mouse models of mucopolysaccharidoses I and IIIB. Proc Natl Acad Sci U S A 100:1902–1907
Olivier J-C, Huertas R, Lee HJ, Calon F, Pardridge WM (2002) Synthesis of pegylated immunonanoparticles. Pharm Res 19:1137–1143
Pardridge WM (2012) Drug transport across the blood–brain barrier. J Cereb Blood Flow Metab 32:1959–1972
Pardridge WM, Boado RJ (2012) Reengineering biopharmaceuticals for targeted delivery across the blood–brain barrier. Methods Enzymol 503:269–292
Pardridge WM, Buciak JL, Friden PM (1991) Selective transport of an anti-transferrin receptor antibody through the blood–brain barrier in vivo. J Pharmacol Exp Ther 259:66–70
Pardridge WM, Kang YS, Buciak JL, Yang J (1995) Human insulin receptor monoclonal antibody undergoes high affinity binding to human brain capillaries in vitro and rapid transcytosis through the blood–brain barrier in vivo in the primate. Pharm Res 12:807–816
Ponce R, Abad L, Amaravadi L, Gelzleichter T, Gore E et al (2009) Immunogenicity of biologically-derived therapeutics: assessment and interpretation of nonclinical safety studies. Regul Toxicol Pharmacol 54:164–182
Rudick RA, Zirretta DK, Herndon RM (1982) Clearance of albumin from mouse subarachnoid space: a measure of CSF bulk flow. J Neurosci Methods 6:253–259
Ruth L, Eisenberg D, Neufeld EF (2000) Alpha-L-iduronidase forms semi-crystalline spherulites with amyloid-like properties. Acta Crystallogr D Biol Crystallogr 56:524–528
Salahuddin TS, Johansson BB, Kalimo H, Olsson Y (1988) Structural changes in the rat brain after carotid infusions of hyperosmolar solutions: a light microscopic and immunohistochemical study. Neuropathol Appl Neurobiol 14:467–482

Schlachetzki F, Zhu C, Pardridge WM (2002) Expression of the neonatal Fc receptor (FcRn) at the blood–brain barrier. J Neurochem 81:203–206

Skarlatos S, Yoshikawa T, Pardridge WM (1995) Transport of [^{125}I]transferrin through the rat blood-brain barrier. Brain Res 683:164–171

Tong PY, Kornfeld S (1989) Ligand interactions of the cation-dependent mannose 6-phosphate receptor. J Biol Chem 264:7970–7975

Triguero D, Buciak JB, Pardridge WM (1990) Capillary depletion method for quantifying blood–brain barrier transcytosis of circulating peptides and plasma proteins. J Neurochem 54:1882–1888

Weinberg WC, Frazier-Jessen MR, Wu WJ, Weir A, Hartsough M et al (2005) Development and regulation of monoclonal antibody products: challenges and opportunities. Cancer Metastasis Rev 24:569–584

Wraith JE, Scarpa M, Beck M, Bodamer OA, De Meirleir L et al (2008) Mucopolysaccharidosis type II (Hunter syndrome): a clinical review and recommendations for treatment in the era of enzyme replacement therapy. Eur J Pediatr 167:267–277

Wu D, Yang J, Pardridge WM (1997) Drug targeting of a peptide radiopharmaceutical through the primate blood–brain barrier in vivo with a monoclonal antibody to the human insulin receptor. J Clin Invest 100:1804–1812

Yamada K, Kinoshita A, Kohmura E, Sakaguchi T, Taguchi J et al (1991) Basic fibroblast growth factor prevents thalamic degeneration after cortical infarction. J Cereb Blood Flow Metab 11:472–478

Yan Q, Matheson C, Sun J, Radeke MJ, Feinstein SC et al (1994) Distribution of intracerebral ventricularly administered neurotrophins in rat brain and its correlation with trk receptor expression. Exp Neurol 127:23–36

Zhang Y, Pardridge WM (2001) Mediated efflux of IgG molecules from brain to blood across the blood–brain barrier. J Neuroimmunol 114:168–172

Zhou Q-H, Boado RJ, Lu JZ, Hui EK-W, Pardridge WM (2011) Chronic dosing of mice with a transferrin receptor monoclonal antibody-GDNF fusion protein. Drug Metab Dispos 39:1149–1154

Zhou QH, Boado RJ, Pardridge WM (2012a) Selective plasma pharmacokinetics and brain uptake in the mouse of enzyme fusion proteins derived from species-specific receptor-targeted antibodies. J Drug Target 20:715–719

Zhou QH, Boado RJ, Lu JZ, Hui EK, Pardridge WM (2012b) Brain-penetrating IgG-iduronate 2-sulfatase fusion protein for the mouse. Drug Metab Dispos 40:329–335

Ziegler RJ, Salegio EA, Dodge JC, Bringas J, Treleaven CM et al (2011) Distribution of acid sphingomyelinase in rodent and non-human primate brain after intracerebroventricular infusion. Exp Neurol 231:261–271

Novel Methods for Addressing Immunogenicity of Therapeutic Enzymes

Leslie P. Cousens, Leonard Moise, and Anne S. De Groot

The Immunogenicity Problem in Protein Therapeutics

Therapeutic protein products encompass monoclonal antibodies, human cytokines, cellular growth factors, hormones, clotting factors, enzymes, anticoagulants, and fusion proteins. They offer the advantages of increased specificity and reduced toxicity compared to small molecule drugs. However, when administered to patients, protein-based drugs have the potential to elicit anti-therapeutic protein immune responses (Rosenberg 2003; Barbosa 2011). These immune responses may have a number of clinical outcomes ranging in severity. Some immune responses may have no observed clinical impact at all, while others may, to varying degrees, impact drug safety and efficacy (Eser et al. 2013; Tatarewicz et al. 2014). Rarely, therapeutic protein-induced immune responses may target an endogenous protein counterpart of the therapeutic and result in life-threatening complications (Haselbeck 2003; Eckardt and Casadevall 2003).

Drug-induced immunogenicity is typically measured as a function of the anti-drug antibody (ADA) level. ADA can be measured by a number of a well-established, low-cost, high-throughput platform technologies, including ELISA, radioimmune precipitation, surface plasmon resonance, and electrochemiluminescence (Koren et al. 2008; Shankar et al. 2008, 2014). Patient samples can be easily obtained from small volumes of blood, thus both longitudinal sampling and long-term storage are feasible. Direct consequences of ADA on pharmacokinetics (PK) and

L.P. Cousens
EpiVax, Inc., 146 Clifford St., Providence, RI 2903, USA

L. Moise • A.S. De Groot, M.D. (✉)
EpiVax, Inc., 146 Clifford St., Providence, RI 2903, USA

Institute of Immunology and Informatics, University of Rhode Island, Kingston, RI, USA
e-mail: annied@epivax.com

A. Rosenberg, B. Demeule (eds.), *Biobetters*, AAPS Advances
in the Pharmaceutical Sciences Series 19, DOI 10.1007/978-1-4939-2543-8_5

pharmacodynamics (PD) may be demonstrated using a number of in vitro methods (Shankar et al. 2008; Pendley and Shankar 2011). Transient immune responses can be observed to many protein therapeutics, usually driven by innate immune triggers such as aggregates and process contaminants that drive non-specific, low titer anti-drug immune responses. However, it is important to consider that the development of a persistent ADA response represents the culmination of a complex "adaptive immune response" process, involving antigen presenting cells (APC), T cells, secreted cytokines, and B cells. Each component of this adaptive immune response presents unique opportunities to intervene and mitigate therapeutic protein immunogenicity.

T cell recognition of peptide epitopes derived from immunogenic protein therapeutics is an early event in the immunologic process that drives adaptive anti-drug responses. In order for a therapeutic protein to be detected by the T cell arm of the adaptive immune system, it must first be taken up by APC, and processed into small peptides. Some of these peptides will stably bind to major histocompatibility complex (MHC) class II molecules (human leukocyte antigen, HLA, in humans) to form a peptide/MHC complex that may activate an effector T cell through interactions with a T cell receptor at the APC/T cell interface, as depicted in Fig. 1 (Weaver et al. 2008). Since humans express a number of different HLA class II alleles, each of which may bind a different epitope, the development of immune responses will vary from individual to individual. In individuals who have one or more HLA that are able to present therapeutic protein-derived T cell epitopes, a cascade of molecular and cellular events may follow that can lead to B cell activation, isotype switching (IgM to IgG), affinity maturation, and B cell memory development that are manifest as ADA.

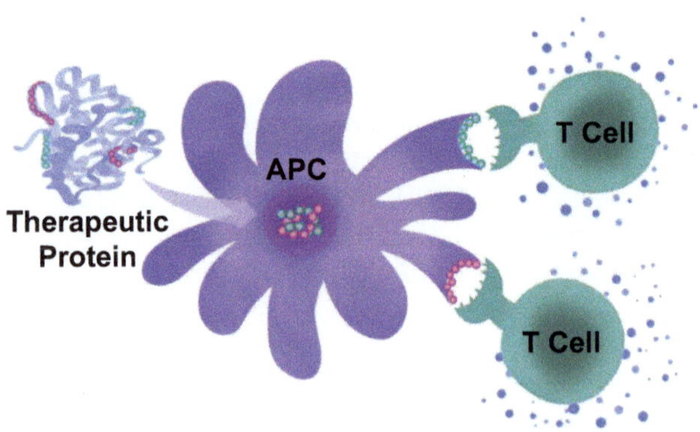

Fig. 1 Processing and presentation of therapeutic proteins. Therapeutic proteins can be taken up by antigen-presenting cells, processed by endosomal and lysosomal proteases, and presented in complex with MHC class II molecules on the cell surface to stimulate antigen-specific T cell functions, such as cytokine production

Immunogenicity for Replacement Proteins: Pompe Disease

Immune responses to endogenous proteins are considered to be unlikely, due to immunological tolerance that is established during fetal development and maintained throughout life. In contrast, immune responses to therapeutic proteins are expected when the protein is foreign, either because its sequence is derived from a different species or because patients lack the natural analog as a result of genetic deletion or modification, as can be observed in Hemophilia A (lacking FVIII) and Pompe disease (lacking lysosomal acid alpha-glucosidase (GAA) enzyme). The development and phenotype of immune responses to FVIII and GAA, for instance, may depend on several factors which include: (1) the extent of the genetic deletion (minimal, partial, complete), (2) the prevalence of the protein in cells and circulation, and (3) the HLA alleles expressed by the patient (Opdenakker et al. 2003; Haribhai et al. 2003; Link et al. 2014; McLachlan et al. 2012; Romball and Weigle 1999; Bachmann et al. 1994).

For example, in Pompe disease, patients have genetic deficiencies in GAA that result in accumulation of glycogen in lysosomes of cells throughout the body. These deficiencies are manifest across a full spectrum of protein expression, ranging from a complete lack of GAA protein expression to various degrees of partial protein expression. Treatment with fully human recombinant GAA can trigger high titer Anti-Drug Antibody (ADA) that interfere with drug efficacy. The severity of immune responses to recombinant enzyme replacement therapy is inversely proportional to the amount of endogenous GAA produced by the patient. Individuals that do not produce any natural GAA are categorized as cross-reactive immunologic material (CRIM)-negative, and those with partial GAA protein expression, are categorized as CRIM-positive. The interdependent relationship between incidence of ADA and CRIM-status in Pompe patients leads to the following dilemma: the less GAA expressed, the more severe the disease, the greater the dependence on the replacement protein, but the greater the risk and severity of ADA.

For CRIM-negative Pompe infants (who have complete GAA deficiency), repeated dosing of replacement enzyme is truly a life-saving treatment. But relief for many of these children is transient, curtailed by ADA against the "foreign" protein, after which they succumb quickly to disease (Joseph et al. 2008; Mendelsohn et al. 2009; Garman et al. 2004). High ADA titers are clearly correlated with poor outcomes. Thus, in spite of having an effective treatment for Pompe disease, there remains a critical unmet medical need for a less immunogenic GAA, especially for CRIM-negative babies. In this case "a biobetter" drug would be clearly defined as less immunogenic. Development of a less-immunogenic GAA replacement would also have broad-reaching implications in other replacement-protein therapies, and generally for many other protein therapeutics. For example, a number of other enzyme replacement therapies (ERTs) and therapeutic protein products are limited by ADA to some extent (e.g. Fabryzyme, Factor VIII, Cerezyme, monoclonal antibodies; (Deegan 2012; Saint-Remy et al. 2004; Starzyk et al. 2007; Maneiro et al. 2013)).

Approaches to Mitigating Immunogenicity

Drug-Induced Immunosuppression

The development of experimental immune tolerance regimens to inhibit ADA against life-saving enzyme replacement therapies is an active area of investigation. Current approaches for mitigating GAA-ADA are based on treatment with methotrexate (MTX) to inhibit the proliferation of lymphocytes, Rituximab (Rituxan) to deplete antibody (Ab)-producing and antigen-presenting B cells (Joseph et al. 2008; Garman et al. 2004; Rohrbach et al. 2010), and intravenous immunoglobulin (IVIg) as both immunomodulatory agent, as well as an anti-infective (Messinger et al. 2012). This approach has had excellent outcomes in inducing immune tolerance to the GAA enzyme replacement therapy (ERT) when applied prophylactically at the commencement of ERT. However, this treatment strategy is not as effective in eliminating entrenched immune responses mediated by long-lived plasma cells. Thus the timing of intervention in patients experiencing an ADA response is critical. In such scenarios, additional pharmacological agents that target long-lived plasma cells have been shown to be of use: Bortezomib, a drug currently in use for plasma cell leukemia and multiple myeloma, has been shown to reverse entrenched high titer antibody responses (Banugaria et al. 2013; Moran et al. 2012). However, both the short term and the long-term risks of Rituximab and Bortezomib, pertaining to their broad and non-specific effects on the immune system, warrant rigorous study (Gea-Banacloche 2010). Alternative strategies to immune tolerance induction (i.e. "treat the therapeutic protein") are presented below.

Deimmunization

The ability of a specific HLA molecule to present certain epitopes from a vaccine antigen, contributing to the development of a B cell response via provision of T cell help, is a key principle in rational vaccine design. Thus the presence of T cell epitopes tends to be correlated with the development of antibody responses. To escape immune responses, tumor cells (Scanlan and Jager 2001) and pathogens (Mullbacher 1992; Hill et al. 1997) can accumulate mutations that alter T cell epitope sequences (Vossen et al. 2002), rendering the tumor or pathogen invisible to the immune response. It stands to reason, then, that the deletion of T cell epitopes from an immunogen is a promising strategy to reduce detrimental immune responses (Celis et al. 1998; Min et al. 1996; Zuckerman 1996; McDermott et al. 1999). Deliberate removal of T cell epitopes to deimmunize biologic products is feasible, and has been the focus of several previous reports and reviews (Parker et al. 2013; Jones et al. 2009; De Groot and Martin 2009; Moise et al. 2012; Mazor et al. 2012; De Groot et al. 2005). The first step in targeting T cell epitopes for deimmunization is to identify the location and identity of T cell epitopes within a protein's amino acid sequence. Towards this goal, a variety of approaches can be applied, including immunoinformatics tools designed to predict T cell epitopes based on amino acid sequence (Paul et al. 2013), or in vitro cultures of PBMC that can be exposed to overlapping peptide fragments of the therapeutic protein that will elicit particular T cell function (Li Pira et al. 2010). While tools to predict T cell epitopes are few in number, their application in the context of immunogenicity screening has been validated in the literature (Wang et al. 2008; Jawa et al. 2013).

As noted above, a number of protein therapeutics have been de-immunized by removing T cell epitopes. One of the first efforts in this regard was the deimmunization of staphylokinase (Collen et al. 1996), where modification of specific amino acids that contributed to HLA binding led to a reduction in T cell responses. In the case of human FVIII, an epitope 15 amino acids in length was determined to bind to particular HLA class II alleles. Modification of the amino acid sequence of this epitope reduced its potential to bind to MHC class II; this molecule was subsequently determined to be less immunogenic by in vitro assays (Jones et al. 2005; Gilles et al. 1999). Epitope modification has also been applied to other proteins in studies performed by Hellendoorn et al. (2004), Tangri et al. (2005), Yeung et al. (2004) and others, using a variety of approaches.

We have used an in silico algorithm for T cell epitope mapping in conjunction with a design tool that iteratively searches optimal substitutions for any given amino acid within a given epitope based on the contribution each makes to HLA binding. This approach is designed to minimize the number of sequence changes to one or two key amino acids per epitope, thus reducing the potential impact on protein structure, function, and clinical performance. One of the key aspects of this approach is to target promiscuous epitopes since a single amino acid alteration can have an impact across HLA alleles, by lowering potential binding to multiple HLAs. As described by Moise et al. for FVIII (Moise et al. 2012), in silico design was followed with in vitro testing of the modified FVIII epitopes and in vivo validation of lowered immunogenicity. In additional (unpublished) collaborations, we have employed this strategy to map and modify the amino acid sequence of several therapeutic proteins. In multiple cases (unpublished studies), deimmunization by OptiMatrix lowered immunogenicity (measured in vitro and/or in vivo) while retaining function of the protein, supporting the feasibility of this approach.

We recently applied this approach to GAA (De Groot et al. 2003). First, we analyzed the full amino acid sequence of human GAA for T cell epitope content. The GAA sequence was parsed into 9-mer frames, where each 9-mer overlapped the last by eight amino acids. Each 9-mer was scored for predicted binding affinity to a panel of Class II HLA alleles (De Groot et al. 2010). Eight 15 to 20-amino acid long peptides predicted to bind multiple HLA types were selected for synthesis and the binding predictions were validated in HLA Class II binding assays (Steere et al. 2006). As shown in Fig. 2, each of the peptides (GAA_1 through 8) was bound to multiple HLA, as predicted. While confirmation using human T cells from exposed patients is still required, the high binding affinity of the peptides predicted indicates that there exist within the GAA sequence several potential promiscuous epitopes that may contribute to its immunogenicity. We consider these promiscuous epitopes to be

	HLA DR Alleles			
	*0101	*0401	*1101	*1501
GAA_1	Very Strong	Moderate	Moderate	Weak
GAA_2	Moderate	Moderate	Moderate	Moderate
GAA_3	Moderate	Non-binder	Non-binder	Strong
GAA_4	Very Strong	Strong	Moderate	Strong
GAA_5	Very Strong	Moderate	Non-binder	Strong
GAA_6	Moderate	Weak	Strong	Moderate
GAA_7	Strong	Moderate	Strong	Very Strong
GAA_8	Weak	Moderate	Moderate	Moderate

Very Strong Binders (IC_{50} < 100 nM)
Strong Binders (IC_{50} 100 – 10,00 nM)
Moderate Binders (IC_{50} 1,000 – 10,000 nM)
Weak Binders (IC_{50} 10,000 – 100,000 nM)
Non-binder (No concentration-dependent inhibition)

Fig. 2 GAA epitope prediction and validation. In silico analysis of the GAA amino acid sequence identified eight distinct peptides predicted to bind to different HLA alleles. Results were confirmed by in vitro HLA binding assay to determine the concentration of test peptide to inhibit 50 % of reference peptide binding (IC_{50}), as summarized above

the best candidates for deimmunizing modifications, since a single amino acid change in these epitopes may reduce binding to, and presentation by, multiple HLAs.

There are several important considerations in the deimmunization of therapeutic proteins. Among them are the numbers of potentially immunogenic epitopes within the protein sequence and the question of how many will be necessary to sufficiently deimmunize. Further, when immunogenic regions overlap with functional domains involved in ligand binding or enzymatic activity, sequence changes can impact critical quality attributes of the final product. Recently, King et al. described a protein design method for reducing immunogenicity that combines targeting of known and predicted T-cell epitopes with maximizing human sequence content to protect protein structure and function (King et al. 2014). As proof-of-concept, deimmunized variants of sfGFP and PE38 were designed, expressed and the function of each protein was validated in vitro. Successful deimmunization of sfGFP protein was demonstrated in vivo in mice as measured by reduced antigen-specific T cells detected by tetramer analysis. PE38 peptide variants were less immunogenic, demonstrated by reduced IL-2 production, when cultured in vitro with human PBMC.

In our own efforts to address these concerns, we have combined iterative deimmunization of epitope clusters and assessment of the impact that these changes have on the overall stability of the tertiary protein structure. This tool, developed in collaboration with Chris Bailey-Kellogg of Dartmouth College, is called "Epi-3D"; it is based on a previously-published tool called EpiSweep (Parker et al. 2013) but uses EpiMatrix instead of online epitope-mapping tools. Use of the tool provides several solutions, each of which may contain several sequence modifications designed to perturb HLA/epitope binding while maintaining protein stability. Once these peptides and their parent proteins are produced, they can then be tested in vitro (in HLA binding and functional assays) and in vivo (immunization studies). The Epi-3D tool is being used to deimmunize lysostaphin.

Antigen Specific Immune Tolerance Induction

Immune responses to self-proteins are controlled by immunologic tolerance mechanisms, certain of which have the potential to be exploited in the context of protein therapeutics. In the immunologic sense, discrimination of self and non-self is initially established in thymus during the latter stages of T cell development. Medullary epithelial cells present a wide range of tissue-specific self-protein epitopes in the context of MHC to immature T cells. T cells bearing T cell receptors (TCR) with either a high affinity for self-peptide:MHC complexes are deleted or deviated into regulatory (i.e. suppressor) T cells. T cells bearing TCR with low or moderate affinity for self-peptide:MHC complexes may escape deletion, complete their maturation program, and exit into the periphery where they may function as either effector cells or regulatory T cells (Treg) (Bluestone and Abbas 2003).

"Adaptive" tolerance develops and is maintained in the periphery where, upon antigen-specific activation through the TCR in the presence of cytokines such as IL-10 and TGF-β, mature T cells differentiate into 'adaptive' Treg. Adaptive Treg induction can be facilitated in a contact-independent manner through the cytokine milieu and/or contact-dependent interactions with APC bearing a regulatory phenotype (Bluestone and Abbas 2003). The role of these 'adaptive' Treg cells may be to dampen effector immune responses (following the primary, vigorous immune reaction, as a means of controlling inflammation), or possibly to facilitate co-existence with some symbiotic bacteria and viruses. Adaptive Treg induction is associated with sustained tolerance (to grafts, allergens and autologous proteins) and may require the existence of Treg cells with the same antigen-specificity as the self-reactive T cells.

In addition to strategies aimed at inhibiting proliferation or eliminating lymphocyte subsets participating in the ADA response, are those aimed at modulating the immune system to become tolerant to a therapeutic protein. IVIG, a pooled blood product from 1,000 or more donors, is used to induce tolerance in clinical settings (Toubi and Etzioni 2005; Kaveri et al. 2011). IVIG has been shown to be associated with modulation of the regulatory T cell axis, including induction of natural (n) Tregs (Ephrem et al. 2008); reduction of IL-17 (Maddur et al. 2011), and by enhancing the suppressive function of Tregs (Kessel et al. 2007). It has thus been applied with much success in a number of autoimmune diseases (Kaveri et al. 2011; Katz et al. 2011). A recent report describes positive clinical outcomes in two Pompe patients who had received prolonged methotrexate and Rituximab therapy for ADA who were also placed on chronic IVIG in an effort to decrease infectious complications. The addition of IVIG may have provided an additional immunomodulatory benefit in promoting tolerance to the GAA therapy through a mechanism that has yet to be defined (Messinger et al. 2012), such as induction of tolerance by exposure to Tregitopes (see below).

Tregitopes

In 2008, evidence was published indicating that highly conserved, promiscuous T cell epitopes are located in the Fc region and framework of the Fab region of IgG (De Groot et al. 2008). We hypothesized that these epitopes, that we named Tregitopes, were regulatory T cell epitopes. Tregitopes have the following four characteristics: (1) Their sequences are highly conserved in immunoglobulin molecules; (2) Almost all contain single 9-mer frames predicted by EpiMatrix to react with multiple HLA Class II alleles; (3) They stimulate Treg cells ($CD4^{+}CD25+FoxP3^{+}$) to proliferate and to produce IL-10; and (4) Co-incubation of Tregitopes with immunogenic peptides (e.g. pre-proinsulin) inhibits effector T cell (Teff) proliferation and suppresses secretion of effector cytokines in response to antigens in vitro (Cousens et al. 2013a). The presence of Tregitopes in immunoglobulins may explain why these unique proteins that undergo somatic hypermutation in the periphery do not appear to trigger immune responses directed at the new 'foreign' hypervariable sequences (Soukhareva et al. 2006). The presence of Tregitopes has a very practical application in the assessment of mAb immunogenicity. A retrospective immunogenicity analysis of 21 human and chimeric mAbs was carried out using in silico tools to predict T cell epitopes. The in silico immunogenicity score was a significant predictor of observed clinical immunogenicity; adjustment for Tregitope content of the mAb increased the predictive power (De Groot and Martin 2009).

Tregitopes may provide insight into mechanisms by which IVIG exerts its therapeutic effects. IVIG has been shown to induce expansion of Tregs and IL-10 secretion in vivo in both animal models and humans (Katz et al. 2011; Lopez et al. 2006; Maddur et al. 2010). IVIG is an effective therapy for Immune Thrombocytopenic Purpura (ITP), Kawasaki Syndrome (KS), polymyositis, dermatomyositis, neurological syndromes such as Guillain-Barré and Chronic Inflammatory Demyelinating Polyneuropathy (CIDP), in cases of severe steroid dependent asthma, and many others (Mazer et al. 2005; Orange et al. 2006). The anti-inflammatory activity of IVIG is attributed to multiple mechanisms including binding of the IgG Fc domain to Fc-gamma receptors, blockade of the FcRn receptor, blockade of Fas-FasL interactions, or interaction of sialylated Fc with a novel macrophage receptor DC-SIGN (Anthony and Ravetch 2010). However, these models have never fully accounted for the observed modulation in T cell phenotype upon co-stimulation with IgG and antigen (Kirschbaum et al. 2006) and the increase of CD4+/CD25+ Tregs after IVIG treatment in skin transplants (Tha-In et al. 2010) or experimental autoimmune encephalitis (EAE) (Toubi and Etzioni 2005). Others have described the immunosuppressive effects of specific IgG-derived peptides subsequently identified as Tregitopes (Maddur et al. 2011; Sharabi et al. 2006; Hahn et al. 2005), providing

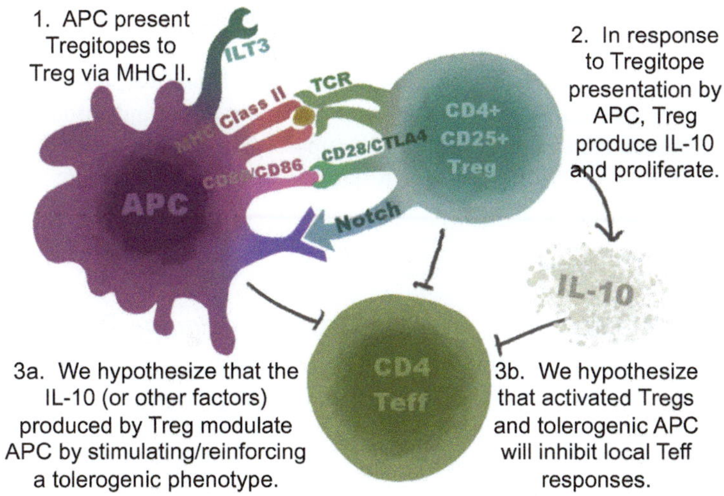

Fig. 3 Proposed Tregitope mechanism of action

independent confirmation of our work. Validation and introduction of Tregitopes as alternatives to IVIG would have a dramatic impact on the use of IVIG for immune modulation therapy and would certainly diminish potential infectious risks from a human blood derived product (Maddur et al. 2010).

Tregitope Mechanism of Action

Since 2008, we have significantly advanced our understanding of the Tregitope mechanism of action, leading to our working hypothesis of the step-wise process that is illustrated in Fig. 3: (1) Tregitope effects are contingent upon APC/MHC II-mediated presentation to a T cell; (2) Tregs recognize Tregitopes presented in the context of MHC II, are activated, and proliferate; (3) these activated Tregs produce IL-10 and interact with the APC to reinforce the development of a tolerogenic APC phenotype; and (4) these tolerogenic APCs and/or Treg act on adjacent antigen-specific effector T cells to suppress their effector responses and convert antigen specific T effectors into antigen-specific Treg (Maddur et al. 2011; Kessel et al. 2007; Cousens et al. 2012, 2013, 2013b).

Applying Tregitopes to tolerance induction in the context of protein therapeutic immunogenicity involves co-presentation of Tregitopes with a biologic protein. While the concept of Tregitope-mediated tolerance induction is relatively novel, this approach has emerged from studies carried out by our laboratory, and those of our collaborators, demonstrating that these specific, highly conserved and promiscuous T-cell epitopes derived from human immunoglobulins activate Treg cells (Maddur et al. 2011) and suppress immune responses in vitro and in vivo (Tha-In et al. 2010).

The corresponding murine epitopes are also effective in murine models (Kessel et al. 2007; van der Marel et al. 2012). The Tregitope discovery has been further validated in in vivo models of autoimmune diseases. Moreover, co-administration of antigens with Tregitopes in vivo and in vitro leads to the induction of antigen-specific tolerance (Kessel et al. 2007) and suppression of both antibody-mediated (Su et al. 2013) and cellular immune responses (Sharabi et al. 2006; Hahn et al. 2005) to co-administered antigens.

Tregitope-Mediated Tolerance Induction for Pompe Patients

Anti-therapeutic antibodies can have a dramatic effect on the safety and efficacy of a protein therapeutic product. Humanization and de-immunization have been applied in therapeutic protein development to reduce immunogenicity with variable success. However these approaches present significant drawbacks in the context of GAA immunogenicity and Pompe disease. Recombinant GAA protein is a fully human sequence and thus is not amenable to humanization. It appears foreign, in part, because these patients are afflicted by a number of mutations within the GAA that are translated into site-specific, truncated, or total gene deletion mutants. The resulting immunological phenotypes are heterogeneous with different GAA epitopes activating variable T cell responses in each individual. Relying on the deimmunization approach alone may be inadequate. Such may be the case where multiple epitope modifications are necessary in order to produce a therapeutic that is deimmunized for the majority of the patient population, but where the ability to implement these changes are constrained by structure or function. One means of inducing protein-specific tolerance to GAA may be by stimulating natural mechanisms of tolerance through the introduction of human regulatory T-cell epitopes into the biologic sequence of GAA (Hahn et al. 2005). This strategy would provide a localized tolerogenic signal for Tregitope-specific T cell induction in the context of GAA-specific T effector activation to facilitate conversion of effector T cells to antigen-specific Tregs. The approach could be further modified to provide co-administration of Tregitope and epitopes that might be presented by the CRIM-negative patients' HLA, potentially further improving the efficacy of the tolerance induction program. Experiments at EpiVax are underway to validate this approach.

Conclusions and Future Directions

Much progress has been made in the field of enzyme-replacement therapy (ERT) over the past decade. New treatments are benefiting patients, and approaches that lead to the reduction of ERT-ADA are under evaluation. Despite these significant advances, the long-term effects of current immune tolerance induction regimens have yet to be defined. Proof of the principle that the combination of deimmunization and Tregitope introduction would mitigate ADA in the context of Pompe disease would have an immediate impact on the field of enzyme replacement therapy and could accelerate adaptation of Tregitope therapy to facilitate effective treatments for these patients. Tregitopes merit consideration if only for reduced dependence on immunosuppressive drugs. However, ongoing studies appear to indicate that Tregitopes elicit antigen-specific tolerance (Kessel et al. 2007), augmenting the potential benefits of this approach in mitigating ADA. Thus, further proof-of-principle studies with Tregitopes are likely to have a far-reaching impact on the clinical development of an entire range of biologic therapies, including enzyme replacement therapy for Pompe disease.

Therapies that safely and permanently harness the immune system to induce long-lasting and specific tolerance in Pompe disease children will address a critical unmet medical need with broad-reaching implications for other replacement-protein therapies that are also limited by ADA (the Lysosomal Storage Disorders, Hemophilia A and B, etc. (Saint-Remy et al. 2004; Starzyk et al. 2007; Deegan 2011)). Clearly, these modified products would have to be produced in cell culture and tested in the appropriate laboratory assays, but in our experience, changes that minimize the perturbation of the three-dimensional structure can be rationally selected a priori. This tolerization approach may accelerate development of a new generation of protein therapeutics, providing an effective solution to the problem of immunogenicity in this field.

References

Anthony RM, Ravetch JV (2010) A novel role for the IgG Fc glycan: the anti-inflammatory activity of sialylated IgG Fcs. J Clin Immunol 30(Suppl 1):S9–S14

Bachmann MF, Hengartner H, Zinkernagel RM (1994) Immunization with recombinant protein: conditions for cytotoxic T cell and/or antibody induction. Med Microbiol Immunol 183(6):315–324

Banugaria SG, Prater SN, McGann JK, Feldman JD, Tannenbaum JA, Bailey C, Gera R, Conway RL, Viskochil D, Kobori JA, Rosenberg AS, Kishnani PS (2013) Bortezomib in the rapid reduction of high sustained antibody titers in disorders treated with therapeutic protein: lessons learned from Pompe disease. Genet Med 15(2):123–131. doi:10.1038/gim.2012.110

Barbosa MD (2011) Immunogenicity of biotherapeutics in the context of developing biosimilars and biobetters. Drug Discov Today 16:345–353

Bluestone JA, Abbas AK (2003) Natural versus adaptive regulatory T cells. Nat Rev Immunol 3:253–257

Celis E, Ou D, Otvos L (1998) Recognition of hepatitis B surface antigen by human T lymphocytes. Proliferative and cytotoxic responses to a major antigenic determinant defined by synthetic peptides. J Immunol 140:1808–1815

Collen D, Bernaerts R, Declerck P, De Cock F, Demarsin E, Jenné S, Laroche Y, Lijnen HR, Silence K, Verstreken M (1996) Recombinant staphylokinase variants with altered immunoreactivity. I: construction and characterization. Circulation 94:197–206

Cousens LP, Mingozzi F, van der Marel S, Su Y, Garman R, Ferreira V, Martin W, Scott DW, De Groot AS (2012) Teaching tolerance: new approaches to enzyme replacement therapy for Pompe disease. Hum Vaccin Immunother 8(10): 1459–1464. PMID: 23095864. http://tinyurl.com/De-Groot-Tregitope-Pompe

Cousens LP, Tassone R, Mazer BD, Ramachandiran V, Scott DW, De Groot AS (2013) Tregitope update: mechanism of action parallels IVIg. Autoimmun Rev 12(3):436–443. PMID: 22944299. http://tinyurl.com/Cousens-Tregitope-Autoimmunity. Epub 2012 Aug 28

Cousens LP, Su Y, McClaine E, Li X, Terry F, Smith R, Lee J, Martin W, Scott DW, De Groot AS (2013a) Application of IgG-derived natural Treg epitopes (IgG Tregitopes) to antigen-specific tolerance induction in a murine model of type 1 diabetes. J Diabetes Res 2013. Article ID 621693. doi:10.1155/2013/621693

Cousens LP, Najafian N, Mingozzi F, Elyaman W, Mazer B, Moise L, Messitt TJ, Su Y, Sayegh M, High K, Khoury SJ, Scott DW, De Groot AS (2013b) In vitro and in vivo studies of IgG-derived Treg epitopes (Tregitopes): a promising new tool for tolerance induction and treatment of autoimmunity. J Clin Immunol 33(1): 43–49. PMID: 22941509. http://www.ncbi.nlm.nih.gov/pmc/articles/PMC3538121/

De Groot AS, Martin W (2009) Reducing risk, improving outcomes: bioengineering less immunogenic protein therapeutics. Clin Immunol 131(2):189–201. doi:10.1016/j.clim.2009.01.009

De Groot AS et al (2003) Immunogenicity of therapeutic biological products. In: Developments in biologicals, vol 112. Karger, Basel, pp 71–80

De Groot AS, Knopp PM, Martin W (2005) De-immunization of therapeutic proteins by T-cell epitope modification. Dev Biol (Basel) 122:171–194

De Groot AS, Moise L, McMurry JA, Wambre E, Van Overtvelt L, Moingeon P, Scott DW, Martin W (2008) Activation of natural regulatory T cells by IgG Fc-derived peptide "Tregitopes". Blood 112(8):3303–3311

De Groot AS et al (2010) Methods in microbiology: immunology of infection, 3rd edn. Elsevier, London, pp 35–66

Deegan PB (2011) Fabry disease, enzyme replacement therapy and the significance of antibody responses. J Inherit Metab Dis. PMID: 22037707. http://www.springerlink.com/content/6517140513818126/

Deegan PB (2012) Fabry disease, enzyme replacement therapy and the significance of antibody responses. J Inherit Metab Dis 35(2):227–243, Epub 2011 Oct 25

Eckardt KU, Casadevall N (2003) Pure red-cell aplasia due to anti-erythropoietin antibodies. Nephrol Dial Transplant 18:865–869

Ephrem A, Chamat S, Miquel C, Fisson S, Mouthon L, Caligiuri G et al (2008) Expansion of CD4+CD25+ regulatory T cells by intravenous immunoglobulin: a critical factor in controlling experimental autoimmune encephalomyelitis. Blood 111(2):715–722, PMID: 17932250

Eser A, Primas C, Reinisch W (2013) Drug monitoring of biologics in inflammatory bowel disease. Curr Opin Gastroenterol 29(4):391–396

Garman RD, Munroe K, Richards SM (2004) Methotrexate reduces antibody responses to recombinant human alpha-galactosidase A therapy in a mouse model of Fabry disease. Clin Exp Immunol 137(3):496–502

Gea-Banacloche JC (2010) Rituximab-associated infections. Semin Hematol 47(2):187–198, PMID: 20350666

Gilles JG, Lavend'homme R, Peerlinck K, Jacquemin MG, Hoylaerts M, Jorieux S, Mazurier C, Vermylen J, Saint-Remy JM (1999) Some factor VIII (FVIII) inhibitors recognise a FVIII epitope(s) that is present only on FVIII-vWF complexes. Thromb Haemost 82:40–45

Hahn BH, Singh RP, La Cava A, Ebling FM (2005) Tolerogenic treatment of lupus mice with consensus peptide induces Foxp3-expressing, apoptosis-resistant, TGF-beta-secreting CD8+ T cell suppressors. J Immunol 175:7728–7737

Haribhai D et al (2003) A threshold for central T cell tolerance to an inducible serum 870 protein. J Immunol 170(6):3007–3014

Haselbeck A (2003) Epoetins: differences and their relevance to immunogenicity. Curr Med Res Opin 19(5):430–432

Hellendoorn K, Jones T, Watkins J, Baker M, Hamilton A, Carr F (2004) Limiting the risk of immunogenicity by identification and removal of T-cell epitopes (DeImmunisation™). Association for Immunotherapy of Cancer: Cancer Immunotherapy – 2nd annual meeting Mainz, Germany, vol 4 (Suppl 1). Cancer Cell International, p S20, 6–7 May 2004

Hill AV, Jepson A, Plebanski M, Gilbert SC (1997) Genetic analysis of host-parasite coevolution in human malaria. Philos Trans R Soc Lond B Biol Sci 352(1359):1317–1325

Jawa V, Cousens LP, Awwad M, Wakshull E, Kropshofer H, De Groot AS (2013) T-cell dependent immunogenicity of protein therapeutics: preclinical assessment and mitigation. Clin Immunol 149(3):534–555. doi:10.1016/j.clim.2013.09.006, Epub 2013 Sep 25

Jones TD, Phillips WJ, Smith BJ, Bamford CA, Nayee PD, Baglin TP, Gaston JS, Baker MP (2005) Identification and removal of a promiscuous CD4+ T cell epitope from the C1 domain of factor VIII. J Thromb Haemost 3:991–1000

Jones TD, Crompton LJ, Carr FJ, Baker MP (2009) Deimmunization of monoclonal antibodies. Methods Mol Biol 525:405–423. doi:10.1007/978-1-59745-554-1_21

Joseph A, Munroe K, Housman M, Garman R, Richards S (2008) Immune tolerance induction to enzyme-replacement therapy by co-administration of short-term, low-dose methotrexate in a murine Pompe disease model. Clin Exp Immunol 152(1):138–146

Katz U, Shoenfeld Y, Zandman-Goddard G (2011) Update on intravenous immunoglobulins (IVIg) mechanisms of action and off- label use in autoimmune diseases. Curr Pharm Des 17(29):3166–3175

Kaveri SV, Maddur MS, Hegde P, Lacroix-Desmazes S, Bayry J (2011) Intravenous immunoglobulins in immunodeficiencies: more than mere replacement therapy. Clin Exp Immunol 164(Suppl 2):2–5

Kessel A, Ammuri H, Peri R, Pavlotzky ER, Blank M, Shoenfeld Y et al (2007) Intravenous immunoglobulin therapy affects T regulatory cells by increasing their suppressive function. J Immunol 179(8):5571–5575, PMID: 17911644

King C, Garza EN, Mazor R, Linehan JL, Pastan I, Pepper M, Baker D (2014) Removing T-cell epitopes with computational protein design. Proc Natl Acad Sci U S A 111(23):8577–8582. doi:10.1073/pnas.1321126111, Epub 2014 May 19

Kirschbaum J, Forschner K, Rasche C, Worm M (2006) Modulation of lymphocyte phenotype and function by immunoglobulins. Br J Dermatol 154:225–230

Koren E, Smith HW, Shores E, Shankar G, Finco-Kent D, Rup B, Barrett YC, Devanarayan V, Gorovits B, Gupta S, Parish T, Quarmby V, Moxness M, Swanson SJ, Taniguchi G, Zuckerman LA, Stebbins CC, Mire-Sluis A (2008) Recommendations on risk-based strategies for detection and characterization of antibodies against biotechnology products. J Immunol Methods 333(1–2):1–9

Li Pira G, Ivaldi F, Moretti P, Manca F (2010) High throughput T epitope mapping and vaccine development. J Biomed Biotechnol 2010:325720. doi:10.1155/2010/325720, Epub 2010 Jun 15

Link J, Lundkvist Ryner M, Fink K, Hermanrud C, Lima I, Brynedal B, Kockum I, Hillert J, Fogdell-Hahn A (2014) Human leukocyte antigen genes and interferon beta preparations influence risk of developing neutralizing anti-drug antibodies in multiple sclerosis. PLoS One 9(3):e90479

Lopez M, Clarkson MR, Albin M, Sayegh MH, Najafian N (2006) A novel mechanism of action for anti-thymocyte globulin: induction of CD4+CD25+Foxp3+ regulatory T cells. J Am Soc Nephrol 17:2844–2853

Maddur MS, Othy S, Hegde P, Vani J, Lacroix-Desmazes S, Bayry J, Kaveri SV (2010) Immunomodulation by intravenous immunoglobulin: role of regulatory T cells. J Clin Immunol 30(Suppl 1):S4–S8

Maddur MS, Kaveri SV, Bayry J (2011) Comparison of different IVIg preparations on IL-17 production by human Th17 cells. Autoimmun Rev 10(12):809–810, PMID: 21376142

Maneiro JR, Salgado E, Gomez-Reino JJ (2013) Immunogenicity of monoclonal antibodies against tumor necrosis factor used in chronic immune-mediated inflammatory conditions: systematic review and meta-analysis. JAMA Intern Med 173(15):1416–1428

Mazer BD, Al-Tamemi S, Yu JW, Hamid Q (2005) Immune supplementation and immune modulation with intravenous immunoglobulin. J Allergy Clin Immunol 116:941–944

Mazor R, Vassall AN, Eberle JA, Beers R, Weldon JE, Venzon DJ, Tsang KY, Benhar I, Pastan I (2012) Identification and elimination of an immunodominant T-cell epitope in recombinant immunotoxins based on Pseudomonas exotoxin A. Proc Natl Acad Sci USA 109(51):E3597–E3603

McDermott AB, Cohen SBA, Zuckerman JN, Madrigal JA (1999) Human leukocyte antigens influence the immune response to a pre-S/S hepatitis B vaccine. Vaccine 17:330–339

McLachlan SM, Aliesky HA, Chen CR, Chong G, Rapoport B (2012) Adjuvant as a 'double edged sword'. PLoS One 7(9):e43517. doi:10.1371/journal.pone.0043517

Mendelsohn NJ, Messinger YH, Rosenberg AS, Kishnani PS (2009) Elimination of antibodies to recombinant enzyme in Pompe's disease. Correspondence. N Engl J Med 360(2):194–195

Messinger YH, Mendelsohn NJ, Rhead W, Dimmock D, Hershkovitz E, Champion M, Jones SA, Olson R, White A, Wells C, Bali D, Case LE, Young SP, Rosenberg AS, Kishnani PS (2012) Successful immune tolerance induction to enzyme replacement therapy in CRIM-negative infantile Pompe disease. Genet Med 14(1):135–142

Min WP, Kamikawaji N, Mineta M, Tana T, Kashiwagi S, Sasazuki T (1996) Identification of an epitope for T-cells correlated with antibody response to hepatitis B surface antigen in vaccinated humans. Hum Immunol 1996(46):93–99

Moise L, Song C, Martin WD, Tassone R, De Groot AS, Scott DW (2012) Effect of HLA DR epitope de-immunization of factor VIII in vitro and in vivo. Clin Immunol 142(3):320–331. doi:10.1016/j.clim.2011.11.010

Moran E, Carbone F, Augusti V, Patrone F, Ballestrero A, Nencioni A (2012) Proteasome inhibitors as immunosuppressants: biological rationale and clinical experience. Semin Hematol 49(3):270–276, PMID: 22726551

Mullbacher A (1992) Viral escape from immune recognition: multiple strategies of adenoviruses. Immunol Cell Biol 70(Pt 1):59–63

Opdenakker G, Van den Steen PE, Laureys G, Hunninck K, Arnold B (2003) Neutralizing antibodies in gene-defective hosts. Trends Immunol 24(2):94–100

Orange JS, Hossny EM, Weiler CR, Ballow M, Berger M, Bonilla FA, Buckley R, Chinen J, El-Gamal Y, Mazer BD et al (2006) Use of intravenous immunoglobulin in human disease: a review of evidence by members of the Primary Immunodeficiency Committee of the American Academy of Allergy, Asthma and Immunology. J Allergy Clin Immunol 117:S525–S553

Parker AS, Choi Y, Griswold KE, Bailey-Kellogg C (2013) Structure-guided deimmunization of therapeutic proteins. J Comput Biol 20(2):152–165. doi:10.1089/cmb.2012.0251

Paul S, Kolla RV, Sidney J, Weiskopf D, Fleri W, Kim Y, Peters B, Sette A (2013) Evaluating the immunogenicity of protein drugs by applying in vitro MHC binding data and the immune epitope database and analysis resource. Clin Dev Immunol 2013:467852. doi:10.1155/2013/467852

Pendley C, Shankar G (2011) Bioanalytical interferences in immunoassays for antibody biotherapeutics. Bioanalysis 3(7):703–706. doi:10.4155/BIO.11.53

Rohrbach M, Klein A, Köhli-Wiesner A, Veraguth D, Scheer I, Balmer C et al (2010) CRIM-negative infantile Pompe disease: 42-month treatment outcome. J Inherit Metab Dis 33(6):751–757, PMID: 20882352

Romball CG, Weigle WO (1999) Cytokines in the induction and circumvention of peripheral tolerance. J Interferon Cytokine Res 19(6):671–678

Rosenberg AS (2003) Immunogenicity of biological therapeutics: a hierarchy of concerns. Dev Biol (Basel) 112:15–21

Saint-Remy JM, Lacroix-Desmazes S, Oldenburg J (2004) Inhibitors in haemophilia: pathophysiology. Haemophilia 10(Suppl 4):146–151

Scanlan MJ, Jager D (2001) Challenges to the development of antigen-specific breast cancer vaccines. Breast Cancer Res 3(2):95–98

Shankar G, Devanarayan V, Amaravadi L, Barrett YC, Bowsher R, Finco-Kent D, Fiscella M, Gorovits B, Kirschner S, Moxness M, Parish T, Quarmby V, Smith H, Smith W, Zuckerman LA, Koren E (2008) Recommendations for the validation of immunoassays used for detection of host antibodies against biotechnology products. J Pharm Biomed Anal 48(5):1267–1281. doi:10.1016/j.jpba.2008.09.020

Shankar G, Arkin S, Cocea L, Devanarayan V, Kirshner S, Kromminga A, Quarmby V, Richards S, Schneider CK, Subramanyam M, Swanson S, Verthelyi D, Yim S (2014) Assessment and reporting of the clinical immunogenicity of therapeutic proteins and peptides-harmonized terminology and tactical recommendations. AAPS J 16(4):658–673. doi:10.1208/s12248-014-9599-2

Sharabi A, Zinger H, Zborowsky M, Sthoeger ZM, Mozes E (2006) A peptide based on the complementarity-determining region 1 of an autoantibody ameliorates lupus by up-regulating CD4+CD25+ cells and TGF-beta. Proc Natl Acad Sci U S A 103(23):8810–8815, Epub 2006 May 30

Soukhareva N, Jiang Y, Scott DW (2006) Treatment of diabetes in NOD mice by gene transfer of Ig-fusion proteins into B cells: role of T regulatory cells. Cell Immunol 240:41–46

Starzyk K, Richards S, Yee J, Smith SE, Kingma W (2007) The long-term international safety experience of imiglucerase therapy for Gaucher disease. Mol Genet Metab 90(2):157–163

Steere AC, Klitz W, Drouin EE, Falk BA, Kwok WW, Neopm GT, Baxter-Lowe LA (2006) Antibiotic-refractory Lyme arthritis is associated with HLA-DR molecules that bind a Borrelia borgdoferi peptide. J Exp Med 203(4):961–971

Su Y, Rossi R, De Groot AS, Scott DW (2013) Regulatory T cell epitopes (Tregitopes) in IgG induce tolerance in vivo and lack immunogenicity per se. J Leukoc Biol 94:377–383, Epub 2013 May 31

Tangri S, Mothe BR, Eisenbraun J, Sidney J, Southwood S, Briggs K, Zinckgraf J, Bilsel P, Newman M, Chesnut R, Licalsi C, Sette A (2005) Rationally engineered therapeutic proteins with reduced immunogenicity. J Immunol 174(6):3187–3196

Tatarewicz SM, Mytych DT, Manning MS, Swanson SJ, Moxness MS, Chirmule N (2014) Strategic characterization of anti-drug antibody responses for the assessment of clinical relevance and impact. Bioanalysis 6(11):1509–1523

Tha-In T, Metselaar HJ, Bushell AR, Kwekkeboom J, Wood KJ (2010) Intravenous immunoglobulins promote skin allograft acceptance by triggering functional activation of CD4+Foxp3+ T cells. Transplantation 89:1446–1455

Toubi E, Etzioni A (2005) Intravenous immunoglobulin in immunodeficiency states: state of the art. Clin Rev Allergy Immunol 29(3):167–172

van der Marel S, Majowicz A, Kwikkers K, van Logtenstein R, te Velde AA, De Groot AS, Meijer SL, van Deventer SJ, Petry H, Hommes DW, Ferreira V (2012) Adeno-associated virus mediated delivery of Tregitope 167 ameliorates experimental colitis. World J Gastroenterol 18(32): 4288–4299. PMID: 22969191. http://www.ncbi.nlm.nih.gov/pubmed/22969191

Vossen MT, Westerhout EM, Soderberg-Naucler C, Wiertz EJ (2002) Viral immune evasion: a masterpiece of evolution. Immunogenetics 54(8):527–542

Wang P, Sidney J, Dow C, Mothé B, Sette A, Peters B (2008) A systematic assessment of MHC class II peptide binding predictions and evaluation of a consensus approach. PLoS Comput Biol 4(4):e1000048. doi:10.1371/journal.pcbi.1000048

Weaver JM, Lazarski CA, Richards KA, Chaves FA, Jenks SA, Menges PR, Sant AJ (2008) Immunodominance of CD4 T cells to foreign antigens is peptide intrinsic and independent of molecular context: implications for vaccine design. J Immunol 181:3039–3048

Yeung VP, Chang J, Miller J, Barnett C, Stickler M, Harding FA (2004) Elimination of an immunodominant CD4+ T cell epitope in human IFN-beta does not result in an in vivo response directed at the subdominant epitope. J Immunol 172(11):6658–6665

Zuckerman JN (1996) Nonresponse to hepatitis B vaccines and the kinetics of anti-HBs production. J Med Virol 50:283–288

Part II
Monoclonal Antibodies: Degradation Mechanisms and Potential Improvements

Structure of Monoclonal Antibodies

Balakrishnan S. Moorthy, Bo Xie, Ehab M. Moussa, Lavanya K. Iyer, Saradha Chandrasekhar, Jainik P. Panchal, and Elizabeth M. Topp

Abbreviations

ADCC	Antibody-dependent cellular cytotoxicity
CDC	Complement-dependent cytotoxicity
CDR	Complementarity-determining region
C_H	Constant heavy chain
C_L	Constant light chain
Fab	Antigen binding fragment
Fc	Fragment crystallizable
Ig	Immunoglobulin
mAbs	Monoclonal antibodies
MBL	Mannan binding lectin
MR	Mannose receptor
SpA	Staphylococcal protein A
SpG	Streptococcal protein G
V_H	Variable heavy chain
V_L	Variable light chain

B.S. Moorthy • B. Xie • E.M. Moussa • L.K. Iyer • S. Chandrasekhar
J.P. Panchal • E.M. Topp (✉)
Department of Industrial and Physical Pharmacy, Purdue University,
Robert E. Heine Pharmacy Building, 575 Stadium Mall Drive,
West Lafayette, IN 47907-2091, USA
e-mail: sbalakr@purdue.edu; xie108@purdue.edu; emoussa@purdue.edu; iyerl@purdue.edu; chandra6@purdue.edu; jpanchal@purdue.edu; topp@purdue.edu

A. Rosenberg, B. Demeule (eds.), *Biobetters*, AAPS Advances
in the Pharmaceutical Sciences Series 19, DOI 10.1007/978-1-4939-2543-8_6

Introduction

Monoclonal antibodies (mAbs) are the fastest growing sector of the biopharmaceutical industry, with nearly 30 mAbs currently on the market and hundreds more in development. mAbs are large protein drugs and susceptible to many types of physical and chemical degradation. The most commonly observed degradation pathways are aggregation, deamidation, isomerization, oxidation, disulfide rearrangement and peptide bond cleavage. In addition to these modifications, heterogeneity due to multiple glycoforms can variably affect mAb stability and function. Since partially degraded proteins have been associated with altered drug potency and an increased potential for immunogenic side effects, understanding the mechanisms of these degradation pathways is important to designing improved mAbs. In this introductory chapter, the structure of mAbs is explained, with a focus on the functionally important regions that are susceptible to chemical modification. The mechanisms of degradation such as aggregation, deamidation, isomerization, oxidation, glycan hydrolysis, disulfide rearrangement and peptide bond hydrolysis are presented in subsequent chapters, together with available evidence for the effects of these modifications on mAb function. Subsequent chapters also address the effects of in vivo environments on these degradation pathways and the potential to make "biobetter" mAbs by reducing susceptibility to degradation.

Chemical Structure of IgG Antibody

The chemical structure of an antibody is complex and has been reviewed in detail elsewhere (Alzari et al. 1988; Burton and Woof 1992; Davies and Metzger 1983; Harris et al. 1997, 1998; Burton 1985; Terry et al. 1968; Saphire et al. 2001). Briefly, mAbs are comprised of four polypeptide chains, two identical heavy chains and two identical light chains, with a total molecular weight of ~150kD. The chains are held together by disulfide linkages and fold to form a "Y" shaped tetramer (Figs. 1 and 2a). With regard to biological function, the amino- and carboxy-terminal halves of an antibody chain are subdivided into variable and constant regions, respectively. The variable region shows great variability in the amino acid sequence among antibodies and serves as an antigen binding site, while the constant region determines antibody isotype and effector functions.

Based on differences in the amino acid sequences in the constant region of the heavy chains, immunoglobulins can be divided into five different classes: IgG, IgM, IgA, IgD and IgE with heavy chains denoted by the Greek letters γ, μ, α, δ and ε respectively. The structural differences in these isotypes contribute to their diverse specificity and effector functions. All immunoglobulins within a given isotype have very similar heavy chain constant regions. Based on differences in amino acid sequences in the constant region, the light chains in immunoglobulins can be classified into two types, kappa (κ) and lambda (λ). Distribution of these isotypes differs

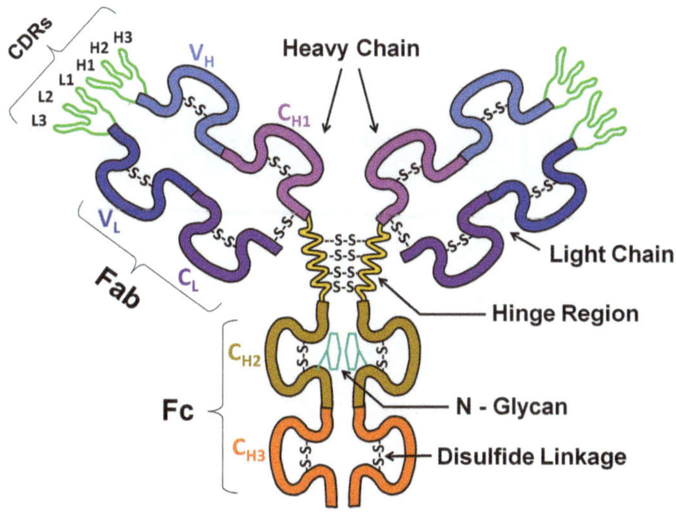

Fig. 1 Schematic diagram of typical IgG structure. The "Y" shaped tetrameric IgG molecule is comprised of two identical light and two identical heavy chains. The light chain consists of V_L (*dark blue*) and C_L (*dark violet*) domains and the heavy chain of V_H (*light blue*), C_H1 (*light violet*), C_H2 (*olive*) and C_H3 (*orange*) domains. The region that connects the C_H1 and C_H2 domains is referred to as the hinge region (*yellow*). The CDRs in the variable domains and the *N*-glycan in the C_H2 domain are represented as *green loops* and *cyan rings*, respectively

in all immunoglobulin classes, subclasses and among different species. However, for any given antibody, the two light chains are always identical and consist of either of κ or λ chains. Unlike heavy chain isotypes, immunoglobulins with κ or λ light chains have no known functional differences.

The remainder of this chapter focuses exclusively on IgG class antibodies, since most therapeutic antibodies today are IgG based. The overall structure of IgG differs from the other antibody classes in the length of the heavy chain and/or the number of structural domains. The IgG heavy chain has one variable heavy (V_H) and three constant heavy (C_H1, C_H2, and C_H3) domains, while the light chain has one variable light (V_L) and one constant light (C_L) domain (Figs. 1 and 2a). The sequence connecting the C_H1 and C_H2 domains forms the hinge region. Based on the length of the hinge region, the IgG class can be further divided into IgG1, IgG2, IgG3 and IgG4 subclasses (see section "Hinge region").

Proteolytic digestion of IgG with papain cleaves the molecule in the hinge region, producing two identical antigen binding fragments (Fab) and one crystallizable fragment (Fc). Each Fab fragment contains one of each of the V_L, C_L, V_H and C_H1 domains from a light and heavy chain. The amino acid sequence in the Fab regions varies among antibodies, so that each can bind specifically to a defined epitope of an antigen. The Fc fragment is comprised of the C_H2 and C_H3 domains from the two heavy chains (Figs. 1 and 2). The Fc domain defines the effector functions of an antibody and the pathway by which the antigen is cleared through different antibody mediated immune responses.

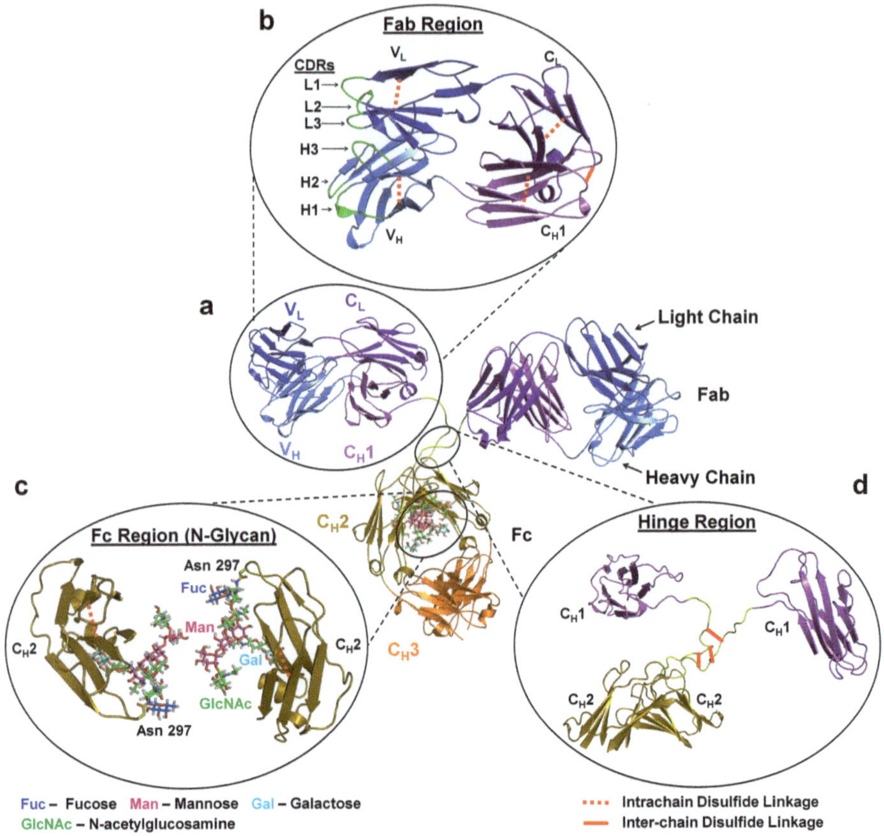

Fig. 2 Structure of monoclonal antibody. (**a**) Cartoon representation of the crystal structure of monoclonal antibody 231 (murine IgG2aκ) (PDB: 1IGT) (Harris et al. 1997). The color codes for domain representation are as shown in Fig. 1. Each domain consists of seven to nine β-strands in two anti-parallel β-sheets forming a β-barrel structure. (**b**) Fab region; the CDR loops (L1, L2, L3, H1, H2 and H3) that connect the β-sheets in V_L and V_H domains are shown in *green*. The intra- and inter-chain disulfide bonds are shown as *red dashed* and *solid lines*, respectively. (**c**) Fc region (*N*-glycan); the highly conserved Asn297 from the two C_H2 domains bear the biantennary *N*-glycan. (**d**) Hinge region; the three inter-chain disulfide bonds that connect the two heavy chains near hinge region are shown as *red lines*

The constant and variable domains are highly structurally homologous among antibodies, with 70–110 amino acids per domain. Each domain is comprised of seven to nine β-strands in two anti-parallel β-sheets forming a β-barrel structure with one intra-chain disulfide bond (Fig. 2b). The orientation of hydrophobic side chains towards the interior of the barrel and the intra-chain disulfide bonds stabilize the constant and variable domains. The V_H/V_L, C_H1/C_L and C_H3/C_H3 domains are paired through hydrophobic interactions. The hydrophobic surface in the C_H2 domain is covered by complex *N*-linked biantennary oligosaccharides, which limit C_H2/C_H2 hydrophobic interactions (Figs. 1 and 2c). In addition to the hydrophobic interactions, each light chain is connected to a heavy chain through a disulfide bond and each heavy chain is connected to another heavy chain through two to eleven disulfide bonds in the hinge region, with the number of hinge disulfides varying with IgG subtype (see section "Hinge region") (Figs. 1 and 2d). The whole IgG molecule is thus stabilized by the pairing of domains through non-covalent hydrophobic interactions and by covalent intra- and inter-chain disulfide bridges that connect the β-sheets within and between domains. The structural complexities of functionally important regions in IgGs are discussed below.

Fab—Complementarity Determining Region (CDRs)

The loops connecting β-strands within each variable domain are arranged non-consecutively into three small sub-regions forming the antigen binding CDRs. The three CDR loops are generally referred as CDR1, CDR2 and CDR3. However, to be specific, the loops formed by the V_L domain are referred to as L1, L2 and L3 and the loops in the V_H domain are referred as H1, H2 and H3 (Figs. 1 and 2b). The CDR loops show the greatest sequence variability between different antibodies and so are sometimes referred as the hypervariable region. The CDR H3 is the most variable of the CDR loops, both in amino acid sequence and length, and is generally the most flexible. The six CDR loops from one Fab domain form a single antigen binding site that can collectively come into contact with an antigen. The recognition of protein antigens by Fab generally utilizes all six of the CDR loops. The crystal structure of the first Fab-hapten complex (Padlan et al. 1973; Segal et al. 1974) showed that the main site of interaction for antibody with antigen is the CDR loops of the Fab domains. The chemical nature and orientation of amino acid residues on the CDR surface determine the specificity and affinity of CDRs for an antigen, and chemical modifications in the CDRs greatly affect these interactions. For example, in vivo deamidation at an Asn residue in the H2 CDR loop may show significant decrease in antigen binding (Huang et al. 2005). Similarly, tryptophan oxidation in the H3 CDR loop of humanized mAb against respiratory syncytial virus (RSV) showed loss of antigen binding and biological function (Wei et al. 2007). Nearly 15–20 % of endogenous IgG bears *N*-glycans in the V_L and V_H regions (Mimura et al. 2007) and the influence of IgG-Fab glycosylation on antigen binding has been the subject of several reports (Kato et al. 1993; Khurana et al. 1997; Tachibana et al. 1993; Wallick et al. 1988; Coloma et al. 1999). *N*-glycan in the L1 CDR loop of human monoclonal antibody specific to lung adenocarcinoma (HB4C5) interfered with antibody activity and an increase in antigen binding of HB4C5 was observed following L1 deglycosylation (Kato et al. 1993; Tachibana et al. 1993). Glycosylation close to the CDR region of an anti-gonadotrophin releasing hormone (GnRH) monoclonal antibody is important for antigen binding, and the presence of sialic acid containing mannose-rich carbohydrate is critical for defining the specificity of antibody (Khurana et al. 1997). An anti-dextran monoclonal antibody with potential N-linked glycosylation site in the H2 loop showed a 15-fold increase in binding affinity to antigen (Wallick et al. 1988; Coloma et al. 1999).

Effector Functions Mediated by FcR-IgGFc Interactions

The Fc receptors (FcRs) and complement binding regions of IgGFc are located in the hinge proximal region of the C_H2 domain. The binding sites for neonatal Fc receptor (FcRn), staphylococcal protein A (SpA), streptococcal protein G (SpG), serum protein mannan binding lectin (MBL) and cellular mannose receptor (MR) are all found in the C_H2/C_H3 interface.

N-glycosylation is arguably the most important co-translational modification in antibodies. Each C_H2 domain of human IgG carries a single covalently linked biantennary *N*-glycan at the highly conserved Asn297 residue (Fig. 2c). The glycan features a heptasaccharide biantennary core and can vary by the addition of fucose to the core *N*-acetylglucosamine, by the addition of *N*-acetylglucosamine to the bisecting mannose, or by extending the arms with the addition of galactose and sialic acid (Fig. 3). The α1-6 arm occupies the pocket formed in the C_H2 domain and the α1-3 arm projects into the interface of the C_H2 and C_H3 domains. The role of glycan is not limited to the protection of hydrophobic surfaces in the C_H2 domain. Glycans also keep the Fc region accessible to the various FcRs that selectively activate different effector pathways.

IgG Fc mediated stimulation of FcRs results in the activation of effector functions, including antibody-dependent cellular cytotoxicity (ADCC), phagocytosis, oxidative burst, and the release of inflammatory mediators. Similarly, complement functions through a variety of mechanisms, including cell lysis, opsonization and phagocytosis, and the production of anaphylatoxins. The nature of the amino acids in the C_H2 domain and in the hinge region determines the ability of different IgG isotypes to activate complement. Interactions between IgGFc and a specialized FcR, FcRn, play an important role in the metabolism and elimination of IgG. IgG-FcRn interactions are essential for endocytic rescue and transcytosis of internalized IgG, mediating the increased plasma half-life.

Any chemical modifications in the Fc region or the *N*-glycans can influence the effector functions of IgG. For example, deglycosylation or aglycosylation has been shown to abolish the binding of IgG with FcRs and complement (Nose and Wigzell 1983; Pound et al. 1993), and removal of a fucose group from the oligosaccharides enhanced mAb ADCC activity (Satoh et al. 2006). Disruption of intramolecular inter-

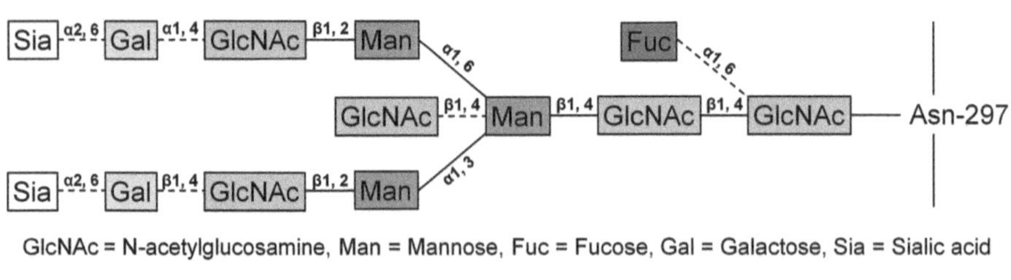

Fig. 3 Structure of *N*-linked glycans in IgG. The *N*-glycans in IgG are composed of a common heptasaccharide biantennary core structure (bonds shown as *solid lines*) and can be heterogeneously modified by the addition of sugar residues (bonds shown as *dashed lines*). The modifications include the addition of fucose to the core *N*-acetylglucosamine, addition of *N*-acetylglucosamine to the bisecting mannose or extension of arms by the addition of galactose and sialic acid

actions between residues in the C_H2 domain and *N*-glycan leads to the exposure of aggregation-prone regions and aggregation (Voynov et al. 2009). Similarly, increased sialylation of Fc glycans has been shown to decrease antibody binding to FcRs and to diminish the corresponding IgG ADCC activity (Scallon et al. 2007). The terminal galactose content of IgG affects complement-dependent cytotoxicity (CDC), but not ADCC activity. Deamidation in Fc domains occurs over time for mAbs administered intravenously (Liu et al. 2009); however the effect of Fc deamidation on antibody function in vivo has not been reported. Methionine oxidation in the Fc region has been shown to decrease the binding affinity of IgGs to FcRs (Pan et al. 2009; Bertolotti-Ciarlet et al. 2009) and increase the deamidation rate (Liu et al. 2008).

Hinge Region

In human IgG, the hinge region is formed by a stretch of 12 amino acids for IgG2 and IgG4, 15 amino acids for IgG1 and 62 amino acids for IgG3 (Redpath et al. 1998). The hinge region links the two heavy chains together by one or more inter-chain disulfide bonds (Figs. 1 and 2d). The number of these bonds and their positions in the hinge region vary both from class to class and within the subclasses of each class. The number of disulfide bonds in human IgG subtypes is two for IgG1 and IgG4, four for IgG2 and eleven for IgG3. In addition to the different number of Cys residues available for forming disulfide bonds, the hinge region is rich in Pro residues which promote a random coil structure, conferring flexibility. The hinge region serves as a spacer that separates the Fab-arms of the molecule from the Fc. This allows each Fab domain to move relative to the other during antigen binding. Along with the C_H2 region, the hinge region plays an important role in complement activation. X-ray crystallographic analysis of a complex between IgG-Fc and a soluble form of FcR has demonstrated that the FcR binds to the lower hinge and the hinge proximal regions of the two C_H2 domains asymmetrically with a 1:1 stoichiometry (Radaev et al. 2001; Sondermann et al. 2000). Since the hinge region is the least structured and most solvent-exposed, it is often the most susceptible to peptide bond hydrolysis and disulfide rearrangement (Smith et al. 1996; Wang et al. 2011). Incorrect disulfide bonding has been shown to affect the biological functions of IgG (Liu and May 2012).

Summary

This introductory chapter has provided an overview of the structure of an antibody and the functionally important regions that can undergo posttranslational modifications. As discussed above, chemical changes in the CDR, in the CH_2–CH_3 domain interface, and in the hinge region are expected to have direct effects on mAb in vivo performance. Chemical changes at other sites may also alter performance, however, albeit indirectly. The next chapter reviews the physicochemical stability of mAbs, including detailed descriptions of various antibody degradations pathways and their effect on in vivo properties.

References

Alzari PM, Lascombe MB, Poljak RJ (1988) Three-dimensional structure of antibodies. Annu Rev Immunol 6:555–580

Bertolotti-Ciarlet A, Wang W, Lownes R, Pristatsky P, Fang Y, Mckelvey T, Li Y, Drummond J, Prueksaritanont T, Vlasak J (2009) Impact of methionine oxidation on the binding of human IgG1 to Fc Rn and Fc gamma receptors. Mol Immunol 46:1878–1882

Burton DR (1985) Immunoglobulin G: functional sites. Mol Immunol 22:161–206

Burton DR, Woof JM (1992) Human antibody effector function. Adv Immunol 51:1–84

Coloma MJ, Trinh RK, Martinez AR, Morrison SL (1999) Position effects of variable region carbohydrate on the affinity and in vivo behavior of an anti-(1→6) dextran antibody. J Immunol 162:2162–2170

Davies DR, Metzger H (1983) Structural basis of antibody function. Annu Rev Immunol 1:87–117

Harris LJ, Larson SB, Hasel KW, Mcpherson A (1997) Refined structure of an intact IgG2a monoclonal antibody. Biochemistry 36:1581–1597

Harris LJ, Skaletsky E, Mcpherson A (1998) Crystallographic structure of an intact IgG1 monoclonal antibody. J Mol Biol 275:861–872

Huang L, Lu J, Wroblewski VJ, Beals JM, Riggin RM (2005) In vivo deamidation characterization of monoclonal antibody by LC/MS/MS. Anal Chem 77:1432–1439

Kato M, Mochizuki K, Hashizume S, Tachibana H, Shirahata S, Murakami H (1993) Activity enhancement of a lung cancer-associated human monoclonal antibody HB4C5 by N-deglycosylation. Hum Antibodies Hybridomas 4:9–14

Khurana S, Raghunathan V, Salunke DM (1997) The variable domain glycosylation in a monoclonal antibody specific to GnRh modulates antigen binding. Biochem Biophys Res Commun 234:465–469

Liu H, May K (2012) Disulfide bond structures of IgG molecules: structural variations, chemical modifications and possible impacts to stability and biological function. mAbs 4:17–23

Liu D, Ren D, Huang H, Dankberg J, Rosenfeld R, Cocco MJ, Li L, Brems DN, Remmele RL Jr (2008) Structure and stability changes of human IgG1 Fc as a consequence of methionine oxidation. Biochemistry 47:5088–5100

Liu YD, Van Enk JZ, Flynn GC (2009) Human antibody Fc deamidation in vivo. Biologicals 37:313–322

Mimura Y, Ashton PR, Takahashi N, Harvey DJ, Jefferis R (2007) Contrasting glycosylation profiles between Fab and Fc of a human IgG protein studied by electrospray ionization mass spectrometry. J Immunol Methods 326:116–126

Nose M, Wigzell H (1983) Biological significance of carbohydrate chains on monoclonal antibodies. Proc Natl Acad Sci U S A 80:6632–6636

Padlan EA, Segal DM, Spande TF, Davies DR, Rudikoff S, Potter M (1973) Structure at 4.5 A resolution of a phosphorylcholine-binding fab. Nat New Biol 245:165–167

Pan H, Chen K, Chu L, Kinderman F, Apostol I, Huang G (2009) Methionine oxidation in human IgG2 Fc decreases binding affinities to protein A and FcRn. Protein Sci 18:424–433

Pound JD, Lund J, Jefferis R (1993) Aglycosylated chimaeric human IgG3 can trigger the human phagocyte respiratory burst. Mol Immunol 30:233–241

Radaev S, Motyka S, Fridman WH, Sautes-Fridman C, Sun PD (2001) The structure of a human type III Fcgamma receptor in complex with Fc. J Biol Chem 276:16469–16477

Redpath S, Michaelsen TE, Sandlie I, Clark MR (1998) The influence of the hinge region length in binding of human IgG to human Fcgamma receptors. Hum Immunol 59:720–727

Saphire EO, Parren PW, Pantophlet R, Zwick MB, Morris GM, Rudd PM, Dwek RA, Stanfield RL, Burton DR, Wilson IA (2001) Crystal structure of a neutralizing human IgG against HIV-1: a template for vaccine design. Science 293:1155–1159

Satoh M, Iida S, Shitara K (2006) Non-fucosylated therapeutic antibodies as next-generation therapeutic antibodies. Expert Opin Biol Ther 6:1161–1173

Scallon BJ, Tam SH, Mccarthy SG, Cai AN, Raju TS (2007) Higher levels of sialylated Fc glycans in immunoglobulin G molecules can adversely impact functionality. Mol Immunol 44: 1524–1534

Segal DM, Padlan EA, Cohen GH, Rudikoff S, Potter M, Davies DR (1974) The three-dimensional structure of a phosphorylcholine-binding mouse immunoglobulin fab and the nature of the antigen binding site. Proc Natl Acad Sci U S A 71:4298–4302

Smith MA, Easton M, Everett P, Lewis G, Payne M, Riveros-Moreno V, Allen G (1996) Specific cleavage of immunoglobulin G by copper ions. Int J Pept Protein Res 48:48–55

Sondermann P, Huber R, Oosthuizen V, Jacob U (2000) The 3.2-A crystal structure of the human IgG1 Fc fragment-Fc gammaRIII complex. Nature 406:267–273

Tachibana H, Seki K, Murakami H (1993) Identification of hybrid-type carbohydrate chains on the light chain of human monoclonal antibody specific to lung adenocarcinoma. Biochim Biophys Acta 1182:257–263

Terry WD, Matthews BW, Davies DR (1968) Crystallographic studies of a human immunoglobulin. Nature 220:239–241

Voynov V, Chennamsetty N, Kayser V, Helk B, Forrer K, Zhang H, Fritsch C, Heine H, Trout BL (2009) Dynamic fluctuations of protein-carbohydrate interactions promote protein aggregation. PLoS One 4:E8425

Wallick SC, Kabat EA, Morrison SL (1988) Glycosylation of a VH residue of a monoclonal antibody against alpha (1-6) dextran increases its affinity for antigen. J Exp Med 168:1099–1109

Wang X, Kumar S, Singh SK (2011) Disulfide scrambling in IgG2 monoclonal antibodies: insights from molecular dynamics simulations. Pharm Res 28:3128–3144

Wei Z, Feng J, Lin HY, Mullapudi S, Bishop E, Tous GI, Casas-Finet J, Hakki F, Strouse R, Schenerman MA (2007) Identification of a single tryptophan residue as critical for binding activity in a humanized monoclonal antibody against respiratory syncytial virus. Anal Chem 79:2797–2805

Prediction of Aggregation In Vivo by Studies of Therapeutic Proteins in Human Plasma

Tudor Arvinte, Emilie Poirier, and Caroline Palais

Abbreviations

EM	Electron microscopy
i.a.	Intra-arterial
i.m.	Intramuscular
i.v.	Intravenous
NTA	Nanoparticle tracking analysis
PFI	Particle-flow imaging
s.c.	Subcutaneous

Introduction

In the last decade major progress had been achieved in the detection of protein aggregates and in the understanding of aggregation mechanisms. Using different new analytical methods, it is possible to characterize a variety of protein aggregates in aqueous solutions, ranging from loose oligomers to more rigid, larger aggregates of nanometer to micrometer diameter. The goal of aggregation studies in the pharmaceutical industry is to develop stable formulations for therapeutic protein

T. Arvinte (✉)
Therapeomic Inc., Bio Park Rosental, Mattenstrasse 22, WRO-1055, Basel 4002, Switzerland

Department of Pharmaceutical Sciences, University of Geneva–University of Lausanne,
30 Quai Ernest Ansermet, 1205, Geneva, Switzerland
e-mail: tudor.arvinte@unibas.ch

E. Poirier • C. Palais
Therapeomic Inc., Bio Park Rosental, Mattenstrasse 22, WRO-1055, Basel 4002, Switzerland
e-mail: emilie.poirier@therapeomic.com; caroline.palais@therapeomic.com

A. Rosenberg, B. Demeule (eds.), *Biobetters*, AAPS Advances
in the Pharmaceutical Sciences Series 19, DOI 10.1007/978-1-4939-2543-8_7

products that contain, in general, the smallest possible number of aggregates in the solution prior to in vivo administration. However, after injection of a non-aggregated protein solution into organs or tissues, aggregation can occur at the injection site or in specialized microenvironments in tissues. This local in vivo aggregation has been used to develop slow-release formulations for marketed products such as the long-acting insulin analog, Lantus® (http://products.sanofi.us/lantus/lantus.html) and peptidic gonadotropin-releasing hormone receptor blocker, Firmagon® (www.accessdata.fda.gov/drugsatfda_docs/label/2013/022201s002lbl.pdf).

In vivo aggregation of therapeutic proteins after administration may influence the absorption/bioavailability properties of the drugs (Filipe et al. 2012; Demeule et al. 2009a). These aggregation phenomena are not generally addressed in pharmacokinetic studies of biopharmaceuticals administered by s.c. and i.v. routes (Zheng et al. 2012; Wang et al. 2012; Richter et al. 2012). The local in vivo aggregation of therapeutic proteins after administration can be evaluated by initially analyzing the aggregation that occurs in mixtures of human serum or plasma with clear protein solutions rather than in formulation buffer alone (Arvinte et al. 2013). For example, large aggregates, up to 10 μm diameter, were formed when human plasma was mixed with 5 % dextrose solutions of Herceptin® (trastuzumab) and Avastin® (bevacizumab), but not Remicade® (infliximab); no aggregates formed when the antibodies were formulated in 0.9 % NaCl (Arvinte et al. 2013). For Herceptin® and Avastin® the prescribing information states that 0.9 % NaCl should be used as the i.v. infusion solution and the use of 5 % dextrose is prohibited (www.accessdata.fda.gov/drugsatfda_docs/label/2010/103792s5250lbl.pdf, www.accessdata.fda.gov/drugsatfda_docs/label/2011/125085s225lbl.pdf). These new observations show that antibodies may form aggregates in vivo when inappropriate infusion solutions are used and indicate that these aggregates may be associated with immunogenic and toxic effects (Arvinte et al. 2013). The presented antibody cases further demonstrate how a formulation issue (and not the underlying protein) may cause aggregation in vivo with potential negative effects on the patients. To minimize such risks, we proposed that the development of protein formulations for new drugs, biosimilars and biobetters should include studies of their compatibility in different formulations with human plasma under conditions as near as possible to those planned to be used in human studies. Such studies are of direct relevance to the drugs intended for i.v. administration. Additionally biopharmaceuticals that have s.c. as the desired route for delivery often require i.v. clinical Phase I or Phase II studies. The s.c. and i.m. spaces contain interstitial fluid which has a similar composition to plasma, containing also lipoproteins (Lundberg et al. 2013). Since interstitial fluid is difficult to collect and to have in large volumes (Lundberg et al. 2013), human plasma can be used as a first model for aggregation properties in the s.c. compartment.

In this chapter we present new structural and size distribution analyses of the aggregates formed when 5 % dextrose solutions of Herceptin® and Avastin® are mixed with human plasma and propose a model for these aggregation phenomena. Comparison of the plasma aggregation properties of an originator protein with its biosimilar and biobetter protein products is proposed as a parameter that can be used to improve the quality of these products.

Materials

Bevacizumab (Avastin®, Roche), trastuzumab (Herceptin®, Roche) and insulin glargine (Lantus®, Sanofi-Aventis) were purchased. All proteins were analyzed prior to the expiration date. Herceptin® is a lyophilized antibody formulation; after reconstitution with water, the solution contains 21 mg/mL trastuzumab in 20 mg/mL (52.9 mM) α,α-trehalose dihydrate, 0.5 mg/mL (2.4 mM) L-histidine HCl, 0.32 mg/mL (2.1 mM) L-histidine and 0.09 mg/mL (0.009 % w/v) polysorbate 20, pH 6. Avastin® is a liquid antibody formulation of 25 mg/mL bevacizumab in 60 mg/mL (158.6 mM) α,α-trehalose dihydrate, 5.8 mg/mL (42 mM) monobasic sodium phosphate monohydrate, 1.2 mg/mL (8.5 mM) dibasic sodium phosphate anhydrous and 0.4 mg/mL (0.04 % w/v) polysorbate 20, pH 6.2. Lantus® consists of 3.64 mg/mL insulin glargine, 30 μg/mL zinc, 2.7 mg/mL m-cresol and 20 mg/mL glycerol 85 %, pH 4. Human plasma bags were purchased from Blutspendezentrum SRK Beider Basel (Basel, Switzerland). Dextrose and NaCl were purchased as analytical grade powders from Fluka (Switzerland). All other chemicals were of analytical grade.

Light Microscopy

Herceptin® stock solution of 21 mg/mL was diluted to 1 mg/mL with 0.9 % NaCl or 5 % dextrose. 2.5 μL of the 1 mg/mL Herceptin® solutions (in 0.86 % sodium chloride or 4.8 % dextrose in water) or 2.5 μL of undiluted Lantus®, were placed inside microscope slides and then 2.5 μL of human plasma were added. Two types of microscope slides were used: FastRead 102™ slides (Immune Systems, Paignton, UK) and KOVA® Glasstic® slides (Hycor, Garden Grove, USA). A Leica DMRXE microscope (Leica Microsystems GmbH, Wetzlar, Germany) with a 5X/0.15na objective (Leica Microsystems GmbH, Wetzlar, Germany) was used. The images were acquired with a Sony NEX-5 camera and its firmware, and processed using the ImageJ version 1.43u software (National Institutes of Health, Bethesda, Maryland, USA). The experiments were performed at room temperature; the micrographs were taken within 1 min after addition of human plasma.

Particle-Flow Imaging (PFI)

Particle-flow imaging measurements were performed at room temperature with the Occhio FC200S particle counter (Occhio, Angleur, Belgium). Presented data are averages of three different measurements: each measurement was performed on 250 μL of solution. Herceptin® (21 mg/mL) and Avastin® (25 mg/mL) stock solutions were diluted to 1 mg/mL in 5 % dextrose in water (final 4.8 % dextrose). Prior to the measurements the 1 mg/mL antibody solutions in 4.8 % dextrose were mixed 1:1 (v:v) with filtered (0.22 μm) human plasma.

Nanoparticle Tracking Analysis (NTA)

NTA measurements were performed with a NanoSight LM20 (Malvern, Worcestershire, United Kingdom) equipped with a sample chamber with a 650 nm laser, using the NTA 2.3 software for capturing and analyzing the data. The samples were measured at room temperature for 120 s with manual shutter and gain adjustments; presented data are the averages of four measurements. 10 μL of antibody stock solutions (21 mg/mL Herceptin® and 25 mg/mL Avastin®) were diluted into 1.5 mL of 5 % dextrose in water (final concentrations: 0.14 mg/mL Herceptin® and 0.17 mg/mL Avastin®). 5 μL of filtered (0.22 μm) human plasma were then added to the 1.5 mL antibody-dextrose solution prior to measurement.

Electron Microscopy (EM)

Herceptin® (21 mg/mL) and Avastin® (25 mg/mL) stock solutions were diluted to 1 mg/mL in 5 % dextrose in water. Prior to the measurements the 1 mg/mL antibody solutions in 4.8 % dextrose were mixed 1:1 (v:v) with human plasma at room temperature. 5 μL aliquots of each sample were incubated for 30 s on fresh glow-discharged Cu grids (mesh 400) covered with parlodium/carbon film. The grids were stained with fresh diluted 2 % uranyl acetate for 10 s. Excess of stain on the grids was absorbed by filter paper and the grids were consequently air-dried. The grids were analyzed using a transmission electron microscope Philips CM-100 with an acceleration voltage of 80 kV.

Results and Discussion

Intra-arterial and i.v. routes are often used for the delivery of drugs. Different in vitro and in vivo studies show that the solution leaving the infusion needle does not mix immediately with the human blood: a stream of infused solution forms and flows through the blood vessels. In vitro models using conventional catheters showed significant streaming of the infused solutions for i.a. infusion at the commonly used infusion rates (2–4 mL/min) (Lutz et al. 1986, 2002; Arasa and Aldridge 2013). Streaming may contribute to non-uniform distribution of the drug in the body, resulting sometimes in up to a fivefold higher concentration of the drug in peripheral blood vessels, which was proposed to be one origin of focal toxicity in vivo (Lutz et al. 1986). Streaming of the i.a. infused solution was shown to exist in vivo and this streaming was associated with heterogeneous chemotherapy drug deposition and side effects (Blacklock et al. 1986; Saris et al. 1991). Besides poor mixing and non-uniform drug distribution, a large surface of contact forms between the infused stream and the blood during streaming (Lutz et al. 1986). At this interface surface the infused therapeutic protein may aggregate with blood components, similar to the aggregation observed in vitro by light microscopy when antibody solutions come in contact with human plasma (Arvinte et al. 2013). The aggregates which formed at the interface between plasma and Herceptin® in 5 % dextrose solution, Fig. 1b, had sizes between 0.5 and 9 μm, with a mean diameter of 4 μm (Arvinte et al. 2013). These aggregates in plasma were much larger than the aggregates present in the clear solution of Herceptin® in 5 % dextrose, which were up to 200 nm, Fig. 1a and (Demeule et al. 2009b).

An instantaneous and strong aggregation was also observed at the interface between human plasma and Lantus®, a marketed slow-release formulation of insulin glargine known to aggregate in vivo at the injection site. The aggregated zone formed at the Lantus®–plasma interface was more compact and dense compared to the aggregated zone formed when Herceptin® in 5 % dextrose was mixed with human plasma, Fig. 1c and d, white arcs. These observations show that in vitro plasma aggregation studies may be predictive for the in vivo aggregation of therapeutic proteins.

In the case of peptide and small protein drugs such as Lantus®, the plasma aggregation model may be used for the development of s.c. or i.m. slow-release formulations. For example, in the case of Lantus® we measured solubilization of aggregates formed upon mixture with human plasma (similar appearance to Fig. 1d) within 24 h of incubation at 37°C (data not shown). Nonetheless, unintended aggregation at s.c. injection site may result in loss of efficacy, local irritation or other immunogenic reactions and such adverse events may be more pronounced with larger protein aggregates.

Careful use of the proposed plasma model is advised; otherwise it may not detect aggregation and incompatibility of the injected biopharmaceuticals in all cases. One important element appears to be that aggregation may occur not immediately after s.c. administration. In some cases we observed that clear protein–plasma mixtures were aggregated only after 10 min (data not shown). Thus for each protein we recommend performing detailed plasma interaction studies and, if possible, to correlate them with in vivo results.

The aggregates formed when Herceptin® and Avastin® in 5 % dextrose are mixed with human plasma were characterized by nanoparticle tracking analysis (NTA), light microscopy, particle-flow imaging (PFI) and electron microscopy. For these proteins, all methods showed that the aggregation occurred rapidly, within seconds, after mixing.

PFI experiments showed that the plasma–dextrose–Herceptin® sample contained particles in the 3–10 μm region, Fig. 2a, as well as aggregates larger than 10 μm, Fig. 2f and g. The plasma–dextrose–Avastin® sample contained less aggregates than the Herceptin® sample, Fig. 2a, d, and e. The control plasma–dextrose–sample had a very small number of aggregates, Fig. 2a, b and c. In both Herceptin® and Avastin® mixtures, fibrils between 20 and 60 μm were observed, Fig. 2d–g. The PFI results are in agreement with the published light microscopy data, Fig. 3a, and b also showing that there are more particles in the Herceptin® sample (Arvinte et al. 2013). Light microscopy showed that the particles formed in plasma–Herceptin®–5 %

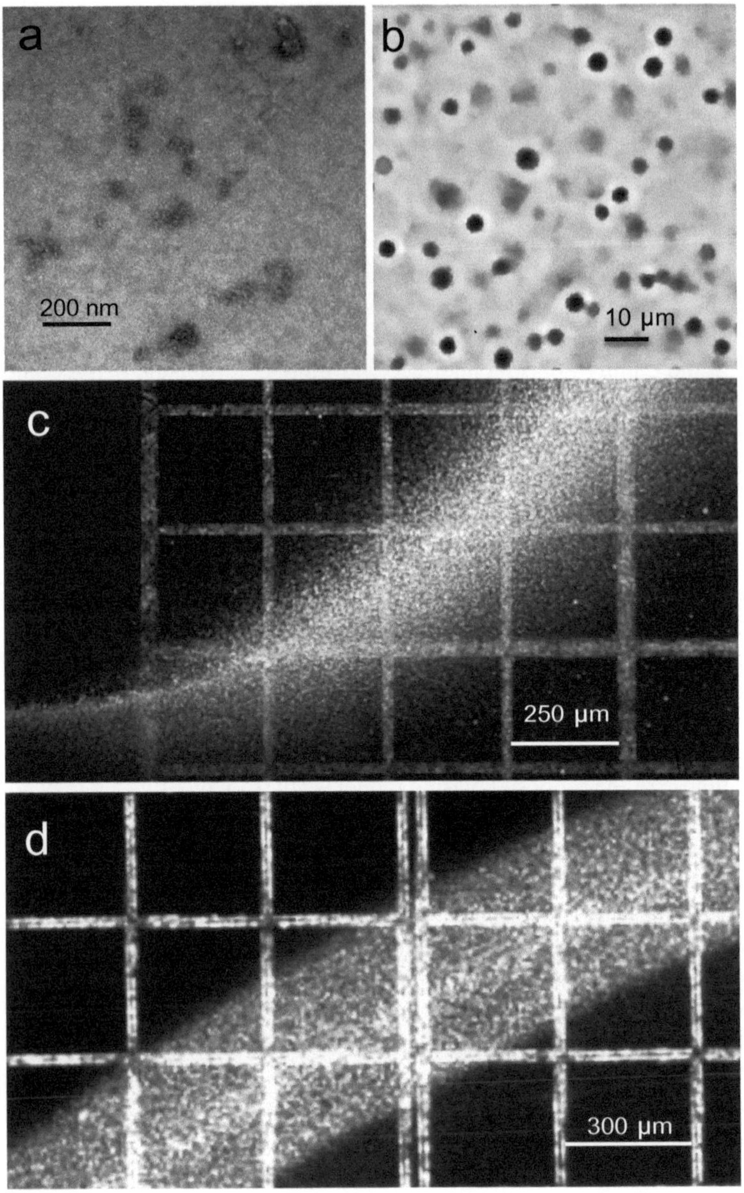

Fig. 1 Transmission electron microscopy picture of 1.28 mg/mL trastuzumab in 5 % dextrose (**a**); experimental conditions described in Demeule et al. (2009b). Light microscopy pictures (**b**) and (**c**) of Herceptin® diluted in 5 % dextrose mixed with human plasma (from Arvinte et al. 2013) and (**d**) the interface between Lantus® solution and human plasma

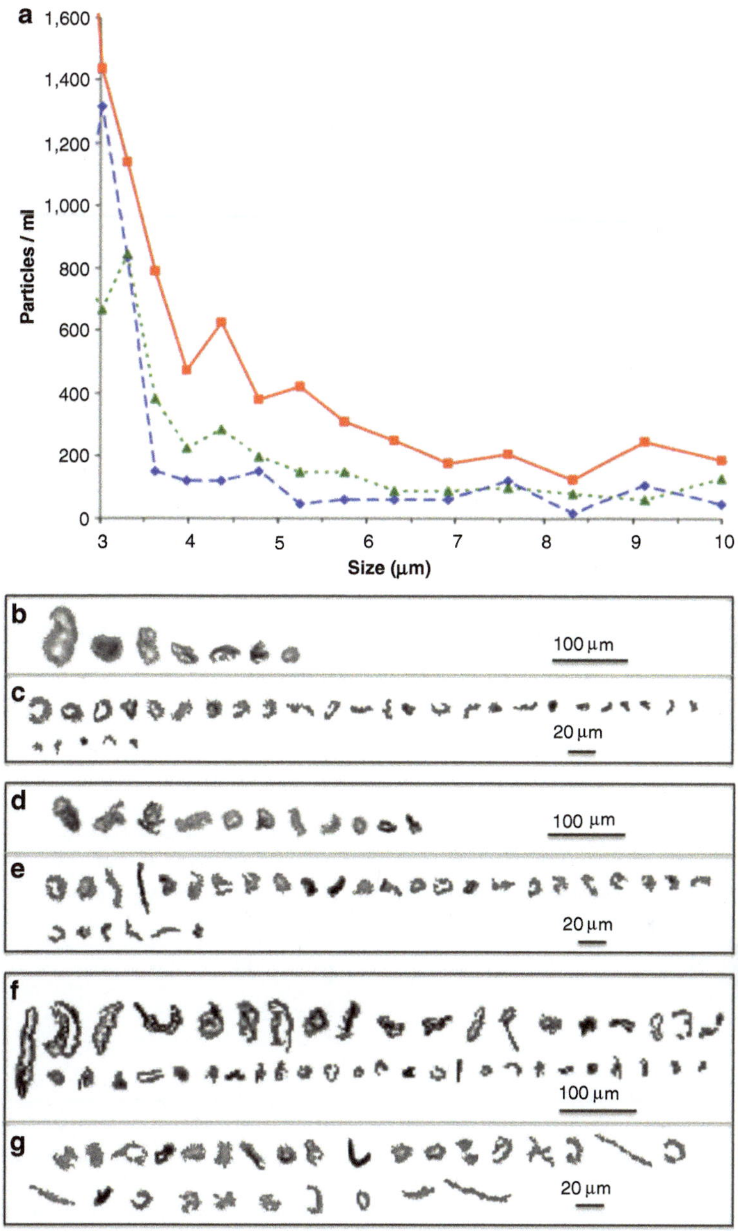

Fig. 2 (**a**) Particle size distributions measured from particle-flow imaging experiments: (-◆-) human plasma mixed with 5 % dextrose, (-▲-) 0.5 mg/mL Avastin® in 2.5 % dextrose and human plasma and (-■-) 0.5 mg/mL Herceptin® in 2.5 % dextrose and human plasma. (**b–g**) Pictures of the aggregates detected by particle-flow imaging: (**b, c**) human plasma in 5 % dextrose; (**d, e**) 0.5 mg/mL Avastin® in 2.5 % dextrose and human plasma and (**f, g**) 0.5 mg/mL Herceptin® in 2.5 % dextrose and human plasma

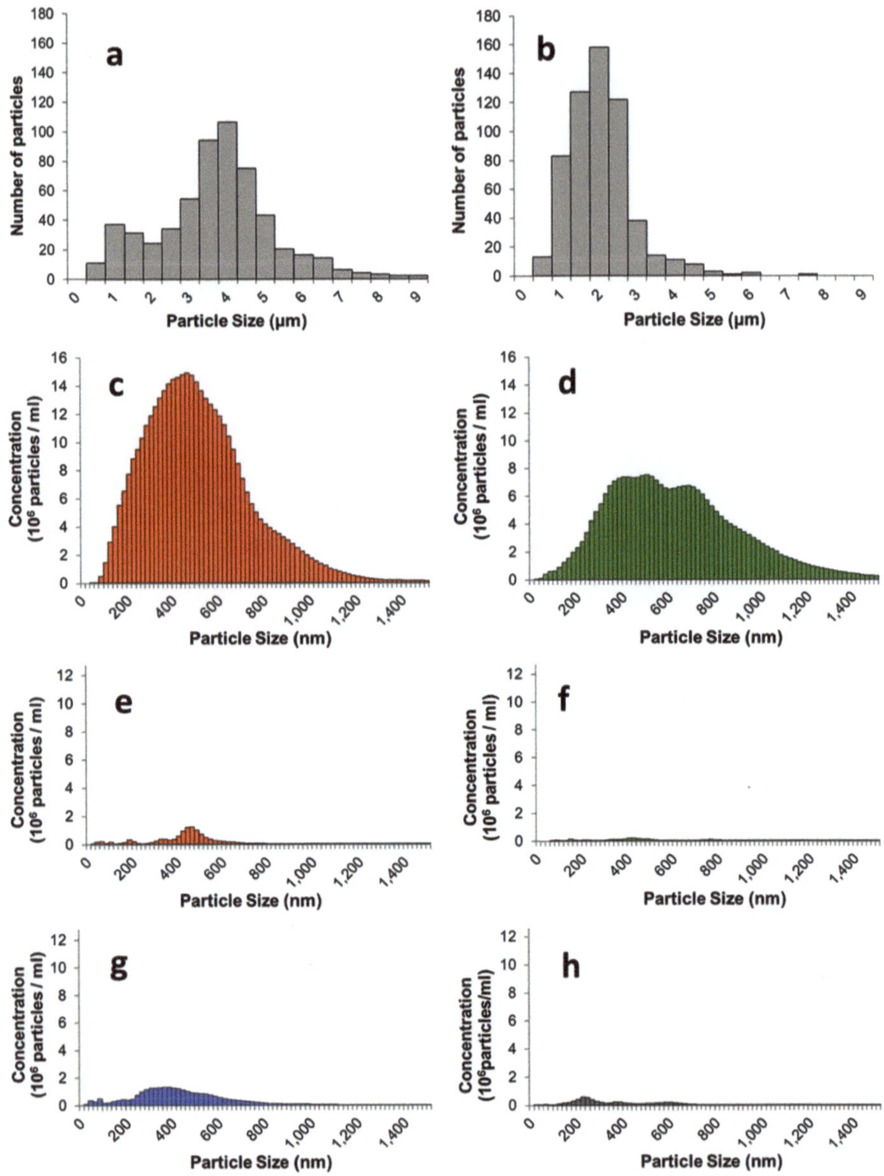

Fig. 3 (**a**) Particle size distributions of the aggregates observed by light microscopy at the contact region between Herceptin® diluted in 5 % dextrose and human plasma and (**b**) between Avastin® diluted in 5 % dextrose and human plasma. Particle size distributions measured by NTA of: (**c**) Herceptin® in 5 % dextrose and human plasma; (**d**) Avastin® in 5 % dextrose and human plasma; (**e**) Herceptin® in 5 % dextrose; (**f**) Avastin® in 5 % dextrose; (**g**) human plasma mixed with 5 % dextrose and (**h**) 5 % dextrose solution

dextrose and in plasma–Avastin®–5 % dextrose solutions had diameters between 0.5 and 9 μm, with a mean diameter of 4 μm for Herceptin® and 2 μm for Avastin®, Fig. 3a and b (Arvinte et al. 2013).

Aggregates in the 100–1,500 nm size range were studied by NTA using diluted solutions which permitted a good particle analysis, 0.17 mg/mL for Avastin®, 0.14 mg/mL for Herceptin® and 300 times diluted human plasma (5 μl of plasma added to 1.5 mL of 5 % dextrose). In these very diluted antibodies and human plasma solutions, many aggregates were observed, Fig. 3c and d, suggesting that the aggregates are non-dissociable. More aggregates were observed in the plasma–dextrose–Herceptin® sample: the size distribution showed one main peak around 500 nm and a shoulder at 900 nm, Fig. 3c. The aggregate distribution in the plasma–dextrose–Avastin® sample had a population with peaks around 400, 500 and 700 nm, Fig. 3d. Herceptin® was shown to aggregate in 5 % dextrose (Demeule et al. 2009b). NTA analysis showed that Herceptin® aggregates in 5 % dextrose are between 50 and 600 nm with the main peak at 500 nm, Fig. 3e. Fewer aggregates were measured in Avastin®–5 % dextrose solution, Fig. 3f.

Plasma diluted in 5 % dextrose contained a small number of particles, with sizes between 100 and 900 nm, with a peak at about 400 nm, Fig. 3g. These particles are very likely lipoproteins, the peak at 400 nm having a size distribution similar to that of the very-low-density human lipoproteins (Bierman et al. 1966). The other lipoproteins found in human plasma have smaller diameters, 20–200 nm (Zhang et al. 2013). Dextrose alone contained few aggregates, Fig. 3h.

Electron micrographs of human plasma mixed with 5 % dextrose showed the presence of globular lipoprotein structures, Fig. 4a and b, similar to published electron micrographs of lipoproteins (Bierman et al. 1966; Zhang et al. 2013). Larger and more numerous globular structures were observed in plasma–dextrose–Avastin® (Fig. 4c and d) and in plasma–dextrose–Herceptin® samples (Fig. 4e and f). For both antibodies, large globular structures in the micrometer range were also observed, Fig. 4d and f. These large globular structures appear to form by agglomeration of smaller globular structures similar in appearance to the lipoproteins.

In Fig. 5 we propose a model for the aggregation phenomena that occur when dextrose solutions of Herceptin® and Avastin® are mixed with human plasma. The spherical shape of the structures observed by light and electron microscopy suggests that the aggregates are formed through an agglomeration of antibodies and plasma lipoproteins. At present we do not have data which explains the molecular origin of the aggregates and support our model of an apparent affinity of plasma lipoproteins with Herceptin® and Avastin® when formulated in dextrose, but not in NaCl. In this context it is interesting to note that for other proteins such as Neupogen®, dextrose is recommended and NaCl is prohibited.

In 5 % dextrose, Herceptin® forms loose aggregates; such aggregates do not form in 0.9 % NaCl (Demeule et al. 2009b). These antibody aggregates have sizes up to 500 nm in diameter, Fig. 5a (based on data in (Demeule et al. 2009b)). In our model we propose that these loose antibody aggregates, when mixed with human plasma, Fig. 5b, bind to lipoproteins and induce the agglomeration of lipoproteins forming the larger globular structures observed by EM (about 1,500 nm) and NTA

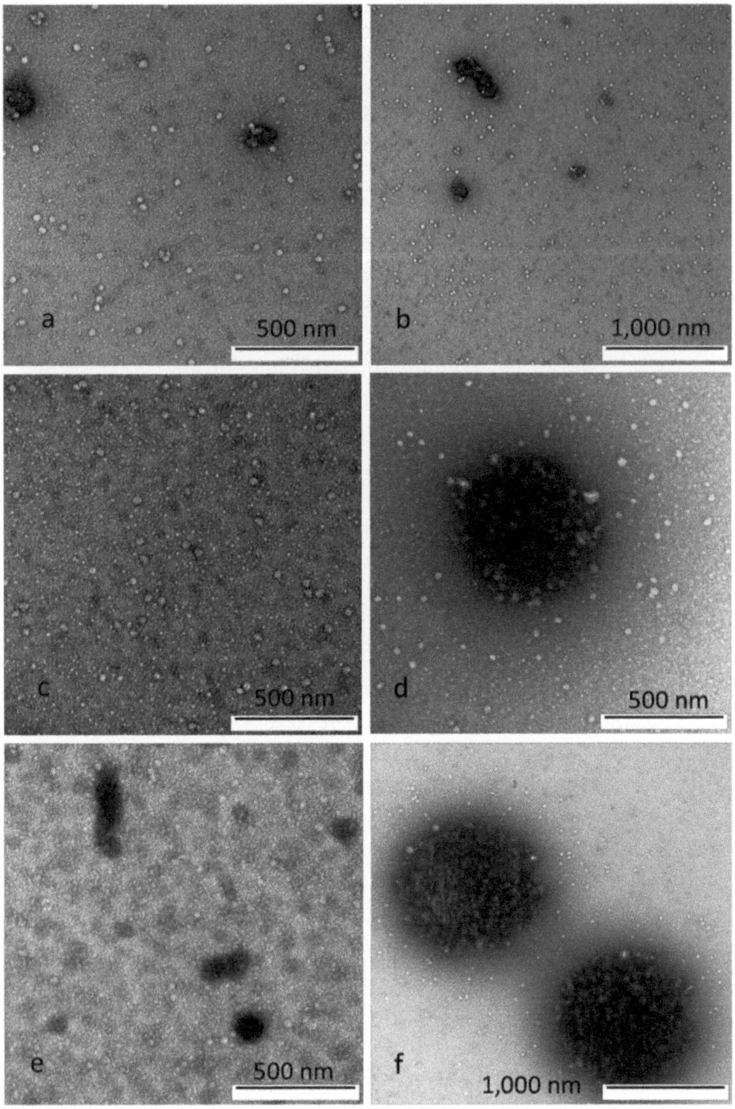

Fig. 4 (**a**, **b**) Transmission electron microscopy pictures of human plasma in 5 % dextrose; (**c**, **d**) Avastin® mixed in 5 % dextrose and human plasma; and (**e**, **f**) Herceptin® mixed in 5 % dextrose and human plasma

(500–1,000 nm), Fig. 5c. These globular structures continue to grow and bind one to another, forming the 1–10 μm aggregates observed by light microscopy, Fig. 5d, and the aggregates and fibrils up to 200 μm detected by PFI, Fig. 5e. The model does not explain why aggregation in human plasma occurred for 5 % dextrose formulations of Herceptin® and Avastin® and why no aggregation was observed when the antibodies were formulated in 0.9 % NaCl.

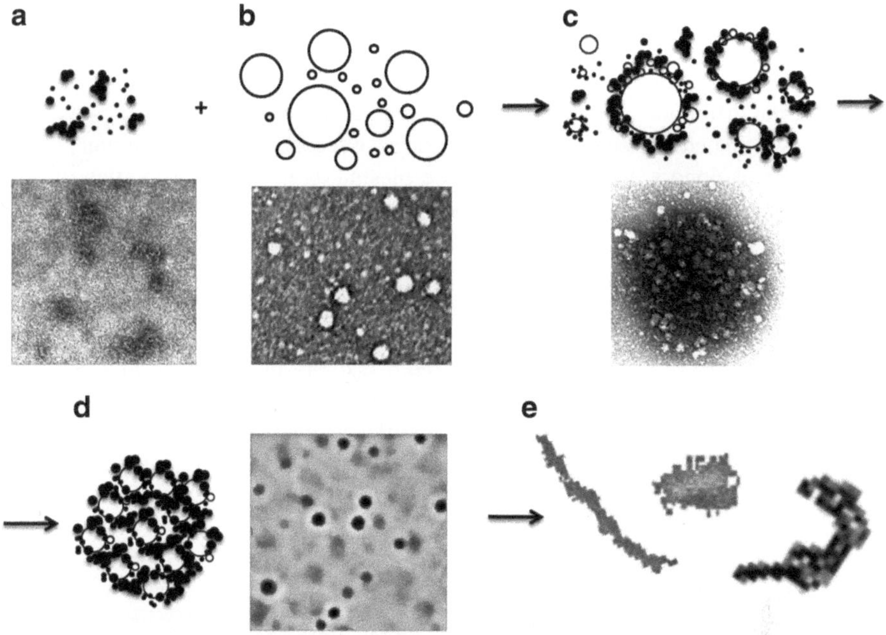

Fig. 5 Model for the aggregation phenomena observed when dextrose formulations of Herceptin® and Avastin® antibodies are mixed with human plasma. Aggregates of antibodies in 5 % dextrose (**a**) interact with plasma lipoproteins (**b**) forming large aggregates (**c**). These aggregates coalesce into large globular structures (**d**) and fibrils (**e**)

Formation of aggregates after infusion of protein drugs in vivo may be one origin of observed "infusion reactions" during the clinical studies of biopharmaceuticals (FDA Guidance for Industry 2014; Gamarra et al. 2006). The aggregates that form when human plasma is mixed with 5 % dextrose–Avastin® solutions could be one origin of the increased severe side effects (such as arterial thromboembolic events, myocardial infarction and angina) observed for Avastin® in combination with chemotherapy, compared to chemotherapy alone (FDA 2005). The formation of such aggregates is likely to be facilitated by the streaming phenomena previously described to be the origin of focal toxicity in vivo (Lutz et al. 1986). Human studies showed that reduced streaming can be obtained by using a pulsatile infusion (Saris et al. 1991). Theoretical studies showed that changing the geometry of the infusion catheter tip can also reduce the streaming phenomena (Arasa and Aldridge 2013).

Plasma aggregation of some biopharmaceuticals may also be dependent on the human donors. The same plasma aggregation behavior was observed when Herceptin®, Avastin® and Remicade® in 5 % dextrose and 0.9 % NaCl formulations were incubated with human plasma from three different donors (Arvinte et al. 2013). However, in other studies, we observed differences in aggregation when plasma from different healthy donors was used. In one research study, where two IgG antibodies of the same subtype were produced for the same target, the first antibody candidate formed aggregates when mixed with plasma from five healthy volunteers and did not form aggregates in plasma from another five volunteers (data not shown). The second antibody candidate did not aggregate in plasma from any of the ten donors, and it was therefore selected for further development. The origin of the difference in the aggregation properties of the two antibodies was not further investigated. Based on such case studies we recommend using human plasma from different donors when evaluating the plasma aggregation properties of biopharmaceuticals in different formulations.

In the development of biosimilar and biobetter proteins, it is also important to compare their plasma aggregation properties with those of the originator. For biosimilars the aim should be to have formulations that have very similar or more favorable aggregation properties compared to the originator. In some cases we have found that originator products that are administered by s.c. or i.m. routes, form aggregates when mixed with human plasma. In these cases the biosimilar formulation should be adapted to have similar aggregation properties as the originator product, Fig. 6, unless such aggregation is correlated to significant immunogenicity or adverse

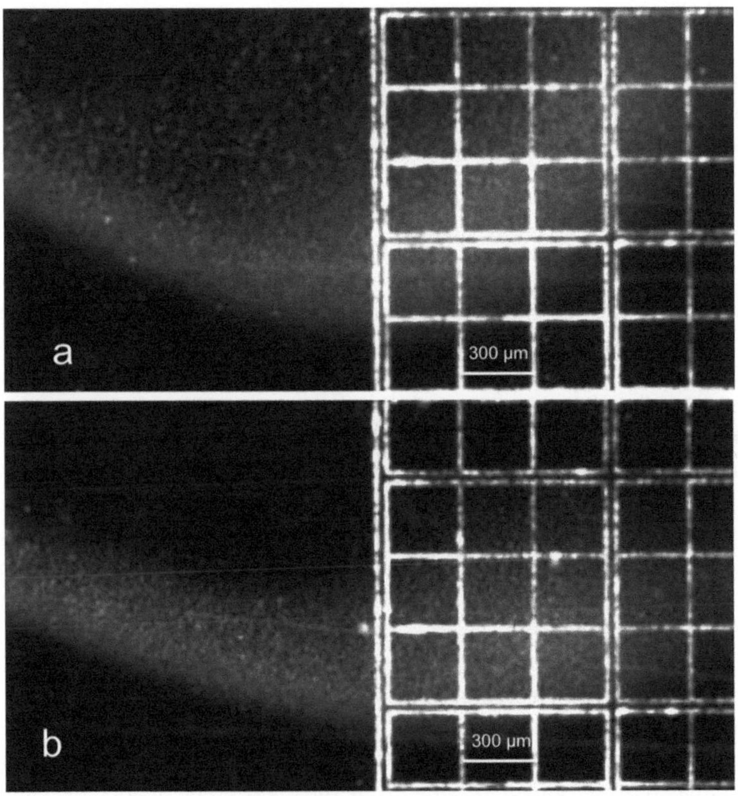

Fig. 6 Light microscopy pictures of the aggregates that formed in the mixing region of human plasma of a biosimilar protein drug formulation (**a**) and the originator product (**b**). The aggregates are small and diffuse forming the *white arc*

events. In the case of biobetters that are administered by s.c or i.m. injection, plasma aggregation can be used to improve the product. One possibility is to develop formulations that do not aggregate at the injection site (in case the originator aggregates) aiming for a better dose–response and dose reduction, less immunogenicity and a reduction of adverse events. In other situations, developing slow-release biobetter formulations that aggregate at the injection site may also have beneficial effects on pharmacokinetic and pharmacodynamic properties compared to the originator, in addition to being more user-friendly with potentially reduced side effects.

The antibody plasma aggregation phenomena presented in this chapter have different implications for research and development of new therapeutic protein products as well as of biosimilars and biobetters. Thus, the in vivo proof of concept for the protein drug may depend on the formulation used in animal and human studies. In early phases, compatibility of new protein drugs with plasma and serum can be selection criteria in research (for example for monoclonal antibodies, Demeule et al. 2009b). For animal proof of concept and toxicity studies, the compatibility of the formulation with the plasma of the animal model should be examined since aggregation can be different in animal plasma and in human plasma. Autologous animal proteins studies provided very helpful information on immunogenicity profiles and generation of antibodies against biotechnology products (Koren et al. 2008). In one of our projects with a human protein drug, the autologous animal protein was produced biotechnologically to study its toxicological effects in the same animal model. Unexpectedly, the animal protein therapeutic was found to aggregate strongly in the animal plasma but did not aggregate in human plasma; the human protein did not aggregate in either human or animal plasma. Had the toxicology evaluations been performed in animals using the animal protein, the observed toxicology would likely have been misleading (Therapeomic Inc., non-published data).

In summary, for the formulation development of new protein drugs, biosimilars and biobetters we propose that the best formulation found in physical and chemical stability studies should also be tested for its potential to generate protein aggregation in human plasma. For the intended i.v. administration of therapeutic proteins, the plasma compatibility studies should be performed under conditions as close as possible to those planned for use in the clinic. Human plasma compatibility can also be used as a model for local aggregation after s.c. and i.m. administration, complementing in vivo toxicity studies of local injection site reactions.

References

Arasa K, Aldridge RC (2013) Computational analysis of catheter-tip geometries for optimizing drug infusion in arterial blood flow. Am J Biomed Eng 3(4):91–98

Arvinte T, Palais C, Green-Trexler E, Gregory S, Mach HG, Narasimhan C, Shameem M (2013) Aggregation of biopharmaceuticals in human plasma and human serum. Implications for drug research and development. mAbs 5(3):491–500

Bierman EL, Hayen TL, Hawkins JN, Ewing AM, Lindgren FT (1966) Particle-size distribution of very low density plasma lipoproteins during fat absorption in man. J Lipid Res 7:65–72

Blacklock JB, Wright DC, Dedrick RL, Blasberg RG, Lutz RJ, Doppmann JL, Oldfield EH (1986) Drug streaming during intra-arterial chemotherapy. J Neurosurg 64:284–291

Demeule B, Shire SJ, Liu J (2009a) A therapeutic antibody and its antigen form different complexes in serum than in phosphate-buffered saline: a study by analytical ultracentrifugation. Anal Biochem 388:279–287

Demeule B, Palais C, Machaidze G, Gurny R, Arvinte T (2009b) New methods allowing the detection of protein aggregates: a case study on trastuzumab. mAbs 1:142–150

FDA (2005) "Important Drug Warning" Letter to Genentech, January 5, 2005 on the "increased risk of arterial thromboembolic events associated with the use of AVASTIN™ (Bevacizumab) in combination with chemotherapy". www.fda.gov/downloads/Safety/MedWatch/SafetyInformation/SafetyAlertsforHumanMedicalProducts/UCM164188.pdf

FDA Guidance for Industry (2014) Immunogenicity assessment for therapeutic protein products http://www.fda.gov/downloads/drugs/guidancecomplianceregulatoryinformation/guidances/ucm338856.pdf

Filipe V, Poole R, Oladunjoye O, Braeckmans K, Jiskoot W (2012) Detection and characterization of subvisible aggregates of monoclonal IgG in serum. Pharm Res 29:2202–2212

Gamarra RM, McGraw SD, Drelichman VS, Maas LC (2006) Serum sickness-like reactions in patients receiving intravenous infliximab. J Emerg Med 30(1):41–44

Koren E et al (2008) Recommendations on risk-based strategies for detection and characterization of antibodies against biotechnology products. J Immunol Methods 333(1–2):1–9

Lundberg J, Rudling M, Angelin B (2013) Interstitial fluid lipoproteins. Curr Opin Lipidol 24(4):327–331

Lutz RJ, Dedrick RL, Bortos JW, Oldfield EH, Blacklock JB, Doppman JL (1986) Mixing studies during intracarotid artery infusions in an in vitro model. J Neurosurg 64:277–283

Lutz RJ, Warren K, Balis F, Patronas N, Dedrick RL (2002) Mixing during intravertebral arterial infusions in an in vitro model. J Neurooncol 58:95–106

Richter WF, Bhansali SG, Morris ME (2012) Mechanistic determinants of biotherapeutics absorption following SC administration. AAPS J 14(3):559–570

Saris SC, Blasberg RG, Carson RE, deVroom HL, Lutz R, Dedrick RL, Pethgrew K, Chang R, Doppman J, Wright DC, Herscovitch P, Oldfield EH (1991) Intravascular streaming during carotid artery infusions: demonstration in humans and reduction using diastole-phased pulsatile administration. J Neurosurg 74:763–772

Wang W et al (2012) Lymphatic transport and catabolism of therapeutic proteins after subcutaneous administration to rats and dogs. Drug Metab Dispos 40(5):952–962

Zhang L, Tong H, Garewal M, Ren G (2013) Optimizing negative-staining electron microscopy for lipoprotein studies. Biochim Biophys Acta 1830:2150–2159

Zheng Y et al (2012) Minipig as a potential translatable model for monoclonal antibody pharmacokinetics after intravenous and subcutaneous administration. mAbs 4(2):243–255

Effect of Hydrolytic Degradation on the In Vivo Properties of Monoclonal Antibodies

Balakrishnan S. Moorthy, Bo Xie, Ehab M. Moussa, Lavanya K. Iyer, Saradha Chandrasekhar, Jainik P. Panchal, and Elizabeth M. Topp

Abbreviations

AChR	Acetylcholine receptors
ADCC	Antibody-dependent cellular cytotoxicity
CDC	Complement-dependent cytotoxicity
CDR	Complementarity-determining region
CH	Constant heavy chain
CL	Constant light chain
DKP	Diketopiperazine
EndoF1-3	Endo-β-*N*-acetylgluccosaminidases
ER	Endoplasmic reticulum
Fab	Antigen binding fragment
Fc	Fragment crystallizable
GSH	Glutathione
Ig	Immunoglobulin
mAbs	Monoclonal antibodies
MBL	Mannan binding lectin
MG	Myasthenia gravis
MR	Mannose receptor
PDI	Protein disulfide isomerase
PML	Progressive multifocal leucoencephalopathy

B.S. Moorthy • B. Xie • E.M. Moussa • L.K. Iyer • S. Chandrasekhar
J.P. Panchal • E.M. Topp (✉)
Department of Industrial and Physical Pharmacy, Purdue University,
Robert E. Heine Pharmacy Building, 575 Stadium Mall Drive,
West Lafayette, IN 47907-2091, USA
e-mail: sbalakr@purdue.edu; xie108@purdue.edu; emoussa@purdue.edu;
iyerl@purdue.edu; chandra6@purdue.edu; jpanchal@purdue.edu; topp@purdue.edu

A. Rosenberg, B. Demeule (eds.), *Biobetters*, AAPS Advances
in the Pharmaceutical Sciences Series 19, DOI 10.1007/978-1-4939-2543-8_8

scFv Single chain fragment variable
SLE Systemic lupus erythematosus
SpA Staphylococcal protein A
SpG Streptococcal protein G
VH Variable heavy chain
VL Variable light chain

Introduction

Hydrolysis is the cleavage of a covalent bond by reaction with water. Hydrolysis constitutes one of two major classes of chemical degradation reactions in mAbs and other protein drugs, the second being oxidation. Hydrolysis can occur in any of the aqueous environments that mAbs encounter, including upstream and downstream processing, fill-finish operations, in the shipping and shelf-storage of solution drug products, and in the bloodstream and tissues following administration. Less obviously, hydrolysis can also occur in solid forms of mAbs, such as lyophilized powders, when residual moisture is present. Hydrolytic reactions can change the composition of amino acid side chains, break peptide bonds in the light and heavy chains, alter glycosylation patterns and disrupt the disulfide bonds responsible for mAb quaternary structure, all of which have the potential to affect both safety and efficacy. To develop biosimilar and "biobetter" mAbs in a rational way, an understanding of hydrolytic reactions and their effects on in vivo performance is necessary. When hydrolysis has little effect on safety or efficacy, minor differences in hydrolytic stability between innovator and follow-on products can perhaps be tolerated. Conversely, hydrolytic reactions or reaction sites with demonstrable effects on safety and efficacy deserve increased industry and regulatory attention.

This chapter reviews four of the most common hydrolytic reactions in mAbs: deamidation, Fc-glycan hydrolysis, peptide bond hydrolysis (proteolysis) and non-oxidative disulfide rearrangement. While non-oxidative disulfide rearrangements are not formally hydrolytic, they are included because they involve proton transfer and are not oxidative. For each reaction, the mechanism is first presented, followed by a discussion of the intrinsic and extrinsic factors known to affect its rate. Available evidence for the effects of the reaction on mAb performance is then presented and discussed. Ideally, such evidence would be based on in vivo studies in animal and/or human subjects documenting a relationship between a specific type of degradation and altered efficacy, safety, pharmacokinetics or biodistribution. Such evidence is scarce. The potential for in vivo effects of hydrolytic degradation can also be inferred from in vitro studies such as receptor binding or cell-based bioactivity assays; the available in vitro evidence for the effects of each hydrolytic reaction is also presented. The chapter concludes with a discussion of approaches to mitigating the effects of hydrolytic reactions in mAbs, including antibody engineering and formulation approaches.

Deamidation

Deamidation is one of the most common routes of protein degradation, and occurs in vivo as part of aging as well as in therapeutic protein formulations. Of the 20 naturally occurring amino acids, Asn and Gln can undergo non-enzymatic deamidation to form Asp (or isoAsp) or glutamyl products. The loss of the amide functional group results in the formation of charged, more acidic products, which are often responsible for the so-called "charge variants" detected in anion or cation exchange chromatography. The subsections below describe the chemical mechanism of deamidation, factors influencing the reaction, and evidence for the effects of the reaction on the in vivo properties of mAbs.

Mechanism of Protein Deamidation

Deamidation occurs most commonly at Asn residues. At neutral to alkaline pH, the backbone N+1 amide nitrogen becomes deprotonated and initiates nucleophilic attack on the carbonyl group of the Asn side-chain. This generates a tetrahedral intermediate, which gains a proton and subsequently degrades to form a succinimide intermediate with the loss of ammonia. The rate-limiting step is the formation of the succinimide, and hence pH is an important factor affecting deamidation rates. In the presence of water, the succinimide hydrolyzes to form a mixture of Asp and isoAsp products (Fig. 1). The ratio of isoAsp/Asp is typically on the order of 3:1 (Geiger and Clarke 1987). At acidic pH, the side chain of Asn can undergo

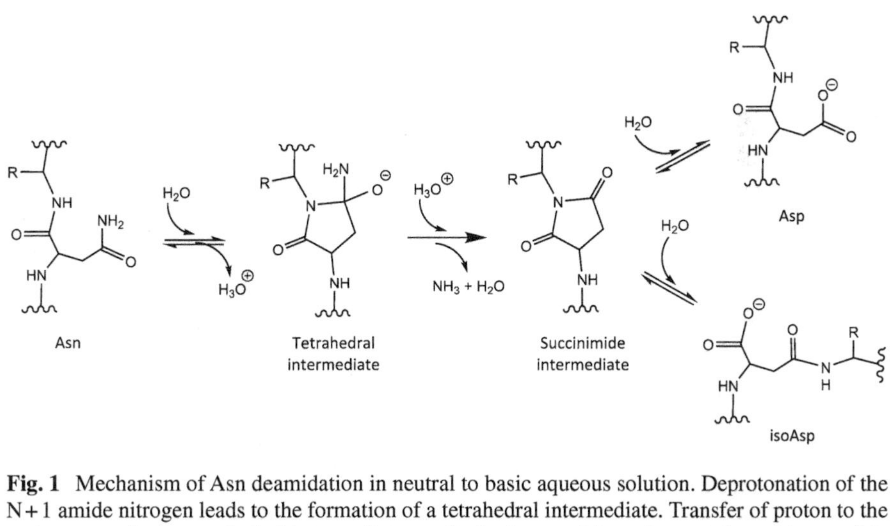

Fig. 1 Mechanism of Asn deamidation in neutral to basic aqueous solution. Deprotonation of the N+1 amide nitrogen leads to the formation of a tetrahedral intermediate. Transfer of proton to the amino group in the tetrahedral intermediate results in the expulsion of ammonia and irreversible formation of a succinimide intermediate. Hydrolysis of the succinimide at either of the carbonyl groups to form Asp or isoAsp variants of the original sequence. Asp can undergo isomerization to form isoAsp through the succinimide intermediate

direct hydrolysis to form an Asp derivative without the formation of isoAsp. Two related routes of degradation are Asp isomerization and Asp racemization. Isomerization refers to the dehydration-mediated conversion of Asp to isoAsp via the succinimide intermediate (Fig. 1) (Manning et al. 2010). Racemization is a parallel degradation pathway that results in the formation of D-Asp and D-isoAsp products from a racemized tetrahedral intermediate, with transient formation of D-Asn (Li et al. 2003).

Deamidation of Gln forms glutamyl and isoglutamyl products. Deamidation is less common at Gln than at Asn, since Gln deamidation requires the formation of a six-membered ring intermediate that is thermodynamically unfavorable, as opposed to the energetically favored five-membered succinimide ring formed in Asn deamidation (Manning et al. 2010). Hence, deamidation rates for Asn typically are much faster than for Gln, and the products of Asn deamidation are much more commonly observed as product variants in protein drug products.

Factors Affecting Deamidation

Several intrinsic (protein structure-related) and extrinsic (environment-related) factors affect the rate and propensity for deamidation:

Intrinsic Factors. The amino acid immediately C-terminal to Asn (i.e., N + 1 amino acid) affects deamidation rates to a greater extent than that to the N-terminal side (i.e., N − 1 amino acid). Unstructured peptides with Gly, Ser or Thr C-terminal to Asn (i.e., XAsnGlyY, XAsnSerY and XAsnThrY) show faster deamidation than sequences with other amino acids in this position (Brennan and Clarke 1994; Manning et al. 2010; Chelius et al. 2005). The small side-chain of Gly and the hydroxyl group-containing side chains of Ser and Thr offer little steric hindrance and favor nucleophilic reactions, resulting in faster deamidation rates. Isomerization and racemization tendencies have been observed for AspGly and AspAsp sequences (Geiger and Clarke 1987; Zhang et al. 2011). However, not all susceptible sequences are labile, since higher order structure also affects deamidation rates. Kosky et al. observed slower Asn deamidation rates for peptides with greater α-helix content (Kosky et al. 1999). In another study, the hinge region of mAbs, although relatively flexible, showed higher deamidation rates only at elevated temperatures (Hambly et al. 2009). Solvent exposure and local conformational flexibility of the deamidating or isomerizing residue determine whether the formation of the cyclic intermediate is energetically favorable (Capasso and Salvadori 1999; Stotz et al. 2004; Wakankar et al. 2007). For example, although trastuzumab (Herceptin®) has six AsnGly and AsnSer sequences which are potential sites of deamidation, the reaction is observed only at Asn30 and Asn55, which are followed by Thr and Gly, respectively (Harris et al. 2001). Cacia et al. observed that despite similar solvent exposure, an AsnGly sequence in the light chain CDR region of an anti-IgE mAb undergoes isomerization, while the same sequence in the heavy chain CDR region does not (Cacia et al. 1996). Another mAb with two AsnGly sequences in the

Fc region (Asn315 and Asn384) is deamidated predominantly at Asn384 under physiological conditions in vitro, but showed deamidation at both sites under denaturing conditions (Liu et al. 2009). This behavior has been attributed to stabilization of the primary sequence by side-chain hydrogen bonds and reverse turns, which impose structural constraints on isoAsp formation. While conformational changes can affect deamidation propensity, deamidation can also induce local structural changes, notably through the introduction of an extra methylene group in the peptide backbone when the isoAsp product is formed. For example, deamidation of Asn33 in an AsnGly sequence in the light chain CDR1 region of an IgG antibody produced Asp and isoAsp variants (Vlasak et al. 2009). The deamidated isoAsp species was susceptible to papain proteolysis in an adjacent CDR2 loop, suggesting that local conformational changes had occurred as a result of deamidation.

Extrinsic Factors. Extrinsic factors that accelerate the rate of deamidation include high pH, high temperature and, to a lesser extent, high solvent dielectric constant (Zheng and Janis 2006; Kroon et al. 1992; Brennan and Clarke 1993). These factors affect the rate of deprotonation of the backbone nitrogen and the stability of the nucleophilic nitrogen anion. In an 8-week stability study, Zheng and Janis showed that the deamidation rate of Asn55 in the heavy chain of a mAb was a function of pH, buffer components and temperature (Zheng and Janis 2006). In addition to affecting reaction kinetics directly, these factors may also affect protein conformation and the deamidation propensity of otherwise protected residues. Deamidation rates were higher in phosphate buffer than in citrate, succinate and tartrate, especially at higher temperature and pH (Zheng and Janis 2006). However, this chemical instability did not translate to physical instability over the period studied, as the loss of monomer and formation of soluble and insoluble aggregates was insignificant. Similar chemical degradation via deamidation and absence of significant physical degradation was observed for a humanized IgG1 antibody, even after storage at 40 °C for 6 months (Liu et al. 2006). In addition to three Asn deamidation sites, one Gln deamidation site was found in these studies. Although the mAb has four AsnGly sequences in the Fc region, none showed detectable deamidation, indicating that they may be protected from solvent exposure. Of 62 Gln residues, the only deamidated residue was close to the N-terminus of the heavy chain and hence may have had greater solvent exposure.

Enzymes are also known to influence deamidation. Glutaminase-mediated deamidation of Gln occurs in several plant proteins such as wheat gluten and soy protein (Suppavorasatit and Cadwallader 2012; Yong et al. 2006). Enzymatic deamidation of Asn in vivo is not known. However, enzymes appear to be involved in the repair of isomerized Asp. The enzyme isoaspartyl carboxyl O-methyltransferase (PIMT) methylates the deamidated isoAsp residue and catalyzes its conversion to Asp (Aswad et al. 2000). The commercially available ISOQUANT kit detects isoAsp using recombinant PIMT (Aswad et al. 2000; Vlasak et al. 2009). For additional information on the mechanism and factors affecting deamidation and isomerization, the reader is referred to several excellent reviews (Robinson and Robinson 2004; Aswad 1994).

Evidence for the Effects of Deamidation on mAb In Vivo Performance

While in vivo evidence is lacking, there is in vitro evidence that deamidation can affect antigen binding affinity and the potency of mAbs. When Asp isomerization occurred in the CDR region of an anti-IgE antibody, Cacia et al. observed significantly reduced binding affinity of the isoAsp species towards IgE compared to the native form (Cacia et al. 1996). Similarly, reduced antigen binding was observed for deamidated Asn33 in the light chain CDR region of an IgG antibody (Vlasak et al. 2009). In an unpublished BIAcore study, deamidation of Asn55 in the heavy chain CDR region of a mAb caused a 14-fold reduction in binding affinity for the antigen (Huang et al. 2005). In another case, deamidation of Asn30 to Asp in the light chain CDR region of trastuzumab caused a 30 % loss in potency, while isomerization of Asp102 to isoAsp in the heavy chain CDR region caused >80 % loss of potency (Harris et al. 2001).

Deamidation in the CDR region and Fc regions of an antibody can affect antigen binding and Fc receptor binding, respectively. Deamidation at other sites may not directly affect avidity for receptors or antigens, but may alter the antibody structure sufficiently to allow further chemical and physical degradation. Endogenous IgG antibodies show ~23 % deamidation in vivo, as observed by Liu et al. using deamidation at Asn384 in the Fc region (Liu et al. 2009). In addition, deamidation of a recombinant IgG2 antibody was studied in vitro under conditions that mimicked physiological conditions as well as in vivo. The antibody was administered intravenously to three human subjects and serum collected at various time points for pharmacokinetic studies. The fraction of deamidated Asn384 increased at a rate of 0.73 % per day, in agreement with the in vitro deamidation rate of 0.76 % per day, with a serum half-life of ~102 days. The results showed that in IgG2, in vitro deamidation rates in which the in vitro conditions mimic key features of the in vivo environment are comparable to in vivo rates.

Deamidation may produce proteins with altered potency and immunogenicity. Although evidence for increased immunogenicity of deamidated mAbs is scant, Mamula et al. demonstrated that injection of isoAsp variants of a peptide derived from murine cytochrome c, an endogenous protein, elicited a stronger immune response in mice than the normal Asp peptide (Mamula et al. 1999). Moreover, the antibodies against the isoAsp antigen also cross-reacted on the endogenous unmodified protein. Similar antigenicity has been observed with deamidated tumor antigen (Doyle et al. 2006). This raises concerns about potential autoimmune reactions that may arise from deamidated protein therapeutics either in formulations or, to a greater extent, in vivo. In addition to compromising safety, such immune responses may also affect efficacy by altering pharmacokinetics. The development of antibodies to therapeutic mAbs is not uncommon. Such responses can affect efficacy, but the origin of such responses has not been elucidated (Baert et al. 2003; Bartelds et al. 2007, 2011; Colombel et al. 2010; Garcês et al. 2013).

The formation and clearance of a deamidated mAbwas monitored following administration of a mAb in monkeys (Huang et al. 2005). As similar clearance rates were observed for Asp and isoAsp products formed by deamidation at Asn55 in the heavy chain CDR region, the deamidation half-life (i.e., the time at which 50 % of the mAb is deamidated) of the chosen mAb was calculated to be 6 days. Moreover, the route of administration (intravenous vs. subcutaneous) did not affect the deamidation rate. Khawli et al. compared in vitro and in vivo properties of a mAb that showed acidic and basic charge variants after purification (Khawli et al. 2010). In vitro binding studies using the acidic variants, which included deamidated product, showed slightly decreased binding to rat FcRn when compared to basic variants and wild-type protein. However, in vitro potency remained equivalent for all three species indicating lack of effect on the CDR region recognizing antigen. No significant pharmacokinetic differences were observed when these charge variants were administered to rats intravenously. Administration of the three fractions subcutaneously too did not show differences between the fractions, in that similar exposure as measured by AUC and Tmax values was observed.

Racemization of proteins is known to occur in vivo during aging. Although there is no evidence that the in vivo activity and immunogenicity are altered in racemized mAbs, they may be present as impurities in the formulation. They may also accumulate in the body as the enzyme PIMT has little to no ability to methylate D-Asp and D-isoAsp (Clarke 1999; Lowenson and Clarke 1992).

Fc *N*-Glycan Hydrolysis

Glycan hydrolysis is not a major route of degradation for mAbs in vivo. The *N*-linked glycans in the Fc domain undergo little lysosomal hydrolysis until the protein is extensively degraded (Kuranda and Aronson 1986; Fisher and Aronson 1992; Brassart et al. 1987; Aronson and Kuranda 1989). Similarly, the stability of *N*-glycans during formulation, manufacturing and storage of mAbs is seldom a concern. However, variations in mAb glycoform structure introduced during upstream processing can have a profound effect on in vivo performance, and removal of even a single terminal monosaccharide can significantly affect receptor binding and biodistribution. These variations may be the result of enzymatic degradation, changes in growth medium composition or drift in cell-line performance. This section describes the mechanisms of enzymatic hydrolysis of Fc *N*-glycans and the evidence for the effects of glycoform variability on the in vivo performance of mAbs.

Mechanism of Glycan Hydrolysis

Glycan hydrolysis includes complete glycan release (deglycosylation) by in vitro enzymatic cleavage; semi-complete release via in vivo digestion by glycosidases (e.g., from human bacterial pathogens); and the sequential removal of terminal

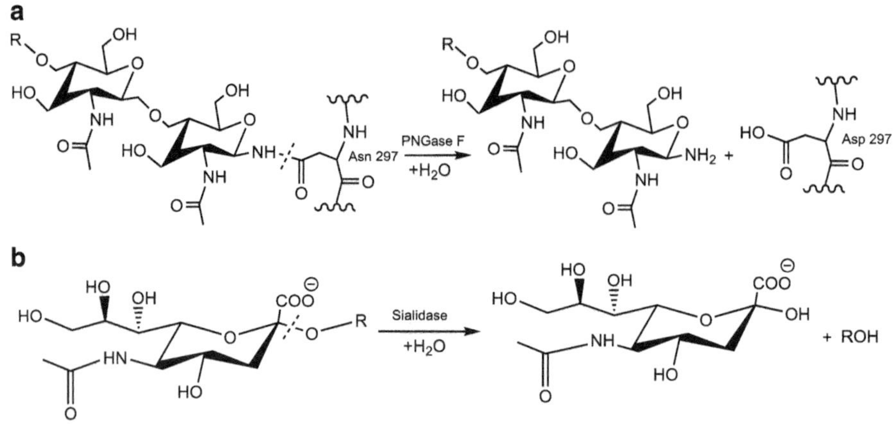

Fig. 2 Mechanism of *N*-glycan hydrolysis. (**a**) PNGase F catalyzed *N*-glycan hydrolysis at Asn297 releases *N*-glycan and converts Asn297 to Asp297 (R represents *N*-glycan groups appended to the two GlcNAc groups shown). (**b**) Sialidase catalyzed sialic acid hydrolysis removes the terminal sialic acid from the rest of the glycan structure (R represents the *N*-glycan groups). The *dashed line* in both reaction mechanisms indicates the cleavage site by the enzyme

glycans by glycosidases in vitro. A number of enzymes catalyze these reactions, each with its own specific mechanisms. Two are provided here as examples of enzymatic glycan hydrolysis (Huhn et al. 2009; Mechref et al. 2005; Omtvedt et al. 2006; Qian et al. 2007).

Complete Glycan Hydrolysis. Peptide-*N*-(*N*-acetyl-β-glucosaminyl) asparagine amidase (PNGase F) catalyzes the cleavage of *N*-linked oligosaccharides from the protein chain (Fig. 2). At *N*-glycosylation site Asn297, the glycosylamine linkage between glycans and protein is hydrolyzed to release a 1-amino oligosaccharide, while Asn297 is deamidated to Asp. The 1-amino-oligosaccaride then undergoes hydrolysis to a di-*N*-acetylchitobiose unit as the reducing end (Tarentino and Plummer 1994).

Terminal Monosaccharide Removal. Sialidases (or neuraminidases) catalyze the hydrolysis of the glycosidic linkage between the terminal monosaccharide and the rest of the glycan moiety, e.g., the terminal α2,6 sialic acid (Fig. 2). The reaction mechanism consists of four steps including binding, formation of the endocyclic sialosyl cation transition-state intermediate, and the formation and release of sialic acid (Taylor and Vonitzstein 1994). This terminal hydrolysis occurs by a standard double displacement mechanism with overall retention of the anomeric configuration (Koshland 1953; Sinnott 1990; Mccarter and Withers 1994; Zechel and Withers 2000; Vasella et al. 2002; Taylor and Vonitzstein 1994).

Factors Affecting Glycan Hydrolysis

Intrinsic Factors. The location of Fc glycan at Asn297 protects the carbohydrate from solvent exposure and enzymatic accessibility, since the glycans are buried in the pocket formed by the two C_H2 domains. This structural protection of Fc *N*-glycan is a major contributor to mAb stability, structure and effector functions. Denaturation is usually necessary to expose the glycan for cleavage by enzymes. Glycan hydrolysis also depends on enzymatic specificities to the types of *N*-glycan substrates such as high-mannose, hybrid or complex, as well as on particular glycosidic linkages.

Extrinsic Factors. Since glycans can only be hydrolyzed enzymatically, the enzymes themselves are the major extrinsic factors affecting the reaction. PNGase F is a well-known hydrolase that specifically cleaves carbon–nitrogen (C–N) bonds (Fig. 2a). It hydrolyzes all types of *N*-glycans in mAbs after protein denaturation and releases the intact glycan, and is often used in glycan and protein analysis (Plummer and Tarentino 1991; Tretter et al. 1991; Maley et al. 1989; Tarentino et al. 1985; Elder and Alexander 1982). Sialidases are "retaining" glycoside hydrolases (i.e., with retention of glycan stereochemistry) found in many organisms. They hydrolyze and remove α2,3, α2,6 and α2,8 linked terminal sialic acid. β-Galactosidase, as an essential retaining enzyme in the human lysosome, selectively catalyzes the hydrolysis of β1,4 linked terminal galactose via a double displacement mechanism (Asp and Dahlqvist 1972; Alpers 1969; Zhang et al. 1994; McCarter et al. 1997; Distler and Jourdian 1973). Other enzymes specifically release many other terminal monosaccharides, such as endo-*N*-acetyl-β-D-glucosaminidase for bifurcated *N*-acetylglucosamine and α-mannosidase for terminal mannose groups (Karamanos 1997; Li 1967; Li and Li 1968; Kimura et al. 1999). All these enzymes are employed for in vitro research.

Bacterial glycosidases (or glycoside hydrolases) from human bacterial pathogens often target the IgG of human adaptive immune system, hydrolyze its *N*-glycan, destroy the immune response and even colonize the host. This also promotes nutrient acquisition for the survival and persistence of pathogens (Burnaugh et al. 2008; Roberts et al. 2000). Since the effects of these pathogens provide insights into the importance of glycosylation, they are introduced briefly here. EndoS is a well-characterized bacterial glycosidase discovered from *Streptococcus pyogenes*. It is highly specific for human IgG and cleaves the β1,4 linkage between two GlcNAc groups from the non-reducing end of the core glycan, releasing the chitobiose core and leaving a GlcNAc with a α1,6 fucose attached (Collin and Olsen 2001b; Collin et al. 2008). EndoS can hydrolyze all the classes of *N*-linked glycans from native IgG without denaturation, while the other glycosidases only hydrolyze denatured glycoproteins (Tarentino and Plummer 1994; Collin and Olsen 2001a; Collin et al. 2008). Molecular analysis of EndoS indicates that catalysis involves Glu235 and Trp residues (Allhorn et al. 2008b). The N-terminus of EndoS is processed by the cysteine proteinase SpeB, which also is secreted from *Streptococcus pyogenes* and may regulate EndoS activity, while the C-terminus is necessary for direct interactions with IgG, though the details have not yet been elucidated (Kimura et al. 1999).

EndoS2, EndoC and EndoE have activity similar to EndoS and cleave the IgG glycan with similar mechanisms (Collin and Olsen 2003; Sjogren et al. 2011; Garbe and Collin 2012). Endo-β-*N*-acetylgluccosaminidases (EndoF1-3), expressed from *E. Meningoseptica*, are similar to EndoS in activity, but each EndoF is specific for particular glycan types (Waddling et al. 2000; Trimble and Tarentino 1991; Elder and Alexander 1982). Endo-β-*N*-acetylgluccosaminidase (EndoD) from *S. Pneumoniae*, often works together with EndoF1-3 to release complex oligosaccharides from IgG (Muramatsu et al. 1978, 2001). All these glycosidases found in the aforementioned human bacterial pathogens can hydrolyze the *N*-linked glycan of IgG, gaining nutrients from the released glycans to promote growth and infectivity. The functional effects of such deglycosylation follow in the next section.

Evidence for the Effects of Fc N-Glycan Hydrolysis on mAb In Vivo Performance

As noted above, Fc *N*-glycans are stable and hydrolysis rarely occurs spontaneously in vivo. Enzymatic removal of the *N*-glycans allows the two C_H2 domains to approach each other to form a closed Fc structure leading to the failure of corresponding effector functions (Krapp et al. 2003). In vitro analysis of the binding affinity of various truncated glycoforms of IgG Fc demonstrated a progressive decrease in the affinity of binding to FcγR (Krapp et al. 2003; Mimura et al. 2000, 2001; Zheng et al. 2011). As affinity between IgG Fc and FcγR is controlled by Fc *N*-glycosylation, glycosidases may alter IgG effector functions through diminished binding to Fc receptors and reduced activation of the complement pathway. Modification of glycosylation by in vivo enzymatic hydrolysis is a promising approach to modulating antibody effector functions, and the research findings regarding pathogen-derived immunomodulatory molecules are being applied to develop treatments for antibody-mediated pathological diseases such as autoimmune disorders (Collin and Fischetti 2004). For example, EndoS has been shown to release all the *N*-glycans from four human IgG subclasses in vitro in purified plasma as well as in human whole blood (Allhorn et al. 2008a). As expected, the affinity of IgG for FcγRs decreased dramatically following EndoS exposure, with the exception of the IgG2 subclass, which showed increased binding to FcγRIIB. Furthermore, the immunomodulatory ability of EndoS was demonstrated in vitro using samples from patients with systemic lupus erythematosus (SLE), an autoimmune disease with chronic or episodic inflammation (Lood et al. 2012). In studies of EndoS-treated immune complexes (ICs) purified from SLE patients or RNA-containing ICs formed in vitro, all pro-inflammatory properties of the ICs were abolished after EndoS treatment. EndoS treatment also decreased IC size and glycosylation. These findings support the potential development of EndoS as a treatment for SLE as well as the potential for inhibition of the deglycosylating enzymes as an anti-microbial strategy. The results also provide evidence for the in vivo effects of minor changes in IgG glycoforms such as those catalyzed by EndoS.

Peptide Bond Hydrolysis

Peptide bond hydrolysis is the cleavage of a peptide bond due to reaction with water, resulting in free carboxy- and amino-terminal groups. The reaction is also called 'clipping', particularly when it occurs near the N- or C-termini. Although all peptide bonds are potential sites for hydrolysis, most are sufficiently stabilized by resonance that hydrolysis is not observed. This section describes the mechanisms of peptide bond hydrolysis and the amino acid sequences that are particularly prone to the reaction, together with evidence for in vivo effects of peptide bond hydrolysis in mAbs.

Mechanism of Peptide Bond Hydrolysis

Peptide bond hydrolysis is a nucleophilic reaction that occurs by several different mechanisms, including:

Direct Hydrolysis. In direct hydrolysis, the polarized carbonyl carbon serves as a site for attack by water or other suitable electron donating species (nucleophile) (Fig. 3a). A tetrahedral intermediate is formed with the incoming nucleophile and carbon. Subsequently, the C–N bond of CO–NH is cleaved to form COOH and NH_2 groups.

Hydrolysis with Diketopiperazine (DKP) Formation. This reaction occurs when a free N-terminal amino group attacks the carbonyl of the penultimate peptide bond. A DKP is formed due to cyclization of the two N-terminal amino groups, and a clipped protein is produced. This reaction occurs readily in solution, especially for proteins containing Pro in the penultimate position near the N-terminus (e.g.,Pro-Gly-NH_2). Pro can exist in a *cis* conformation that is required for the reaction, although DKP formation is also observed with Ala and Gly (Fig. 3b). DKP cyclization is catalyzed by both acids and bases, and the type of buffer has an effect on DKP formation (Goolcharran and Borchardt 1998; Stein 1993).

β-Elimination Mediated Hydrolysis. Hydrolysis of the Ser-Cys bond in IgG1 heavy chain can occur via β-elimination (Cordoba et al. 2005; Cohen et al. 2007). In β-elimination, hydroxide ion attack on the Cα proton of a Cys generates dehydroalanine and persulfide (Trivedi et al. 2009) (Fig 3c). Dehydroalanine is reactive and can participate in electrophilic addition reactions. An N-terminal amide and C-terminal peptide fragments with terminal amide Ser and pyruvoyl groups, respectively, are produced upon dehydroalanine hydrolysis. The extent of Ser-Cys peptide bond hydrolysis has been shown to increase with pH as hydroxide ion mediated deprotonation of Cys α-carbon is favored under alkaline conditions.

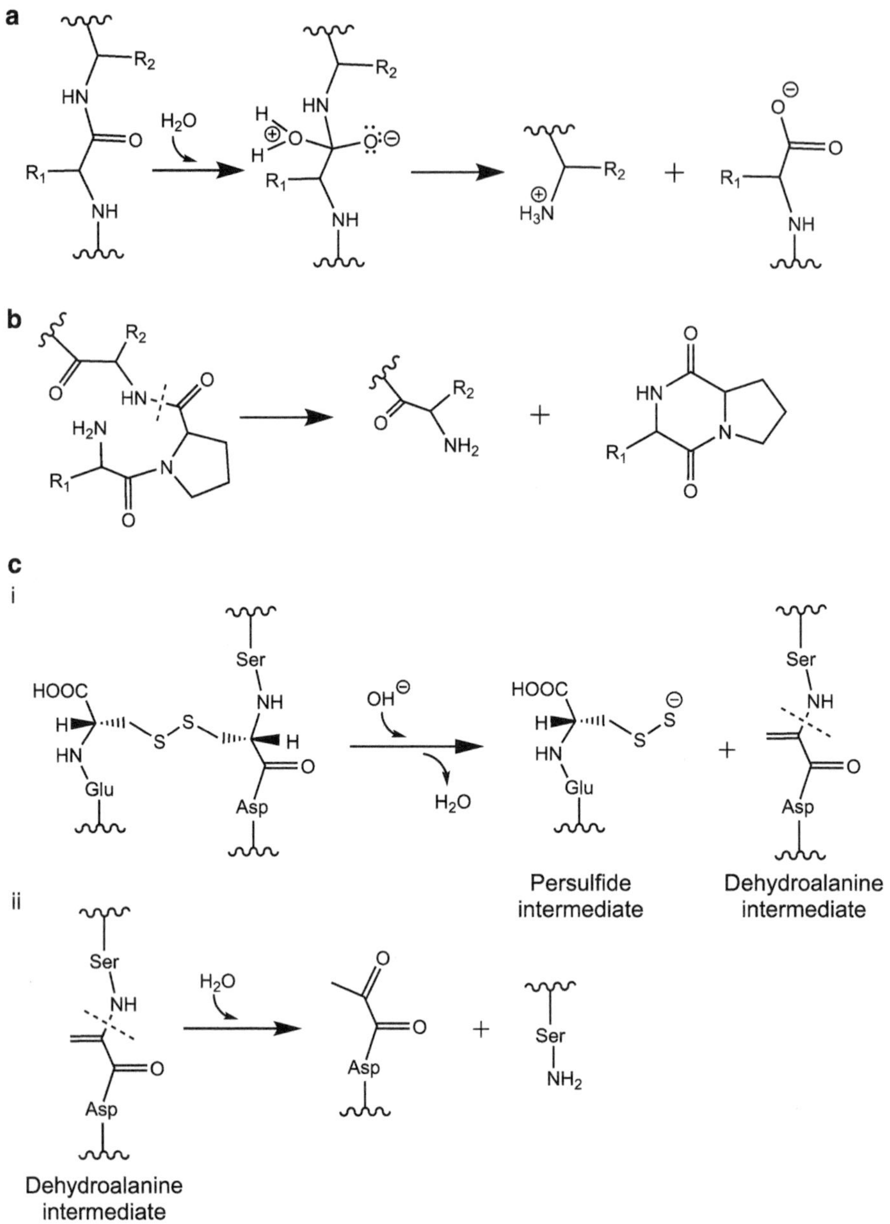

Fig. 3 Mechanism of peptide bond hydrolysis. (**a**) Direct hydrolysis with nucleophilic attack forming a tetrahedral intermediate which loses water to give a clipped peptide. (**b**) DKP formation due to cyclization of two N-terminal amino acids with peptide bond cleavage. (**c**) Beta elimination mediated hydrolysis, in which dehydroalanine and persulfide intermediates are formed (**c**-i), the Ser-Cys bond is hydrolyzed via dehydroalanine intermediate (**c**-ii)

Factors Affecting Peptide Bond Hydrolysis

Intrinsic Factors. Studies of peptides show that Asp-Ser, Asp-Tyr and Asp-Pro sequences are susceptible to proteolysis (Liu et al. 2006; Manning et al. 2010). While the C_H2 and C_H3 domains are structured, the interface between them is relatively flexible and prone to enzymatic hydrolysis at low pH (Ellerson et al. 1976). In addition, the hinge region is highly susceptible to hydrolysis and deserves special mention. Hinge region hydrolysis has been reported for a number of mAbs (Rao and Kroon 1993; Liu et al. 2006; Cordoba et al. 2005). In a month-long stability study of four recombinant IgG1 antibodies at pH 5.2 and different temperatures, Cordoba et al. observed Fab and Fab + Fc fragments. Little to no fragmentation was observed at low temperature. Hydrolysis was detected in the heavy chain upper hinge region sequence Ser-Cys-Asp-Lys-Thr-His-Thr for all the mAbs studied, with similar extents of fragmentation. Within this sequence, the major hydrolysis reactions were at Asp-Lys and His-Thr. The authors suggest that the local conformational flexibility of the hinge region may lower the activation energy for hydrolysis and favor the formation of Fab and Fab + Fc fragments.

Extrinsic Factors. pH plays an important role in peptide bond hydrolysis. β-elimination is accelerated at basic pH. Hinge region hydrolysis is increased at both acidic and basic pH (Xiang et al. 2007; Usami et al. 1996). A stability study of a recombinant humanized IgG1 antibody at 40 °C showed cleavage between hinge region residues Asp-Lys and His-Thr at neutral, acid and basic pH (Xiang et al. 2007). The rate of degradation of a model peptide corresponding to the hinge region sequence was much faster than that of the intact antibody or F(ab)2 fragments, suggesting a protective effect against hinge region hydrolysis by the Fab region. However, hinge region hydrolysis rates for intact antibody and F(ab)2 fragments were similar, indicating that the Fc region does not protect the hinge region against direct hydrolysis. Similar observations were made by Kamerzell et al. (2010). pH also affects the site of fragmentation, possibly by affecting antibody conformation. Stability studies at various pH conditions showed that the major hinge region hydrolysis site shifted towards the C-terminus with decreasing pH (Gaza-Bulseco and Liu 2008). At pH 4, the hydrolysis site was no longer in the hinge region, but was located in the C_H2 region. The authors proposed that structural rearrangement of the antibody caused the Fab fragments to move closer to the N-terminus of the hinge region, thus shielding it from hydrolysis. This is consistent with a study in which hydrolysis increased the interaction between C_H1 and C_H2 regions in rabbit IgG (Kravchuk et al. 1994).

Enzymes catalyze hydrolysis and are regularly used to obtain mAb fragments, and enzymes such as papain and pepsin are used to digest proteins in vitro. The effect of host cell enzymes was studied by Cordoba et al. wherein a mAb was recovered at various stages of production from CHO cell culture (Cordoba et al. 2005). Although clipping was detected, enzymatic proteolysis was not observed with increasing levels of host cell proteases, suggesting that fragmentation was not enzymatically mediated.

Evidence for the Effects of Peptide Bond Hydrolysis on mAb In Vivo Performance

In general, peptide bond hydrolysis is likely to be detrimental to mAb binding and effector functions, and can promote further chemical instability (Voorter et al. 1988). However, when appropriately designed, antibody fragments may provide benefits in terms of serum stability and tissue uptake. In vivo studies using xenograft tumor models have shown that the uptake of Fab and F(ab)2 fragments into tumors displaying the corresponding antigen is better than the uptake of the intact antibody. Moreover, the fragments are cleared more rapidly from non-tumor tissue, resulting in higher tumor concentrations (Herlyn et al. 1983; Buchegger et al. 1983; Mather et al. 1987). However, unintended in vivo hydrolysis is more likely to render therapy ineffective due to the loss of the effector arm of the antibody or a reduction in binding affinity. Brezski et al. showed cleaved antibodies were not as effective as intact protein and the loss of activity correlated with decreased affinity to Fc gamma receptors (Brezski et al. 2009). As a general rule, IgG fragments have better distribution and shorter half-life than intact IgG. The half-life of intact IgG is known to be extended by its interaction with Fc receptors. In a study using radiolabelled intact IgG, F(ab′)2 and Fab in murine model, Bazin-Redureau et al. showed that the in vivo distribution of IgG was dependent on molecular weight. Fab and F(ab′)2 were cleared more quickly than intact IgG. The elimination half-life of intact IgG was tenfold longer than that of the fragments (Bazin-Redureau et al. 1997). Smith et al. showed differences in both pharmacokinetics and immunogenicity between intact IgG and Fab fragments using rabbit and baboon models (Smith et al. 1979). They demonstrated that Fab fragments had considerably higher intravascular distribution but significantly weaker immunogenic response when administered intramuscularly as compared to intact IgG, suggesting that hydrolysis of mAbs may reduce the potential for immunogenic response. Whether these results with IgG fragments can be extrapolated to partially degraded mAbs in humans is not known, however.

Interestingly, some serum antibodies are capable of catalyzing peptide bond hydrolysis. It has also been shown that the proteolytic fragments of proteolytic serum IgG are also active. Li et al. characterized serum IgG preparations and found two proteolytic species with approximate masses of 50 and 150 kDa. The light chain dimers (50 kDa) were shown to be primarily responsible for proteolytic activity along with some contribution from full-length tetrametic IgG (Li et al. 2000). Proteolytic fragments may have different immunoreactivity, depending on the enzyme that cleaves the antibody. Superior immune response and better tumor localization were observed in a murine xenograft model following bromelain-mediated hydrolysis of an antibody, compared to hydrolysis with papain and pepsin (Milenic et al. 1989). Such fragments can be used for diagnostic purposes as they retain the selectivity towards the site of interest while decreasing Fc mediated non-selective binding. These limited studies suggested that the activity, biodistribution and immunogenicity of mAbs may be increased or decreased by hydrolysis in vivo, and that additional investigation is warranted.

Disulfide Rearrangement

Disulfide bonds stabilize proteins by restricting structural flexibility. However, with a dissociation energy 0.4-fold lower than that of C–C and C–H bonds, they are more prone to undergo cleavage (Correia 2010). This section reviews the mechanisms of hydrolytic reactions involving disulfide bonds as well as the evidence for the effects of these reactions on the performance of mAbs in vivo.

Mechanism of Disulfide Rearrangement

Disulfide bonds can undergo rearrangement via multiple pathways. Reaction mechanisms for the dominant hydrolytic pathways are discussed here; oxidative pathways are discussed elsewhere in this volume. Disulfide scrambling, also called "disulfide shuffling", proceeds via disulfide bond reduction by direct attack of hydroxyl ion in basic media and the subsequent generation of a thiolate anion (RS^-) and sulfenic acid (RSOH) (Fig. 4a). Disruption of the disulfide bond is facilitated by hydroxide ion deprotonation; α/β-elimination produces thiolate/thioaldehyde or dehydroalanine/persulfide that are reactive and can participate in different reaction pathways (Trivedi et al. 2009). RS^- is a reactive species that can then initiate scrambling by attacking a native disulfide (R′SSR″) or reacting with RSOH (Fig. 4b) to form non-native disulfides (Gilbert 1995). A related reaction, thiol-disulfide exchange, involves the nucleophilic attack of RS^- on a disulfide bond (R′SSR″) followed by expulsion of the thiol group with a lower pK_a (Summa et al. 2007) (Fig. 4c).

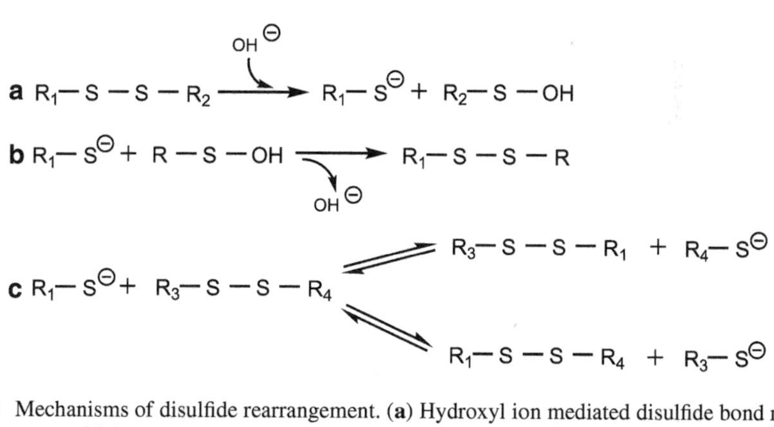

Fig. 4 Mechanisms of disulfide rearrangement. (**a**) Hydroxyl ion mediated disulfide bond reduction generates thiolate anion and sulfenic acid. (**b**) Thiolate anion generated in (**a**) can react with another sulfenic acid to form a non-native disulfide bond. (**c**) Two mechanistically equivalent and reversible thiol-disulfide exchange reactions. Nucleophilic attack of thiolate anion from (**a**) on a native disulfide results in the formation of scrambled disulfides with the expulsion of a thiolate anion

Factors Affecting Disulfide Rearrangement

Intrinsic Factors. The C-terminal amino acid of the light chain influences disulfide bond susceptibility to reduction in IgG1λ (Liu et al. 2011). The presence of either a Ser or Ala C-terminal to Cys on the light chain increased the susceptibility of the inter light-heavy chain disulfide bond to reduction, but not the inter-heavy chain or intra-light chain bonds (Liu et al. 2011). The contribution of the C-terminal amino acid to bond reduction could be due to a conformational or redox potential change, or both, as opposed to steric hindrance from an additional residue, which would be expected to protect the disulfide bond. The effect of a C-terminal Ser on disulfide rearrangement is different in IgG2. Three disulfide isomers of IgG2 have been identified in antibodies isolated from myeloma plasma and normal human serum: IgG2-A, IgG2-A/B and IgG2-B (Wypych et al. 2008). The conversion reaction follows the pathway: A↔A/B↔B. In IgG2-A/B, one of the Fab-arms is linked to the hinge region by disulfide bonds and the other arm is independent, as in IgG2-A. In the B form, both Fab-arms are linked to the hinge region by disulfide bonds. IgG2s with a λ LC have low levels of A/B and B isoforms compared to IgG2 antibodies with a κ LC. This difference could be due to steric hindrance, differences in the kinetics or thermodynamics of the disulfide shuffling reaction or differences in flexibility (Liu et al. 2008; Dillon et al. 2008).

Extrinsic Factors. Factors that promote disulfide bond reduction and/or generation of a thiolate anion and increase the rate of disulfide scrambling include pH, temperature, ionic strength, enzymes and low molecular weight thiols (Szajewski and Whitesides 1980; Snyder et al. 1981; Liu et al. 2008; Park and Ryu 1995). Disulfide scrambling is accelerated under basic conditions. Following administration, the pH in blood and tissue fluids is sufficiently basic to promote the reaction relative to rates in stabilized formulations, as is the higher body temperature. The presence of low molecular weight thiols in vivo, such as glutathione and cysteine, may also promote reactivity.; The reaction shows Arrhenius behavior and activation energies in the range 30–70 kJ/mol have been reported (Wiita et al. 2006).

Protein disulfide isomerase (PDI) is found in the endoplasmic reticulum (ER) of cells and plays an important role in disulfide bond formation in proteins. Park et al. studied the kinetics of PDI catalyzed mAb refolding and assembly (Park and Ryu 1995). Differences observed in the rate of mAb assembly in the absence and presence of PDI demonstrated the role of enzyme catalysis in disulfide bond formation. The reaction pathway is not altered in the presence of PDI, but the interconversions of some intermediates that form the final mAb structure are favored. Since PDI is found inside cells and mAbs interact externally with cell surface receptors, enzyme catalyzed disulfide shuffling typically would not be expected to occur after administration in vivo. Changes in the concentrations of disulfide isoforms over time in serum have been attributed to the presence of reactive thiols such as those in albumin, Cys, cystine and glutathione (GSH) (Liu et al. 2008) and not due to the presence of enzymes. The concentration of thiols in serum is 13–20 μM; this is not very different from cell culture medium conditions in which disulfide rearrangement is known to occur. Disulfide shuffling in serum has been observed for IgG1, IgG2 and IgG4 (Correia 2010) and under conditions mimicking human serum in vitro (e.g., whole blood or PBS with low molecular weight thiols; see also below) (Dillon et al. 2008). The redox environment in blood is thus favorable for mediating disulfide rearrangement reactions without the need for enzyme catalysis.

Evidence for the Effects of Disulfide Rearrangement on mAb In Vivo Performance

In serum, IgG molecules adopt a compact structure due to the presence of hundreds of other proteins and show enhanced association with antigens, other IgGs and low molecular weight thiols (Correia 2010; Demeule et al. 2009). Disulfide bonds can participate in multiple reactions that can disrupt native mAb conformation. These covalent modifications include hinge region fragmentation, the formation of dimers/oligomers, trisulfide variants and thioether linkage, and Fab-arm exchange.

Cohen et al. observed hinge region fragmentation following LC-HC disulfide bond cleavage via β-elimination (Cohen et al. 2007). The reaction mechanism is discussed in the peptide bond hydrolysis section. Yoo et al. demonstrated that IgG2 forms disulfide linked dimers/oligomers facilitated by Cys in the hinge region, both in cell culture and human serum (Yoo et al. 2003). IgG2 is produced in response to carbohydrate antigens in humans; these antibodies usually bind to their targets with lower affinity than those that bind protein antigens. Thus the formation of covalently linked polymers may promote more effective binding in the presence of multiple variable regions. Trisulfide variants formed by the insertion of a sulfur atom occur in all IgG subclasses and trisulfide levels as high as 40 % have been detected in a non-clinical recombinant IgG1 (Gu et al. 2010). Binding affinity to antigens was not affected by the presence of trisulfides, however, even for an IgG1 with ~39 % trisulfide content. In vivo, IgG1 containing 20 % trisulfide converted completely into disulfides following intraperitoneal injection in rats after 24 h (Gu et al. 2010). Although loss of therapeutic efficacy was not reported, the quantitation of such modifications may be important in determining the quality of drug product.

Significant differences in potency have been observed for antibodies with incomplete disulfide bond formation in the heavy chain variable domain. Cell-based binding assays and ELISA showed no difference in the binding affinity and potency of an IgG1 that had unpaired Cys in the V_H domain of the heavy chain at Cys22 and Cys96, residues which typically form an intra-chain disulfide bond (Zhang et al. 2012). However, in studies with omalizumab, an anti-IgE antibody that contains one sulfhydryl/mol of antibody, a reduction in potency was reported for Fab fragments that had unpaired Cys residues (Harris 2005; Ouellette et al. 2010; Shire et al. 2010).

The formation of thioether (also called lanthionine) linkages involves the removal of sulfur from –C–S–S–C– to form –C–S–C–. Thioethers have been detected in mAbs during production and storage (Tous et al. 2005). Although the mechanism needs further investigation, Cohen et al. observed that alkaline pH favored the formation of a thioether-linked antibody (Cohen et al. 2007). Zhang et al. monitored thioether level changes in IgG1 over time in blood following single-dose intravenous injection in humans (Zhang et al. 2013). Conversion rates were faster for IgG1 with a λ LC than for those with a κ LC, a difference attributed to dehydrogenation impairment with κ LC. Thioether formation was also detected in endogenous antibodies at the level of 5 % in IgG1κ and 11 % in IgG1λ. The effect of thioether formation on the safety and efficacy of IgG1 is unknown, but may affect Fab orientation and overall activity.

Dillon et al. also investigated the effect of the disulfide bonding pattern on the biological activity of IgG2 (Dillon et al. 2008). The bioassay used to determine binding properties was inhibition of IL-1β- induced IL-6 production in whole blood and human chondrocytes. IgG2-A and IgG2-B isoforms showed significant differences in inhibiting Il-6 production. In other biological activity studies with IgG2 mAbs with different therapeutic targets, the isoforms exhibited differences in potency depending on antigen accessibility. IgG2-A showed greater potency when directed against a cell surface receptor while the two isoforms A and B showed no difference in potency against a non-membrane bound antigen. Liu et al. reported that in serum samples collected from patients after intravenous administration of both 1,000 and 300 mg doses of IgG2, the relative concentration of the three isoforms (A, A/B and B) changed over time. Isoform conversion in vivo can thus be used as a marker for antibody age and changes in activity in blood. The effect of disulfide heterogeneity on biological activity of IgG2 in vivo is unknown.

Therapeutic IgG4 antibodies undergo Fab-arm exchange with endogenous human IgG4 in vivo facilitated by the hinge and C_H3 regions (Labrijn et al. 2009). The wild-type IgG4 hinge region has the sequence Cys-Pro-Ser-Cys that is thought to promote intra-chain disulfide bonds instead of inter-heavy chain bonds (Burton and Wilson 2007). Fab-arm exchange results in the formation of a hybrid antibody that is bispecific but binds only monovalently, even to repeating antigens. Fab-arm exchange in IgG4 is promoted by GSH concentrations as low as 0.5 mM in vitro (van der Neut Kolfschoten et al. 2007). IgG4 antibodies; IgG4-EGFR, natalizumab and TGN1412 formed bispecific antibodies with IgG4-CD20 when added in equimolar amounts, and in the presence of 0.5 mM GSH after incubation at 37 °C for 24 h (Labrijn et al. 2009). Bispecific antibodies formed by Fab-arm exchange lose their ability to cross-link antigen so that signaling and its downstream consequences are dampened (Fig. 5). Thus, bispecific IgG4s show potential as therapeutic agents for immunotherapeutic applications where the immune system's effector functions are undesired (Labrijn et al. 2011). Van der Neut Kolfschoten et al. have shown that Fab-arm exchange can be useful in conferring anti-inflammatory properties to IgG4 in monkey models of myasthenia gravis (MG) (van der Neut Kolfschoten et al. 2007). As an example, IgG1-637, an IgG1 patient-derived acetylcholine receptor (AChR) specific antibody, triggers AChR degradation and induced MG disease in rhesus monkeys. While studies in vitro with rhabdomyosarcoma TE671 cells expressing AchR showed that both IgG4-637 and IgG1-637

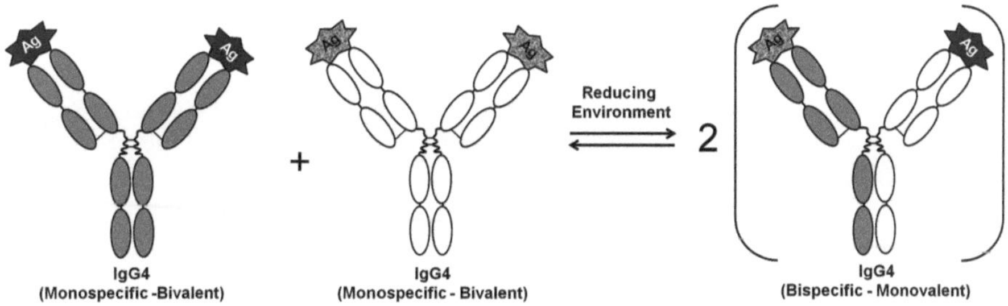

Fig. 5 In vivo Fab arm exchange in human monoclonal IgG4 antibodies. IgG4 molecules are dynamic and can undergo Fab arm exchange to form bispecific antibodies. The heavy chain—light chain pair of one IgG4 molecule exchanges with heavy chain—light chain pair from another IgG4 to form antibody with two distinct Fab arms. The bispecific antibodies lose their ability to cross-link identical antigens and to form immune complexes (van der Neut Kolfschoten et al. 2007). *Ag* = antigen

have similar reactivity towards AChR after administration in vivo, the cross-linking activity of IgG4-637 towards AChR decreased due to Fab-arm exchange, offering protection against IgG1-637 induced MG. The results demonstrate that Fab-arm exchange influences immune response, with beneficial effects in this model of MG.

Fab-arm exchange may also have adverse effects. Labrijn et al. have shown that Fab-arm exchange occurs in natalizumab (Tysabri®) and TGN1412, both containing a wild-type hinge region (Labrijn et al. 2009). Natalizumab is used in the treatment of multiple sclerosis. TGN1412, a CD-28 specific IgG4 based therapeutic was developed for rheumatoid arthritis and B cell chronic lymphocytic leukemia. The kinetics of Fab-arm exchange for TGN1412 were found to be comparable to the onset of adverse effects following treatment in a clinical study that ultimately led to its discontinuation. However, the correlation between this reaction and the cytokine storm post TGN1412 administration is not evidence of causation, and the in vivo effects of Fab-arm exchange require further study.

Mitigating the Effects of Hydrolytic Reactions

Various high resolution biophysical and biochemical methods are used to identify chemical modifications in mAbs and to study their effects on stability, antigen binding, effector functions and pharmacokinetics (Wang et al. 2005; Huang et al. 2005; Boswell et al. 2010; Pace et al. 2013). The knowledge gained is being applied to improve the stability and functions of antibody through careful formulation design. In addition, recent advances in molecular engineering have made it possible to eliminate many sites of degradation (Presta 2008). This section discusses formulation and protein engineering approaches to develop hydrolytically stable mAbs.

Ex Vivo Approaches to Mitigation: Antibody Formulation

One of the simplest approaches to mitigating the effects of hydrolytic reactions is to minimize reaction rate by controlling the composition of the formulation. As discussed above, hydrolytic reactions such as deamidation, Asp isomerization, peptide bond hydrolysis and thiol-disulfide exchange are sensitive to pH, and the appropriate selection of buffer in solution formulations can be sufficient to control these reactions in some cases. For example, a study of the effect of pH on human mAb subclasses IgG1, IgG2, and IgG4 showed stability and minimal heat-induced degradation at pH 5.0–5.5 (Ishikawa et al. 2010). Excipients such as sugars, polymers, surfactants and amino acids can also have stabilizing effects and are commonly included in mAb solution formulations. For example, optimized IgG formulations containing 20 % sorbitol and 1 M Gly prevented fragmentation induced by various stresses (Mueller et al. 2013). The mechanisms by which these excipients exert their effects are generally not as well understood as those of buffers, however, and the approach to selecting and optimizing such excipients remains largely a matter of trial-and-error.

Water of course plays a key role in hydrolytic degradation reactions. Though solution formulations are often desirable for administration, stability can often be improved by removing water to produce dry powders that are reconstituted. These powders typically are produced by lyophilization (freeze-drying) or spray-drying. Although mAbs in lyophilized solids are usually more stable than in solution, it is still necessary to protect antibodies during freeze-drying, storage and reconstitution (Daugherty and Mrsny 2006). As with solution formulations, careful selection of buffers and other excipients can improve stability in lyophilized formulations (Cleland et al. 1993; Meyer et al. 2009). For example, sucrose at high concentration in combination with Gly has been shown to reduce deamidation in a lyophilized antibody formulation in an accelerated stability study (Meyer et al. 2009). Similarly, specific molar ratios of the stabilizers sucrose, trehalose and mannitol have effectively prevented lyophilization induced structural alterations and deamidation in a mAb (Cleland et al. 2001). Lyophilization-induced structural perturbations have also been associated with increases in intermolecular disulfide formation and aggregation (Costantino et al. 1998). Inclusion of carbohydrate-like trehalose or sucrose in antibody formulations has been shown to preserve the antibody native structure, thereby reducing thiol-disulfide exchange mediated aggregation during storage (Andya et al. 2003). Water cannot be completely removed from these solid formulations, and the residual water that remains can have both stabilizing and destabilizing effects (Breen et al. 2001; Chang et al. 2005).

Antibody Engineering Approaches to Mitigate Hydrolytic Reactions *In Vivo*

Antibodies that have been engineered to improve the properties of an existing product have been called "biobetter" antibodies (Beck 2011). The extent of such alterations varies from mutating a single amino acid to the deletion of multiple

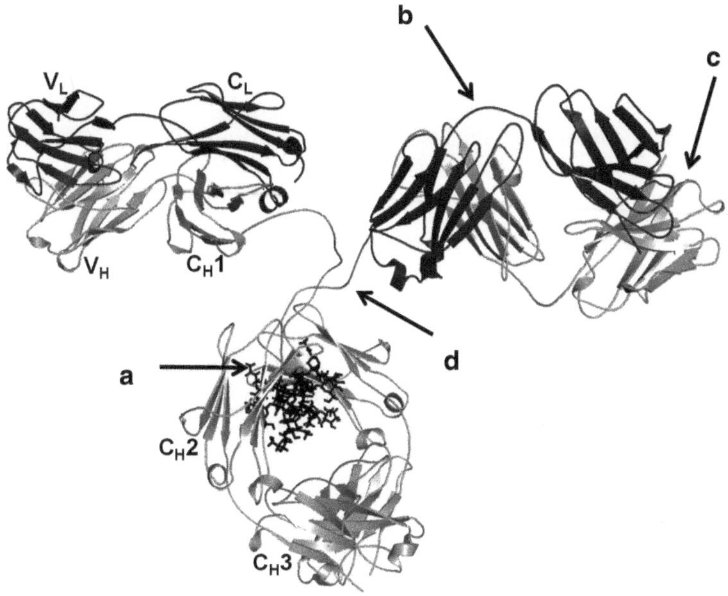

Fig. 6 Antibody engineering to produce hydrolytically stable mAbs. (**a**) Glycoengineering to control the number of glycoforms. (**b**) Introduction of additional disulfide bonds to improve Fab stability. (**c**) Mutations to eliminate deamidation in the CDR regions. (**d**) Mutations to avoid disulfide scrambling and proteolysis in the hinge region

domains from the structure without functional loss (Nakano et al. 2010; Holliger and Hudson 2005; Filpula 2007). Examples of such modifications in specific mAb domains are provided below (see also Fig. 6).

Antibody Engineering: Fab Region

Asn deamidation and Asp isomerization in the CDR loops are among the most common and deleterious types of antibody degradation. An approach to eliminating these reactions is to mutate the Asn or Asp residue to another amino acid without affecting antigen binding (Igawa et al. 2011). When mutating the Asn/Asp residue adversely affects mAb function, the N+1 residue can be altered instead, exploiting the sensitivity of these reactions to the neighboring amino acid. For example, in an antibody containing the deamidation-prone Asn-Gly sequence, the N+1 Gly residue was substituted with Arg in an attempt to eliminate the deamidating site (Nakano et al. 2010).

The introduction of non-native disulfide bonds has been used to improve the stability and the strength of interaction between domains in antibody. The interactions between V_L and V_H domains in scFv determine the mutual stabilization of the two domains and their functions (Rothlisberger et al. 2005; Ewert et al. 2003).

A second disulfide linkage between Cys54 and Cys78 in scFv increased the thermostability of V_H domains by 14–18 °C (Kim et al. 2012). In another study, introducing an additional disulfide in a V_{HH} nanobody by mutating Ala/Gly54Cys and Ile78Cys residues produced a significant increase in thermal stability and resistance to proteolysis with only minor perturbations in target binding affinities (Hussack et al. 2011). Changing the positions of inter-light and inter-heavy chain disulfide bonds (C_L–C_H1) by mutating different residues in the C-terminal region of the C_H1 domain led to the formation of thermostable IgG4 (Peters et al. 2012). Attempts have also been made to prevent disulfide scrambling and Fab-arm exchange in IgG though disulfide engineering. Replacing a single disulfide bond between C_L and C_H1 domains with a more stable thioether bridge increased the serum permanence time in normal mice threefold and did not affect the pharmacokinetics (Rodrigues et al. 1993). It has also been proposed that Fab-arm exchange in human IgG4 can be prevented by careful mutagenesis (Labrijn et al. 2009).

Antibody Engineering: Fc and Hinge Regions

Engineering the *N*-linked carbohydrates in the C_H2 domain of the mAb Fc region, or "glycoengineering", has been used to alter stability, pharmacokinetic and pharmacodynamic properties. Afucosylated glycoforms of antibodies used to treat rheumatoid arthritis (Rituximab®) and breast cancer (Trastuzumab®) showed 40- to 100-fold increases in ADCC activity with decreased plasma half-life, respectively (Beck et al. 2010; Beck 2011; Junttila et al. 2010). Glycoengineering has also been used to evaluate various glycan structures for conferring resistance of the protein backbone to papain digestion. The G0 glycoform was shown to confer twice the resistance to papain digestion as the G2 and G2S2 glycoforms, suggesting that terminal sugars of Fc glycans may be important in antibody stability and in conferring resistance to proteases, in addition to influencing effector functions (Raju and Scallon 2007). Recently, a number of aglycosylated therapeutic antibodies have entered clinical trials, with the absence of glycan compensated by mutations in the Fc region. Eliminating glycosylation eliminates the need to control glycan isoforms. In addition, the greater flexibility of the aglycosylated Fc region can be utilized to produce antibodies with unique FcR selectivity and ADCC activity (Keymeulen et al. 2005; Jung et al. 2011). Intermolecular disulfide exchange between the heavy chains of IgG4 creates a mixture of bispecific antibodies. This has been prevented by a single point mutation of a Ser residue to Pro (Ser241Pro) resulting in a normal disulfide bonded homodimer (Angal et al. 1993). Adding engineered inter- and intra-domain disulfide bonds at the C-terminus of the C_H3 domain increased the thermostability of human IgG1 without modulating its function (Wozniak-Knopp et al. 2012; Wozniak-Knopp and Ruker 2012). A single point mutation His229Tyr improved the wild type antibody by inhibiting hinge cleavage by 98 % and suggested that the redox active Tyr did not replicate the ability of His to facilitate radical induced degradation (Yan et al. 2012).

Conclusions

mAbs are the most rapidly growing class of biopharmaceuticals and efforts are being made to improve their stability and biological functions. The mechanisms of common hydrolytic degradation reactions, their effects on mAb in vivo functions and the various approaches to mitigate the effects of these reactions have been discussed in this chapter. While considerable progress has been made in understanding the mechanisms of these reactions, the consequences for pharmacodynamic and pharmacokinetic properties are less well understood. Controlling hydrolytic reactions during formulation, manufacturing and storage deserves continued attention, particularly in cases in which in vivo efficacy and safety are compromised. Additional research is needed to improve our understanding of the effects of hydrolytic degradation on antigen binding, FcR interactions, effector functions and immunogenicity. Such information will support the rational design of biobetter mAbs with improved stability, safety and efficacy.

References

Allhorn M, Olin AI, Nimmerjahn F, Collin M (2008a) Human IgG/Fc gamma R interactions are modulated by streptococcal IgG glycan hydrolysis. PLoS One 3:e1413

Allhorn M, Olsen A, Collin M (2008b) Endos from streptococcus pyogenes is hydrolyzed by the cysteine proteinase SpeB and requires glutamic acid 235 and tryptophans for IgG glycan-hydrolyzing activity. BMC Microbiol 8:3

Alpers DH (1969) Separation and isolation of rat and human intestinal beta-galactosidases. J Biol Chem 244:1238–1246

Andya JD, Hsu CC, Shire SJ (2003) Mechanisms of aggregate formation and carbohydrate excipient stabilization of lyophilized humanized monoclonal antibody formulations. AAPS PharmSci 5:E10

Angal S, King DJ, Bodmer MW, Turner A, Lawson AD, Roberts G, Pedley B, Adair JR (1993) A single amino acid substitution abolishes the heterogeneity of chimeric mouse/human (IgG4) antibody. Mol Immunol 30:105–108

Aronson NN, Kuranda MJ (1989) Lysosomal degradation of Asn-linked glycoproteins. FASEB J 3:2615–2622

Asp NG, Dahlqvist A (1972) Human small intestine -galactosidases: specific assay of three different enzymes. Anal Biochem 47:527–538

Aswad DW (ed) (1994) Deamidation and isoaspartate formation in peptides and proteins. CRC, Ann Arbor

Aswad DW, Paranandi MV, Schurter BT (2000) Isoaspartate in peptides and proteins: formation, significance, and analysis. J Pharm Biomed Anal 21:1129–1136

Baert F, Noman M, Vermeire S, Van Assche G, D'Haens G, Carbonez A, Rutgeerts P (2003) Influence of Immunogenicity on the long-term efficacy of infliximab in Crohn's disease. N Engl J Med 348:601–608

Bartelds GM, Wijbrandts CA, Nurmohamed MT, Stapel S, Lems WF, Aarden L, Dijkmans BA, Tak PP, Wolbink GJ (2007) Clinical response to adalimumab: relationship to anti-adalimumab antibodies and serum adalimumab concentrations in rheumatoid arthritis. Ann Rheum Dis 66:921–926

Bartelds GM, Krieckaert CL, Nurmohamed MT, Van Schouwenburg PA, Lems WF, Twisk JW, Dijkmans BA, Aarden L, Wolbink GJ (2011) Development of antidrug antibodies against adalimumab and association with disease activity and treatment failure during long-term follow-up. JAMA 305:1460–1468

Bazin-Redureau MI, Renard CB, Scherrmann JM (1997) Pharmacokinetics of heterologous and homologous immunoglobulin G, F(ab′)2 and Fab after intravenous administration in the rat. J Pharm Pharmacol 49:277–281

Beck A (2011) Biosimilar, biobetter and next generation therapeutic antibodies. MAbs 3:107–110

Beck A, Cochet O, Wurch T (2010) Glycofi's technology to control the glycosylation of recombinant therapeutic proteins. Expert Opin Drug Discov 5:95–111

Boswell CA, Tesar DB, Mukhyala K, Theil FP, Fielder PJ, Khawli LA (2010) Effects of charge on antibody tissue distribution and pharmacokinetics. Bioconjug Chem 21:2153–2163

Brassart D, Baussant T, Wieruszeski JM, Strecker G, Montreuil J, Michalski JC (1987) Catabolism of N-glycosylprotein glycans - evidence for a degradation pathway of sialylglyco-asparagines resulting from the combined action of the lysosomal aspartylglucosaminidase and endo-N-acetyl-beta-D-glucosaminidase – a 400-mhz H-1-NMR study. Eur J Biochem 169:131–136

Breen ED, Curley JG, Overcashier DE, Hsu CC, Shire SJ (2001) Effect of moisture on the stability of a lyophilized humanized monoclonal antibody formulation. Pharm Res 18:1345–1353

Brennan TV, Clarke S (1993) Spontaneous degradation of polypeptides at aspartyl and asparaginyl residues: effects of the solvent dielectric. Protein Sci 2:331–338

Brennan TV, Clarke S (1994) The effects of sequence and solution environment. In: Aswad DW (ed) Deamidation and isoaspartate formation in pepties and proteins. CRC, Ann Arbor

Brezski RJ, Vafa O, Petrone D, Tam SH, Powers G, Ryan MH, Luongo JL, Oberholtzer A, Knight DM, Jordan RE (2009) Tumor-associated and microbial proteases compromise host IgG effector functions by a single cleavage proximal to the hinge. Proc Natl Acad Sci U S A 106:17864–17869

Buchegger F, Haskell CM, Schreyer M, Scazziga BR, Randin S, Carrel S, Mach JP (1983) Radiolabeled fragments of monoclonal antibodies against carcinoembryonic antigen for localization of human colon carcinoma grafted into nude mice. J Exp Med 158:413–427

Burnaugh AM, Frantz LJ, King SJ (2008) Growth of streptococcus pneumoniae on human glycoconjugates is dependent upon the sequential activity of bacterial exoglycosidases. J Bacteriol 190:221–230

Burton DR, Wilson IA (2007) Immunology. Square-dancing antibodies. Science 317:1507–1508

Cacia J, Keck R, Presta LG, Frenz J (1996) Isomerization of an aspartic acid residue in the complementarity-determining regions of a recombinant antibody to human IgE: identification and effect on binding affinity. Biochemistry 35:1897–1903

Capasso S, Salvadori S (1999) Effect of the three-dimensional structure on the deamidation reaction of ribonuclease A. J Pept Res 54:377–382

Chang LL, Shepherd D, Sun J, Tang XC, Pikal MJ (2005) Effect of sorbitol and residual moisture on the stability of lyophilized antibodies: implications for the mechanism of protein stabilization in the solid state. J Pharm Sci 94:1445–1455

Chelius D, Rehder DS, Bondarenko PV (2005) Identification and characterization of deamidation sites in the conserved regions of human immunoglobulin G antibodies. Anal Chem 77:6004–6011

Clarke S (1999) A protein carboxyl methyltransferase that recognizes age-damaged peptides and proteins and participates in their repair. In: Cheng X, Blumenthal RM (eds) S-adenosylmethionine-dependent methyltransferases: structures and functions. World Scientific, Singapore, pp 123–148

Cleland JL, Powell MF, Shire SJ (1993) The development of stable protein formulations: a close look at protein aggregation, deamidation, and oxidation. Crit Rev Ther Drug Carrier Syst 10:307–377

Cleland JL, Lam X, Kendrick B, Yang J, Yang TH, Overcashier D, Brooks D, Hsu C, Carpenter JF (2001) A specific molar ratio of stabilizer to protein is required for storage stability of a lyophilized monoclonal antibody. J Pharm Sci 90:310–321

Cohen SL, Price C, Vlasak J (2007) Beta-elimination and peptide bond hydrolysis: two distinct mechanisms of human IgG1 hinge fragmentation upon storage. J Am Chem Soc 129: 6976–6977

Collin M, Fischetti VA (2004) A novel secreted endoglycosidase from Enterococcus faecalis with activity on human immunoglobulin G and ribonuclease B. J Biol Chem 279:26802

Collin M, Olsen A (2001a) Effect of SpeB and EndoS from streptococcus pyogenes on human immunoglobulins. Infect Immun 69:7187–7189

Collin M, Olsen A (2001b) Endos, a novel secreted protein from streptococcus pyogenes with endoglycosidase activity on human IgG. EMBO J 20:3046–3055

Collin M, Olsen A (2003) Extracellular enzymes with immunomodulating activities: variations on a theme in streptococcus pyogenes. Infect Immun 71:2983–2992

Collin M, Shannon O, Bjorck L (2008) IgG glycan hydrolysis by a bacterial enzyme as a therapy against autoimmune conditions. Proc Natl Acad Sci U S A 105:4265–4270

Colombel JF, Sandborn WJ, Reinisch W, Mantzaris GJ, Kornbluth A, Rachmilewitz D, Lichtiger S, D'haens G, Diamond RH, Broussard DL, Tang KL, Van Der Woude CJ, Rutgeerts P, SONIC Study Group (2010) Infliximab, azathioprine, or combination therapy for Crohn's disease. N Engl J Med 362:1383–1395

Cordoba AJ, Shyong BJ, Breen D, Harris RJ (2005) Non-enzymatic hinge region fragmentation of antibodies in solution. J Chromatogr B Analyt Technol Biomed Life Sci 818:115–121

Correia IR (2010) Stability of IgG isotypes in serum. MAbs 2:221–232

Costantino HR, Schwendeman SP, Langer R, Klibanov AM (1998) Deterioration of lyophilized pharmaceutical proteins. Biochemistry (Mosc) 63:357–363

Daugherty AL, Mrsny RJ (2006) Formulation and delivery issues for monoclonal antibody therapeutics. Adv Drug Deliv Rev 58:686–706

Demeule B, Shire SJ, Liu J (2009) A therapeutic antibody and its antigen form different complexes in serum than in phosphate-buffered saline: a study by analytical ultracentrifugation. Anal Biochem 388:279–287

Dillon TM, Ricci MS, Vezina C, Flynn GC, Liu YD, Rehder DS, Plant M, Henkle B, Li Y, Deechongkit S, Varnum B, Wypych J, Balland A, Bondarenko PV (2008) Structural and functional characterization of disulfide isoforms of the human IgG2 subclass. J Biol Chem 283:16206–16215

Distler JJ, Jourdian GW (1973) The purification and properties of beta-galactosidase from bovine testes. J Biol Chem 248:6772–6780

Doyle HA, Zhou J, Wolff MJ, Harvey BP, Roman RM, Gee RJ, Koski RA, Mamula MJ (2006) Isoaspartyl post-translational modification triggers anti-tumor T and B lymphocyte immunity. J Biol Chem 281:32676–32683

Elder JH, Alexander S (1982) Endo-beta-N-acetylglucosaminidase F: endoglycosidase from flavobacterium meningosepticum that cleaves both high-mannose and complex glycoproteins. Proc Natl Acad Sci U S A 79:4540–4544

Ellerson JR, Yasmeen D, Painter RH, Dorrington KJ (1976) Structure and function of immunoglobulin domains. III. Isolation and characterization of a fragment corresponding to the Cgamma2 homology region of human immunoglobin G1. J Immunol 116:510–517

Ewert S, Honegger A, Pluckthun A (2003) Structure-based improvement of the biophysical properties of immunoglobulin VH domains with a generalizable approach. Biochemistry 42:1517–1528

Filpula D (2007) Antibody engineering and modification technologies. Biomol Eng 24:201–215

Fisher KJ, Aronson NN (1992) Cloning and expression of the cDNA sequence encoding the lysosomal glycosidase di-N-acetylchitobiase. J Biol Chem 267:19607–19616

Garbe J, Collin M (2012) Bacterial hydrolysis of host glycoproteins - powerful protein modification and efficient nutrient acquisition. J Innate Immun 4:121–131

Garcês S, Demengeot J, Benito-Garcia E (2013) The immunogenicity of anti-TNF therapy in immune-mediated inflammatory diseases: a systematic review of the literature with a meta-analysis. Ann Rheum Dis 72:1947–1955

Gaza-Bulseco G, Liu H (2008) Fragmentation of a recombinant monoclonal antibody at various PH. Pharm Res 25:1881–1890

Geiger T, Clarke S (1987) Deamidation, isomerization, and racemization at asparaginyl and aspartyl residues in peptides. Succinimide-linked reactions that contribute to protein degradation. J Biol Chem 262:785–794

Gilbert HF (1995) Thiol/disulfide exchange equilibria and disulfide bond stability. Methods Enzymol 251:8–28

Goolcharran C, Borchardt RT (1998) Kinetics of diketopiperazine formation using model peptides. J Pharm Sci 87:283–288

Gu S, Wen D, Weinreb PH, Sun Y, Zhang L, Foley SF, Kshirsagar R, Evans D, Mi S, Meier W, Pepinsky RB (2010) Characterization of trisulfide modification in antibodies. Anal Biochem 400:89–98

Hambly DM, Banks DD, Scavezze JL, Siska CC, Gadgil HS (2009) Detection and quantitation of IgG 1 hinge aspartate isomerization: a rapid degradation in stressed stability studies. Anal Chem 81:7454–7459

Harris RJ (2005) Heterogeneity of recombinant antibodies: linking structure to function. Dev Biol (Basel) 122:117–127

Harris RJ, Kabakoff B, Macchi FD, Shen FJ, Kwong M, Andya JD, Shire SJ, Bjork N, Totpal K, Chen AB (2001) Identification of multiple sources of charge heterogeneity in a recombinant antibody. J Chromatogr B Biomed Sci Appl 752:233–245

Herlyn D, Powe J, Alavi A, Mattis JA, Herlyn M, Ernst C, Vaum R, Koprowski H (1983) Radioimmunodetection of human tumor xenografts by monoclonal antibodies. Cancer Res 43:2731–2735

Holliger P, Hudson PJ (2005) Engineered antibody fragments and the rise of single domains. Nat Biotechnol 23:1126–1136

Huang L, Lu J, Wroblewski VJ, Beals JM, Riggin RM (2005) In vivo deamidation characterization of monoclonal antibody by LC/MS/MS. Anal Chem 77:1432–1439

Huhn C, Selman MH, Ruhaak LR, Deelder AM, Wuhrer M (2009) IgG glycosylation analysis. Proteomics 9:882–913

Hussack G, Hirama T, Ding W, Mackenzie R, Tanha J (2011) Engineered single-domain antibodies with high protease resistance and thermal stability. PLoS One 6:E28218

Igawa T, Tsunoda H, Kuramochi T, Sampei Z, Ishii S, Hattori K (2011) Engineering the variable region of therapeutic IgG antibodies. MAbs 3:243–252

Ishikawa T, Ito T, Endo R, Nakagawa K, Sawa E, Wakamatsu K (2010) Influence of PH on heat-induced aggregation and degradation of therapeutic monoclonal antibodies. Biol Pharm Bull 33:1413–1417

Jung ST, Kang TH, Kelton W, Georgiou G (2011) Bypassing glycosylation: engineering aglycosylated full-length IgG antibodies for human therapy. Curr Opin Biotechnol 22:858–867

Junttila TT, Parsons K, Olsson C, Lu Y, Xin Y, Theriault J, Crocker L, Pabonan O, Baginski T, Meng G, Totpal K, Kelley RF, Sliwkowski MX (2010) Superior in vivo efficacy of afucosylated trastuzumab in the treatment of Her2-amplified breast cancer. Cancer Res 70:4481–4489

Kamerzell TJ, Li M, Arora S, Ji JA, Wang YJ (2010) The relative rate of immunoglobulin gamma 1 fragmentation. J Pharm Sci 100(4):1341–1349

Karamanos Y (1997) Endo-N-acetyl-beta-D-glucosaminidases and their potential substrates: structure/function relationships. Res Microbiol 148:661–671

Keymeulen B, Vandemeulebroucke E, Ziegler AG, Mathieu C, Kaufman L, Hale G, Gorus F, Goldman M, Walter M, Candon S, Schandene L, Crenier L, De Block C, Seigneurin JM, De Pauw P, Pierard D, Weets I, Rebello P, Bird P, Berrie E, Frewin M, Waldmann H, Bach JF, Pipeleers D, Chatenoud L (2005) Insulin needs after CD3-antibody therapy in new-onset type 1 diabetes. N Engl J Med 352:2598–2608

Khawli LA, Goswami S, Hutchinson R, Kwong ZW, Yang J, Wang X, Yao Z, Sreedhara A, Cano T, Tesar D, Nijem I, Allison DE, Wong PY, Kao YH, Quan C, Joshi A, Harris RJ, Motchnik P (2010) Charge variants in IgG1: isolation, characterization, in vitro binding properties and pharmacokinetics in rats. MAbs 2:613–624

Kim DY, Kandalaft H, Ding W, Ryan S, Van Faassen H, Hirama T, Foote SJ, Mackenzie R, Tanha J (2012) Disulfide linkage engineering for improving biophysical properties of human VH domains. Protein Eng Des Sel 25:581–589

Kimura Y, Hess D, Sturm A (1999) The N-glycans of jack bean alpha-mannosidase. Structure, topology and function. Eur J Biochem 264:168–175

Koshland DE (1953) Stereochemistry and the mechanism of enzymatic reactions. Biol Rev Camb Philos Soc 28:416–436

Kosky AA, Razzaq UO, Treuheit MJ, Brems DN (1999) The effects of alpha-helix on the stability of Asn residues: deamidation rates in peptides of varying helicity. Protein Sci 8:2519–2523

Krapp S, Mimura Y, Jefferis R, Huber R, Sondermann P (2003) Structural analysis of human IgG-Fc glycoforms reveals a correlation between glycosylation and structural integrity. J Mol Biol 325:979–989

Kravchuk ZI, Vlasov AP, Liakhnovich GV, Martsev SP (1994) A stable conformer of IgG, prepared by an acidic influence: study by calorimetry, binding of the C1q complement component, and monospecific anti-IgG. Biokhimiia 59:1458–1477

Kroon DJ, Baldwin-Ferro A, Lalan P (1992) Identification of sites of degradation in a therapeutic monoclonal antibody by peptide mapping. Pharm Res 9:1386–1393

Kuranda MJ, Aronson NN (1986) A di-N-acetylchitobiase activity is involved in the lysosomal catabolism of asparagine-linked glycoproteins in Rat-liver. J Biol Chem 261:5803–5809

Labrijn AF, Buijsse AO, Van Den Bremer ET, Verwilligen AY, Bleeker WK, Thorpe SJ, Killestein J, Polman CH, Aalberse RC, Schuurman J, Van De Winkel JG, Parren PW (2009) Therapeutic IgG4 antibodies engage in Fab-arm exchange with endogenous human IgG4 in vivo. Nat Biotechnol 27:767–771

Labrijn AF, Rispens T, Meesters J, Rose RJ, Den Bleker TH, Loverix S, Van Den Bremer ET, Neijssen J, Vink T, Lasters I, Aalberse RC, Heck AJ, Van De Winkel JG, Schuurman J, Parren PW (2011) Species-specific determinants in the IgG CH3 domain enable Fab-arm exchange by affecting the noncovalent CH3-CH3 interaction strength. J Immunol 187:3238–3246

Li YT (1967) Studies on the glycosidases in jack bean meal. I. Isolation and properties of alpha-mannosidase. J Biol Chem 242:5474–5480

Li YT, Li SC (1968) Studies on the glycosidases in jack bean meal. II. Sepation of various glycosidases by isoelectric focusing. J Biol Chem 243:3994–3996

Li L, Kalaga R, Paul S (2000) Proteolytic components of serum IgG preparations. Clin Exp Immunol 120:261–266

Li B, Borchardt RT, Topp EM, Vandervelde D, Schowen RL (2003) Racemization of an asparagine residue during peptide deamidation. J Am Chem Soc 125:11486–11487

Liu H, Gaza-Bulseco G, Sun J (2006) Characterization of the stability of a fully human monoclonal IgG after prolonged incubation at elevated temperature. J Chromatogr B Analyt Technol Biomed Life Sci 837:35–43

Liu YD, Chen X, Enk JZ, Plant M, Dillon TM, Flynn GC (2008) Human IgG2 antibody disulfide rearrangement in vivo. J Biol Chem 283:29266–29272

Liu YD, Van Enk JZ, Flynn GC (2009) Human antibody Fc deamidation in vivo. Biologicals 37:313–322

Liu H, Zhong S, Chumsae C, Radziejewski C, Hsieh CM (2011) Effect of the light chain C-terminal serine residue on disulfide bond susceptibility of human immunoglobulin G1lambda. Anal Biochem 408:277–283

Lood C, Allhorn M, Lood R, Gullstrand B, Olin AI, Ronnblom L, Truedsson L, Collin M, Bengtsson AA (2012) IgG glycan hydrolysis by endoglycosidase S diminishes the proinflammatory properties of immune complexes from patients with systemic lupus erythematosus: a possible new treatment? Arthritis Rheum 64:2698–2706

Lowenson JD, Clarke S (1992) Recognition of D-aspartyl residues in polypeptides by the erythrocyte L-isoaspartyl/D-aspartyl protein methyltransferase. Implications for the repair hypothesis. J Biol Chem 267:5985–5995

Maley F, Trimble RB, Tarentino AL, Plummer TH (1989) Characterization of glycoproteins and their associated oligosaccharides through the use of endoglycosidases. Anal Biochem 180:195–204

Mamula MJ, Gee RJ, Elliott JI, Sette A, Southwood S, Jones PJ, Blier PR (1999) Isoaspartyl post-translational modification triggers autoimmune responses to self-proteins. J Biol Chem 274:22321–22327

Manning MC, Chou DK, Murphy BM, Payne RW, Katayama DS (2010) Stability of protein pharmaceuticals: an update. Pharm Res 27:544–575
Mather SJ, Durbin H, Taylor-Papadimitriou J (1987) Identification of immunoreactive monoclonal antibody fragments for improved immunoscintigraphy. J Immunol Methods 96:255–264
Mccarter JD, Withers SG (1994) Mechanisms of enzymatic glycoside hydrolysis. Curr Opin Struct Biol 4:885–892
Mccarter JD, Burgoyne DL, Miao SC, Zhang SQ, Callahan JW, Withers SG (1997) Identification of Glu-268 as the catalytic nucleophile of human lysosomal beta-galactosidase precursor by mass spectrometry. J Biol Chem 272:396–400
Mechref Y, Muzikar J, Novotny MV (2005) Comprehensive assessment of N-glycans derived from a murine monoclonal antibody: a case for multimethodological approach. Electrophoresis 26:2034–2046
Meyer JD, Nayar R, Manning MC (2009) Impact of bulking agents on the stability of a lyophilized monoclonal antibody. Eur J Pharm Sci 38:29–38
Milenic DE, Esteban JM, Colcher D (1989) Comparison of methods for the generation of immunoreactive fragments of a monoclonal antibody (B72.3) reactive with human carcinomas. J Immunol Methods 120:71–83
Mimura Y, Church S, Ghirlando R, Ashton PR, Dong S, Goodall M, Lund J, Jefferis R (2000) The influence of glycosylation on the thermal stability and effector function expression of human IgG1-Fc: properties of a series of truncated glycoforms. Mol Immunol 37:697–706
Mimura Y, Sondermann P, Ghirlando R, Lund J, Young SP, Goodall M, Jefferis R (2001) Role of oligosaccharide residues of IgG1-Fc in Fc gamma RIIb binding. J Biol Chem 276: 45539–45547
Mueller M, Loh MQ, Tee DH, Yang Y, Jungbauer A (2013) Liquid formulations for long-term storage of monoclonal IgGs. Appl Biochem Biotechnol 169:1431–1448
Muramatsu T, Koide N, Maeyama K (1978) Further-studies on "endo-beta-N-acetylglucosaminidase-D". J Biochem 83:363–370
Muramatsu H, Tachikui H, Ushida H, Song X, Qiu Y, Yamamoto S, Muramatsu T (2001) Molecular cloning and expression of endo-beta-N-acetylglucosaminidase D, which acts on the core structure of complex type asparagine-linked oligosaccharides. J Biochem 129:923–928
Nakano K, Ishiguro T, Konishi H, Tanaka M, Sugimoto M, Sugo I, Igawa T, Tsunoda H, Kinoshita Y, Habu K, Orita T, Tsuchiya M, Hattori K, Yamada-Okabe H (2010) Generation of a humanized anti-glypican 3 antibody by CDR grafting and stability optimization. Anticancer Drugs 21:907–916
Omtvedt LA, Royle L, Husby G, Sletten K, Radcliffe CA, Harvey DJ, Dwek RA, Rudd PA (2006) Glycan analysis of monoclonal antibodies secreted in deposition disorders indicates that subsets of plasma cells differentially process IgG glycans. Arthritis Rheum 54:3433–3440
Ouellette D, Alessandri L, Chin A, Grinnell C, Tarcsa E, Radziejewski C, Correia I (2010) Studies in serum support rapid formation of disulfide bond between unpaired cysteine residues in the VH domain of an immunoglobulin G1 molecule. Anal Biochem 397:37–47
Pace AL, Wong RL, Zhang YT, Kao YH, Wang YJ (2013) Asparagine deamidation dependence on buffer type, PH, and temperature. J Pharm Sci 102:1712–1723
Park SH, Ryu DD (1995) Protein disulfide isomerase reaction kinetics in endoplasmic reticulum for monoclonal antibody refolding and assembly. Ann N Y Acad Sci 750:291–299
Peters SJ, Smales CM, Henry AJ, Stephens PE, West S, Humphreys DP (2012) Engineering an improved IgG4 molecule with reduced disulfide bond heterogeneity and increased Fab domain thermal stability. J Biol Chem 287:24525–24533
Plummer TH Jr, Tarentino AL (1991) Purification of the oligosaccharide-cleaving enzymes of flavobacterium meningosepticum. Glycobiology 1:257–263
Presta LG (2008) Molecular engineering and design of therapeutic antibodies. Curr Opin Immunol 20:460–470
Qian J, Liu T, Yang L, Daus A, Crowley R, Zhou QW (2007) Structural characterization of N-linked oligosaccharides on monoclonal antibody cetuximab by the combination of orthogonal matrix-assisted laser desorption/ionization hybrid quadrupole-quadrupole time-of-flight tandem mass spectrometry and sequential enzymatic digestion. Anal Biochem 364:8–18

Raju TS, Scallon B (2007) Fc glycans terminated with N-acetylglucosamine residues increase antibody resistance to papain. Biotechnol Prog 23:964–971

Rao PE, Kroon DJ (1993) Orthoclone OKT3. Chemical mechanisms and functional effects of degradation of a therapeutic monoclonal antibody. Pharm Biotechnol 5:135–158

Roberts G, Tarelli E, Homer KA, Philpott-Howard J, Beighton D (2000) Production of an endo-beta-N-acetylglucosaminidase activity mediates growth of enterococcus faecalis on a high-mannose-type glycoprotein. J Bacteriol 182:882–890

Robinson NE, Robinson AB (2004) Molecular clocks: deamidation of asparaginyl and glutaminyl residues in peptides and proteins. Althouse Press, Cave Junction, Or

Rodrigues ML, Snedecor B, Chen C, Wong WL, Garg S, Blank GS, Maneval D, Carter P (1993) Engineering Fab′ fragments for efficient F(ab)2 formation in Escherichia coli and for improved in vivo stability. J Immunol 151:6954–6961

Rothlisberger D, Honegger A, Pluckthun A (2005) Domain interactions in the Fab fragment: a comparative evaluation of the single-chain Fv and Fab format engineered with variable domains of different stability. J Mol Biol 347:773–789

Shire SJ, Gombotz W, Bechtold-Peters K, Andya J (eds) (2010) Current trends in monoclonal antibody and manufacturing. Springer, New York

Sinnott ML (1990) Catalytic mechanisms of enzymatic glycosyl transfer. Chem Rev 90:1171–1202

Sjogren J, Okumura CYM, Collin M, Nizet V, Hollands A (2011) Study of the IgG endoglycosidase EndoS in group a streptococcal phagocyte resistance and virulence. BMC Microbiol 11:120

Smith TW, Lloyd BL, Spicer N, Haber E (1979) Immunogenicity and kinetics of distribution and elimination of sheep digoxin-specific IgG and Fab fragments in the rabbit and baboon. Clin Exp Immunol 36:384–396

Snyder GH, Cennerazzo MJ, Karalis AJ, Field D (1981) Electrostatic influence of local cysteine environments on disulfide exchange kinetics. Biochemistry 20:6509–6519

Stein RL (1993) Mechanism of enzymatic and nonenzymatic prolyl cis-trans isomerization. Adv Protein Chem 44:1–24

Stotz CE, Borchardt RT, Middaugh CR, Siahaan TJ, Vander Velde D, Topp EM (2004) Secondary structure of a dynamic type I'beta-hairpin peptide. J Pept Res 63:371–382

Summa D, Spiga O, Bernini A, Venditti V, Priora R, Frosali S, Margaritis A, Di Giuseppe D, Niccolai N, Di Simplicio P (2007) Protein-thiol substitution or protein dethiolation by thiol/disulfide exchange reactions: the albumin model. Proteins 69:369–378

Suppavorasatit I, Cadwallader KR (2012) Effect of enzymatic deamidation of soy protein by protein-glutaminase on the flavor-binding properties of the protein under aqueous conditions. J Agric Food Chem 60:7817–7823

Szajewski RP, Whitesides GM (1980) Rate constants and equilibrium-constants for thiol-disulfide interchange reactions involving oxidized glutathione. J Am Chem Soc 102:2011–2026

Tarentino AL, Plummer TH (1994) Enzymatic deglycosylation of asparagine-linked glycans – purification, properties, and specificity of oligosaccharide-cleaving enzymes from flavobacterium-meningosepticum. Methods Enzymol 230:44–57

Tarentino AL, Gomez CM, Plummer TH (1985) Deglycosylation of asparagine-linked glycans by peptide - N-glycosidase-F. Biochemistry 24:4665–4671

Taylor NR, Vonitzstein M (1994) Molecular modeling studies on ligand-binding to sialidase from influenza-virus and the mechanism of catalysis. J Med Chem 37:616–624

Tous GI, Wei Z, Feng J, Bilbulian S, Bowen S, Smith J, Strouse R, Mcgeehan P, Casas-Finet J, Schenerman MA (2005) Characterization of a novel modification to monoclonal antibodies: thioether cross-link of heavy and light chains. Anal Chem 77:2675–2682

Tretter V, Altmann F, Marz L (1991) Peptide-N4-(N-acetyl-beta-glucosaminyl)asparagine amidase-F cannot release glycans with fucose attached alpha-1-]3 to the asparagine-linked N-acetylglucosamine residue. Eur J Biochem 199:647–652

Trimble RB, Tarentino AL (1991) Identification of distinct endoglycosidase (endo) activities in flavobacterium-meningosepticum - endo-F1, endo-F2, and endo-F3 - endo-F1 and endo-H hydrolyze only high mannose and hybrid glycans. J Biol Chem 266:1646–1651

Trivedi MV, Laurence JS, Siahaan TJ (2009) The role of thiols and disulfides on protein stability. Curr Protein Pept Sci 10:614–625

Usami A, Ohtsu A, Takahama S, Fujii T (1996) The effect of PH, hydrogen peroxide and temperature on the stability of human monoclonal antibody. J Pharm Biomed Anal 14:1133–1140

Van Der Neut Kolfschoten M, Schuurman J, Losen M, Bleeker WK, Martinez-Martinez P, Vermeulen E, Den Bleker TH, Wiegman L, Vink T, Aarden LA, De Baets MH, Van De Winkel JG, Aalberse RC, Parren PW (2007) Anti-inflammatory activity of human IgG4 antibodies by dynamic Fab arm exchange. Science 317:1554–1557

Vasella A, Davies GJ, Bohm M (2002) Glycosidase mechanisms. Curr Opin Chem Biol 6:619–629

Vlasak J, Bussat MC, Wang S, Wagner-Rousset E, Schaefer M, Klinguer-Hamour C, Kirchmeier M, Corvaia N, Ionescu R, Beck A (2009) Identification and characterization of asparagine deamidation in the light chain CDR1 of a humanized IgG1 antibody. Anal Biochem 392: 145–154

Voorter CE, De Haard-Hoekman WA, Van Den Oetelaar PJ, Bloemendal H, De Jong WW (1988) Spontaneous peptide bond cleavage in aging alpha-crystallin through a succinimide intermediate. J Biol Chem 263:19020–19023

Waddling CA, Plummer TH, Tarentino AL, Van Roey P (2000) Structural basis for the substrate specificity of endo-beta-N-acetylglucosaminidase F-3. Biochemistry 39:7878–7885

Wakankar AA, Borchardt RT, Eigenbrot C, Shia S, Wang YJ, Shire SJ, Liu JL (2007) Aspartate isomerization in the complementarity-determining regions of two closely related monoclonal antibodies. Biochemistry 46:1534–1544

Wang L, Amphlett G, Lambert JM, Blattler W, Zhang W (2005) Structural characterization of a recombinant monoclonal antibody by electrospray time-of-flight mass spectrometry. Pharm Res 22:1338–1349

Wiita AP, Ainavarapu SR, Huang HH, Fernandez JM (2006) Force-dependent chemical kinetics of disulfide bond reduction observed with single-molecule techniques. Proc Natl Acad Sci U S A 103:7222–7227

Wozniak-Knopp G, Ruker F (2012) A C-terminal interdomain disulfide bond significantly stabilizes the Fc fragment of IgG. Arch Biochem Biophys 526:181–187

Wozniak-Knopp G, Stadlmann J, Ruker F (2012) Stabilisation of the Fc fragment of human IgG1 by engineered intradomain disulfide bonds. PLoS One 7:E30083

Wypych J, Li M, Guo A, Zhang Z, Martinez T, Allen MJ, Fodor S, Kelner DN, Flynn GC, Liu YD, Bondarenko PV, Ricci MS, Dillon TM, Balland A (2008) Human IgG2 antibodies display disulfide-mediated structural isoforms. J Biol Chem 283:16194–16205

Xiang T, Lundell E, Sun Z, Liu H (2007) Structural effect of a recombinant monoclonal antibody on hinge region peptide bond hydrolysis. J Chromatogr B Analyt Technol Biomed Life Sci 858:254–262

Yan B, Boyd D, Kaschak T, Tsukuda J, Shen A, Lin Y, Chung S, Gupta P, Kamath A, Wong A, Vernes JM, Meng GY, Totpal K, Schaefer G, Jiang G, Nogal B, Emery C, Vanderlaan M, Carter P, Harris R, Amanullah A (2012) Engineering upper hinge improves stability and effector function of a human IgG1. J Biol Chem 287:5891–5897

Yong YH, Yamaguchi S, Matsumura Y (2006) Effects of enzymatic deamidation by protein-glutaminase on structure and functional properties of wheat gluten. J Agric Food Chem 54:6034–6040

Yoo EM, Wims LA, Chan LA, Morrison SL (2003) Human IgG2 can form covalent dimers. J Immunol 170:3134–3138

Zechel DL, Withers SG (2000) Glycosidase mechanisms: anatomy of a finely tuned catalyst. Acc Chem Res 33:11–18

Zhang SQ, Mccarter JD, Okamuraoho Y, Yaghi F, Hinek A, Withers SG, Callahan JW (1994) Kinetic mechanism and characterization of human beta-galactosidase precursor secreted by permanently transfected Chinese-hamster ovary cells. Biochem J 304:281–288

Zhang J, Yip H, Katta V (2011) Identification of isomerization and racemization of aspartate in the Asp-Asp motifs of a therapeutic protein. Anal Biochem 410:234–243

Zhang T, Zhang J, Hewitt D, Tran B, Gao X, Qiu ZJ, Tejada M, Gazzano-Santoro H, Kao YH (2012) Identification and characterization of buried unpaired cysteines in a recombinant monoclonal IgG1 antibody. Anal Chem 84:7112–7123

Zhang Q, Schenauer MR, Mccarter JD, Flynn GC (2013) IgG1 thioether bond formation in vivo. J Biol Chem 288:16371–16382

Zheng JY, Janis LJ (2006) Influence of PH, buffer species, and storage temperature on physicochemical stability of a humanized monoclonal antibody LA298. Int J Pharm 308:46–51

Zheng K, Bantog C, Bayer R (2011) The impact of glycosylation on monoclonal antibody conformation and stability. MAbs 3:568–576

Oxidation of Proteins in the In Vivo Environment: What We Know; What We Need to Study and Potential Mitigation Strategies

Christian Schöneich

Introduction

The oxidation of proteins in vivo continues to be an important focus of biomedical research (Davies 2005) as elevated levels of oxidized proteins have been associated with a large number of pathologies (Guttmann and Ghoshal 2011; Martinez et al. 2010; Nakamura et al. 2012) as well as biological aging (Stadtman 1988, 1992; Oliver et al. 1987). Protein oxidation can lead to changes in activity, conformation, protein–protein interactions, and half-life (Stadtman and Oliver 1991; Levine et al. 1981; Bota and Davies 2002; Davies and Lin 1988; Davies et al. 1987a, b; Davies and Delsignore 1987; Davies 1987), and trigger autophagy and/or apoptosis (Chan et al. 2012; Dunlop et al. 2011). An important consequence of oxidative (and other) covalent protein modifications can be the potential immunogenicity of the modified proteins (Eggleton et al. 2013), especially if antibodies directed against neo-epitopes cross-react with native, unmodified proteins, breaking immune tolerance (Griffiths 2008; Omersel et al. 2008, 2011; van Beers et al. 2011; Sauerborn et al. 2010; Jiskoot et al. 2009; Schellekens and Jiskoot 2006). Here, autoantibody generation may lead to autoimmune disorders. Analogous to endogenous proteins, therapeutic proteins may be subject to oxidation in vivo (i.e., after administration to the patient) though potential oxidation reactions of therapeutic proteins in vivo will likely be restricted to the extracellular space. The potential for immunogenicity is, therefore, not only an issue for endogenous proteins but also for the development and safe delivery of protein therapeutics (van Beers et al. 2011; Sauerborn et al. 2010; Jiskoot et al. 2009; Schellekens and Jiskoot 2006).

C. Schöneich (✉)
Department of Pharmaceutical Chemistry, The University of Kansas,
2095 Constant Avenue, Lawrence, KS 66047, USA
e-mail: schoneic@ku.edu

A. Rosenberg, B. Demeule (eds.), *Biobetters*, AAPS Advances
in the Pharmaceutical Sciences Series 19, DOI 10.1007/978-1-4939-2543-8_9

In the past, the chemical characterization of oxidized proteins from tissue or biological fluids required careful isolation and/or fractionation of sufficient quantities, mandated by the nature and/or detection limits of analytical methods such as Edman sequencing, amino acid analysis and the available mass spectrometry equipment. Today, modern proteomic/mass spectrometric methods allow protein characterization to be performed with significantly lower quantities of material (Witze et al. 2007; Dalle-Donne et al. 2005). In addition, the screening for and identification of post-translational modifications on a large set of peptides and proteins has been significantly accelerated by the development of bioinformatics tools (Cappadona et al. 2012).

Proteomic techniques are applied to an increasing number of tissues and biological fluids in order to achieve a more exhaustive molecular characterization of such tissues in disease states or assessing the outcome of therapeutic interventions (Witze et al. 2007; Dalle-Donne et al. 2005; Lam et al. 2013; Lalowski et al. 2013; Lai et al. 2013; Indovina et al. 2013; Honda et al. 2013). These tools are also applied to carefully characterize the chemical integrity (Manning et al. 2010) of all protein therapeutics during the manufacturing and storage of drug substance and drug product (Demartini et al. 2013). Interestingly, only a limited number of studies have focused on the chemical, specifically oxidative, fate of protein therapeutics *after* administration to the patient, i.e. the chemical integrity of protein therapeutics in vivo (Yin et al. 2013; Zhang et al. 2013). Results of such studies may have significant consequences for the development of formulations and for protein engineering. For example, if a specific degradation product affects potency and/or immunogenicity, its administration or in vivo formation could have therapeutic consequences (i.e. it would be considered a critical product quality attribute (CQA)) and consideration should be given to mechanisms to preclude such degradation in vitro (by formulation control) or in vivo (by protein engineering). In contrast, if a protein therapeutic contains significant levels of a specific chemical degradation product but the degraded product (1) maintains potency and (2) does not show any negative side effects or enhanced immunogenicity, then one could argue that the degradation product is not a CQA (Yin et al. 2013) and, to the extent that this is fully evaluated, not require very tight control of levels or require protein engineering to preclude its generation. Of course, wide variation in levels of the degradation product from lot to lot speaks to potential problems in maintaining the consistency of manufacturing and thus, a reasonable level of control of the levels are appropriate.

In the following, the available information on in vivo oxidation of protein pharmaceuticals will be summarized with specific emphasis on antibodies. Prior to that, a brief review on the generation of oxidants and potential oxidation sites in proteins will be given.

The Generation of Oxidants

Multiple enzymatic and non-enzymatic pathways can generate a variety of radical and non-radical oxidants in vivo. The extent to which these oxidants are formed in or released into the extracellular space will depend, in part, on the physiological

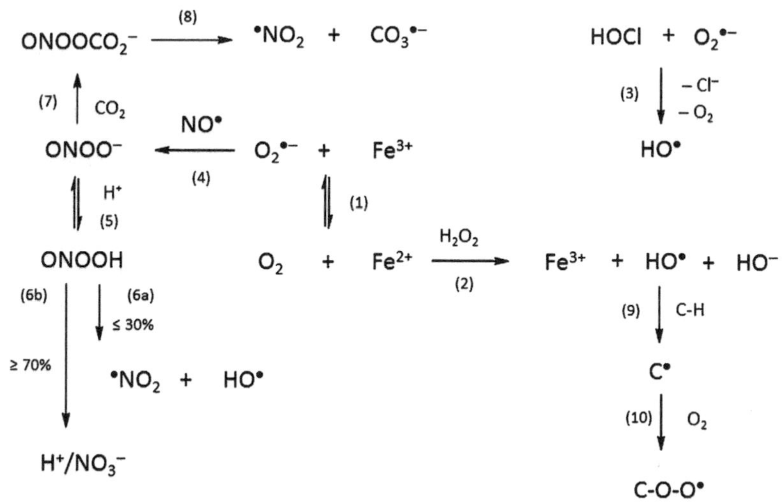

Scheme 1 Mechanisms of generation of radical and non-radical oxidants

conditions and/or activities of the organism (Rees et al. 2008; Nikolaidis and Jamurtas 2009). For example, inflammatory conditions or exercise will affect the number and/or activity of leukocytes and macrophages, which can release superoxide ($O_2^{\bullet-}$), hydrogen peroxide (H_2O_2), hypochlorous acid (HOCl) and/or peroxynitrite ($ONOO^-$) (Rees et al. 2008; Nikolaidis and Jamurtas 2009). The reactions of these oxidants with proteins will be summarized in more detail below. Additional oxidants may be formed via the reactions of superoxide, hydrogen peroxide, hypochlorous acid and peroxynitrite with redox-active transition metals (e.g., iron or copper) or CO_2. For example, the reaction of hydrogen peroxide with ferrous iron (Fe^{II}) yields highly reactive hydroxyl radicals ($HO^{\bullet}$) in the classic Fenton reaction (Scheme 1, reaction 2), where Fe^{II} may be provided by reaction of ferric iron (Fe^{III}) with superoxide (reaction 1) (Koppenol 1993; Goldstein et al. 1993). When generated through metal-catalyzed processes, hydroxyl radicals may either diffuse from the metal complexes to react as "free" hydroxyl radicals or remain bound to the metal and react as "complexed" hydroxyl radicals; both species are very reactive. Hydroxyl radical formation has also been reported for the reaction of hypochlorous acid with superoxide (reaction 3) (Candeias et al. 1993) and for the homolytic (i.e., the cleavage of a covalent bond into two radicals) dissociation of peroxynitrite in acidic solution, i.e. of peroxynitrous acid (ONOOH) (reaction 6a) (Kissner et al. 2003; Koppenol and Kissner 1998; Merenyi et al. 1998; Merenyi and Lind 1998). However, the yield of the homolytic dissociation is still a matter of considerable debate. While the concentrations of free iron and copper will be very low and are tightly controlled under normal physiologic conditions, this situation can change as a result of certain pathologies such as, e.g., iron overload or neurodegenerative diseases (Taba 2013; Parker et al. 2013; Litwin et al. 2013; Langkammer et al. 2013; Williams et al. 2012). Moreover, metal-catalyzed reactions of oxidants (e.g., hydrogen peroxide or peroxynitrite) easily proceed with protein-bound metals (Stadtman

Scheme 2 The fragmentation of an alkoxyl radical

$$R{-}CH_2{-}O^\bullet \longrightarrow R^\bullet + H_2C{=}O \qquad (11)$$

and Oliver 1991; Szweda and Stadtman 1992; Stadtman and Berlett 1991; Stadtman 1990; Zhao et al. 1997). In the case of heme-proteins, these reactions lead to the formation of ferryl species, which are strong oxidants on their own (Boccini and Herold 2004; Herold and Shivashankar 2003; Mehl et al. 1999; Battistuzzi et al. 2010). In the case of non-heme protein-bound metals, the reaction with oxidants frequently leads to site-specific protein oxidation of amino acids involved in metal-binding or located adjacent to the metal-binding site (Stadtman and Oliver 1991; Szweda and Stadtman 1992; Stadtman and Berlett 1991; Stadtman 1990; Zhao et al. 1997). Whenever hydroxyl radicals are generated, their efficient reaction with C–H bonds (reaction 9) can generate carbon-centered radicals, most of which will react in a diffusion-controlled manner, i.e. in a reaction which is associated with little to no activation energy, with molecular oxygen to yield peroxyl radicals ($ROO^\bullet$) (reaction 10) (Davies 2005). Peroxyl radicals are good oxidants, which react via hydrogen transfer, one-electron or two-electron oxidation (Ingold 1969), where two-electron oxidation may proceed via oxygen transfer. Specifically, hydrogen transfer and one-electron oxidation generate an organic peroxide (ROOH), which is a moderate oxidant itself. In contrast, the oxygen transfer results in the formation of an alkoxyl radical ($RO^\bullet$), a highly reactive species, which can either react via hydrogen- or electron transfer, or via α–β fragmentation (Scheme 2, reaction 11), generating a carbonyl product and a carbon-centered radical, i.e. a source for an additional peroxyl radical.

The reaction of hydrogen peroxide with CO_2 produces peroxymonocarbonate ($HOC(O)OO^-$), an oxidant which reacts about 100-times faster with methionine (Met) as compared to hydrogen peroxide alone (Richardson et al. 2003). At physiological concentrations of CO_2 in blood, this would translate into a ca. twofold acceleration of Met oxidation by hydrogen peroxide. In contrast, the reaction of peroxynitrite with CO_2 (HCO_3^-) ultimately generates a pair of radicals, nitrogen dioxide ($^\bullet NO_2$) and carbonate radical anion ($^\bullet CO_3^-$), both of which will further react with various amino acids (Scheme 1, reactions 8 and 9) (Pryor et al. 1997; Uppu et al. 1996; Lymar et al. 1996; Lymar and Hurst 1996).

In general, plasma has been reported to contain ≈ 0.25 μM hydrogen peroxide (Nikolaidis and Jamurtas 2009). In addition to leukocytes, erythrocytes represent a viable source for oxidant generation (Nikolaidis and Jamurtas 2009). Hydrogen peroxide readily crosses cell membranes, so that even the intracellular generation of hydrogen peroxide can elevate extracellular hydrogen peroxide levels (Rees et al. 2008). Nitric oxide (NO), a precursor of peroxynitrite, diffuses across cell membranes, and even the diffusion of peroxynitrite across lipid and erythrocyte membranes has been reported, which can be facilitated by ion channels (Carballal et al. 2013; Denicola et al. 1998; Marla et al. 1997). It is important to note, though, that antioxidant and scavenging reactions inside the cell may limit the actual fraction of oxidants diffusing into the extracellular space, as will reactions of these oxidants with proteins and other biomolecules inside the cell and within the membrane.

Several isoforms of NADPH oxidase, located on the plasma membrane, directly release hydrogen peroxide into the extracellular space (Rees et al. 2008; Nikolaidis and Jamurtas 2009). In fact, the generation of reactive oxygen species at the plasma membrane has been identified as a mechanism for the unmasking or masking of autoantibodies (Crane and Low 2008; McIntyre et al. 2006), a mechanism relevant for our discussion of the in vivo oxidation of therapeutic proteins. Here, the term "masking" refers to rendering the autoantibody "unreactive" towards its usual epitope and the term "unmasking" refers to a reversal of this situation. The various NADPH oxidase isoforms are expressed in a tissue- and cell-dependent manner, and, therefore, the release of hydrogen peroxide from the plasma membrane to the extracellular space is not restricted to blood (Crane and Low 2008).

Target Sites for Protein Oxidation

It appears that protein therapeutics may encounter a variety of reactive oxygen species depending on the physiologic conditions of the patients (Nikolaidis and Jamurtas 2009). Consequently, the potential oxidation targets of therapeutic proteins need to be discussed in view of the manifold species that may react. Frequently, the oxidative degradation of therapeutic proteins is monitored via Met oxidation to methionine sulfoxide (MetO). Specifically two Met residues in the Fc domain of antibodies, Met-252 and Met-428, are critical for binding to the neonatal Fc receptor (FcRn), where oxidation of these two Met residues can significantly reduce binding (Bertolotti-Ciarlet et al. 2009), affect the conformation of the C_H2 domain (containing Met-252) (Liu et al. 2008) and reduce serum half-life (Correia 2010; Wang et al. 2011). It can be expected that oxidation of a single Met-252 or single Met-428 within the dimeric antibody structure has a significantly lower effect of FcRn binding as compared to oxidation of both Met residues in both chains. In addition, oxidation of these Met residues affects binding to the Fcγ receptor 131H allele, FcγRIIa-131H (Bertolotti-Ciarlet et al. 2009), protein A and protein G (Gaza-Bulseco et al. 2008). In fact, protein A chromatography has been developed into an analytical tool to screen for Met-252 and Met-428 oxidation in antibodies (Loew et al. 2012). Of the reactive oxygen species introduced in the previous section, hydrogen peroxide, organic hydroperoxides, peroxynitrite, hypochlorous acid, and peroxyl radicals convert Met into MetO and have, therefore, to be considered as sources for Met oxidation of antibodies in vivo (Schoneich 2005). However, specifically, peroxynitrite, hypochlorous acid and peroxyl radicals will generate additional oxidation products via reaction with other amino acids. For example, peroxynitrite will nitrate tryptophan (Trp) to nitro-Trp (nTrp; possible products include 1-, 2-, 4-, 5-, 6- and 7-nTrp) (Nuriel et al. 2011) and tyrosine (Tyr) to 3-nitro-Tyr (3-nTyr) (Carballal et al. 2013; Radi 2013; Beckman and Koppenol 1996; Ischiropoulos et al. 1992), either via metal-catalyzed nitration or via the formation of carbonate radical anion and nitrogen dioxide in the presence of CO_2. Hypochlorous acid will chlorinate Tyr to 3-chloro-Tyr (3-cTyr) (Bergt et al. 2004;

Himmelfarb et al. 2001; Hazen and Heinecke 1997), cysteine (Cys) to sulfenyl chloride (R-S-Cl) (Fu et al. 2002), and free amines to N-chloramines (R_2N-Cl) (Pattison et al. 2011; Hawkins and Davies 1998, 1999). Here, the sulfenyl chloride serves as a precursor for the formation of sulfenic acid, disulfide, or sulfonamide, and chloramines serve as precursors for highly reactive aminyl radicals. Peroxyl radicals will attack protein Trp, Tyr and Cys (Ma et al. 1999; Viner et al. 1997) yielding a series of oxidation products: the main oxidation products generated from Trp are N-formylkynurenine, hydroxyl-Trp and/or oxyindolealanine (Steinmann et al. 2012). Tyr is converted into 3,4-dihydroxyphenylalanine (DOPA), its ortho-quinone oxidation product, and additional cross-linked products, generated through nucleophilic reaction of amines and amides with the ortho-quinone (Steinmann et al. 2012). An important intermediate formed during Tyr oxidation is the tyrosyl radical, which can couple with adventitious radicals such as nitrogen dioxide (to 3-nTyr) (Carballal et al. 2013; Bartesaghi et al. 2010), superoxide (to hydroperoxy and hydroxyl-derivatives of Tyr) (Nagy et al. 2009; Winterbourn et al. 2004; Moller et al. 2012), additional tyrosyl radicals (to dityrosine) (Giulivi et al. 2003; Giulivi and Davies 1994) and peroxyl radicals (Shchepin et al. 2010). The latter generates a para-coupling product amenable to *intra*molecular Diels-Alder reaction when conjugated dienes are present, as expected for peroxyl radicals generated from polyunsaturated fatty acids. The reaction of peroxyl radicals with Cys (Hildenbrand and Schulte-Frohlinde 1997) is expected to generate thiyl radicals ($RS^•$) and/or sulfenic acid, depending on the propensity of the peroxyl radicals for one-electron oxidation/hydrogen transfer vs. oxygen transfer. Sulfenic acids will either react with thiol/thiolate to generate disulfide or oxidize further to sulfinic (RSO_2H) or sulfonic acid (RSO_3H) (Ratnayake et al. 2013). In contrast, thiyl radicals may enter into a series of *intra*molecular hydrogen transfer reactions, either within the Cys residue (i.e., a 1,2- or 1,3-hydrogen transfer) (Nauser et al. 2012; Mozziconacci et al. 2012) or with αC–H and/or side chain C–H bonds of adjacent amino acids (Nauser et al. 2004; Nauser and Schoneich 2003; Mozziconacci et al. 2010a, 2011). Specifically the 1,3-hydrogen transfer within Cys (Scheme 3, reaction 13) leads to the elimination of $HS^•$ (reaction 14), generating dehydroalanine (Dha), an electrophile amenable for covalent cross-link formation with nucleophiles such as the mercapto group from Cys or the amino group from Lys. In contrast, the hydrogen transfer reactions with adjacent amino acids lead to the formation of carbon-centered radicals (reaction 15). The addition of oxygen to these carbon-centered radicals provides another source for peroxyl radicals, and peroxyl radicals generated at the αC-position of protein amino acids serve as precursors for fragmentation (reactions 16–18) (Garrison 1987).

Specifically the presence of transition metals promotes the formation of hydroxyl radicals or "complexed" hydroxyl radicals. Hydroxyl radicals will react with all amino acids though preferentially with aromatic and sulfur-containing amino acids (Davies 2005). When antibodies where exposed to metal-catalyzed oxidation, i.e. reaction conditions consistent with hydroxyl radical generation, mass spectrometry analysis confirmed the oxidation of histidine (His), Met, Trp, Tyr and phenylalanine (Phe) (Luo et al. 2011; Zhou et al. 2013).

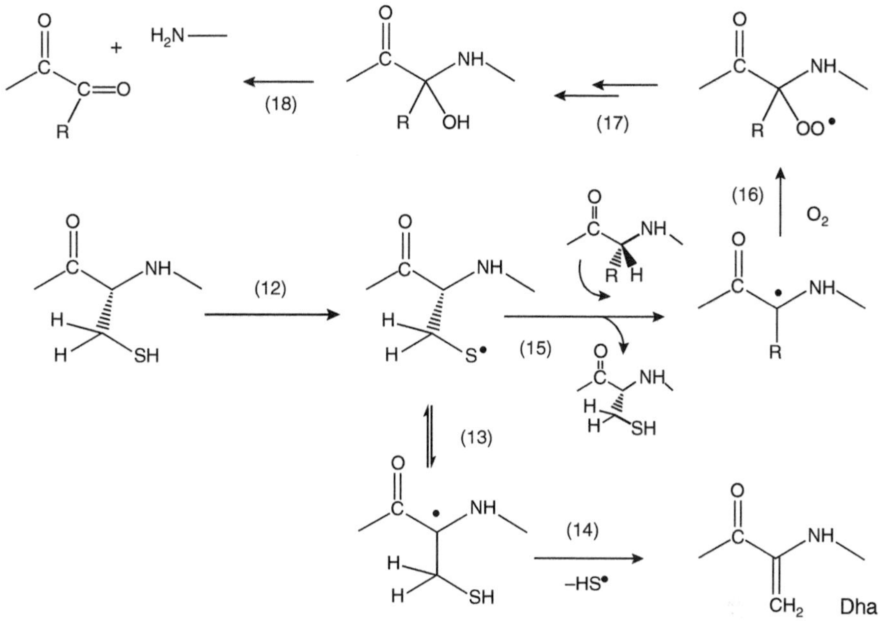

Scheme 3 Thiyl radical-induced reactions in peptides and proteins

Among the amino acid targets discussed above, not only the oxidation of Met but also of Trp can have biological consequences. For example, the oxidation of Trp-32 within the complementarity determining region 1 (CDR1) of the light chain in IgG1 reduces binding of antigen (Hensel et al. 2011). Theoretically, antibodies do not contain reduced Cys, as all Cys residues are expected to be part of disulfide bonds. However, in practice pharmaceutical antibodies contain a fraction of reduced Cys residues (Brych et al. 2010; Yoo et al. 2003; Zhang and Czupryn 2002), amenable to the oxidation reactions introduced above. To date, biological consequences of His, Phe and Tyr oxidation in pharmaceutical antibodies are unknown. Oxidized antibodies can be pro-inflammatory, and can display reduced binding not only to the FcRn and FcγRIIa-131H (Bertolotti-Ciarlet et al. 2009), but also to the C1q component of the C1 complex and to antigens (Hensel et al. 2011; Griffiths and Lunec 1996). The latter suggest a chemical and/or conformational modification of one or more of the CDRs, and in this regard the oxidation-sensitivity reported for Trp-32 located within the CDR1 of the light chain is important (Hensel et al. 2011). A characteristic feature of antibodies is the high prevalence of Tyr in antigen binding sites, where Tyr constitutes ca. 10 % of the CDR and is involved in ca. 25 % of the antigen contacts (Ye et al. 2008; Birtalan et al. 2008; Fellouse et al. 2005). Hence, any oxidation, nitration and/or chlorination of Tyr within the CDR may lead to significant changes in antigen recognition. Representative studies with human interferon-β have revealed a high potential for metal-catalyzed oxidation to cause immunogenicity (van Beers et al. 2011). Target amino acids oxidized under these conditions were

Met, His, Tyr and Trp (van Beers et al. 2011; Torosantucci et al. 2013), suggesting that the potential in vivo oxidation of these amino acids may also lead to immunogenicity of other therapeutic proteins, including antibodies.

Oxidation of Pharmaceutical Proteins In Vivo

Only a few studies have reported in vivo oxidation of therapeutic proteins. An interesting example is the conversion of a disulfide bond to a thioether, which does not represent a *direct* oxidation but includes an oxidation step in the mechanism (Friedman 1999). When comparing the stability of IgG1 in different environments, Yin et al. recognized that the formation of a thioether (lanthionine) between the heavy and light chains occurred faster in rat plasma (0.7 %/day) and rats in vivo (0.3 %/day) as compared to the reaction in PBS (0.1 %/day) (Yin et al. 2013). This thioether bond replaced an original disulfide bond between heavy chain (HC) and light chain (LC). The replacement of a disulfide by a thioether can have significant conformational consequences as the distance between the two Cys residues initially providing the disulfide bond is shortened by a distance equivalent to one sulfur–sulfur bond in the thioether. Complementary studies by Zhang et al. on endogenous antibodies in humans localized thioether formation in vivo to LC-Cys-214 and HC-Cys-220 with a formation rate of ca. 0.1 %/day in healthy patients (Zhang et al. 2013) thus indicating that this is likely to occur in therapeutic proteins. One mechanism of thioether formation is initiated by β-elimination of a disulfide, generating dehydroalanine (Dha) and perthiol (RSSH), where the perthiol subsequently eliminates elemental sulfur (S) (Friedman 1999) and the resulting thiol adds to Dha via Michael addition. Technically, in this reaction sequence the release of elemental sulfur from the perthiol represents an *oxidation* step. In vitro, alternative mechanisms for thioether formation exist such as the photo-induced homolytic cleavage of a disulfide, followed by disproportionation of a thiyl radical pair, dithiohemiacetal formation, and photolytic conversion of the dithiohemiacetal to thioether (Mozziconacci et al. 2010b). However, although these reactions have been observed in vitro, such reaction sequence is unlikely to play a role for thioether formation in vivo.

Interestingly, Yin et al. did not detect any significant in vivo oxidation of IgG1 in rat blood at Met and Trp (Yin et al. 2013), i.e., amino acids, which have been reported to be oxidation-sensitive in vitro. In contrast, human growth hormone (hGH) undergoes significant Met oxidation in rat blood in vivo, specifically of Met-14 (Battersby et al. 1995). An important difference between these two studies is the timeframe of protein recovery from the blood, i.e. 5–45 min for hGH (Battersby et al. 1995) and 2–10 days for IgG1 (Yin et al. 2013). If oxidized IgG1 were rapidly turned over then oxidized IgG1 may be not be detectable in the recovered IgG1. However, previous results on serum half-life of IgG1 in human FcRn mice showed that only significant but not mild Met oxidation resulted in a reduction of serum half-life from 56 ± 11 to 9.4 ± 0.3 h (Wang et al. 2011). Hence, if the lack of oxidized

IgG1 in the study by Yin et al. (2013) were to be rationalized by rapid oxidation-dependent turnover, oxidative modifications other than of Met (or even other, non-oxidative post-translational modifications or combinations thereof) would need to account for such rapid turnover. Clearly, more research on the effect(s) of post-translational modifications on IgG1 turnover is needed.

Approaches to Mitigate Protein Oxidation In Vivo

While only limited information on the in vivo oxidation of therapeutic proteins is available (see above), literally no studies have addressed the mitigation of specific in vivo oxidation problems. However, lessons can be learned from in vitro oxidation studies (i.e., accelerated stability studies and/or real-time storage conditions) where protein engineering can been used to make conservative substitutions of specific chemically labile amino acids while maintaining potency of the therapeutic protein(s). In this regard, consideration should be given to expand accelerated stability studies such as to include oxidation by oxidants which are generated during inflammatory conditions (i.e., hypochlorous acid, peroxynitrite and/or nitrogen dioxide; see above).

Conclusion and Outlook

Considering the variety of oxidative post-translational modifications, which have been detected on endogenous proteins, surprisingly few studies have addressed the potential for oxidative modifications of therapeutic proteins in vivo. Specifically for the development of proteins with long circulation half-lives in humans, such as antibodies, the potential for oxidative modifications becomes an issue, especially when these proteins can encounter inflammatory conditions. Additional complexity may be expected for antibody-drug conjugates (ADCs), where oxidation cannot only target the protein but also the linker and the conjugated drug(s).

References

Bartesaghi S, Wenzel J, Trujillo M, Lopez M, Joseph J, Kalyanaraman B, Radi R (2010) Lipid peroxyl radicals mediate tyrosine dimerization and nitration in membranes. Chem Res Toxicol 23:821–835. doi:10.1021/tx900446r

Battersby JE, Mukku VR, Clark RG, Hancock WS (1995) Affinity purification and microcharacterization of recombinant DNA-derived human growth hormone isolated from an in vivo model. Anal Chem 67:447–455

Battistuzzi G, Bellei M, Bortolotti CA, Sola M (2010) Redox properties of heme peroxidases. Arch Biochem Biophys 500:21–36. doi:10.1016/j.abb.2010.03.002

Beckman JS, Koppenol WH (1996) Nitric oxide, superoxide, and peroxynitrite: the good, the bad, and ugly. Am J Physiol 271:C1424–C1437

Bergt C, Fu X, Huq NP, Kao J, Heinecke JW (2004) Lysine residues direct the chlorination of tyrosines in YXXK motifs of apolipoprotein A-I when hypochlorous acid oxidizes high density lipoprotein. J Biol Chem 279:7856–7866. doi:10.1074/jbc.M309046200

Bertolotti-Ciarlet A, Wang W, Lownes R, Pristatsky P, Fang Y, McKelvey T, Li Y, Drummond J, Prueksaritanont T, Vlasak J (2009) Impact of methionine oxidation on the binding of human IgG1 to Fc Rn and Fc gamma receptors. Mol Immunol 46:1878–1882. doi:10.1016/j.molimm.2009.02.002

Birtalan S, Zhang Y, Fellouse FA, Shao L, Schaefer G, Sidhu SS (2008) The intrinsic contributions of tyrosine, serine, glycine and arginine to the affinity and specificity of antibodies. J Mol Biol 377:1518–1528. doi:10.1016/j.jmb.2008.01.093

Boccini F, Herold S (2004) Mechanistic studies of the oxidation of oxyhemoglobin by peroxynitrite. Biochemistry 43:16393–16404. doi:10.1021/bi0482250

Bota DA, Davies KJ (2002) Lon protease preferentially degrades oxidized mitochondrial aconitase by an ATP-stimulated mechanism. Nat Cell Biol 4:674–680. doi:10.1038/ncb836

Brych SR, Gokarn YR, Hultgen H, Stevenson RJ, Rajan R, Matsumura M (2010) Characterization of antibody aggregation: role of buried, unpaired cysteines in particle formation. J Pharm Sci 99:764–781. doi:10.1002/jps.21868

Candeias LP, Patel KB, Stratford MR, Wardman P (1993) Free hydroxyl radicals are formed on reaction between the neutrophil-derived species superoxide anion and hypochlorous acid. FEBS Lett 333:151–153

Cappadona S, Baker PR, Cutillas PR, Heck AJ, van Breukelen B (2012) Current challenges in software solutions for mass spectrometry-based quantitative proteomics. Amino Acids 43:1087–1108. doi:10.1007/s00726-012-1289-8

Carballal S, Bartesaghi S, Radi R (2013) Kinetic and mechanistic considerations to assess the biological fate of peroxynitrite. Biochim Biophys Acta. doi:10.1016/j.bbagen.2013.07.005

Chan SW, Dunlop RA, Rowe A, Double KL, Rodgers KJ (2012) L-DOPA is incorporated into brain proteins of patients treated for Parkinson's disease, inducing toxicity in human neuroblastoma cells in vitro. Exp Neurol 238:29–37. doi:10.1016/j.expneurol.2011.09.029

Correia IR (2010) Stability of IgG isotypes in serum. MAbs 2:221–232

Crane FL, Low H (2008) Reactive oxygen species generation at the plasma membrane for antibody control. Autoimmun Rev 7:518–522. doi:10.1016/j.autrev.2008.04.004

Dalle-Donne I, Scaloni A, Giustarini D, Cavarra E, Tell G, Lungarella G, Colombo R, Rossi R, Milzani A (2005) Proteins as biomarkers of oxidative/nitrosative stress in diseases: the contribution of redox proteomics. Mass Spectrom Rev 24:55–99. doi:10.1002/mas.20006

Davies KJ (1987) Protein damage and degradation by oxygen radicals. I. General aspects. J Biol Chem 262:9895–9901

Davies MJ (2005) The oxidative environment and protein damage. Biochim Biophys Acta 1703:93–109. doi:10.1016/j.bbapap.2004.08.007

Davies KJ, Delsignore ME (1987) Protein damage and degradation by oxygen radicals. III. Modification of secondary and tertiary structure. J Biol Chem 262:9908–9913

Davies KJ, Lin SW (1988) Oxidatively denatured proteins are degraded by an ATP-independent proteolytic pathway in Escherichia coli. Free Radic Biol Med 5:225–236

Davies KJ, Lin SW, Pacifici RE (1987a) Protein damage and degradation by oxygen radicals. IV. Degradation of denatured protein. J Biol Chem 262:9914–9920

Davies KJ, Delsignore ME, Lin SW (1987b) Protein damage and degradation by oxygen radicals. II. Modification of amino acids. J Biol Chem 262:9902–9907

Demartini DR, Pasquali G, Carlini CR (2013) An overview of proteomics approaches applied to biopharmaceuticals and cyclotides research. J Proteomics. doi:10.1016/j.jprot.2013.06.009

Denicola A, Souza JM, Radi R (1998) Diffusion of peroxynitrite across erythrocyte membranes. Proc Natl Acad Sci U S A 95:3566–3571

Dunlop RA, Brunk UT, Rodgers KJ (2011) Proteins containing oxidized amino acids induce apoptosis in human monocytes. Biochem J 435:207–216. doi:10.1042/BJ20100682

Eggleton P, Nissim A, Ryan BJ, Whiteman M, Winyard PG (2013) Detection and isolation of human serum autoantibodies that recognize oxidatively modified autoantigens. Free Radic Biol Med 57:79–91. doi:10.1016/j.freeradbiomed.2012.11.006

Fellouse FA, Li B, Compaan DM, Peden AA, Hymowitz SG, Sidhu SS (2005) Molecular recognition by a binary code. J Mol Biol 348:1153–1162. doi:10.1016/j.jmb.2005.03.041

Friedman M (1999) Chemistry, biochemistry, nutrition, and microbiology of lysinoalanine, lanthionine, and histidinoalanine in food and other proteins. J Agric Food Chem 47:1295–1319

Fu X, Mueller DM, Heinecke JW (2002) Generation of intramolecular and intermolecular sulfenamides, sulfinamides, and sulfonamides by hypochlorous acid: a potential pathway for oxidative cross-linking of low-density lipoprotein by myeloperoxidase. Biochemistry 41: 1293–1301

Garrison WM (1987) Reaction-mechanisms in the radiolysis of peptides, polypeptides, and proteins. Chem Rev 87:381–398

Gaza-Bulseco G, Faldu S, Hurkmans K, Chumsae C, Liu H (2008) Effect of methionine oxidation of a recombinant monoclonal antibody on the binding affinity to protein A and protein G. J Chromatogr B Analyt Technol Biomed Life Sci 870:55–62. doi:10.1016/j.jchromb.2008.05.045

Giulivi C, Davies KJ (1994) Dityrosine: a marker for oxidatively modified proteins and selective proteolysis. Methods Enzymol 233:363–371

Giulivi C, Traaseth NJ, Davies KJ (2003) Tyrosine oxidation products: analysis and biological relevance. Amino Acids 25:227–232. doi:10.1007/s00726-003-0013-0

Goldstein S, Meyerstein D, Czapski G (1993) The Fenton reagents. Free Radic Biol Med 15:435–445

Griffiths HR (2008) Is the generation of neo-antigenic determinants by free radicals central to the development of autoimmune rheumatoid disease? Autoimmun Rev 7:544–549. doi:10.1016/j.autrev.2008.04.013

Griffiths HR, Lunec J (1996) The C1q binding activity of IgG is modified in vitro by reactive oxygen species: implications for rheumatoid arthritis. FEBS Lett 388:161–164

Guttmann RP, Ghoshal S (2011) Thiol-protease oxidation in age-related neuropathology. Free Radic Biol Med 51:282–288. doi:10.1016/j.freeradbiomed.2011.04.017

Hawkins CL, Davies MJ (1998) Hypochlorite-induced damage to proteins: formation of nitrogen-centred radicals from lysine residues and their role in protein fragmentation. Biochem J 332(Pt 3):617–625

Hawkins CL, Davies MJ (1999) Hypochlorite-induced oxidation of proteins in plasma: formation of chloramines and nitrogen-centred radicals and their role in protein fragmentation. Biochem J 340(Pt 2):539–548

Hazen SL, Heinecke JW (1997) 3-Chlorotyrosine, a specific marker of myeloperoxidase-catalyzed oxidation, is markedly elevated in low density lipoprotein isolated from human atherosclerotic intima. J Clin Invest 99:2075–2081. doi:10.1172/JCI119379

Hensel M, Steurer R, Fichtl J, Elger C, Wedekind F, Petzold A, Schlothauer T, Molhoj M, Reusch D, Bulau P (2011) Identification of potential sites for tryptophan oxidation in recombinant antibodies using tert-butylhydroperoxide and quantitative LC-MS. PLoS One 6:e17708. doi:10.1371/journal.pone.0017708

Herold S, Shivashankar K (2003) Metmyoglobin and methemoglobin catalyze the isomerization of peroxynitrite to nitrate. Biochemistry 42:14036–14046. doi:10.1021/bi0350349

Hildenbrand K, Schulte-Frohlinde D (1997) Time-resolved EPR studies on the reaction rates of peroxyl radicals of poly(acrylic acid) and of calf thymus DNA with glutathione. Re-examination of a rate constant for DNA. Int J Radiat Biol 71:377–385

Himmelfarb J, McMenamin ME, Loseto G, Heinecke JW (2001) Myeloperoxidase-catalyzed 3-chlorotyrosine formation in dialysis patients. Free Radic Biol Med 31:1163–1169

Honda K, Ono M, Shitashige M, Masuda M, Kamita M, Miura N, Yamada T (2013) Proteomic approaches to the discovery of cancer biomarkers for early detection and personalized medicine. Jpn J Clin Oncol 43:103–109. doi:10.1093/jjco/hys200

Indovina P, Marcelli E, Pentimalli F, Tanganelli P, Tarro G, Giordano A (2013) Mass spectrometry-based proteomics: the road to lung cancer biomarker discovery. Mass Spectrom Rev 32:129–142. doi:10.1002/mas.21355

Ingold KU (1969) Peroxy radicals. Acc Chem Res 2:1–9
Ischiropoulos H, Zhu L, Chen J, Tsai M, Martin JC, Smith CD, Beckman JS (1992) Peroxynitrite-mediated tyrosine nitration catalyzed by superoxide dismutase. Arch Biochem Biophys 298:431–437
Jiskoot W, van Schie RM, Carstens MG, Schellekens H (2009) Immunological risk of injectable drug delivery systems. Pharm Res 26:1303–1314. doi:10.1007/s11095-009-9855-9
Kissner R, Nauser T, Kurz C, Koppenol WH (2003) Peroxynitrous acid–where is the hydroxyl radical? IUBMB Life 55:567–572
Koppenol WH (1993) The centennial of the Fenton reaction. Free Radic Biol Med 15:645–651
Koppenol WH, Kissner R (1998) Can O=NOOH undergo homolysis? Chem Res Toxicol 11:87–90. doi:10.1021/tx970200x
Lai ZW, Nice EC, Schilling O (2013) Glycocapture-based proteomics for secretome analysis. Proteomics 13:512–525. doi:10.1002/pmic.201200414
Lalowski M, Magni F, Mainini V, Monogioudi E, Gotsopoulos A, Soliymani R, Chinello C, Baumann M (2013) Imaging mass spectrometry: a new tool for kidney disease investigations. Nephrol Dial Transplant 28:1648–1656. doi:10.1093/ndt/gft008
Lam SW, Jimenez CR, Boven E (2013) Breast cancer classification by proteomic technologies: current state of knowledge. Cancer Treat Rev. doi:10.1016/j.ctrv.2013.06.006
Langkammer C, Ropele S, Pirpamer L, Fazekas F, Schmidt R (2013) MRI for iron mapping in Alzheimer's disease. Neurodegener Dis. doi:10.1159/000353756
Levine RL, Oliver CN, Fulks RM, Stadtman ER (1981) Turnover of bacterial glutamine synthetase: oxidative inactivation precedes proteolysis. Proc Natl Acad Sci U S A 78:2120–2124
Litwin T, Gromadzka G, Szpak GM, Jablonka-Salach K, Bulska E, Czlonkowska A (2013) Brain metal accumulation in Wilson's disease. J Neurol Sci 329:55–58. doi:10.1016/j.jns.2013.03.021
Liu D, Ren D, Huang H, Dankberg J, Rosenfeld R, Cocco MJ, Li L, Brems DN, Remmele RL Jr (2008) Structure and stability changes of human IgG1 Fc as a consequence of methionine oxidation. Biochemistry 47:5088–5100. doi:10.1021/bi702238b
Loew C, Knoblich C, Fichtl J, Alt N, Diepold K, Bulau P, Goldbach P, Adler M, Mahler HC, Grauschopf U (2012) Analytical protein a chromatography as a quantitative tool for the screening of methionine oxidation in monoclonal antibodies. J Pharm Sci 101:4248–4257. doi:10.1002/jps.23286
Luo Q, Joubert MK, Stevenson R, Ketchem RR, Narhi LO, Wypych J (2011) Chemical modifications in therapeutic protein aggregates generated under different stress conditions. J Biol Chem 286:25134–25144. doi:10.1074/jbc.M110.160440
Lymar SV, Hurst JK (1996) Carbon dioxide: physiological catalyst for peroxynitrite-mediated cellular damage or cellular protectant? Chem Res Toxicol 9:845–850. doi:10.1021/tx960046z
Lymar SV, Jiang Q, Hurst JK (1996) Mechanism of carbon dioxide-catalyzed oxidation of tyrosine by peroxynitrite. Biochemistry 35:7855–7861. doi:10.1021/bi960331h
Ma YS, Chao CC, Stadtman ER (1999) Oxidative modification of glutamine synthetase by 2,2′-azobis(2- amidinopropane) dihydrochloride. Arch Biochem Biophys 363:129–134. doi:10.1006/abbi.1998.1076
Manning MC, Chou DK, Murphy BM, Payne RW, Katayama DS (2010) Stability of protein pharmaceuticals: an update. Pharm Res 27:544–575. doi:10.1007/s11095-009-0045-6
Marla SS, Lee J, Groves JT (1997) Peroxynitrite rapidly permeates phospholipid membranes. Proc Natl Acad Sci U S A 94:14243–14248
Martinez A, Portero-Otin M, Pamplona R, Ferrer I (2010) Protein targets of oxidative damage in human neurodegenerative diseases with abnormal protein aggregates. Brain Pathol 20:281–297. doi:10.1111/j.1750-3639.2009.00326.x
McIntyre JA, Wagenknecht DR, Faulk WP (2006) Redox-reactive autoantibodies: detection and physiological relevance. Autoimmun Rev 5:76–83. doi:10.1016/j.autrev.2005.07.009
Mehl M, Daiber A, Herold S, Shoun H, Ullrich V (1999) Peroxynitrite reaction with heme proteins. Nitric Oxide 3:142–152. doi:10.1006/niox.1999.0217
Merenyi G, Lind J (1998) Free radical formation in the peroxynitrous acid (ONOOH)/peroxynitrite (ONOO-) system. Chem Res Toxicol 11:243–246. doi:10.1021/tx980026s

Merenyi G, Lind J, Goldstein S, Czapski G (1998) Peroxynitrous acid homolyzes into *OH and *NO2 radicals. Chem Res Toxicol 11:712–713. doi:10.1021/tx980043h

Moller MN, Hatch DM, Kim HY, Porter NA (2012) Superoxide reaction with tyrosyl radicals generates para-hydroperoxy and para-hydroxy derivatives of tyrosine. J Am Chem Soc 134:16773–16780. doi:10.1021/ja307215z

Mozziconacci O, Kerwin BA, Schoneich C (2010a) Reversible hydrogen transfer between cysteine thiyl radical and glycine and alanine in model peptides: covalent H/D exchange, radical-radical reactions, and L- to D-Ala conversion. J Phys Chem B 114:6751–6762. doi:10.1021/jp101508b

Mozziconacci O, Kerwin BA, Schoneich C (2010b) Photolysis of an intrachain peptide disulfide bond: primary and secondary processes, formation of H2S, and hydrogen transfer reactions. J Phys Chem B 114:3668–3688. doi:10.1021/jp910789x

Mozziconacci O, Kerwin BA, Schoneich C (2011) Reversible hydrogen transfer reactions of cysteine thiyl radicals in peptides: the conversion of cysteine into dehydroalanine and alanine, and of alanine into dehydroalanine. J Phys Chem B 115:12287–12305. doi:10.1021/jp2070453

Mozziconacci O, Williams TD, Schoneich C (2012) Intramolecular hydrogen transfer reactions of thiyl radicals from glutathione: formation of carbon-centered radical at Glu, Cys, and Gly. Chem Res Toxicol 25:1842–1861. doi:10.1021/tx3000494

Nagy P, Kettle AJ, Winterbourn CC (2009) Superoxide-mediated formation of tyrosine hydroperoxides and methionine sulfoxide in peptides through radical addition and intramolecular oxygen transfer. J Biol Chem 284:14723–14733. doi:10.1074/jbc.M809396200

Nakamura T, Cho DH, Lipton SA (2012) Redox regulation of protein misfolding, mitochondrial dysfunction, synaptic damage, and cell death in neurodegenerative diseases. Exp Neurol 238:12–21. doi:10.1016/j.expneurol.2012.06.032

Nauser T, Schoneich C (2003) Thiyl radicals abstract hydrogen atoms from the (alpha)C-H bonds in model peptides: absolute rate constants and effect of amino acid structure. J Am Chem Soc 125:2042–2043. doi:10.1021/ja0293599

Nauser T, Pelling J, Schoneich C (2004) Thiyl radical reaction with amino acid side chains: rate constants for hydrogen transfer and relevance for posttranslational protein modification. Chem Res Toxicol 17:1323–1328. doi:10.1021/tx049856y

Nauser T, Koppenol WH, Schoneich C (2012) Reversible hydrogen transfer reactions in thiyl radicals from cysteine and related molecules: absolute kinetics and equilibrium constants determined by pulse radiolysis. J Phys Chem B 116:5329–5341. doi:10.1021/jp210954v

Nikolaidis MG, Jamurtas AZ (2009) Blood as a reactive species generator and redox status regulator during exercise. Arch Biochem Biophys 490:77–84. doi:10.1016/j.abb.2009.08.015

Nuriel T, Hansler A, Gross SS (2011) Protein nitrotryptophan: formation, significance and identification. J Proteomics 74:2300–2312. doi:10.1016/j.jprot.2011.05.032

Oliver CN, Ahn BW, Moerman EJ, Goldstein S, Stadtman ER (1987) Age-related changes in oxidized proteins. J Biol Chem 262:5488–5491

Omersel J, Jurgec I, Cucnik S, Kveder T, Rozman B, Sodin-Semrl S, Bozic B (2008) Autoimmune and proinflammatory activity of oxidized immunoglobulins. Autoimmun Rev 7:523–529. doi:10.1016/j.autrev.2008.04.005

Omersel J, Avbersek-Luznik I, Grabnar PA, Kveder T, Rozman B, Bozic B (2011) Autoimmune reactivity of IgM acquired after oxidation. Redox Rep 16:248–256. doi:10.1179/174329211X13190184351680

Parker SJ, Koistinaho J, White AR, Kanninen KM (2013) Biometals in rare neurodegenerative disorders of childhood. Front Aging Neurosci 5:14. doi:10.3389/fnagi.2013.00014

Pattison DI, O'Reilly RJ, Skaff O, Radom L, Anderson RF, Davies MJ (2011) One-electron reduction of N-chlorinated and N-brominated species is a source of radicals and bromine atom formation. Chem Res Toxicol 24:371–382. doi:10.1021/tx100325z

Pryor WA, Lemercier JN, Zhang H, Uppu RM, Squadrito GL (1997) The catalytic role of carbon dioxide in the decomposition of peroxynitrite. Free Radic Biol Med 23:331–338

Radi R (2013) Protein tyrosine nitration: biochemical mechanisms and structural basis of functional effects. Acc Chem Res 46:550–559. doi:10.1021/ar300234c

Ratnayake S, Dias IH, Lattman E, Griffiths HR (2013) Stabilising cysteinyl thiol oxidation and nitrosation for proteomic analysis. J Proteomics. doi:10.1016/j.jprot.2013.06.019

Rees MD, Kennett EC, Whitelock JM, Davies MJ (2008) Oxidative damage to extracellular matrix and its role in human pathologies. Free Radic Biol Med 44:1973–2001. doi:10.1016/j.freeradbiomed.2008.03.016

Richardson DE, Regino CA, Yao H, Johnson JV (2003) Methionine oxidation by peroxymonocarbonate, a reactive oxygen species formed from CO2/bicarbonate and hydrogen peroxide. Free Radic Biol Med 35:1538–1550

Sauerborn M, Brinks V, Jiskoot W, Schellekens H (2010) Immunological mechanism underlying the immune response to recombinant human protein therapeutics. Trends Pharmacol Sci 31:53–59. doi:10.1016/j.tips.2009.11.001

Schellekens H, Jiskoot W (2006) Erythropoietin-associated PRCA: still an unsolved mystery. J Immunotoxicol 3:123–130. doi:10.1080/15476910600845567

Schoneich C (2005) Methionine oxidation by reactive oxygen species: reaction mechanisms and relevance to Alzheimer's disease. Biochim Biophys Acta 1703:111–119. doi:10.1016/j.bbapap.2004.09.009

Shchepin R, Moller MN, Kim HY, Hatch DM, Bartesaghi S, Kalyanaraman B, Radi R, Porter NA (2010) Tyrosine-lipid peroxide adducts from radical termination: para coupling and intramolecular Diels-Alder cyclization. J Am Chem Soc 132:17490–17500. doi:10.1021/ja106503a

Stadtman ER (1988) Protein modification in aging. J Gerontol 43:B112–B120

Stadtman ER (1990) Metal ion-catalyzed oxidation of proteins: biochemical mechanism and biological consequences. Free Radic Biol Med 9:315–325

Stadtman ER (1992) Protein oxidation and aging. Science 257:1220–1224

Stadtman ER, Berlett BS (1991) Fenton chemistry. Amino acid oxidation. J Biol Chem 266:17201–17211

Stadtman ER, Oliver CN (1991) Metal-catalyzed oxidation of proteins. Physiological consequences. J Biol Chem 266:2005–2008

Steinmann D, Ji JA, Wang YJ, Schoneich C (2012) Oxidation of human growth hormone by oxygen-centered radicals: formation of Leu-101 hydroperoxide and Tyr-103 oxidation products. Mol Pharm 9:803–814. doi:10.1021/mp3001028

Szweda LI, Stadtman ER (1992) Iron-catalyzed oxidative modification of glucose-6-phosphate dehydrogenase from Leuconostoc mesenteroides. Structural and functional changes. J Biol Chem 267:3096–3100

Taba P (2013) Metals and movement disorders. Curr Opin Neurol 26:435–441. doi:10.1097/WCO.0b013e3283629beb

Torosantucci R, Sharov VS, van Beers M, Brinks V, Schoneich C, Jiskoot W (2013) Identification of oxidation sites and covalent cross-links in metal catalyzed oxidized interferon Beta-1a: potential implications for protein aggregation and immunogenicity. Mol Pharm 10:2311–2322. doi:10.1021/mp300665u

Uppu RM, Squadrito GL, Pryor WA (1996) Acceleration of peroxynitrite oxidations by carbon dioxide. Arch Biochem Biophys 327:335–343. doi:10.1006/abbi.1996.0131

van Beers MM, Sauerborn M, Gilli F, Brinks V, Schellekens H, Jiskoot W (2011) Oxidized and aggregated recombinant human interferon beta is immunogenic in human interferon beta transgenic mice. Pharm Res 28:2393–2402. doi:10.1007/s11095-011-0451-4

Viner RI, Krainev AG, Williams TD, Schoneich C, Bigelow DJ (1997) Identification of oxidation-sensitive peptides within the cytoplasmic domain of the sarcoplasmic reticulum Ca2+–ATPase. Biochemistry 36:7706–7716. doi:10.1021/bi970058z

Wang W, Vlasak J, Li Y, Pristatsky P, Fang Y, Pittman T, Roman J, Wang Y, Prueksaritanont T, Ionescu R (2011) Impact of methionine oxidation in human IgG1 Fc on serum half-life of monoclonal antibodies. Mol Immunol 48:860–866. doi:10.1016/j.molimm.2010.12.009

Williams R, Buchheit CL, Berman NE, LeVine SM (2012) Pathogenic implications of iron accumulation in multiple sclerosis. J Neurochem 120:7–25. doi:10.1111/j.1471-4159.2011.07536.x

Winterbourn CC, Parsons-Mair HN, Gebicki S, Gebicki JM, Davies MJ (2004) Requirements for superoxide-dependent tyrosine hydroperoxide formation in peptides. Biochem J 381:241–248. doi:10.1042/BJ20040259

Witze ES, Old WM, Resing KA, Ahn NG (2007) Mapping protein post-translational modifications with mass spectrometry. Nat Methods 4:798–806. doi:10.1038/nmeth1100

Ye JD, Tereshko V, Frederiksen JK, Koide A, Fellouse FA, Sidhu SS, Koide S, Kossiakoff AA, Piccirilli JA (2008) Synthetic antibodies for specific recognition and crystallization of structured RNA. Proc Natl Acad Sci U S A 105:82–87. doi:10.1073/pnas.0709082105

Yin S, Pastuskovas CV, Khawli LA, Stults JT (2013) Characterization of therapeutic monoclonal antibodies reveals differences between in vitro and in vivo time-course studies. Pharm Res 30:167–178. doi:10.1007/s11095-012-0860-z

Yoo EM, Wims LA, Chan LA, Morrison SL (2003) Human IgG2 can form covalent dimers. J Immunol 170:3134–3138

Zhang W, Czupryn MJ (2002) Free sulfhydryl in recombinant monoclonal antibodies. Biotechnol Prog 18:509–513. doi:10.1021/bp025511z

Zhang Q, Schenauer MR, McCarter JD, Flynn GC (2013) IgG1 thioether bond formation in vivo. J Biol Chem 288:16371–16382. doi:10.1074/jbc.M113.468397

Zhao F, Ghezzo-Schoneich E, Aced GI, Hong J, Milby T, Schoneich C (1997) Metal-catalyzed oxidation of histidine in human growth hormone. Mechanism, isotope effects, and inhibition by a mild denaturing alcohol. J Biol Chem 272:9019–9029

Zhou S, Mozziconacci O, Kerwin BA, Schoneich C (2013) Fluorogenic tagging methodology applied to characterize oxidized tyrosine and phenylalanine in an immunoglobulin monoclonal antibody. Pharm Res 30:1311–1327. doi:10.1007/s11095-012-0970-7

Molecular Assessment of Monoclonal Antibody-Based Therapeutics Enabling Lead Selection for Clinical Development

Vikas K. Sharma and Robert F. Kelley

Introduction

Monoclonal antibodies (mAbs) and mAb-based therapeutics (e.g. antibody drug conjugates, bispecifics, Fab fragments, Fc-fusion proteins, etc.) continue to emerge as an important class of biologics for the treatment of a variety of diseases. Over 35 mAb-based therapeutics have been approved and many hundreds remain in clinical trials in the areas of oncology, cancer-immunology, immunology, infectious diseases, and other acute and chronic ailments (as of October 2014) (http://www.immunologylink.com/FDA-APP-Abs.html). The promise demonstrated by mAb-based therapies in treating unmet medical needs has led several biopharmaceutical companies to invest heavily in their development. Such investment has led to the simultaneous development of several mAb-based drugs for a given indication and/or target within an organization, and almost concurrent development of drugs for similar indications across different biopharmaceutical companies. This is important for patients and the medical community as it ensures that multiple therapeutic options may become available for a given disease and, in some instances, provides a tailored therapeutic regimen for a given disease. However, in order to enhance the chances of clinical and commercial success, companies must ensure early on that the lead molecule (also referred to as "lead candidate") introduced into clinical trials is optimal with respect to the therapeutics' efficacy, safety, pharmacokinetics, manufacturability, stability, and delivery profile.

Biopharmaceutical companies strive to produce "first-in-class" or "best-in-class" efficacious and safe mAb-based therapeutics, primarily driven by their unique mechanism of action or specific targeted biology. Furthermore, there is an effort to

V.K. Sharma (✉) • R.F. Kelley
Pharmaceutical Development, Genentech, Inc., South San Francisco, CA 94030, USA
e-mail: sharma.vikas_k@gene.com

A. Rosenberg, B. Demeule (eds.), *Biobetters*, AAPS Advances
in the Pharmaceutical Sciences Series 19, DOI 10.1007/978-1-4939-2543-8_10

produce "biobetters", or new mAb or mAb-based molecules, which target the same validated target epitope as the original molecule. The biobetter is engineered to have improved efficacy, safety, pharmacokinetics, or manufacturability/delivery (Thayer 2013). For example, a long-acting variant of an existing molecule generated through altered FcRn binding would potentially be considered a biobetter. Along with the improved therapeutic properties offered by a biobetter, an additional benefit to the company may come in the form of prolonged patent protection on the new molecule. There is a definite financial advantage associated with introducing a biobetter since it capitalizes on the already existing market share associated with a specific validated target.

Drug development of a new molecule associated with a novel target or of a biobetter is a prolonged effort that typically spans more than a decade. Technical or clinical delays during this process could potentially put the whole program at risk, requiring significantly more investment and, most importantly, delaying the availability of the therapeutic to the patient. Therefore, many companies have adopted the strategy of approaching first-in-human (FIH) Phase I trials faster because this provides the investigator the opportunity to decide early on whether the drug has the potential to move forward based on its safety/efficacy profile. An efficient early CMC (Chemistry, Manufacturing and Controls) development program (referred to as process development in the remaining text) is highly desirable to achieving this goal, and requires that the lead therapeutic candidate behave well during all of the typical steps of development. These steps include: (1) upstream cell culture processing, (2) purification or downstream processing, (3) drug product processing and filling, (4) storage stability, and (5) ability to deliver via the intended route (intravenous, subcutaneous, ocular, etc.). Given the general similarities in overall conformation and physicochemical properties of mAb-based therapeutics, strategies have been built around developing efficient platform processes for manufacturing and formulation (Shukla et al. 2007; Rathore et al. 2013). These strategies are beneficial for those mAb-based therapeutics that are generally well behaved from a physical and chemical stability point of view. However, if a lead candidate does not behave as expected (i.e., its physical–chemical characteristics do not meet the expectations of the platform processes, it is not adequately stable to provide intended shelf-life in its dosage form, or cannot be manufactured/delivered at high concentrations due to viscosity or precipitation issues), significant delays with additional resource investment are almost certainly likely and therefore to hamper achieving faster FIH trials. The main concern is that with traditional development approaches to biologics (Fig. 1) the information regarding whether the lead candidate is well behaved is often not available until the molecule actually enters the early process development phase.

"Molecular Assessment" (MA) or "Developability Assessment" programs have been designed and put in place by several companies to de-risk such technical development hurdles, alleviate CMC-related development concerns and ensure predictable behavior of mAb-based therapeutics during both early and long-term process development (Ramachanderand and Rathore 2013; Yang et al. 2013). Figure 1 compares the traditional versus "MA-Enabled" approach of early development of biologics. MA studies are carried out prior to deciding whether to bring a molecule into full development and serve the following purposes:

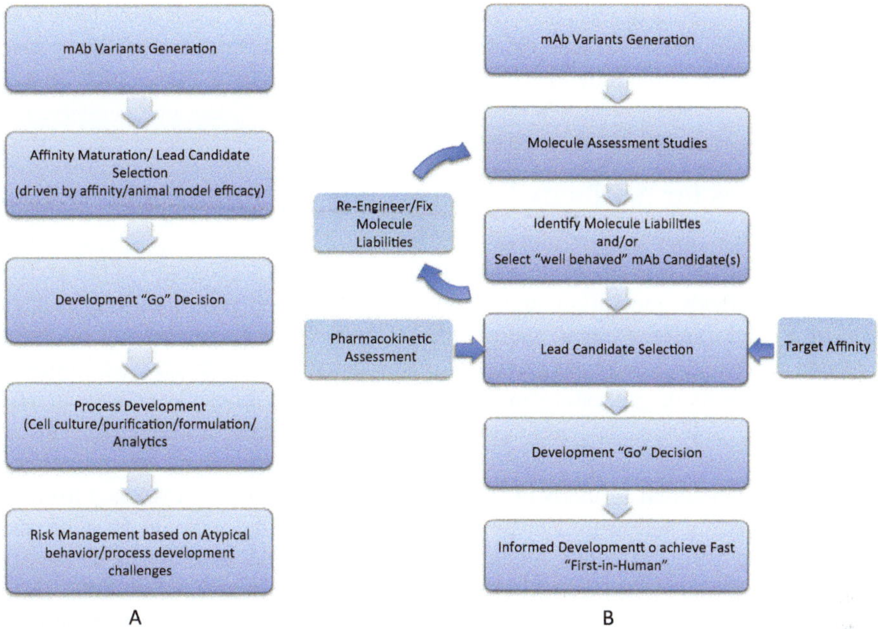

Fig. 1 A schematic of typical steps involved in the early process development phase of biologics related to mAb-based therapeutics from generation of mAb variants through Phase I development. (**a**) Conventional approach. (**b**) Molecule Assessment-enabled approach

1. Enable selection of a lead candidate (out of a number of available molecules) that demonstrates fit-to-platform manufacturing processes and has an optimal physico-chemical profile with respect to stability, manufacturability, and delivery.
2. Generate information about potential technical liabilities associated with a given molecule, in case the lead candidate retains undesirable properties, such that adequate resources/timelines could be allocated once the molecule enters early development
3. For biobetters, ensure that the selected lead is indeed superior to the existing clinical/marketed molecule, especially when pharmaceutical/manufacturability attributes are being improved to attain a certain clinical goal (for example, improved stability to achieve sustained delivery)

Ultimately, information generated through MA studies is used in conjunction with other relevant information, including affinity and pharmacokinetic profile, to make decisions regarding the lead selection. Affinity is typically the top-level criterion in molecule selection, i.e., given the biology and the target antigen, decision making on lead candidate selection in almost all cases will be driven primarily by affinity for the target.

In this chapter we discuss the MA studies in detail beginning with the current technologies available for antibody generation and how these relate to the overall

MA strategy. We then discuss how MA studies address specific concerns related to technical development and lead candidate selection. Specific case studies are also presented to further illustrate the utility of MA studies.

Technologies for Antibody Discovery and Optimization

One of the primary reasons contributing to the value of MA programs is the ability of current antibody generation/engineering technologies to yield multiple variants against a single antigen/target, such that MA studies can be carried out on these multiple candidates. The broad pursuit of mAbs as human therapeutic agents began with the capability of generating antibodies of human or human-like sequence, thus reducing the potential for immunogenic, anti-therapeutic immune responses. Initially, this involved using hydridoma technology (Kohlerand and Milstein 1975) to isolate a monoclonal antibody of desired specificity, affinity, and potency from an immunized rodent followed by "humanization" of the monoclonal (Riechmann et al. 1988; Queen et al. 1989; Carter et al. 1992). Humanization has been hugely successful with ~16 FDA-approved and US-marketed products originating from this process. However, humanization can be labor and time consuming and the final candidate may have many residues of rodent origin, in both complementarity-determining regions (CDR) and framework regions, which can carry risk of immunogenicity in humans. Several technologies have been developed that yield human antibodies and ten such antibodies have received approval for US marketing.

Technologies that have been developed to generate more human-like antibodies with increased speed and efficiency may be classified as employing in vivo or in vitro selection to generate lead candidates. Transgenic animals expressing human immunoglobulins have been developed such that human antibodies can be elicited through immunization and recovered using standard hybridoma methodology (Lonberg 2005; Jakobovits et al. 2007; Murphy et al. 2014). Hybridoma technology is not routinely applied to human B-cells; nonetheless, therapeutic antibodies can sometimes be recovered directly from humans through a B-cell-cloning process (Wrammert et al. 2008; Nakamura et al. 2013). In this procedure, B-cells expressing an antibody of desired specificity are identified from human peripheral blood cells obtained via blood draw from infected or vaccinated donors and the antibody genes recovered using recombinant DNA technology (Nakamura et al. 2013).

In vitro methods for antibody discovery include a variety of display technologies. These may be based on prokaryotic expression, phage (Barbas et al. 1991) display, or eukaryotic expression systems, yeast (Feldhaus et al. 2003) or mammalian (Bowers et al. 2011) cell display. In phage, yeast, and mammalian display, antibodies specific to a target are selected from pools of variants displayed on bacteriophage or cells through binding to antigen in vitro. Specificity can be enhanced via counter-selection either against closely related homologues, general non-specific binding, or both. Since the genetic information for the selected antibody is contained in the particle on which it is displayed, specific antibodies can be identified by sequencing the DNA from cells or phage enriched through the selection process. The variant pool, or "library", can be constructed from a natural, non-immune source (de Haard et al. 1999), or be composed of semi-synthetic (Soderlind et al. 2000; Hoet et al. 2005) or fully synthetic antibody diversity (Lee et al. 2004; Zhai et al. 2011; Rothe et al. 2008). Non-immune libraries can be prepared by cloning antibody heavy and light chain variable domains from mRNA isolated from human peripheral blood cells and typically yield antibodies of modest affinity and specificity. Semi-synthetic libraries often combine donor-derived CDR3 sequences with synthetic variable domains. In this approach, CDR3 sequences from pooled human donors are cloned into acceptor synthetic human framework regions to generate a complete variable domain gene. The synthetic acceptor domains may have diversity in CDRs1,2 designed on natural sequences and built using synthesized DNA. Fully synthetic libraries generate diverse CDRs, including the length and sequence diversity of CDR3, with synthesized DNA and the designs are based on sequences observed in natural human antibodies. A single human framework or multiple human frameworks may be employed in the library. Both the semi- and fully-synthetic libraries usually yield more sequence-diverse, antigen-specific, and higher affinity antibodies than the non-immune libraries.

Merits of Antibody Discovery Technologies Relative to Molecular Assessment

The in vivo and in vitro approaches present contrasting virtues for generation of lead antibody candidates (Table 1). For immunization of normal or human Ig transgenic rodents, the power of the immune system to generate high affinity, specific candidates is harnessed. The process of B cell clonal expansion eliminates many non-specific clones and, coupled with somatic mutation, can produce high affinity antibodies in a somewhat stochastic process. Although highly specific antibodies can be obtained from in vitro methods by employing counter-selection, it is difficult to know a priori what should be used for counter-selection to avoid off-target binding in vivo. Increased off-target binding can lead to atypical, fast pharmacokinetics and adverse events (Hotzel et al. 2012).

Antibodies isolated from human Ig transgenic animals may have properties suitable for clinical development without further engineering. Monoclonals obtained from immunization of wild-type rodents will obviously require humanization, with expected time and labor costs. However, antibodies from both types of animals may not meet potency requirements for a particular indication, or have sequence liabilities, that require further in vitro engineering. For example, a higher affinity antibody may be needed to neutralize a high copy target on a virus as well as enable less frequent, subcutaneous dosing (Pantua et al. 2013). Affinities that can be obtained in vivo may be limited by the selection process, and codon usage, to *KD* values of $\geq$0.1 nM (Li et al. 2014a). In vitro display methods do not have these affinity limitations and can be used to increase the affinity of antibodies derived from natural or

Table 1 Relative merits of in vivo and in vitro selection

Property	In vivo selection	In vitro selection
Technology	Hybridoma: wt or human Ig transgenics; B cell cloning	Phage, yeast, mammalian display
Specificity	Highly specific through clonal selection	Limited off-target counter selection
Affinity	High affinity through somatic mutation with some constraints ($KD \geq 0.1$ nM)	High affinity ($KD \leq 0.1$ nM) possible via secondary library design/selection protocol
Engineering	May require further CDR amino acid changes (assessed in vitro) to increase affinity and/or fix instabilities	May require extensive CDR engineering to achieve affinity/specificity requirements and/or fix instabilities
Cross-reactivity	Cross-reactivity with rodent antigen may be challenging	Rodent cross-reactive often possible
CDR glycosylation sites	Potential sites introduced during somatic mutation	Potential sites can be minimized with library design
Expression	Clonal selection generates antibodies expressable in animals but may not translate to recombinant systems	Frameworks can be pre-selected for high expression
Chemical stability	No elimination of potential unstable sites during selection	Can reduce potential sites with library design
Physical stability	Semi-random combination of frameworks may yield unstable antibodies	Starting frameworks can be designed for thermal stability

synthetic sources. Secondary libraries can be designed to optimize clones obtained from initial screening; however, since the diversity that can be represented in a library is in the range of 10^{11} unique sequences, and potential CDR diversity is 60^{20}, affinity improvement can be labor intensive, especially if there is little information on the structure of the antibody-antigen complex. In some cases, especially for agonist antibodies that stimulate a signaling pathway (Hünig 2012), it may be desirable to limit the affinity improvement so as to avoid any safety consequences for the engineered antibody. In addition, care needs to be taken during the affinity maturation process to avoid the introduction of unintended cross-reactivity, with potentially severe consequences (Linette et al. 2013), or off-target binding, which can result in unexpected fast clearance (Hotzel et al. 2012).

Antibodies obtained from both in vivo and in vitro selections have the potential for N-linked glycosylation sites (Asn-X(not Pro)-Ser/Thr) in the CDR regions. These can be problematic, since, depending on the expression host and culture conditions, there can be variability in glycosylation status resulting in batch-to-batch inconsistency. Occurrence of glycosylation sites in antibodies from in vivo selection is difficult to control, especially since for some germline genes, a single codon change will create a potential glycosylation site (Li et al. 2014a). Such sites may need to be removed through CDR engineering to enable a consistent drug substance.

The potential for glycosylation sites can be reduced for in vitro selections through the design used for synthetic libraries. For example, the potential for N-linked glycosylation sites can be reduced by designing the oligonucleotides used to create the CDR sequences in the library such as to avoid N-X-S/T glycosylation motifs. Furthermore, the potential for known chemical instabilities, such as deamidation and isomerization hot spots, to be present in CDR regions can be evaluated and reduced through judicious library design (Tiller et al. 2013). An immune response in rodents may elicit antibodies with thermally unstable VH/VL pairings that can result in challenges for development. As shown by Tiller et al. (2013), this problem can be avoided in synthetic libraries by limiting the repertoire to combinations known to have stable properties, at least at the germline level. In any case, generation of multiple mAb candidates using the above-mentioned technologies is likely to carry liabilities related to process development. Therefore, MA studies are crucial to further assess mAb variants and de-risk long-term process development.

De-risking Process Development Through MA

As mentioned earlier, an efficient and streamlined early development program is desirable to achieve fast FIH trials. However, challenges are often encountered during these early stages where little is known about the physicochemical behavior of the lead candidate. The question becomes what process development challenges can be de-risked or alleviated through the MA exercise? In our experience MA studies would typically address the following process development concerns:

- Expression during cell line development. If the yield is not above a certain g/L number, the cell culture process may not be economically acceptable.
- Compatibility and fit of the lead candidate into platform purification steps. Additional development studies and alteration to platform steps would necessitate additional resources and prolonged timelines.
- Identification of chemical degradation hotspots in the CDRs antigen binding sites and assessment of overall physicochemical stability (for example, aggregation). Degradation could lead to unacceptable potency loss or present unwanted risk in terms of safety, thereby limiting the shelf-life of the lead candidate as a liquid dosage form.
- Amenability to high concentration formulations with respect to manufacturability and delivery into subcutaneous space through a device. The primary concern here is unacceptable viscosity, solubility and aggregation issues at high protein concentrations
- Compatibility with the subcutaneous environment. The lead candidate should remain soluble upon injection into the subcutaneous space

In this chapter we will primarily focus on how MA studies help de-risk some of the process development challenges as outlined by the bottom three bullets in the preceding paragraph. Additional MA studies may be performed regarding agitation, freeze-thaw, and filtration to support transportation and drug product manufacturing (Ramachanderand and Rathore 2013). In the next section, we delve into specific drug product related issues that are critical concerns during process development.

Site-Specific Chemical Degradation

Chemical modification of amino acid side chains, especially those involved in target/antigen binding, could lead to potency loss and a shortened shelf-life. These chemical modifications may also lead to additional degradation pathways such as aggregation (Philoand and Arakawa 2009), affect immunogenic properties (Hermeling et al. 2004), or potentially alter the pharmacokinetic profile (Iwao et al. 2006) of the lead candidate. Common modifications include the reactivity of side chains with solutes (glycation at lysine side chains, enzymatic glycosylation of certain asparagine side chains, i.e., Asn-X-Ser or Asn-X-Thr, where X is not Pro) or degradation of the side chains (Asn deamidation, Asp isomerization, Met or Trp oxidation, and/or hydrolytic clipping of Asp-Pro or Asn-Pro amino acid pairs) (Manning et al. 2010; Wang et al. 2007). In addition, the presence of free Cys (free thiol) in either the CDR or the variable domain is undesirable as it may lead to intermolecular dimerization via either oxidative formation of disulfides or intermolecular disulfide exchange (Yu et al. 2007). Therefore, as part of the initial sequence assessment during MA, amino acids Asn (D), Asp (N), Met (M), Trp (W), Lys (K), or Cys (C) should be carefully scrutinized.

Physical Instability

As part of the physical stability evaluation, the main concern for mAbs is the formation of either submicron or nanometer aggregates (SEC-based high molecular weight species or oligomers) or visible particulates (greater than 100 μm) (Carpenter et al. 2009; Narhi et al. 2012). The concerns with aggregates regard induction of immune responses, infusion type reactions pertaining to fixation of C′ and decreased or enhanced potency (depending on mechanism of action) (Hermeling et al. 2004; Kumar et al. 2011; Rosenberg 2006). Therefore, effort is made during both process development and stability to ensure that the level of soluble aggregates remains at a safe, preferably minimal, level and that no visible particulates are formed. The concern with soluble aggregates is more elevated for higher molecular weight aggregates than for dimer species (Rosenberg 2006). Therefore, while a few percent of dimer presence is tolerable, even a fraction of percent of very high molecular weight species (formed of multimers) may not be acceptable.

On the other hand, there is little tolerance for the presence of visible, large, particulates as the particulates could potentially lead to capillary occlusion, pulmonary dysfunction, and immunogenicity (Doessegger et al. 2012), as well as needle clogging during administration. Furthermore, such particles are alarming to end-users

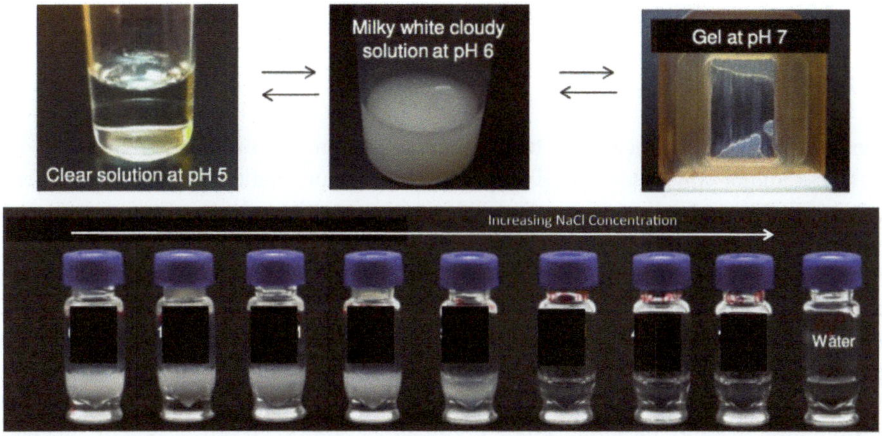

Fig. 2 An example of a Fab molecule with solubility issues observed during early development. The molecule solubility is pH dependent where it transitions from clear solution (pH 5) to turbid solution (pH 6) to a gel (pH 7) (*top panel*). The molecule's solubility also showed salt-dependence exhibiting higher solubility in solutions with high NaCl concentrations (*bottom panel*)

(health care professionals/patients) who are advised to inspect the product for visible particles prior to preparation/administration and not use the product if visible particles are observed. Often times, the presence of visible particles in a drug product is indicative of the fact that "something wrong has happened with product quality". Gross precipitation may actually indicate solubility issues with the lead candidate and therefore directly affect the amount of soluble protein present. Figure 2 shows an example of a Fab, whose solubility was dependent on solution conditions. The Fab molecule was found to be more soluble at lower pH and under high ionic strength conditions. Unfortunately, this Fab also contained an isomerization hot spot and, therefore, could not be developed as a liquid formulation. This molecule illustrates a "perfect storm" example where solubility (and therefore precipitation) can be mitigated by altering pH, salt-type, and ionic strength, but other issues, such as chemical stability, may become more prominent at the newly selected formulation conditions. This molecule was not selected using MA studies and, therefore, its subsequent development required significant effort, time and resources. Balancing product quality for this Fab involved trade-offs on different aspects of molecular stability so that a desirable drug product could be developed. Clearly, it makes sense if such issues are identified prior to lead candidate selection and proper steps taken to select the right lead candidate. Concomitantly, there is an opportunity to develop a biobetter that could improve Fab stability and solubility, while the first generation molecule moves forward to establish a clinical proof of concept and gets introduced into the market due to patients' needs. In such instances, MA can be used to ensure that the "biobetter" is indeed devoid of such stability or solubility issues, with benefits potentially extending to a better stability profile in the in vivo environment.

High Concentration Formulation Considerations

Viscosity

One of the primary concerns of high concentration mAb solutions is increased viscosity at higher concentrations. This presents two main challenges: manufacturability and fitness for delivery in syringes (delivery into the subcutaneous space via a syringe-needle device) (Schmit et al. 2014; Shire et al. 2004; Yadav et al. 2010). An ultrafiltration/diafiltration (UF/DF) step using tangential flow filtration across a membrane is required to concentrate a mAb solution to a desired concentration. High viscosity solutions tend to develop high pressure across the membrane which prevents the ability to reach the target concentration (Narasimhan et al. 2012). Therefore, if the solution viscosity is above a certain limit, high concentration mAb formulation may not be achievable without either additional processing or substantial losses in yield. Similarly, a high viscosity solution may be difficult to expel through the narrow bore needles typically used in pre-filled syringes or auto-injectors as defined by the Poiseuille's law (Adler 2012). mAbs of a similar isotype, for example IgG1, that differ mostly in the CDRs could exhibit a range of viscosity-concentration profiles (Fig. 3). Therefore if the lead candidate mAb entering into early development exhibits high viscosity in its optimal formulation, it is likely to present significant challenges with respect to manufacturing and delivery of the therapeutic into the intended site.

Phase Separation

Phase separation is another manifestation of physical instability often associated with high concentrations of mAbs, wherein two phases of liquid (one low in protein and the second enriched in protein) may exist in equilibrium (Mason et al. 2010; Nishi et al. 2010). Phase separation may affect product quality by causing protein

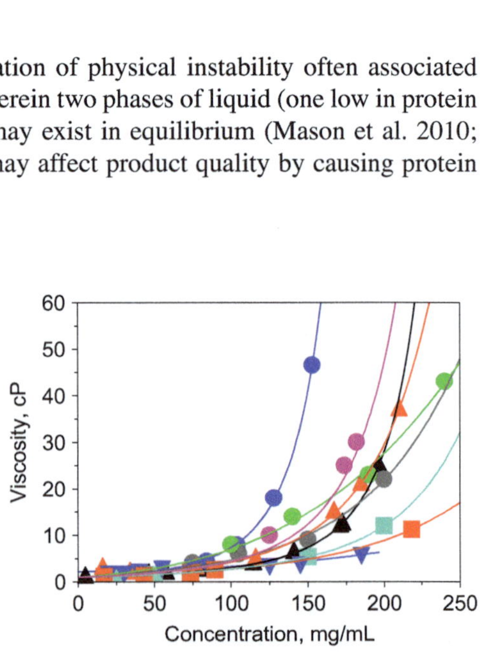

Fig. 3 Viscosity-concentration profiles of a number of IgG1 mAbs that differ with sequence differences in the Fv domain with most differences in the CDR regions. All viscosity measurements were carried out at an identical shear rate of 1,000 s^{-1} under ambient temperature conditions

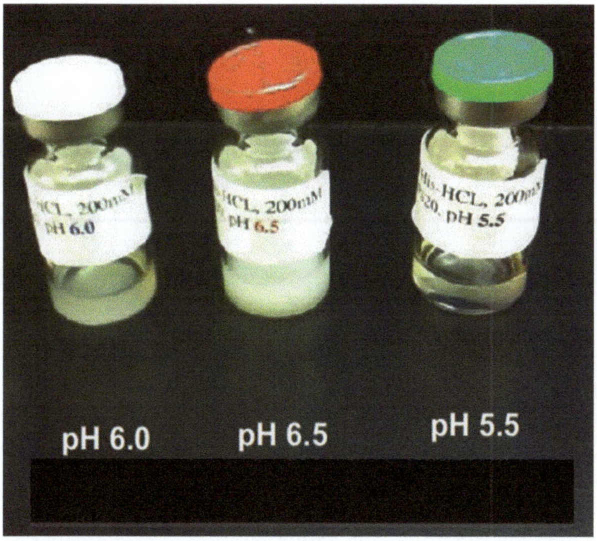

Fig. 4 pH-dependent phase separation of a mAb at high concentration under high salt conditions. The mAb was soluble at pH 5.5 at 150 mg/mL but formed an opaque gel at pH 6.5 following prolonged storage at 5 °C

concentration inhomogeneity within the final drug product container and potentially leading to dose inhomogeneity. Additionally, the end-user may be advised to not use a product if a "gel-like" phase is present. It is likely that prolonged storage of the high concentration phase may lead to gelation, irreversible phase separation or aggregation. Figure 4 illustrates an example of a mAb that exhibits pH-dependent gelation with formation of a gel upon storage at 5 °C. Additional development work will be required for such molecules to ensure that the final selected formulation does not induce phase separation upon prolonged storage of the drug product. Both phase separation and/or gelation should be assessed during MA studies.

Compatibility with the Subcutaneous (s.c.) Space

High concentration formulations are primarily developed for subcutaneous injections in order to accommodate high doses in a small injection volume. Often, mAb-based drugs are formulated within an acidic pH range (pH 5–6), and only exposed to more neutral pH and physiological salt conditions upon injection into the s.c. space. The antibody should stay soluble under physiological conditions. Any mAb insolubility in the form of large particles may cause local injection site reactions, inflammation, or trigger an immune response. Therefore, antibodies with solubility issues may not be the ideal candidates for s.c. delivery. Figure 5 illustrates an example of a mAb that readily precipitates at concentrations as low as 75 mg/mL in buffers that represent physiological conditions. It is no surprise that this mAb was not

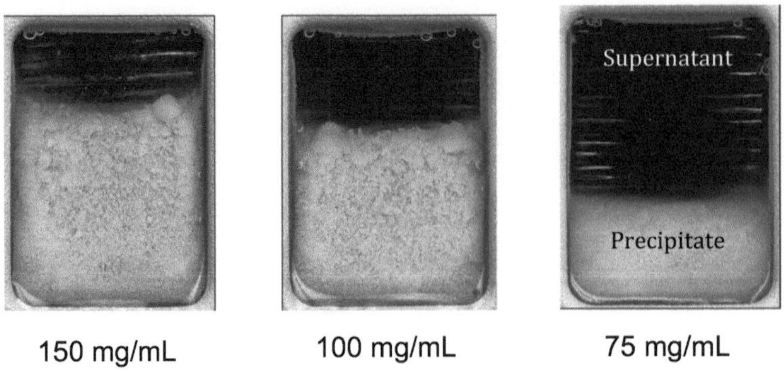

Fig. 5 Concentration-dependent precipitation of a mAb in phosphate buffered saline (PBS) at pH 7.2. The formulated mAb solution was dialysed against PBS at 37 °C for 24 h

a suitable candidate for subcutaneous administration at high concentrations. In fact, injection of this high concentration mAb in non-primate models led to local inflammation at the site of injection indicative of local intolerance. Once again, this presents an opportunity to develop a biobetter that is devoid of the issues related to subcutaneous injection. MA can demonstrate if a biobetter molecule is devoid of such delivery-related concerns.

Experimental Tools for MA and Data Interpretation

The experiments for MA need to be carefully devised to accommodate the typical time and material quantity constraints associated with the timing of early clinical studies (prior to the candidate being selected for IND-enabling toxicological studies). The quality of the material used in MA testing is also critical as impurities or product quality heterogeneity (size variants and charge variants) may affect the outcome of the studies. While material produced using a stable cell line is desirable, a stable cell line may not be developed or available at the time when MA studies are being conducted. Therefore, material from transient cell lines may be used for MA. In general, prior to selecting the cell line, an evaluation should be performed on whether data generated from material using transient cell lines is reliable enough to understand the physicochemical behavior of mAb candidates, enable selection of the lead candidate, or generate information about liabilities associated with final candidate. Various studies that are typically conducted as part of MA, along with typical material requirements are primarily driven by the attribute that is being probed (Table 2). Some of the details of MA studies are discussed in the next section.

Table 2 Typical studies conducted as part of MA along with molecular attributes that are tested, material requirements and typical analytical tools that are used for such studies

MA study	Molecule attribute tested	Material requirement	Analytical tools
Accelerated temperature study (low concentration)	Chemical hotspots	3–5 mg	Size exclusion chromatography (SEC)
	Asp isomerization		Ion exchange chromatography (IEC)/ Isoelectric focusing (IEF)
	Asn deamidation		Peptide Map/LC-MS
	Fragmentation		
	Aggregation		
	Oxidation		
Induced oxidation	Met oxidation	3 mg	Peptide map/LC-MS
	Trp oxidation		
Viscosity-concentration profile	Viscosity	25–75 mg	Rheometer
Accelerated temperature study (high concentration)	Aggregation	25 mg	Visual Assessment
	Phase separation		SEC
	Solubility		IEC/IEF
	Gelation		
In vivo compatibility	Solubility	25 mg	Turbidity
			Visual Assessment
Thermal unfolding assessment	Onset of unfolding	1 mg	Differential scanning calorimetry
			Fluorescence Scanning
			Dynamic Light Scattering

Assessment of Chemical Hotspots

Accelerated Temperature Stability Studies

Similar to the stability studies conducted during typical formulation screening, a stressed temperature study is carried out during MA to assess chemical hot spot liabilities. For most mAbs and mAb-based drugs, 40 °C serves as an appropriate stress temperature, which is close to the normal in vivo temperature of 37 °C. The selection of the stress temperature is often driven by the onset of thermal unfolding of the molecule(s) in consideration (as determined by differential scanning calorimetry or a similar thermal scanning technique). Conducting stability studies at temperatures close to the onset of thermal unfolding can lead to unnecessary degradation that is not representative of the degradation experienced during shelf-life storage conditions. Therefore, if molecular properties (aglycosylation, fusion proteins), pH and/or presence of excipients lower the onset of unfolding, a milder temperature such as 30 °C, may be considered for conducting stressed temperature stability studies. It should be noted that the onset of unfolding is typically pH- and ionic strength-dependent; therefore, a low unfolding temperature in formulation conditions is not necessarily predictive of in vivo instability. For mAbs, the higher in vivo pH (compared to typical formulation conditions) usually increases the unfolding temperature so that a lower onset of unfolding temperature observed under acidic formulation conditions may not be a concern under physiologically relevant solution conditions.

The attributes typically assessed through accelerated temperature studies are Asp isomerization, Asn deamidation, formation of aggregates and fragments, and overall changes in the charge variants. Typically, such studies are carried out using one or more formulation conditions, including the intended platform formulation, to assess the chemical and physical liabilities that are relevant to drug product formulation. Asn deamidation sites are identified using neutral pH or basic pH conditions, since deamidation may not occur at target (acidic) formulation conditions. Neutral pH assessment also allows the understanding of potential degradation pathways that could be relevant under physiological conditions.

The samples generated through these studies are analyzed using (1) peptide mapping to identify site-specific Asp or Asn degradation, (2) size-exclusion chromatography (SEC) to monitor changes in size variants and (3) ion-exchange chromatography (IEC) or image capillary iso-electric focusing (icIEF) to monitor changes in charge-variants (Chirinoand and Mire-Sluis 2004). These studies can be carried out at low mAb concentration since the primary focus is to identify chemical hot spots with the assumption that mAb concentration will not affect the degradation pathways. This concentration-independence is likely true for chemical and charge-variant changes. However, this does not hold for size-variants, where aggregation is often concentration dependent. Nevertheless, this early assessment at low concentration will help highlight if any unusual behavior is observed in size variants upon stressed stability. The use of peptide mapping at this stage also helps to identify any post-translational modifications in the CDRs of the starting sample, such as glycation or glycosylation. Site-specific analysis by peptide map at this stage also allows the identification and possibly removal by mutation of the problematic residues.

The interpretation of stability data from these studies depends on the amount of acceptable mAb degradation from a product quality point of view. This knowledge about degradation limits could come from historical experience with previous molecules or from Arrhenius-based predictions given the degradation mechanism involved. For example, for Asp isomerization an activation energy of 21 kcal/mol/K (Wakankar et al. 2007a) and a site-specific degradation at 40 °C over 1 week of 10 % would translate into a degradation over 2 years at 5 °C of 20 %. This amount of degradation may not be acceptable, especially if such degradation is associated with equivalent potency loss or induction of immune response. The above example uses 21 kcal/mol/K as the activation energy, however, other activation energies, or a range may be considered given prior experience with other molecules. A similar approach can also be adopted for Asn deamidation.

For size-variant and charge-variant changes upon stressed stability, the data generated as part of MA studies could be evaluated against historical data with other molecules. A risk-based assessment could be performed to decide upon what extent of degradation under stressed conditions may be acceptable to yield a desirable shelf-life.

Oxidation Hotspots

Accelerated temperature stability studies may not identify oxidation hotspots involving Met and Trp because several other factors besides thermal stress could lead to oxidation. These factors include presence of metal, exposure to light and levels of reactive oxygen species (Torosantucci et al. 2014). Therefore, additional testing may be required to identify labile Trp or Met sites to capture these additional oxidative mechanisms. One of the ways to address this is use of a known oxidizer, such as *tert*-butyl hydrogen peroxide (TBHP) or 2,2′-Azobis(2-amidinopropane) dihydrochloride (AAPH) (Wakankar et al. 2007a; Hensel et al. 2011; Ji et al. 2009), to mimic presence of peroxide as well as organic free radicals in solution. In addition light sensitivity and metal sensitivity can be tested using appropriate light stress and addition of metal oxidizing system, such as Fe^{2+}/Fe^{3+}, respectively (Mallaney et al. 2014). Appropriate controls, positive (known oxidation sensitive sites) and negative (known oxidation-resistant sites) can be used to gauge the risk factor around new molecules. While such an approach may not be predictive, it could be used to identify potential hotspots that present oxidation risk during the product development cycle from manufacturing to storage.

Viscosity

For high concentration mAb solutions, viscosity is an important attribute for both manufacturability and delivery. Highly viscous mAb solutions may present significant challenges and in some cases limit achieving the target concentration intended for delivery. One of the primary challenges of viscosity measurements during pre-development stages is the requirement for significant quantities of material which may not be readily available. High concentration mAbs are often formulated at ≥100 mg/mL and therefore, the viscosity needs to be assessed at such concentrations. Often, a multi-point viscosity concentration curve is useful for understanding the concentration-dependence due to a non-linear dependence of viscosity on concentration.

The commonly available tool to measure viscosity of mAb solutions is a cone-and-plate based rheometer (Shire et al. 2004; Jezek et al. 2011). Viscosity can be measured using volumes as low as 60 uL, thereby substantially reducing the material requirements. While many mAb solutions exhibit a non-Newtonian rheology, i.e., these solutions exhibit a shear-dependent viscosity profile, measurement at a single shear rate is sufficient to rank order various mAbs in terms of their relative viscosities. Viscosity-concentration curves are generated by concentrating the mAb solution to a high concentration, diluting it to generate 2 or 3 different concentrations and measuring the viscosity at a constant shear rate and temperature. Such experiments requiring multiple concentration measurements for a single mAb variant may use >50 mg total protein. However, if material is limited viscosity measurement at a single concentration can be performed for rank ordering multiple mAb candidates; a single concentration measurement can be performed by using 15 mg or less of the protein.

Other methods available to measure viscosity of mAb solutions include capillary-based rheometers (Hudson et al. 2014). There are also technologies available to measure viscosities of smaller volumes of liquid such as quartz crystal based rheometry (Saluja et al. 2007), micro-fluidics based rheometry (Jezek et al. 2011), particle tracking velocimetry based rheology (Jezek et al. 2011), etc. However, these methods need further evaluation to be adapted for broader use

Solubility Studies

Solubility studies are required to ensure that the mAb candidate remains soluble in formulation buffer at the target concentration upon manufacturing and upon storage. For mAbs that are delivered through the subcutaneous route, it is also necessary to ascertain that the lead candidate is soluble in physiologically relevant conditions. This is because mAb solutions for subcutaneous delivery are high concentration solutions and may have substantial localized residence time (several hours) before complete disposition into the systemic circulation (Porterand and Charman 2000). This concern may be somewhat lowered for intravenous delivery because of typical low protein concentrations used and dilution into the blood stream. However, there may be concerns about aggregation of mAbs in the infusion bags commonly used to infuse such drugs through i.v. route. It has been reported that mAb aggregates could form in an infusion bag depending on the solution type in the infusion bag (for example, dextrose versus NaCl) and/or on the protein formulation composition (Arvinte et al. 2013; Kumru et al. 2012). Therefore, candidates that form aggregates, soluble, subvisible or visible precipitates, in physiologically relevant conditions or infusion solutions may not be the best candidates to move forward with.

Solubility studies in formulation conditions can be done as part of the stability study set up at the intended target concentration. A simple visual assessment of the formulated bulk is often sufficient to inform of either absence or presence of large particulate material. Attention should be paid while formulating and/or concentrating the mAb solution to look for any unusual visual phenomena and ensure that the resulting solution is clear and free of particulate matter. To assess solubility of a mAb under physiological conditions following a s.c. injection, the mAb can be first concentrated to high concentration (since this is mostly relevant for high concentration s.c. injection) and then dialyzed against a physiologically relevant buffer such as phosphate buffered saline at pH 7.2. The dialyzed solution is then assessed for any visible precipitation/formation of particulates. Compatibility of mAbs with infusion bag solutions can be assessed by mixing and diluting mAbs (in their respective formulations) in the infusion bag followed by storage at the appropriate temperatures for a reasonable period of time. Further analysis for soluble, subvisible and/or visible aggregates as well as chemical analysis could be carried out to assess the risk around the compatibility of mAbs with infusion solutions.

Critically, the in vivo solubility should be assessed in the presence of human plasma or serum (or other relevant body fluids), as discussed in Chap. 7 of this volume.

Miscellaneous Studies

Often, additional studies may be conducted to screen candidates early on, especially when multiple variants are available and are still being optimized with respect to biological affinity. These typically include thermal scanning techniques such as differential scanning calorimetry, differential thermal fluorescence scanning and/or thermal scanning using dynamic light scattering. The basic idea here is to remove candidates based on simple response in thermal scanning and to move forward with only those candidates, which are more stable to thermal scanning stress. Thus molecules that show propensity to aggregate or unfold at lower temperatures are not considered for further development. Note that this approach may only be useful when a restricted number of molecules are available against a single target such that later MA studies need not be performed on a large set of molecules.

MA Case Studies

In Silico Engineering of Chemical Hotspots

Table 3 shows the heavy chain CDRs and light chain CDRs of a mAb. Simple examination of the sequence identifies several potential hotspots (shown as underlined). These include isomerization sites (DS and DG), deamidation sites (NS), clipping sites (DP), a glycosylation site (NLS) and oxidation sites (W and M). In this case, certain sites were engineered out prior to conducting any experimental stability studies. The binding affinity of the re-engineered molecule with these changes incorporated was not affected and the potency (as determined by binding affinity) of the molecule was maintained (data not shown). Note that certain other residues, while identified as potential hotspots, were not altered (HC CDR1 W, HC CDR3 M, LC CDR2 NS and LC CDR NP). This was based on the sequence comparison of this particular mAb with other mAbs, which revealed the presence of these residues being conserved and information that these amino acid residues are known not to undergo significant degradation. Additional experimental testing did show that these unaltered residues were stable against thermal stress and, therefore, no further changes were made to the CDR sequence.

Table 3 An example of in silico MA assessment of chemical hot spots in a mAb and engineering approaches to alleviate associated stability risks

CDR sequences	Stability liability	Re-engineered sites
Heavy Chain CDRs:		
DSIXXXXYWN	Isomerization, Oxidation	**ES**IXXXXYWN
YIXXXXXTYY**NLS**LRS	N-Linked Glycosylation	NLS site mutated to remove glycosylation hotspot
ITXXXXAMDY	Oxidation	ITXXXXAMDY
Light Chain CDRs:		
RXXESV**DG**XX**NS**FLH	Isomerization, Deamidation	RXXESV**SG**XX**LS**FLH
LAXXXNS	Deamidation	LAXXXNS
QXXXVDPWT	Hydrolytic Cleavage, Oxidation	QXXXVDPWT

Identified hotspots are underlined and sites that were engineered out to remove stability liabilities are shown in bold and underlined

Examples of Chemical Hotspots

In this section, two separate examples are provided to illustrate the utility of early MA studies for identification of Asp isomerization sites and steps that can be taken to either fix the labile site or pick a more stable candidate.

The first example compares two different mAbs for different antigen targets (same clinical indication) and therefore, both mAbs were considered lead candidates for early development. Table 4 shows the heavy chain CDR 2 sequences of these two mAbs, mAb 1 and mAb 2. As shown by the underlined and bolded residues, MAb 1 contains multiple isomerization sites, DD, DG, and DS, whereas mAb 2 contains a different isomerization motif, DD. Both mAbs were prepared in buffered solution and were subsequently stressed at 40 °C for 2 weeks. The extent of isomerization was measured by monitoring formation of isoAsp using peptide map followed by Extracted-Ion-Chromatogram (EIC)/LC-MS. Only 1 % formation of iso-Asp at the DD/DG site was observed in mAb 1 over 2 weeks at 40 °C, whereas, a 25 % formation of iso-Asp was detected for the first Asp at the DD motif in mAb 2 at similar stressed conditions. It is interesting to note that the extent of isomerization for the DG site is significantly reduced in mAb 1, compared to the first Asp in DD (mAb 1). Based on simple peptide models, DD would have been predicted to be less reactive based on the more bulky D residue at the n + 1 location (Wakankar et al. 2007b). Clearly, the tertiary fold of the antibody and local environment has impact on the Asp isomerization rate in a mAb. A comparison of the sequence between mAb 1 and mAb 2 revealed a DS motif in mAb 1 at the same location where DD was present in mAb 2. From previous studies it was known that this DS motif is stable against isomerization under the given stress conditions. Based on this information, the DDF motif in mAb 2 was switched to a DSV motif and this completely

Table 4 A comparison of isomerization rates (formation of isoAsp) among different mAbs as part of MA to identify chemical hotspots

mAbs	CDR	Isomerization rate at 40 °C (site)
mAb 1	Heavy Chain CDR 2 XXXXP**<u>DDG</u>**DTDYA**<u>DS</u>**VXX	0.5 %/week (DDG)
		Not detected (DS)
mAb2	Heavy Chain CDR 2 XXXXTETGEPTYA**<u>DD</u>**FXX	12.5 %/week (DD)
mAb 3	Heavy Chain (VH1 framework) XXXXTYTGEPTYA**<u>DD</u>**FXX	Not detected (DD)
mAb 4	Heavy Chain (VH3 framework) XXXXTYTGEPTYA**<u>DD</u>**FXX	6 %/week (DD)

mAbs were stressed at 40 °C for 2 weeks and analyzed by enzymatic digestion and peptide mapping followed by LC-MS. The rates are shown for the bold and underlined isomerization hotspots. For DD hotspots, the rates correspond to the first D in the pair

alleviated the isomerization at this specific site. mAb 2 was therefore re-engineered from an isomerization-labile to an isomerization-stable molecule.

The second example compares two mAb candidates for the same target and therefore, only one of these would be selected to move into early development. This is an interesting example because the two mAb variants generated had the exact same six CDRs but had different frameworks for the constant domain. The isomerization site of interest was the first Asp of the DD motif present in the heavy chain CDR2 of both mAbs (Table 4, mAb 3 and 4). One of the mAbs used VH1 as the heavy chain framework construct, while the other used the VH3 heavy chain construct. Similar to the above-mentioned example, both mAbs were prepared in a buffered solution and were stressed at 40 °C for 2 weeks. The extent of isomerization was measured by monitoring formation of isoAsp using peptide mapping followed by Extracted-Ion-Chromatogram (EIC)/LC-MS. The results show that the framework construct can impact the isomerization rates even when the CDRs have the exact same sequence. The DD motif in the VH1 heavy chain CDR2 showed only a 1 % formation of isoAsp over 2 weeks at 40 °C compared to a 12 % formation of isoAsp for the same motif in VH3 heavy chain CDR2. Based on these data, the mAb with the VH1 framework was selected for further development. These data also point to the fact that the tertiary fold and the presence of neighboring residues in the variable domain can affect the reaction rates of residues in the CDRs.

Case Study for Improved Stability

Herein, we present an example of a molecule, Molecule 1, that presented significant stability issues such that the formulation needed to be lyophilized to attain the desired shelf-life. Regardless of shelf life, the stabilized lyophilized formulation proved sufficient for therapeutic use only requiring short half-life in vivo (days). However, Molecule 1 may not be suitable for sustained delivery approaches due to

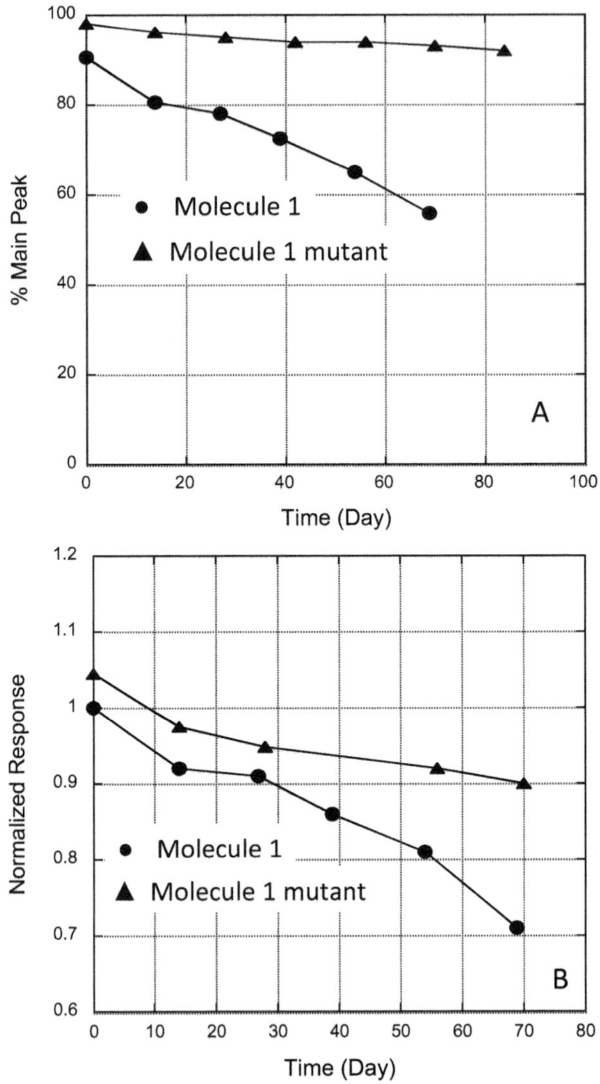

Fig. 6 Case study of a molecule being developed to improve upon the stability where a liquid formulation was not feasible with the existing molecule. Molecule 1 mutant shows reduction in percent main peak loss as measured by IEC in PBS at 37 °C (Panel *a*) corresponding to retention of affinity as measured by Biacore under similar solution conditions as compared to Molecule 1 itself (Panel *b*)

its high rate of Asn deamidation, poor solubility, and aggregation under physiologically relevant conditions Therefore, a subsequent version of Molecule 1 was developed, termed as Molecule 1 mutant. Modifications were made to the Molecule 1 mutant amino acids without much effect on the binding affinity to the target epitope, while overcoming the stability liabilities. Figure 6 shows the percent main peak loss profiles as measured by IEC as well as the antigen binding as measured by Biacore comparing Molecule 1 and Molecule 1 mutant at 37 °C in PBS. As shown, improvement in the stability profile along with a decreased loss in potency was observed in Molecule 1 mutant compared to Molecule 1 itself. In this case MA was valuable in ensuring that the amino acid changes resulted in stability improvements. From a stability point of view, therefore, Molecule 1 mutant could potentially be considered as the biobetter version of the original molecule.

mAbs and Viscosity

As mentioned previously, viscosity is a critical parameter for manufacturing and delivery of high concentration mAb solutions. mAb solutions with high viscosity produce significant challenges during the UF/DF process and may not be injectable through a narrow gauge needle without considerable force. With respect to MA studies, the additional challenge with mAbs is that high viscosity is not easily fixable unlike the chemical hotspot liabilities. That is, "fixing" viscosity problem is not a matter of simply engineering a few residues, since the origins of high viscosity are typically attributed to the non-specific interactions (electrostatic or hydrophobic) that usually extend to a larger molecular surface than encompassed by just a few residues (Yadav et al. 2010; Duand and Klibanov 2011; Kanai et al. 2008). As a consequence it may be difficult to generate multiple variants against a single target that show sufficient variance in the sequence to result in significantly different viscosity profiles without affecting other relevant biological properties such as affinity and/or pharmacokinetics. Therefore, unless antibody engineering enables generation of multiple candidates within a desired affinity range, in many cases it may not be matter of "selection" but more so of "information generation" when it comes to evaluating viscosity as part of the MA studies. This enables the team to make decisions whether platform processes will be acceptable to manufacture a certain mAb or if significant additional development work/alternate strategies will be required to manufacture/deliver a given mAb candidate.

Figure 7a, b shows examples of viscosity profiles of four mAbs, mAb 5, mAb 6, mAb 7 and mAb 8 compared to a high viscosity mAb and a low viscosity mAb in the same-buffered solution as function of concentration. mAb 5 and 6 were two candidates generated against the same target, whereas, mAb 7 and 8 were two different mAbs against different targets. All four mAbs are of the IgG1 isotype and are at least 90 % homologous with respect to sequence comparison. Based on the viscosity profile, mAbs 5 and 6 were considered acceptable with respect to viscosity profile and no additional processing or extensive formulation development work was required (Fig. 7a). On the other hand, mAb 7 was considered to have a viscosity profile that required additional steps to enable the UF/DF process in order to

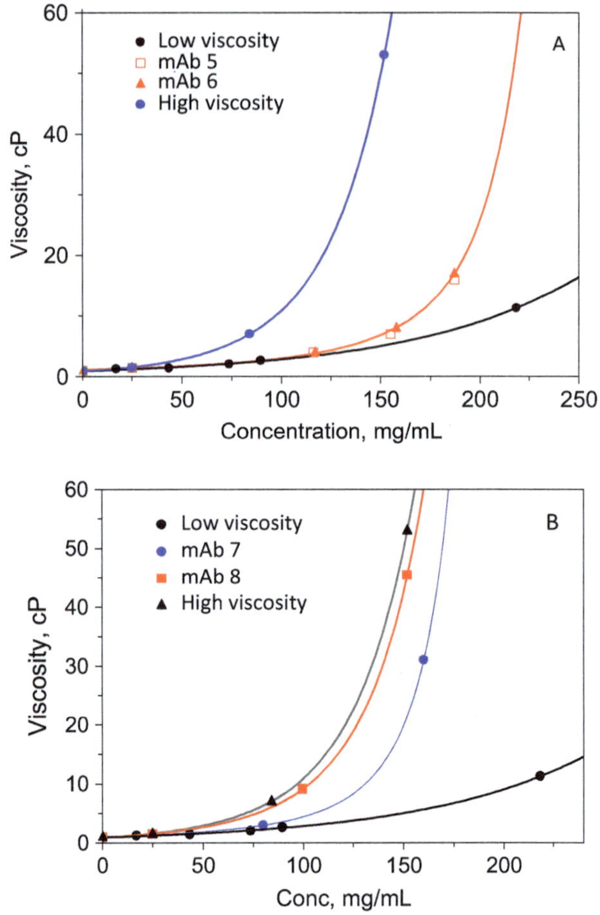

Fig. 7 (**a**) Viscosity profiles of two mAbs as part of molecule assessment studies. Both mAbs are considered low viscosity. (**b**) Viscosity profiles of two mAbs as part of molecule assessment studies. mAb 7 was considered high viscosity and needed further processing modifications to manufacture. mAb 8 was not considered manufacturable at the intended target concentration without significant challenges. The *points* are experimental data and the *line* represents a fit to the equation y = a + bexp(cx)

achieve the desired target concentration (Fig. 7b). mAb 8 was considered to be highly viscous at the initial target concentration and would have posed significant challenges for the UF/DF process to be successful (Fig. 7b) (Ramachanderand and Rathore 2013).

The above examples clearly demonstrate the utility of MA as a critical strategy to enable re-engineering mAbs to attain target stability characteristics, enable selection of the “stable” candidate when multiple options may be available, or simply generate information to develop strategy around process development with the appropriate allocation of time and resources.

Future Directions

While the molecule assessment strategy has been well established in many companies, the process of performing this early assessment continues to evolve concomitantly with antibody engineering technology to produce multiple candidates against a single target. Current approaches are still largely experimentally based and require several milligrams of material to complete the assessment. The future direction is likely to evolve in two pathways: (1) miniaturization techniques to minimize material requirements and (2) in silico analysis to perform risk assessment on candidates with respect to various properties.

Material quantity is always a concern while performing early assessment of mAb-based candidates. The question often asked is "can a certain experiment or test be carried out with less?" or "what's the minimum material needed?" There is an obvious need to do more with less as part of MA studies. The analytical tools play a critical role in defining the material required. While stability studies can be miniaturized using microgram levels of protein, use of newer analytical tools such as Ultra Performance Liquid Chromatography (UPLC) along with sophisticated MS technology can help miniaturize performance of such studies. Other studies such as rheology still demand several mgs of proteins. Advances in microfluidics and micro-rheology may help reduce the material requirements while providing adequate information on the rheological properties of mAb solutions. Another approach is use of dilute solutions to measure properties apparent at high concentrations. One such example is use of diffusion interaction parameter, k_D, measurable at low concentrations to assess viscosity behavior at high concentration (Connolly et al. 2012). It has been shown that k_D measured in dilute solution conditions can be used to provide a rank ordering of viscosity values at higher concentration. While the technique looks promising, a broader data set may be needed for the technique to have high rate of prediction success. Furthermore, kD provides a qualitative rank-ordering and fails to quantitatively predict the viscosity values. Nevertheless, if multiple candidates are available, the technique could still be used to screen out candidates that demonstrate attractive interactions. Overall, with the improvements in analytics and study designs, a reduction in material requirement will facilitate testing more candidates or perform more tests with the available amounts.

In silico assessment of physicochemical properties of mAbs could greatly facilitate screening in order to select or narrow down the number of candidates for final experimental testing, especially when a large pool of candidates may become available. In silico tools could also aid in reducing material requirements if certain decisions regarding molecular liabilities could be made without performing the actual experiment and therefore, eliminating the need for experimental testing. Computational tools have been presented in the literature to assess various molecular properties such as aggregation, isomerization, and viscosity. For example, spatial aggregation propensity parameter has been developed to identify aggregation-prone regions using three-dimensional structures of antibodies (Chennamsetty et al. 2009). As another example, recently, a multi-variable approach has been proposed to predict viscosity of mAbs using several structural properties of Fv region of mAbs including net charge, zeta potential, and pI (Li et al. 2014b). Regarding the chemical stability aspect, a structure-based approach has been proposed to identify Asn or Asp hotspots of deamidation and isomerization, respectively (Sydow et al. 2014). The authors have shown that parameters of conformational flexibility, the size of the C-terminally flanking amino acid residue, and secondary structural parameters can be used to identify such labile hot spots. While these tools are relatively new and their broader use needs to be assessed, they offer promise in being implemented and, along with experimental studies, can aid in molecular assessment studies towards lead candidate selection.

Summary

“Molecular Assessment” or “Developability Assessment” studies have become a necessary part of drug development strategies for biologics, especially for monoclonal antibody based therapeutics. These early studies play a critical role in either selecting the “right” lead candidate with optimal physicochemical properties or generating information about molecule behavior/liabilities that help define the long-term strategy towards technical development. The overall goal, essentially, is to “de-risk” the technical development and achieve faster FIH trials, as well as to enhance the probability of commercial success with minimal unforeseen/unpredictable events. Together with affinity maturation and early pharmacokinetic assessment, MA studies ensure that enough information is available prior to the entry of the lead candidate into the clinical development program and therefore, enhance the likelihood of technical and clinical success.

Furthermore, while MA has become an essential strategy for new lead candidates entering clinical development and targeting novel targets/epitope, MA plays an equally important role in introducing biobetters. In some sense, MA is perhaps even more crucial for biobetters, as these molecules are designed to overcome a certain suboptimal attribute of the existing clinical/marketed molecule (efficacy, safety, pharmacokinetics, manufacturability). Therefore it needs to be ensured that the biobetter indeed exhibits the desired improved attribute and meets the requirements of it being developed as a better therapeutic. One example of such attribute from a technical development point of view is improved stability. A biobetter “stable” molecule could be a better candidate for a preferred dosage form, i.e., liquid compared to a lyophilized drug product or could enable sustained delivery by being more physico-chemically resistant to the in vivo environment. MA can help ensure that such attributes are indeed being exhibited prior to a molecule being selected and designated as a biobetter.

Acknowledgements The authors would like to acknowledge Boyan Zhang, Michael Kim, Li Yi, Benson Gikanga, Sabrina Lo, Rita Wong, Jessica Yang, Ankit Patel, and Bill Galush for contributing data presented in the case studies. The authors acknowledge Jamie Moore, William Galush, Barthélemy Demeule, Karen Rutherford, Samir Sane and Laura Simmons for their careful review of the chapter.

References

Adler M (2012) INJECTABLES-challenges in the development of pre-filled syringes for biologics from a formulation scientist's point of view. Am Pharm Rev 15:96

Arvinte T, Palais C, Green-Trexler E, Gregory S, Mach H, Narasimhan C, Shameem M (2013) Aggregation of biopharmaceuticals in human plasma and human serum. MAbs 5:491–500

Barbas CF 3rd, Kang AS, Lerner RA, Benkovic SJ (1991) Assembly of combinatorial antibody libraries on phage surfaces: the gene III site. Proc Natl Acad Sci U S A 88:7978–7982

Bowers PM, Horlick RA, Neben TY, Toobian RM, Tomlinson GL, Dalton JL, Jones HA, Chen A, Altobell L 3rd, Zhang X, Macomber JL, Krapf IP, Wu BF, McConnell A, Chau B, Holland T, Berkebile AD, Neben SS, Boyle WJ, King DJ (2011) Coupling mammalian cell surface display with somatic hypermutation for the discovery and maturation of human antibodies. Proc Natl Acad Sci U S A 108:20455–20460

Carpenter JF, Randolph TW, Jiskoot W, Crommelin DJA, Middaugh CR, Winter G, Fan Y-X, Kirshner S, Verthelyi D, Kozlowski S, Clouse KA, Swann PG, Rosenberg A, Cherney B (2009) Overlooking subvisible particles in therapeutic protein products: gaps that may compromise product quality. J Pharm Sci 98:1201–1205

Carter P, Presta L, Gorman CM, Ridgway JB, Henner D, Wong WL, Rowland AM, Kotts C, Carver ME, Shepard HM (1992) Humanization of an anti-p185HER2 antibody for human cancer therapy. Proc Natl Acad Sci U S A 89:4285–4289

Chennamsetty N, Voynov V, Kayser V, Helk B, Trout BL (2009) Design of therapeutic proteins with enhanced stability. Proc Natl Acad Sci 106:11937–11942

Chirinoand AJ, Mire-Sluis A (2004) Characterizing biological products and assessing comparability following manufacturing changes. Nat Biotechnol 22:1383–1391

Connolly BD, Petry C, Yadav S, Demeule B, Ciaccio N, Moore JM, Shire SJ, Gokarn YR (2012) Weak interactions govern the viscosity of concentrated antibody solutions: high-throughput analysis using the diffusion interaction parameter. Biophys J 103:69–78

de Haard HJ, van Neer N, Reurs A, Hufton SE, Roovers RC, Henderikx P, de Bruine AP, Arends JW, Hoogenboom HR (1999) A large non-immunized human Fab fragment phage library that permits rapid isolation and kinetic analysis of high affinity antibodies. J Biol Chem 274:18218–18230

Doessegger L, Mahler H-C, Szczesny P, Rockstroh H, Kallmeyer G, Langenkamp A, Herrmann J, Famulare J (2012) The potential clinical relevance of visible particles in parenteral drugs. J Pharm Sci 101:2635–2644

Duand W, Klibanov AM (2011) Hydrophobic salts markedly diminish viscosity of concentrated protein solutions. Biotechnol Bioeng 108:632–636

Feldhaus MJ, Siegel RW, Opresko LK, Coleman JR, Feldhaus JM, Yeung YA, Cochran JR, Heinzelman P, Colby D, Swers J, Graff C, Wiley HS, Wittrup KD (2003) Flow-cytometric isolation of human antibodies from a nonimmune Saccharomyces cerevisiae surface display library. Nat Biotechnol 21:163–170

Hensel M, Steurer R, Fichtl J, Elger C, Wedekind F, Petzold A, Schlothauer T, Molhoj M, Reusch D, Bulau P (2011) Identification of potential sites for tryptophan oxidation in recombinant antibodies using *tert*-butylhydroperoxide and quantitative LC-MS. PLoS One 6:e17708

Hermeling S, Crommelin DA, Schellekens H, Jiskoot W (2004) Structure-immunogenicity relationships of therapeutic proteins. Pharm Res 21:897–903

Hoet RM, Cohen EH, Kent RB, Rookey K, Schoonbroodt S, Hogan S, Rem L, Frans N, Daukandt M, Pieters H, van Hegelsom R, Neer NC, Nastri HG, Rondon IJ, Leeds JA, Hufton SE, Huang L, Kashin I, Devlin M, Kuang G, Steukers M, Viswanathan M, Nixon AE, Sexton DJ, Hoogenboom HR, Ladner RC (2005) Generation of high-affinity human antibodies by combining donor-derived and synthetic complementarity-determining-region diversity. Nat Biotechnol 23:344–348

Hotzel I, Theil FP, Bernstein LJ, Prabhu S, Deng R, Quintana L, Lutman J, Sibia R, Chan P, Bumbaca D, Fielder P, Carter PJ, Kelley RF (2012) A strategy for risk mitigation of antibodies with fast clearance. MAbs 4:753–760

Hudson SD, Sarangapani P, Pathak JA, Migler KB (2014) A microliter capillary rheometer for characterization of protein solutions. J Pharm Sci 104(2):678–685
Hünig T (2012) The storm has cleared: lessons from the CD28 superagonist TGN1412 trial. Nat Rev Immunol 12:317–318
Iwao Y, Anraku M, Hiraike M, Kawai K, Nakajou K, Kai T, Suenaga A, Otagiri M (2006) The structural and pharmacokinetic properties of oxidized human serum albumin, advanced oxidation protein products (AOPP). Drug Metab Pharmacokinet 21:140–146
Jakobovits A, Amado RG, Yang X, Roskos L, Schwab G (2007) From XenoMouse technology to panitumumab, the first fully human antibody product from transgenic mice. Nat Biotechnol 25:1134–1143
Jezek J, Rides M, Derham B, Moore J, Cerasoli E, Simler R, Perez-Ramirez B (2011) Viscosity of concentrated therapeutic protein compositions. Adv Drug Deliv Rev 63:1107–1117
Ji JA, Zhang B, Cheng W, Wang YJ (2009) Methionine, tryptophan, and histidine oxidation in a model protein, PTH: mechanisms and stabilization. J Pharm Sci 98:4485–4500
Kanai S, Liu J, Patapoff T, Shire SJ (2008) Reversible self-association of a concentrated monoclonal antibody solution mediated by Fab-Fab interaction that impacts solution viscosity. J Pharm Sci 97(10):4219–4227
Kohlerand G, Milstein C (1975) Continuous cultures of fused cells secreting antibody of predefined specificity. Nature 256:495–497
Kumar S, Singh S, Wang X, Rup B, Gill D (2011) Coupling of aggregation and Immunogenicity in biotherapeutics: T- and B-cell immune epitopes may contain aggregation-prone regions. Pharm Res 28:949–961
Kumru OS, Liu J, Ji JA, Cheng W, Wang YJ, Wang T, Joshi SB, Middaugh CR, Volkin DB (2012) Compatibility, physical stability, and characterization of an IgG4 monoclonal antibody after dilution into different intravenous administration bags. J Pharm Sci 101:3636–3650
Lee CV, Liang WC, Dennis MS, Eigenbrot C, Sidhu SS, Fuh G (2004) High-affinity human antibodies from phage-displayed synthetic Fab libraries with a single framework scaffold. J Mol Biol 340:1073–1093
Li B, Fouts AE, Stengel K, Luan P, Dillon M, Liang WC, Feierbach B, Kelley RF, Hotzel I (2014a) In vitro affinity maturation of a natural human antibody overcomes a barrier to in vivo affinity maturation. MAbs 6:437–445
Li L, Kumar S, Buck P, Burns C, Lavoie J, Singh S, Warne N, Nichols P, Luksha N, Boardman D (2014b) Concentration dependent viscosity of monoclonal antibody solutions: explaining experimental behavior in terms of molecular properties. Pharm Res 31(11):3161–3167
Linette GP, Stadtmauer EA, Maus MV, Rapoport AP, Levine BL, Emery L, Litzky L, Bagg A, Carreno BM, Cimino PJ, Binder-Scholl GK, Smethurst DP, Gerry AB, Pumphrey NJ, Bennett AD, Brewer JE, Dukes J, Harper J, Tayton-Martin HK, Jakobsen BK, Hassan NJ, Kalos M, June CH (2013) Cardiovascular toxicity and titin cross-reactivity of affinity-enhanced T cells in myeloma and melanoma. Blood 122:863–871
Lonberg N (2005) Human antibodies from transgenic animals. Nat Biotechnol 23:1117–1125
Mallaney M, Wang S-H, Sreedhara A (2014) Effect of ambient light on monoclonal antibody product quality during small-scale mammalian cell culture process in clear glass bioreactors. Biotechnol Prog 30:562–570
Manning M, Chou D, Murphy B, Payne R, Katayama D (2010) Stability of protein pharmaceuticals: an update. Pharm Res 27:544–575
Mason BD, Zhang-van Enk J, Zhang L, Remmele RL Jr, Zhang J (2010) Liquid-liquid phase separation of a monoclonal antibody and nonmonotonic influence of hofmeister anions. Biophys J 99:3792–3800
Murphy AJ, Macdonald LE, Stevens S, Karow M, Dore AT, Pobursky K, Huang TT, Poueymirou WT, Esau L, Meola M, Mikulka W, Krueger P, Fairhurst J, Valenzuela DM, Papadopoulos N, Yancopoulos GD (2014) Mice with megabase humanization of their immunoglobulin genes generate antibodies as efficiently as normal mice. Proc Natl Acad Sci U S A 111:5153–5158
Nakamura G, Chai N, Park S, Chiang N, Lin Z, Chiu H, Fong R, Yan D, Kim J, Zhang J, Lee WP, Estevez A, Coons M, Xu M, Lupardus P, Balazs M, Swem LR (2013) An in vivo human-

plasmablast enrichment technique allows rapid identification of therapeutic influenza A antibodies. Cell Host Microbe 14:93–103
Narasimhan C, Mach H, Shameem M (2012) High-dose monoclonal antibodies via the subcutaneous route: challenges and technical solutions, an industry perspective. Ther Deliv 3:889–900
Narhi LO, Schmit J, Bechtold-Peters K, Sharma D (2012) Classification of protein aggregates. J Pharm Sci 101:493–498
Nishi H, Miyajima M, Nakagami H, Noda M, Uchiyama S, Fukui K (2010) Phase separation of an IgG1 antibody solution under a low ionic strength condition. Pharm Res 27:1348–1360
Pantua H, Diao J, Ultsch M, Hazen M, Mathieu M, McCutcheon K, Takeda K, Date S, Cheung TK, Phung Q, Hass P, Arnott D, Hongo JA, Matthews DJ, Brown A, Patel AH, Kelley RF, Eigenbrot C, Kapadia SB (2013) Glycan shifting on hepatitis C virus (HCV) E2 glycoprotein is a mechanism for escape from broadly neutralizing antibodies. J Mol Biol 425:1899–1914
Philoand JS, Arakawa T (2009) Mechanisms of protein aggregation. Curr Pharm Biotechnol 10:348–351
Porterand CJH, Charman SA (2000) Lymphatic transport of proteins after subcutaneous administration. J Pharm Sci 89:297–310
Queen C, Schneider WP, Selick HE, Payne PW, Landolfi NF, Duncan JF, Avdalovic NM, Levitt M, Junghans RP, Waldmann TA (1989) A humanized antibody that binds to the interleukin 2 receptor. Proc Natl Acad Sci U S A 86:10029–10033
Ramachanderand R, Rathore N (2013) Molecule and manufacturability assessment leading to robust commercial formulation for therapeutic proteins. In: Kolhe P, Shah M, Rathore N (eds) Sterile product development: formulation, process, quality, and regulatory considerations. Springer, New York, pp 33–45
Rathore AS, Godavarti R, Kumar V, Tugcu N (2013) Evolution of the monoclonal antibody purification platform. Biopharm Int 26:32–37
Riechmann L, Clark M, Waldmann H, Winter G (1988) Reshaping human antibodies for therapy. Nature 332:323–327
Rosenberg A (2006) Effects of protein aggregates: an immunologic perspective. AAPS J 8:E501–E507
Rothe C, Urlinger S, Lohning C, Prassler J, Stark Y, Jager U, Hubner B, Bardroff M, Pradel I, Boss M, Bittlingmaier R, Bataa T, Frisch C, Brocks B, Honegger A, Urban M (2008) The human combinatorial antibody library HuCAL GOLD combines diversification of all six CDRs according to the natural immune system with a novel display method for efficient selection of high-affinity antibodies. J Mol Biol 376:1182–1200
Saluja A, Badkar AV, Zeng DL, Kalonia DS (2007) Ultrasonic rheology of a monoclonal antibody (IgG(2)) solution: implications for physical stability of proteins in high concentration formulations. J Pharm Sci 96:3181–3195
Schmit JD, He F, Mishra S, Ketchem RR, Woods CE, Kerwin BA (2014) Entanglement model of antibody viscosity. J Phys Chem B 118:5044–5049
Shire SJ, Shahrokh Z, Liu J (2004) Challenges in the development of high protein concentration formulations. J Pharm Sci 93:1390–1402
Shukla AA, Hubbard B, Tressel T, Guhan S, Low D (2007) Downstream processing of monoclonal antibodies—application of platform approaches. J Chromatogr B 848:28–39
Soderlind E, Strandberg L, Jirholt P, Kobayashi N, Alexeiva V, Aberg AM, Nilsson A, Jansson B, Ohlin M, Wingren C, Danielsson L, Carlsson R, Borrebaeck CA (2000) Recombining germline-derived CDR sequences for creating diverse single-framework antibody libraries. Nat Biotechnol 18:852–856
Sydow JF, Lipsmeier F, Larraillet V, Hilger M, Mautz B, Mølhøj M, Kuentzer J, Klostermann S, Schoch J, Voelger HR, Regula JT, Cramer P, Papadimitriou A, Kettenberger H (2014) Structure-based prediction of asparagine and aspartate degradation sites in antibody variable regions. PLoS One 9:e100736
Thayer AM (2013) Biobetters may be a better bet. Chem Eng News 91:24–25
Tiller T, Schuster I, Deppe D, Siegers K, Strohner R, Herrmann T, Berenguer M, Poujol D, Stehle J, Stark Y, Hessling M, Daubert D, Felderer K, Kaden S, Kolln J, Enzelberger M, Urlinger S

(2013) A fully synthetic human Fab antibody library based on fixed VH/VL framework pairings with favorable biophysical properties. MAbs 5:445–470

Torosantucci R, Schöneich C, Jiskoot W (2014) Oxidation of therapeutic proteins and peptides: structural and biological consequences. Pharm Res 31:541–553

Wakankar AA, Borchardt RT, Eigenbrot CE, Shia S, Wang YJ, Shire SJ, Liu JL (2007a) A spartate isomerization in the complementarity-determining regions of two closely related monoclonal antibodies. Biochemistry 46:1534–1544

Wakankar AA, Borchardt RT, Eigenbrot C, Shia S, Wang YJ, Shire SJ, Liu JL (2007b) Aspartate isomerization in the complementarity-determining regions of two closely related monoclonal antibodies. Biochemistry 46:1534–1544

Wang W, Singh S, Zeng DL, King K, Nema S (2007) Antibody structure, instability, and formulation. J Pharm Sci 96:1–26

Wrammert J, Smith K, Miller J, Langley WA, Kokko K, Larsen C, Zheng NY, Mays I, Garman L, Helms C, James J, Air GM, Capra JD, Ahmed R, Wilson PC (2008) Rapid cloning of high-affinity human monoclonal antibodies against influenza virus. Nature 453:667–671

Yadav S, Liu J, Shire SJ, Kalonia DS (2010) Specific interactions in high concentration antibody solutions resulting in high viscosity. J Pharm Sci 99:1152–1168

Yang X, Xu W, Dukleska S, Benchaar S, Mengisen S, Antochshuk V, Cheung J, Mann L, Babadjanova Z, Rowand J, Gunawan R, McCampbell A, Beaumont M, Meininger D, Richardson D, Ambrogelly A (2013) Developability studies before initiation of process development: improving manufacturability of monoclonal antibodies. MAbs 5:787–794

Yu BL, Vizel A, Young M, Morando A, He B (2007) Impact of degradations on bioactivity: a reflection from a monoclonal antibody. Abstracts of Papers, 234th ACS National Meeting, Boston, MA, United States, August 19–23, 2007:BIOT-136

Zhai W, Glanville J, Fuhrmann M, Mei L, Ni I, Sundar PD, Van Blarcom T, Abdiche Y, Lindquist K, Strohner R, Telman D, Cappuccilli G, Finlay WJ, Van den Brulle J, Cox DR, Pons J, Rajpal A (2011) Synthetic antibodies designed on natural sequence landscapes. J Mol Biol 412:55–71

Part III
New Platforms

Perspectives on Engineering Biobetter Therapeutic Proteins with Greater Stability in Inflammatory Environments

V. Ashutosh Rao

The Need for Biobetters to Overcome Degradation, Aggregation, and Instability in a Proinflammatory Physiological Environment

Rapid progression of genetic engineering technology has accelerated the development and availability of protein-based biopharmaceuticals for clinical use. However, their unique and complex structures render them susceptible to a plethora of post-translational modifications (PTMs) as well as chemical degradative processes they may encounter in distinct in vivo environments. Among the more common degradative reactions that can impact the structure and function of therapeutic proteins are oxidation, proteolysis, phosphorylation, and deamidation. Several PTMs such as oxidative modifications further render modified proteins vulnerable to aggregation or proteolytic degradation by enzymatic or non-enzymatic mechanisms (Torosantucci et al. 2014). Degradation of a therapeutic protein in vivo becomes problematic when the structural modification alters its intended function and safety or efficacy profile (Foye 2008). As a result of protein aggregation or degradation, therapeutic activity can be decreased, increased, or altered to have off-target effects. Protein degradation or aggregation could be facilitated under inflammatory circumstances in diverse clinical settings such as cancer, chronic inflammatory diseases, organ transplants, infectious diseases, and cardiovascular disorders, and can exacerbate an inflammatory response with unintended consequences (Chennamsetty et al. 2009; Hermeling et al. 2004). Therefore, characterizing and controlling the degradation or aggregation profiles for a therapeutic

V.A. Rao (✉)
Division of Biotechnology Review and Research III, Office of Biotechnology Products,
Center for Drug Evaluation and Research, Food and Drug Administration,
10903 New Hampshire Ave, 52-72/2212, Silver Spring, MD 20993, USA
e-mail: Ashutosh.Rao@fda.hhs.gov

A. Rosenberg, B. Demeule (eds.), *Biobetters*, AAPS Advances
in the Pharmaceutical Sciences Series 19, DOI 10.1007/978-1-4939-2543-8_11

protein, especially for indications where the physiological environment presents additional opportunity for instability, is an essential component of a drug manufacturer's control strategy. Evaluating a manufacturer's control strategy is a key risk assessment tool for the regulation of investigational and licensed biologic drugs for human use. Better risk assessment can result from greater characterization of critical structural modifications that can influence therapeutic protein function, application of sensitive and suitable methods to objectively measure protein modifications, and use of relevant preclinical models or human tissue samples for predicting the in vivo impact of the inflammatory environment on such protein alterations (Torosantucci et al. 2014; Foye 2008; Chennamsetty et al. 2009). The increasing number of novel investigational drugs and the simultaneous demand for safer and more effective drugs warrants the need to examine the mechanisms by which therapeutic proteins are modified in vitro and in vivo, as well as apply modern analytical and genetic engineering techniques to design *biobetter* biologic drugs with improved safety and efficacy profiles. This chapter will examine factors known to alter the stability of therapeutic proteins in vivo, potential interactions of susceptible proteins with the inflammatory environment, and review some challenges and potential strategies for designing biobetters.

The Fate of Degraded and Aggregated Therapeutic Proteins in an Inflammatory Environment

Dysfunctional protein aggregation of therapeutic proteins in the body can be triggered by physiologic conditions as they impact on protein dynamics and accelerated by inflammatory physiological conditions. The thermodynamic instability resulting from the conformational changes associated with hydroxyl radical attack and/or metal binding is thought to be responsible for peptide bond cleavage and protein fragmentation, and do not always require a proteolytic enzyme catalyst (Stadtman 2006). Indirect oxidation of protein amino acid side chains occurs through formation of reactive adducts with products of oxidized lipids, amino acids, sugars, and glutathione. The oxidation of granulocyte colony-stimulating factor (G-CSF), interleukin-2, interferons alpha and beta, erythropoietin, growth hormone, insulin, and monoclonal antibodies such as muromonab-CD3 and trastuzumab are well documented and reviewed (Torosantucci et al. 2014). Newer investigational therapeutic proteins with reactive amino acid side chains under appropriate destabilizing conditions are also susceptible to oxidation, degradation, and aggregation (Foye 2008; Kroon et al. 1992; Lam et al. 1997).

Many systemically administered proteins undergo degradation and/or activation via proteolytic enzymes (Kaufman 1998; Zhong and Wright 2013). Others undergo receptor-mediated uptake, non-specific endocytosis, autophagy, target-mediated clearance, and/or formation of immune-complexes followed by complement- or Fc-receptor-mediated clearance mechanisms (Xiao and Gan 2013; Vugmeyster et al. 2012; Rajadhyaksha et al. 2011; Xu and Vugmeyster 2012; Han et al. 2008).

The classical pharmacokinetic paradigms of adsorption, distribution, metabolism, and excretion have unique aspects when it comes to biologic drugs, primarily due to their large, complex structures, and biologic functions. Upon administration, therapeutic proteins can trigger receptor-mediated reactions or interact with endogenous proteins for signaling. Upon entering the blood, a range of hydrolytic enzymes including, but not limited to, carboxypeptidases (cleave C-terminal residues), dipeptidyl carboxypeptidases (C-terminus dipeptides), amino peptidases (N-terminal residues), and amidases (internal peptide bonds) hydrolyze the peptide bonds of the protein and recycle the components for the biogenesis of new proteins (Foye 2008). In some instances, such as with amyloid-β aggregates, the plasmin/plasminogen system is induced and proteolyzes the aggregates (Tucker et al. 2000). Hence, the transport and global tissue distribution of locally administered proteins in their native state is potentially somewhat limited. Within cells, protein aggregates or misfolded proteins can be chaperoned by the heat shock proteins such as Hsp70 and Hsp90 in an ATP-dependent manner to limit their toxicity by sequestering and/or disassembling them (Labbadia et al. 2012). Extracellular chaperones such as clusterin, haptoglobin, $\alpha_2$macroglobulin, αS_2-casein, and β-casein are implicated for the disposal of misfolded proteins in the extracellular space (Almeida and Saraiva 2012; Wyatt et al. 2013; Dabbs et al. 2013). These chaperones can sense and form stable, solubilized high molecular weight complexes with misfolded proteins in an ATP-independent manner (Dabbs et al. 2013; Hochgrebe et al. 2000). Such complexes are then cleared from the extracellular fluids through receptor-mediated endocytosis and subsequent degradation in lysosomes (Wyatt et al. 2013). Current evidence suggests that clusterin and haptoglobins lack the ability to independently refold misfolded proteins after stress (Yerbury et al. 2005; Poon et al. 2000). Therapeutic proteins and their hydrolyzed, aggregated, or fragmented byproducts can be identified as foreign and can trigger immune effector mechanisms that may neutralize therapeutic activity or activate desired or undesired biological mechanisms.

Human biological environments significantly influence the fate of a therapeutic protein in vivo. In turn, a protein modified by in vivo reactions can elicit intended and unintended pharmacodynamic responses. Despite the presence of proteases, catabolic processes and chaperones, unintended protein aggregates are associated with clinical symptoms in humans. The pathophysiological conditions of infection, graft rejection, aging, myocardial infarction, neurodegeneration, diabetes, rheumatoid arthritis, amyotrophic lateral sclerosis (ALS), systemic lupus erythematosus (SLE), other auto-immune disorders, and some cancers represent possible instances in which a proinflammatory environment could accelerate the oxidation and aggregation of a therapeutic protein (Dobson 2006; Rosenberg 2006; Ratanji et al. 2014; Takalo et al. 2013). Misfolding and aggregation of islet amyloid polypeptides have been investigated in the context of diabetes, whereas the deposition of aggregated amyloid proteins in the form of amyloid fibrils is a hallmark of Alzheimer's, Huntington, Parkinson's, and prion-linked diseases such as BSE (Meredith 2005; Calabrese et al. 2006). Hence, it is reasonable to expect that protein aggregates can manifest and persist in vivo, long enough to trigger unintended consequences.

During infection, cytokine activation as well as neutrophil and macrophage congregation at the site of infection would be expected in patients with a robust inflammatory response. Phagocytosis in neutrophils triggers a respiratory burst that generates large amounts of superoxide and hypochlorous acid that can degrade or aggregate proteins (Matheson et al. 1979; Shacter et al. 1988). Altered proteins (either endogenous or therapeutically administered) can trigger macrophages and microglia activation as a first response. These then produce reactive oxygen species (ROS), cytokines, nitric oxide, and prostaglandin E2, and prostaglandin E2, and recruit cells of the adaptive immune system (Amor et al. 2014). Oxidant attack on tyrosine residues of proteins form dityrosyl cross-links, whereas carbonylation and deamidation on lysine, arginine, proline, and threonines can also result in an aldehyde or ketone on the side chain and increase aggregation and degradation (Hazell et al. 1994; Shacter 2000). Histidine residues of proteins are particularly susceptible to oxidation and chelation due to their proximity to the metal binding site resulting in formation of oxo-histidine, asparagine or aspartate. Hypochlorous acid, produced by myeloperoxidase released from activated neutrophils, has been shown to inactivate the enzyme alpha-1-proteinase inhibitor by methionine oxidation. Oxidized alpha-1-proteinase inhibitor has been isolated from synovial fluid of arthritic patients and in lung lavage of smokers (Matheson et al. 1982). Oxidized and aggregated immunoglobulins in rheumatoid arthritis indicate that levels of reactive oxygen species that can modify and aggregate therapeutic proteins can occur in vivo (Jasin 1983). Similarly, elevated levels of reactive oxygen species are present in diabetes patients, at least in part, due to hyperglycemia and protein kinase C-dependent activation of NAD(P) oxidase (Inoguchi et al. 2003).

In addition to reactive oxygen species, reactive nitrogen species (RNS) such as nitric oxide can be formed during inflammation by inducible nitric oxide synthase (iNOS) in macrophages. Nitric oxide in turn induces aggresome formation of nitric oxide synthase and renders the aggregated iNOS inactive (Wang and Xia 2012). While this occurrence is indicative of a feedback loop in iNOS regulation, it suggests that reactive nitrogen species can also trigger protein aggregation in inflammatory environments. In Parkinson's disease, α-synuclein polymers are formed by peroxinitrite-mediated nitration and this forms stable dityrosine cross-linked α-synuclein aggregates in Lewy bodies (Butterfield and Kanski 2001; Giasson et al. 2000).

A protein may also possess inherent structural features that make it susceptible to in vivo aggregation. An interesting example of this possibility is the recent identification of mutant p53 aggregates in cancer (Xu et al. 2011). The p53 protein, which naturally occurs as a tetramer, is proposed to have a prion-like core structure that, upon mutation, has a propensity to form aggregates in vivo (Rangel et al. 2014). The functional DNA-binding domains of p53, where most mutations related to cancer development are found, have the highest propensity to form amyloid-like oligomers and fibrils. While the exact structural mechanisms are still being investigated, the R248Q and R280K mutants have shown to have higher propensity to aggregate (Rangel et al. 2014; Ano Bom et al. 2012) and appear to accelerate the aggregation of wild-type p53. These mutant p53 monomers can even co-aggregate with p63 and p73. A greater degree of tumor invasiveness was found to be correlated with increased mutant p53 aggregation, suggestive of a gain-of-function for mutant p53 via aggregation in tumors (Silva et al. 2014).

Under healthy physiological conditions, antioxidant defense mechanisms scavenge and mitigate excessive ROS. These include the scavenging enzymes superoxide dismutase (SOD), thioredoxin, catalase, glutathione, peroxiredoxins, heme oxygenase, and others which convert excessive superoxide, hydrogen peroxide, and hydroxyl radicals to less reactive products. For example, manganese superoxide dismutase converts superoxide radicals to hydrogen peroxide, which then is converted to water through the glutathione or thioredoxin/peroxiredoxin pathway.

However, in several proinflammatory conditions, elevated ROS coincides with a deficiency of key antioxidant enzymes. For instance, while there is an increased release of ROS in Crohn's disease, there are decreased levels of catalase (but not SOD or glutathione peroxidase) during active disease (Iborra et al. 2011). Inflammatory chronic obstructive pulmonary disease (COPD) and asthma are also associated with increased levels of ROS released from macrophages and an acute loss of superoxide dismutase activity, for which the reasons are not fully clear (Comhair et al. 2000; Kirkham and Rahman 2006). In Alzheimer's and Parkinson's disease, excessive ROS coupled with the high oxygen demand of the brain, lead to a higher propensity for oxidant-induced protein aggregation (Mohsenzadegan and Mirshafiey 2012; Jomova et al. 2010). A markedly low catalase activity in neuronal mitochondria is also in agreement with the finding that these cells rely on other antioxidant systems such as thioredoxin/peroxiredoxin to control the elevated levels of ROS (Sorgato et al. 1974; Lopert et al. 2012). In type 2 diabetes, chronically increased hydrogen peroxide, in the blood and islets, due to a catalase gene mutation, is a proposed risk factor for the disease (Tarnai et al. 2007), with the suggested mechanism being peroxide-mediated damage to pancreatic β-cells (Goth 2008). The aggregation of islet amyloid polypeptide oligomers in pancreatic β-cells has been correlated with the development of type 2 diabetes, in agreement with the hypothesis that autoinflammatory conditions present elevated ROS levels that promote protein oligomerization (Soong et al. 2009).

While autoimmune disorders and the proinflammatory state can present conditions conducive to therapeutic protein oxidation, the converse may also be possible. Oxidative damage to proteins, lipids, and DNA by elevated endogenous free radicals and the resulting byproducts and aggregates can produce highly immunogenic antigens, triggering pathogenic antibodies and platelet activation such as those reported in systemic lupus erythematosus and other diseases (Perl 2013; Shah et al. 2013; Herczenik et al. 2007). Antibodies to superoxide dismutase and catalase are also proposed to be responsible for increased oxidative damage in SLE patients, thereby presenting a cyclical pathogenic state that favors excessive levels of ROS (Mansour et al. 2008). Most non-functional aggregates occurring in physiological systems are controlled with a combination of chaperones and proteases that can be explained by Le Chatelier's dynamic equilibrium principle which suggests that physiological systems are designed to control high concentrations of insoluble aggregates by induction of mechanisms to lower them to levels that are soluble and non-toxic (Gsponer and Babu 2012). However, in SLE this mechanism is overwhelmed by autoantibodies which form against the chaperone proteins (such as Hsp90) which ensure proper folding of other proteins, thereby increasing protein misfolding and aggregation under elevated ROS conditions (Shukla and Pitha 2012). Understanding the physiological environment in the intended patient population, tissue distribution, and target tissue profile can help design smarter drugs that maintain their therapeutic profile while limiting adverse events.

Concomitant administration of redox-active drugs such as doxorubicin, bleomycin, and cisplatin can also transiently modulate the in vivo environment by increasing the release of cytokines and ROS (Wondrak 2009). It is conceivable that such an increase in oxidants and cytokines can facilitate aggregation of coadminstered therapeutic proteins. Increased reactive oxygen species, lipid peroxidation and protein oxidation is reported in breast cancer patients and in animals exposed to clinically-relevant doses of doxorubicin (Amin et al. 2012). Similarly, exposure to therapeutic levels of ionizing or ultraviolet radiation could also elevate highly reactive oxidants (Bossi et al. 2008; Leach et al. 2001). Upregulation of proinflammatory cytokines TNF-α, IL-1β, IL-6, and CCL2, as well as mitochondrial dysfunction and increased ROS/RNS, occur during chemotherapy-induced neuropathy (Areti et al. 2014; Wang et al. 2012). Levels of the antioxidants catalase and glutathione were found diminished in breast carcinoma patients prior to treatment, thus increasing the susceptibility to induced levels of ROS.

Product Features and Physiological Conditions as Critical Factors for Designing Biobetters

Oxidation of methionine and cysteine are the most well-established oxidative modification for therapeutic proteins. However, other types and sites of modifications are also likely (Shacter 2000; Arakawa et al. 1993). Exposure to light (photoinstability), alkaline pH (chemical instability), and conformational changes in higher-order structure (physical instability) present destabilizing conditions that can result in oxidation and denaturation of native proteins. Some preventive measures include lowering pH to minimize sulfhydryl reactions, lowering oxygen tension, use of low molecular weight antioxidants, and inclusion of inert gas layers to mitigate the oxidation and degradation reactions in storage containers. Product-specific formulation and packaging are used to control many of the potentially deleterious conditions a therapeutic protein can encounter during storage and use. Unfortunately, metal leachates or even some excipients can result in unintended oxidation and degradation of proteins after filling of the final drug product during storage and handling. When administered in a proinflammatory environment, such aggregated or degraded protein could exacerbate cytokine release and mediate aggregation of therapeutic or endogenous proteins especially in the context of elevated levels of ROS. The use of excipients such as polysorbate 80 (when autooxidized), citrates, phenols, and EDTA can cause oxidation resulting in degradation or aggregation. This has been observed with G-CSF in the presence of peroxides from expired polysorbates in formulation (Yin et al. 2005; Patel et al. 2011; McGoff and Scher 2000). Similarly, leachates from the container closure system can oxidize proteins. For instance, in the well-documented case of erythropoetin alfa in the presence of tungsten oxide metal leachates resulted in oxidation induced aggregates and induced an erythropoietin-specific immune response in patients (Bennett et al. 2004; Sharma 2007a, b; Seidl et al. 2012). Stressed or accelerated stability studies are particularly useful in characterizing degradation and aggregation profiles as well as for making sure that real-time stability assays are capable of capturing variants of the native protein. However, these measures are not absolute due to the limitations of antioxidants to control for all forms of oxidation, the need for reconstituting or reformulating some final drug products, and the human factor limitations on the in-use conditions that a final drug product encounters during handling and storage. Therefore, there is a need to start with a pharmacologically-active biobetter protein that is less susceptible to oxidation and degradation under stressed conditions. Reduced susceptibility to degradation and aggregation would of course be of further benefit in minimizing the susceptibility to aggregation and chemical degradation in proinflammatory in vivo environments.

Despite the limited tissue distribution of many therapeutic proteins, there are certain tissue sites that might be particularly sensitive to small but sustained exposure to protein aggregates and redox-active variants. For instance, some systemically administered proteins with the propensity to undergo oxidation or the ability to redox cycle and generate reactive free radicals might be toxic to tissue with lowered endogenous levels of antioxidant enzymes such as the heart. In this instance, the damage could result from reactive free radical-mediated damage to endogenous proteins, lipids, and/or DNA. Cardiac tissue is selectively sensitive to oxidative damage because it expresses significantly lower levels of catalase and superoxide dismutase antioxidant enzymes (Doroshow et al. 1980). There are several approved protein therapeutics, including interleukin-2, interferon alpha, and trastuzumab which trigger cardiotoxicity in cancer patients, although a direct link between exposure to oxidized protein and cardiac damage has not been established for these agents (Della Pina et al. 2012). The tissue-selective cardiotoxicity of doxorubicin is linked to the oxidation of cardiac myosin binding C protein and impairment of actin-binding activity (Aryal et al. 2014).

Autophagy, which is normally associated with lysosomal recycling of damaged proteins, is deficient in conditions where misfolded proteins result in a proinflammatory clinical disorder such as Alzheimer's, elevating the risk from unscavenged oxidants and aggregates from therapeutic proteins (Nilsson et al. 2013). Even at tissue sites that are considered relatively immune privileged, the presence of aggregates can still cause inflammation as well as impact the flow of physiological fluid and lead to adverse events. One such example is the obstruction of aqueous outflow by high-molecular weight aggregates of bevacizumab (in a circumstance in which it was repackaged and used off label) with subsequent elevation in intraocular pressure reported in some patients (Kahook et al. 2010). However, focused research on the exact mechanism of oxidative stress, impact of product-related variants, and systemic or tissue-specific toxicity directly resulting from degraded or aggregated therapeutic proteins is needed to better understand the risk.

It may also be pragmatic to develop preclinical models in which in vivo assessments of the native and stressed proteins could be evaluated to better predict stability and immunogenic potential. And while new models might be able to better discriminate risk from a drug product and its variants, it would seem prudent to start by designing biobetter molecules that have a higher threshold for degradation and aggregation to minimize the extent to which they can be degraded and aggregated and thus trigger an immune or toxic response, especially in an inflammatory environment such as in patients with autoimmune disorder (e.g., lupus) or with a propensity for protein misfolding (e.g., Alzheimer's).

Challenges and Potential Strategies for Designing Stable Biobetter Molecules

Protein engineering has already made great strides in designing a new generation of small proteins and peptide therapeutics, such as the development of rapid-acting or long-acting insulins and long acting growth hormone (Phillips et al. 2010; Heikoop et al. 1997). By replacing amino acids, adding glycosylation sites, or conjugating with targeting moieties or stabilizing scaffolds, these newer, engineered versions offer better bioavailability, reduced adverse events, and offer greater convenience to patients. The structural modifications that resulted in a better safety and efficacy profile for some protein therapeutics have been well examined and reviewed elsewhere (Dulaney and Huang 2012; Vigneri et al. 2010; Zinman 2013). While no single intervention can completely eliminate safety risks, currently available protein engineering strategies allow us to prospectively minimize the risk by incorporating structural modifications that might maintain efficacy while lowering the propensity for degradation and aggregation. The structure-based design and development of biobetter large protein therapeutics is challenged by the inherent complexity in structure, range of possible post-translational modifications, different PEGylation or glycosylation chemistries, need for high concentration dosing, and irregular solubility and absorption at high concentrations. A key challenge after modifying the protein molecule to better withstand degradation and aggregation is to confirm the intended biological activity at the intended target site in the patient population. It is also important to proactively characterize any unintended biological activity that might result as a consequence of the structural modifications. Some examples of licensed and investigational biotechnology-derived drugs and their respective protein engineering stabilization strategies are discussed below. This discussion is intended to provide a snapshot of some strategies using one or two examples from the literature and not intended to be a complete listing of all possible strategies or all drugs that could benefit from them. These examples fall under the following general approaches:

- Designing oxidation-resistant or protease-resistant forms of therapeutic proteins
- PEGylation or glycosylation to stabilize or lower antigenicity
- Liposomal encapsulation of therapeutic proteins to delay degradation
- Stabilizing or targeting therapeutic proteins by fusion with proteins or small molecules

The aforementioned structure-based approaches can be considered in combination with local administration of some high-risk proteins and lowering the minimal effective dose of proteins conjugated with targeting antibodies or synergistic small molecules.

Designing Oxidation-Resistant Forms of Therapeutic Proteins

One design strategy to maintain stability and biologic activity involves designing oxidation-resistant forms of therapeutic proteins that are functional but do not have unnecessary methionines, free cysteines, or other reactive side chains. Recombinant human G-CSF is a hematopoietic growth factor wherein oxidation of methionine residues causes a loss in biological activity and receptor binding (Lu et al. 1999; Reubsaet et al. 1998). While there are histidine, tyrosine, tryptophan, and phenylalanine residues in G-CSF that can also be oxidized, the oxidation of one free sulfhydryl on a cysteine residue has been studied for propensity to aggregate G-CSF and four reactive methionine residues have been linked to decreased receptor dimerization, lowered biological activity, and altered serum half-life (Reubsaet et al. 1998; Raso et al. 2005). The oxidation at Met127 and Met122 results in an unstable protein conformation and decreased biological activity. Oxidation at both Met1 and Met138 resulted in greater loss in biological activity than at Met1 alone. Oxidizing all four methionine residues resulted in retention of 3 % of the remaining biological activity of the native protein. The G-CSF molecule also contains a free cysteine at Cys17 and two intramolecular disulfide bonds at Cys36-Cys42 and Cys64-Cys74. Site-directed mutagenesis studies have suggested that Cys17 oxidation can result in structural changes, covalent dimerization, and aggregation (Arakawa et al. 1993; Lu et al. 1992).

Replacing both Met127 and Met138 (but not individually) with Leu, confers greater stability than that of the native G-CSF and retains in vitro biological activity (Lu et al. 1999). In this study, biological activity was measured by cell mitogenesis and G-CSF receptor binding, whereas stability was measured by size-exclusion chromatography of samples held at 37 °C. The substitution of Cys17 with Ala17 had five times the heat stability (at 53 °C) of native G-CSF, whereas Ser17 had comparable stability to the native G-CSF (Ishikawa et al. 1992). The Ala17-containing modified G-CSF, which also included four additional amino acids at the N-terminus, increased neutrophil counts to a higher level with a concomitant slower declining rate in murine and primate animal models following irradiation or cyclophosphamide than did the wild type G-CSF (Jiang et al. 2011). A pharmacokinetic study also reported higher serum half-life, concentration time curve values, and maximum serum concentration than the parent G-CSF in rats. Consistent with its greater stability, the Cys17Ala modified G-CSF showed a lower rate of degradation than the parent G-CSF in whole blood and serum from rats. The increased granulopoiesis and higher bioavailability with the modified G-CSF presents the advantages of greater stability and sustained activity, although the required human dosage would need to be addressed as well. Finally, it should also be mentioned that human growth hormone is another well-studied example where methionine and cysteine residues have been modified for improved stability and activity (Brems et al. 1990; Mulinacci et al. 2013).

Screening for Protease-Resistant Proteins or Peptide Regions

Proteolytic cleavage of proteins is an evolutionarily conserved metabolic process. Members of the protease families of enzymes are involved in hydrolysis of proteins at serine, threonine, cysteine, aspartate, and glutamic acid residues; some metalloproteases require a metal cofactor such as zinc. Substrate-specific proteases reside in almost every major tissue and organ including the gut, skin, and blood. The bioavailability, immunogenicity, and stability of therapeutic proteins, therefore, is dependent on their ability to withstand hydrolysis by proteases long enough to perform their intended biological activity. Fortunately, the sequence specificity for the protease family members, their need for co-factors, and methods to study their enzyme kinetics using purified protein drugs have been relatively well studied. There are also several predictive tools based on molecular modeling and bioinformatics that allow a preliminary and theoretical assessment of the potential cleavage sites in a peptide sequence (Chennamsetty et al. 2009; Boyd et al. 2005; Gasteiger et al. 2005; Song et al. 2011, 2012). Therefore, one potential strategy for biobetter molecules would be to engineer proteins with sequences that are relatively resistant to proteases or contain a scaffold to potentially hide the cleavage sites.

Glycosylation and PEGylation of therapeutic proteins are among the more well-studied modification for engineering protease-resistant peptides (Raju and Scallon 2006; Park et al. 2010). Some of the more novel approaches include the generation of macrocyclic peptides by converting linear peptides into stable macrocycles via allylic substitution catalyzed by palladium (Lawson et al. 2013), adding interchain disulfide bonds to stabilize a coiled coil peptide structure (Tong et al. 2013), and covalently crosslinking hydrocarbons across peptide regions to stabilize α-helical structures (Bird et al. 2010; Braun et al. 2010). In these and other investigations, proteolytic degradation was tested in vitro using α-chymotrypsin, proteinase-K, trypsin, or pepsin at acidic and neutral pH and/or in vivo by pharmacokinetic measurements of a bioavailable drug in an animal model. In the instance of an HIV-1 gp41 specific therapeutic, addition of interchain disulfide bonds conferred greater stability and sustained the product's anti-HIV activity (Tong et al. 2013). Covalently crosslinked hydrocarbons or hydrocarbon double-stapling also appeared to slow down the kinetics of proteolytic degradation while sustaining the proposed antiviral activity of the modified protein (Bird et al. 2010). Overall, there appears to be several established and investigational approaches for engineering a protease-resistant protein; the application of the optimal approach(es) for a given molecule might be dependent on the number of stress points on the molecule, dosage and drug concentration expectations, intended route of administration, and therapeutic index of the drug in its intended target population.

PEGylation, Acylation, or Glycosylation to Stabilize or Lower Antigenicity of Degraded or Native Proteins

Polyethylene glycol (PEG) molecules are flexible and soluble polymers that are conjugated to primary amines on protein structures to increase their size, reduce exposure of the protein to degradative factors, and increase their bioavailability. PEGs of varying size (3.5–40 kDa), structure (linear/branched), and polymer length can be titrated to some extent with the goal of reducing aggregation, immunogenicity, and increasing serum half-life of the therapeutic protein (Beals and Shanafelt 2006). There are several approved products on the market that are PEGylated versions of their native molecule, including growth hormones, insulin, interferons, G-CSF, and L-asparaginase. PEG moieties are usually attached to the protein using *N*-hydroxysuccinamide or aldehyde chemistry and pH modulation (Park et al. 2010; Clark et al. 1996; Luxon et al. 2002; Wang et al. 2010). An advantage of the PEGylation chemistry is the ability to attach the PEG polymer to reactive amino acids such as an unpaired cysteine. The steric hindrance offered by PEGylation to protect the molecule also presents a challenge when considering reductions in receptor binding and site-specific protein–protein interactions from such hindrance. In the case of PEGylated interferon alfa-2a, while there is a significant increase in serum half-life upon PEGylation, there is also a reduction in specific activity (Luxon et al. 2002; Bailon et al. 2001; Wang et al. 2002). Also, while PEGylation can reduce immunogenicity of therapeutic proteins (Veronese and Mero 2008), this may not always be the case and PEG itself may elicit immune responses (Li et al. 2001; Garay et al. 2012), although additional studies are needed to address this concern (Schellekens et al. 2013). Other challenges with PEGylation include the difficulty in predicting the site-specific modification on specific amino acid amines, the coupling efficiency of the PEG, the need for controlling additional quality attributes related to PEG at release and stability, and the need for revisiting patient dosage due to greater bioavailability and/or lower receptor binding for some molecules. The mechanisms of biodegradation for PEG and the fate of chronically administered PEG in patients also warrant further studies. Finally, acylation of protein molecules with a fatty acid conjugate has also been tested with insulin to extend the duration of activity although this technique has not been as well investigated with large therapeutic protein molecules (Beals and Shanafelt 2006). These and other alternative protein scaffolds and antibody constructs are discussed in greater detail in the chapter by M. Gebauer and A. Skerra presented in this book.

More than half of known human proteins are naturally glycosylated and approximately 90 % of these are N-glycosylated. Glycosylation can facilitate protein folding and enhances stability by increasing solubility. Glycosylation is also an important part of the targeting and potency activities of therapeutic enzymes (such as human glucocerebrosidase for Gaucher disease), monoclonal antibodies (cytotoxicity of humanized IgG1 antibodies), cytokines (interferon-beta and -gamma), and clotting factors (recombinant Factor IX). The enzymatic process that attaches glycans to proteins has been applied to the engineering of glycosylated therapeutic proteins with the intent of stabilizing them (Sola and Griebenow 2009). Designing proteins with glycosylation sites and manufacturing them in mammalian cells is intended to result in increasing serum half-life and limiting exposure to stressful physiological conditions. N-linked glycosylation at asparagine residues and O-linked glycosylation at serine or threonine residues using fucose, galactose, mannose, *N*-acetylglucosamine (GlcNAc), *N*-acetylgalactosamine, and sialic acid (*N*-acetylneuraminic acid) residues are among the most prevalent modifications of therapeutic proteins, serving to enhance stability, solubility and reduce aggregation. Glycosylation can reduce pH-mediated denaturation, decrease hydrophobic interactions resulting in improved chemical stability, and reduce aggregation by increasing protein solubility or steric hindrance (Chennamsetty et al. 2009; Sola and Griebenow 2009). Thus, not surprisingly, targeted glycosylation has been used as a tool to optimize clinical performance (Zhong and Wright 2013; Walsh and Jefferis 2006). For example, increasing the sialic acid-containing carbohydrate content of erythropoietin on additional N-linked sites created darbepoetin alpha with an increased circulating half-life and in vivo potency compared to unmodified erythropoietin (Egrie et al. 2003; Elliott et al. 2004; Egrie and Browne 2001). Therefore, glycoengineering these and similar molecules for greater stability requires careful consideration in order to maintain the intended biological activity while preventing unintended "off target" pharmacokinetic and pharmacodynamic outcomes (Zheng et al. 2014). For a more extensive discussion of current strategies for pharmacokinetic optimization, please refer to the chapter by U. Binder and A. Skerra presented in this book.

The common challenges for all biochemical modifications that provide steric protection to large proteins is justifying the need for stabilizing the molecule, understanding the relative stability and dosing differences with the modified version, and characterizing both the in vitro/specific activity and in vivo activity of the modified versions.

Encapsulation of Therapeutic Proteins to Delay Degradation

Microencapsulation of small molecules is a well-established engineering step to delay release and subsequent metabolism. Whether liposomal or other forms of drug delivery systems can minimize a therapeutic protein's degradation and aggregation has not been thoroughly investigated. There have been instances of liposomal delivery of investigational large therapeutic proteins including endonucleases that were topically applied for UV protection, insulins for hyperglycemia, and asparaginase or interleukins for anticancer cytotoxicity (Wolf et al. 2000; Adibzadeh et al. 1992; Gaspar et al. 1996). In vitro models have suggested that liposomes are efficient carriers and that there is biological activity associated with the protein cargo, although confirmatory and follow-up studies have been limited. There have been very few clinical studies in human patients with such agents to better understand their risk to benefit profile (Weiner 1994). There are also other limitations related to manufacturing scale-up and need for sterilization that are challenging with proteins encapsulated in liposomes (Swaminathan and Ehrhardt 2012). An important opportunity with encapsulated therapeutic proteins might be the ability to combine liposomal delivery with intranasal administration to transport therapeutic proteins across the blood brain barrier (Rajadhyaksha et al. 2011). Such strategies might be especially worth investigating for delivering neurotropic factors to treat neurological indications. Investigations have been made in delivering proteins via the nasal epithelial barrier for transcellular transport within the perineural space and ultimately to the cerebrospinal fluid and brain (Rajadhyaksha et al. 2011; Lochhead and Thorne 2012). Milgore et al. showed that intranasal ovalbumin could be delivered to the brain via cationic liposomes by 6 h in rats (Migliore et al. 2010). Similarly, intranasal administration of nerve growth factor in mice has been reported (De Rosa et al. 2005). Potential local and systemic toxicity, especially in cases of chronic administration, need to be considered for such cationic liposomes. A better understanding of the biophysical aspects of the liposomal components, their interaction with the protein cargo, release triggers, and means for scaling-up and sterilization is needed to make optimal use of this potentially promising approach for biobetter protein drugs (Weiner 1994; Ulrich 2002).

Stabilizing or Targeting Therapeutic Proteins by Fusion with Proteins or Small Molecules

Proteins can be fused with other proteins or other longer-lasting conjugates to enhance their effectiveness or stability. The conjugates most widely tested are albumin and the Fc portion of antibodies. Etanercept, a TNF-alpha antagonist for the treatment of rheumatoid arthritis, is one such example where the extracellular domain of p75 TNF receptor and the Fc domain of human IgG1 are fused to increase the serum half-life by increasing its size and mediating endosomal recycling (Goldenberg 1999). Fc fusion also offers the advantage of increasing solubility, secretion, and valency. In cases where a secondary immune function might be desired, Fc fusion proteins can also complement the immune functions of the primary cargo protein by enhancing the effector function (Cines et al. 2008). Other novel forms of antibody-engineered fusion constructs such as Fab (fragment antigen binding) and single chain Fv (fragment, variable) modifications are discussed in detail in other reviews (Jazayeri and Carroll 2008; Gillies et al. 2002). This is discussed in further detail in the chapter by A. Lugovskoy et al. presented in this book. The use of albumin (human serum albumin or rabbit serum albumin) as a fusion protein for prolonging the half-life has also been well-studied. Recombinant Factor VIIa and IX are two coagulation factors whose half-lives have been increased by fusion to albumin using a cleavable peptide linker. In both cases, preclinical studies suggest sustained biological activity to correct bleeding time and blood loss associated with hemophilia B (Schulte 2008, 2009). Interleukin-2 has also been conjugated with human serum albumin to increase half-life and tumor killing in preclinical models (Yao et al. 2004). The conjugation of therapeutic proteins such as TRAIL with transferrin has also been a novel approach for improving tissue targeting of anticancer agents with increased accumulation compared to the parent molecule in preclinical models (Kim et al. 2012). In this study, the authors examined the impact of fusion with both PEG and transferrin in immune-deficient mice, hence studies in an immune proficient preclinical model would be desirable to truly assess the tissue-specific targeting and stabilization of novel fusion proteins. Other tissue targeting opportunities include fusion peptides with a mitochondrial localization signal sequence and lipophilic cations such as triphenylphosphonium are also under development and might offer ways of limiting exposure of the therapeutic protein to oxidation and proteolysis before reaching the intended site of action (Keeney et al. 2009; Reily et al. 2013). It may be advantageous to develop biobetters with stabilizing conjugates that allow tissue-specific targeting or, when possible, consider local administration to minimize systemic exposure and degradation in proinflammatory environments in vivo. Engineering stable, conjugated biobetters may also accelerate clinical development by avoiding the need for titrating greater solubility and stability with excipients (Pavisic et al. 2010).

Closing Thoughts

Protein oxidation and the resulting degradation and/or aggregation can have a profound impact on the stability, immunogenicity, safety, and efficacy of therapeutic proteins. These effects are magnified under disease conditions where a proinflammatory environment and elevated level of ROS/RNS might accelerate protein aggregation and elicit undesired immune responses. The stress points of some large protein molecules that render them susceptible to fragmentation and aggregation have been extensively studied and can be leveraged for designing biobetter therapeutic proteins including monoclonal antibodies. Preclinical models and predictive bioinformatics tools are helpful for preliminary assessments of the risk factors and potential strategies, and should involve an immune-proficient and pharmacologically-relevant test models. Protein engineering strategies that might aid future drug development of biobetters include, but are not limited to, site-directed mutagenesis to obtain oxidation- and protease-resistant molecules, scaffolding the protein or its active moieties to shield it from physiological stress, conjugation with large oligomers or stabilizing peptides and small molecules, encapsulation of the complete protein into liposomes, and local drug administration at the intended site of action (Fig. 1). In conclusion, a more sound strategy for improving the clinical performance of therapeutic proteins is to design biobetter protein molecules that are engineered for stability and sustained biological activity in vivo, rather than to solely augment stability into the final formulation with excipients. Stable biobetters should aim to provide sustained/intended efficacy, reduce immunogenicity, reduce dosing frequency, enhance patient convenience of use, minimize instability during routine handling by health care professionals and patients, and be thoroughly evaluated for off-target or unintended biological activities.

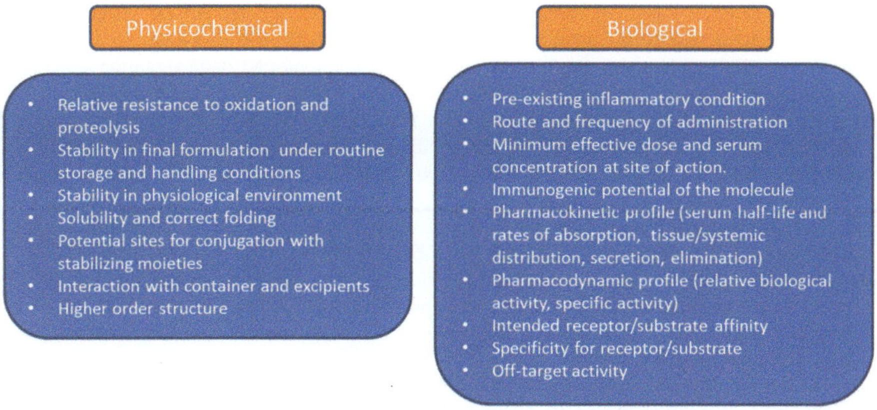

Fig. 1 Physicochemical and biological considerations for designing biobetters

Disclosure The views expressed in this article are those of the author and do not necessarily reflect the official policy or position of the U.S. Food and Drug Administration and the Department of Health and Human Services, nor does mention of trade names, commercial products, or organizations imply endorsement by the U.S. Government.

References

Adibzadeh M et al (1992) Activity of liposomal interleukin-2 in vitro. Mol Biother 4(1):24–28

Almeida MR, Saraiva MJ (2012) Clearance of extracellular misfolded proteins in systemic amyloidosis: experience with transthyretin. FEBS Lett 586(18):2891–2896

Amin KA, Mohamed BM, El-Wakil MA, Ibrahem SO (2012) Impact of breast cancer and combination chemotherapy on oxidative stress, hepatic and cardiac markers. J Breast Cancer 15(3):306–312

Amor S et al (2014) Inflammation in neurodegenerative diseases–an update. Immunology 142(2):151–166

Ano Bom AP et al (2012) Mutant p53 aggregates into prion-like amyloid oligomers and fibrils: implications for cancer. J Biol Chem 287(33):28152–28162

Arakawa T, Prestrelski SJ, Narhi LO, Boone TC, Kenney WC (1993) Cysteine 17 of recombinant human granulocyte-colony stimulating factor is partially solvent-exposed. J Protein Chem 12(5):525–531

Areti A, Yerra VG, Naidu V, Kumar A (2014) Oxidative stress and nerve damage: role in chemotherapy induced peripheral neuropathy. Redox Biol 2:289–295

Aryal B, Jeong J, Rao VA (2014) Doxorubicin-induced carbonylation and degradation of cardiac myosin binding protein C promote cardiotoxicity. Proc Natl Acad Sci U S A 111(5):2011–2016

Bailon P et al (2001) Rational design of a potent, long-lasting form of interferon: a 40 kDa branched polyethylene glycol-conjugated interferon alpha-2a for the treatment of hepatitis C. Bioconjug Chem 12(2):195–202

Beals JM, Shanafelt AB (2006) Enhancing exposure of protein therapeutics. Drug Discov Today Technol 3(1):87–94

Bennett CL et al (2004) Pure red-cell aplasia and epoetin therapy. N Engl J Med 351(14): 1403–1408

Bird GH et al (2010) Hydrocarbon double-stapling remedies the proteolytic instability of a lengthy peptide therapeutic. Proc Natl Acad Sci U S A 107(32):14093–14098

Bossi O et al (2008) UV irradiation increases ROS production via PKCdelta signaling in primary murine fibroblasts. J Cell Biochem 105(1):194–207

Boyd SE, Pike RN, Rudy GB, Whisstock JC, Garcia de la Banda M (2005) PoPS: a computational tool for modeling and predicting protease specificity. J Bioinform Comput Biol 3(3):551–585

Braun CR et al (2010) Photoreactive stapled BH3 peptides to dissect the BCL-2 family interactome. Chem Biol 17(12):1325–1333

Brems DN, Brown PL, Becker GW (1990) Equilibrium denaturation of human growth hormone and its cysteine-modified forms. J Biol Chem 265(10):5504–5511

Butterfield DA, Kanski J (2001) Brain protein oxidation in age-related neurodegenerative disorders that are associated with aggregated proteins. Mech Ageing Dev 122(9):945–962

Calabrese V et al (2006) Redox regulation of cellular stress response in neurodegenerative disorders. Ital J Biochem 55(3–4):263–282

Chennamsetty N, Voynov V, Kayser V, Helk B, Trout BL (2009) Design of therapeutic proteins with enhanced stability. Proc Natl Acad Sci U S A 106(29):11937–11942

Cines DB, Yasothan U, Kirkpatrick P (2008) Romiplostim. Nat Rev Drug Discov 7(11):887–888

Clark R et al (1996) Long-acting growth hormones produced by conjugation with polyethylene glycol. J Biol Chem 271(36):21969–21977

Comhair SA, Bhathena PR, Dweik RA, Kavuru M, Erzurum SC (2000) Rapid loss of superoxide dismutase activity during antigen-induced asthmatic response. Lancet 355(9204):624

Dabbs RA, Wyatt AR, Yerbury JJ, Ecroyd H, Wilson MR (2013) Extracellular chaperones. Top Curr Chem 328:241–268

De Rosa R et al (2005) Intranasal administration of nerve growth factor (NGF) rescues recognition memory deficits in AD11 anti-NGF transgenic mice. Proc Natl Acad Sci U S A 102(10):3811–3816

Della Pina P et al (2012) Biological drugs: classic adverse effects and new clinical evidences. Cardiovasc Toxicol 12(4):285–297

Dobson CM (2006) Protein aggregation and its consequences for human disease. Protein Pept Lett 13(3):219–227

Doroshow JH, Locker GY, Myers CE (1980) Enzymatic defenses of the mouse heart against reactive oxygen metabolites: alterations produced by doxorubicin. J Clin Invest 65(1):128–135

Dulaney SB, Huang X (2012) Strategies in synthesis of heparin/heparan sulfate oligosaccharides: 2000-present. Adv Carbohydr Chem Biochem 67:95–136

Egrie JC, Browne JK (2001) Development and characterization of novel erythropoiesis stimulating protein (NESP). Br J Cancer 84(S1):3–10

Egrie JC, Dwyer E, Browne JK, Hitz A, Lykos MA (2003) Darbepoetin alfa has a longer circulating half-life and greater in vivo potency than recombinant human erythropoietin. Exp Hematol 31(4):290–299

Elliott S et al (2004) Control of rHuEPO biological activity: the role of carbohydrate. Exp Hematol 32(12):1146–1155

Foye WOLTLWDA (2008) Foye's principles of medicinal chemistry. Lippincott Williams & Wilkins, Philadelphia

Garay RP, El-Gewely R, Armstrong JK, Garratty G, Richette P (2012) Antibodies against polyethylene glycol in healthy subjects and in patients treated with PEG-conjugated agents. Expert Opin Drug Deliv 9(11):1319–1323

Gaspar MM, Perez-Soler R, Cruz ME (1996) Biological characterization of L-asparaginase liposomal formulations. Cancer Chemother Pharmacol 38(4):373–377

Gasteiger E et al (2005) Protein Identification and analysis tools on the ExPASy server. In: Walker J (ed) The proteomics protocols handbook. Humana, Totowa, pp 571–607

Giasson BI et al (2000) Oxidative damage linked to neurodegeneration by selective alpha-synuclein nitration in synucleinopathy lesions. Science 290(5493):985–989

Gillies SD et al (2002) Bi-functional cytokine fusion proteins for gene therapy and antibody-targeted treatment of cancer. Cancer Immunol Immunother 51(8):449–460

Goldenberg MM (1999) Etanercept, a novel drug for the treatment of patients with severe, active rheumatoid arthritis. Clin Ther 21(1):75–87, discussion 71–72

Goth L (2008) Catalase deficiency and type 2 diabetes. Diabetes Care 31(12):e93

Gsponer J, Babu MM (2012) Cellular strategies for regulating functional and nonfunctional protein aggregation. Cell Rep 2(5):1425–1437

Han J et al (2008) Involvement of protective autophagy in TRAIL resistance of apoptosis-defective tumor cells. J Biol Chem 283(28):19665–19677

Hazell LJ, van den Berg JJ, Stocker R (1994) Oxidation of low-density lipoprotein by hypochlorite causes aggregation that is mediated by modification of lysine residues rather than lipid oxidation. Biochem J 302(Pt 1):297–304

Heikoop JC, van den Boogaart P, Mulders JW, Grootenhuis PD (1997) Structure-based design and protein engineering of intersubunit disulfide bonds in gonadotropins. Nat Biotechnol 15(7): 658–662

Herczenik E et al (2007) Activation of human platelets by misfolded proteins. Arterioscler Thromb Vasc Biol 27(7):1657–1665

Hermeling S, Crommelin DJ, Schellekens H, Jiskoot W (2004) Structure-immunogenicity relationships of therapeutic proteins. Pharm Res 21(6):897–903

Hochgrebe T, Pankhurst GJ, Wilce J, Easterbrook-Smith SB (2000) pH-dependent changes in the in vitro ligand-binding properties and structure of human clusterin. Biochemistry 39(6): 1411–1419

Iborra M et al (2011) Role of oxidative stress and antioxidant enzymes in Crohn's disease. Biochem Soc Trans 39(4):1102–1106

Inoguchi T et al (2003) Protein kinase C-dependent increase in reactive oxygen species (ROS) production in vascular tissues of diabetes: role of vascular NAD(P)H oxidase. J Am Soc Nephrol 14(8 Suppl 3):S227–S232

Ishikawa M et al (1992) The substitution of cysteine 17 of recombinant human G-CSF with alanine greatly enhanced its stability. Cell Struct Funct 17(1):61–65

Jasin HE (1983) Generation of IgG aggregates by the myeloperoxidase-hydrogen peroxide system. J Immunol 130(4):1918–1923

Jazayeri JA, Carroll GJ (2008) Fc-based cytokines : prospects for engineering superior therapeutics. BioDrugs 22(1):11–26

Jiang Y, Jiang W, Qiu Y, Dai W (2011) Effect of a structurally modified human granulocyte colony stimulating factor, G-CSFa, on leukopenia in mice and monkeys. J Hematol Oncol 4:28

Jomova K, Vondrakova D, Lawson M, Valko M (2010) Metals, oxidative stress and neurodegenerative disorders. Mol Cell Biochem 345(1–2):91–104

Kahook MY et al (2010) High-molecular-weight aggregates in repackaged bevacizumab. Retina 30(6):887–892

Kaufman RJ (1998) Post-translational modifications required for coagulation factor secretion and function. Thromb Haemost 79(6):1068–1079

Keeney PM et al (2009) Mitochondrial gene therapy augments mitochondrial physiology in a Parkinson's disease cell model. Hum Gene Ther 20(8):897–907

Kim TH et al (2012) PEG-transferrin conjugated TRAIL (TNF-related apoptosis-inducing ligand) for therapeutic tumor targeting. J Control Release 162(2):422–428

Kirkham P, Rahman I (2006) Oxidative stress in asthma and COPD: antioxidants as a therapeutic strategy. Pharmacol Ther 111(2):476–494

Kroon DJ, Baldwin-Ferro A, Lalan P (1992) Identification of sites of degradation in a therapeutic monoclonal antibody by peptide mapping. Pharm Res 9(11):1386–1393

Labbadia J et al (2012) Suppression of protein aggregation by chaperone modification of high molecular weight complexes. Brain 135(Pt 4):1180–1196

Lam XM, Yang JY, Cleland JL (1997) Antioxidants for prevention of methionine oxidation in recombinant monoclonal antibody HER2. J Pharm Sci 86(11):1250–1255

Lawson KV, Rose TE, Harran PG (2013) Template-constrained macrocyclic peptides prepared from native, unprotected precursors. Proc Natl Acad Sci U S A 110(40):E3753–E3760

Leach JK, Van Tuyle G, Lin PS, Schmidt-Ullrich R, Mikkelsen RB (2001) Ionizing radiation-induced, mitochondria-dependent generation of reactive oxygen/nitrogen. Cancer Res 61(10):3894–3901
Li J et al (2001) Thrombocytopenia caused by the development of antibodies to thrombopoietin. Blood 98(12):3241–3248
Lochhead JJ, Thorne RG (2012) Intranasal delivery of biologics to the central nervous system. Adv Drug Deliv Rev 64(7):614–628
Lopert P, Day BJ, Patel M (2012) Thioredoxin reductase deficiency potentiates oxidative stress, mitochondrial dysfunction and cell death in dopaminergic cells. PLoS One 7(11):e50683
Lu HS et al (1992) Folding and oxidation of recombinant human granulocyte colony stimulating factor produced in Escherichia coli. Characterization of the disulfide-reduced intermediates and cysteine—serine analogs. J Biol Chem 267(13):8770–8777
Lu HS et al (1999) Chemical modification and site-directed mutagenesis of methionine residues in recombinant human granulocyte colony-stimulating factor: effect on stability and biological activity. Arch Biochem Biophys 362(1):1–11
Luxon BA, Grace M, Brassard D, Bordens R (2002) Pegylated interferons for the treatment of chronic hepatitis C infection. Clin Ther 24(9):1363–1383
Mansour RB et al (2008) Increased levels of autoantibodies against catalase and superoxide dismutase associated with oxidative stress in patients with rheumatoid arthritis and systemic lupus erythematosus. Scand J Rheumatol 37(2):103–108
Matheson NR, Wong PS, Travis J (1979) Enzymatic inactivation of human alpha-1-proteinase inhibitor by neutrophil myeloperoxidase. Biochem Biophys Res Commun 88(2):402–409
Matheson NR, Janoff A, Travis J (1982) Enzymatic oxidation of alpha-1-proteinase inhibitor in abnormal tissue turnover. Mol Cell Biochem 45(2):65–71
McGoff P, Scher DS (2000) Solution formulation of proteins/peptides. In: McNally EJ (ed) Protein formulation and delivery. Marcel Dekker, New York, pp 133–152
Meredith SC (2005) Protein denaturation and aggregation: cellular responses to denatured and aggregated proteins. Ann N Y Acad Sci 1066:181–221
Migliore MM, Vyas TK, Campbell RB, Amiji MM, Waszczak BL (2010) Brain delivery of proteins by the intranasal route of administration: a comparison of cationic liposomes versus aqueous solution formulations. J Pharm Sci 99(4):1745–1761
Mohsenzadegan M, Mirshafiey A (2012) The immunopathogenic role of reactive oxygen species in Alzheimer disease. Iran J Allergy Asthma Immunol 11(3):203–216
Mulinacci F, Poirier E, Capelle MA, Gurny R, Arvinte T (2013) Influence of methionine oxidation on the aggregation of recombinant human growth hormone. Eur J Pharm Biopharm 85(1):42–52
Nilsson P et al (2013) A^2 secretion and plaque formation depend on autophagy. Cell Rep 5(1):61–69
Park JB et al (2010) PEGylation of bacterial cocaine esterase for protection against protease digestion and immunogenicity. J Control Release 142(2):174–179
Patel J, Kothari R, Tunga R, Ritter NM, Tunga BS (2011) Stability considerations for biopharmaceuticals. Part 1: overview of protein and peptide degradation pathways. Bioprocess Int 9:20–31
Pavisic R et al (2010) Recombinant human granulocyte colony stimulating factor pre-screening and screening of stabilizing carbohydrates and polyols. Int J Pharm 387(1–2):110–119
Perl A (2013) Oxidative stress in the pathology and treatment of systemic lupus erythematosus. Nat Rev Rheumatol 9(11):674–686
Phillips NB et al (2010) Supramolecular protein engineering: design of zinc-stapled insulin hexamers as a long acting depot. J Biol Chem 285(16):11755–11759
Poon S, Easterbrook-Smith SB, Rybchyn MS, Carver JA, Wilson MR (2000) Clusterin is an ATP-independent chaperone with very broad substrate specificity that stabilizes stressed proteins in a folding-competent state. Biochemistry 39(51):15953–15960
Rajadhyaksha M, Boyden T, Liras J, El-Kattan A, Brodfuehrer J (2011) Current advances in delivery of biotherapeutics across the blood–brain barrier. Curr Drug Discov Technol 8(2):87–101

Raju TS, Scallon BJ (2006) Glycosylation in the Fc domain of IgG increases resistance to proteolytic cleavage by papain. Biochem Biophys Res Commun 341(3):797–803
Rangel LP, Costa DC, Vieira TC, Silva JL (2014) The aggregation of mutant p53 produces prion-like properties in cancer. Prion 8(1):75–84
Raso SW et al (2005) Aggregation of granulocyte-colony stimulating factor in vitro involves a conformationally altered monomeric state. Protein Sci 14(9):2246–2257
Ratanji KD, Derrick JP, Dearman RJ, Kimber I (2014) Immunogenicity of therapeutic proteins: influence of aggregation. J Immunotoxicol 11(2):99–109
Reily C et al (2013) Mitochondrially targeted compounds and their impact on cellular bioenergetics. Redox Biol 1(1):86–93
Reubsaet JL et al (1998) Oxidation of recombinant methionyl human granulocyte colony stimulating factor. J Pharm Biomed Anal 17(2):283–289
Rosenberg AS (2006) Effects of protein aggregates: an immunologic perspective. AAPS J 8(3):E501–E507
Schellekens H, Hennink WE, Brinks V (2013) The immunogenicity of polyethylene glycol: facts and fiction. Pharm Res 30(7):1729–1734
Schulte S (2008) Use of albumin fusion technology to prolong the half-life of recombinant factor VIIa. Thromb Res 122(Suppl 4):S14–S19
Schulte S (2009) Half-life extension through albumin fusion technologies. Thromb Res 124(Suppl 2):S6–S8
Seidl A et al (2012) Tungsten-induced denaturation and aggregation of epoetin alfa during primary packaging as a cause of immunogenicity. Pharm Res 29(6):1454–1467
Shacter E (2000) Quantification and significance of protein oxidation in biological samples. Drug Metab Rev 32(3–4):307–326
Shacter E, Beecham EJ, Covey JM, Kohn KW, Potter M (1988) Activated neutrophils induce prolonged DNA damage in neighboring cells. Carcinogenesis 9(12):2297–2304
Shah D, Sah S, Wanchu A, Wu MX, Bhatnagar A (2013) Altered redox state and apoptosis in the pathogenesis of systemic lupus erythematosus. Immunobiology 218(4):620–627
Sharma B (2007a) Immunogenicity of therapeutic proteins. Part 1: impact of product handling. Biotechnol Adv 25(3):310–317
Sharma B (2007b) Immunogenicity of therapeutic proteins. Part 3: impact of manufacturing changes. Biotechnol Adv 25(3):325–331
Shukla HD, Pitha PM (2012) Role of hsp90 in systemic lupus erythematosus and its clinical relevance. Autoimmune Dis 2012:728605
Silva JL, De Moura Gallo CV, Costa DC, Rangel LP (2014) Prion-like aggregation of mutant p53 in cancer. Trends Biochem Sci 39(6):260–267
Sola RJ, Griebenow K (2009) Effects of glycosylation on the stability of protein pharmaceuticals. J Pharm Sci 98(4):1223–1245
Song J et al (2011) Bioinformatic approaches for predicting substrates of proteases. J Bioinform Comput Biol 9(1):149–178
Song J et al (2012) PROSPER: an integrated feature-based tool for predicting protease substrate cleavage sites. PLoS One 7(11):e50300
Soong R, Brender JR, Macdonald PM, Ramamoorthy A (2009) Association of highly compact type II diabetes related islet amyloid polypeptide intermediate species at physiological temperature revealed by diffusion NMR spectroscopy. J Am Chem Soc 131(20):7079–7085
Sorgato MC, Sartorelli L, Loschen G, Azzi A (1974) Oxygen radicals and hydrogen peroxide in rat brain mitochondria. FEBS Lett 45(1):92–95
Stadtman ER (2006) Protein oxidation and aging. Free Radic Res 40(12):1250–1258
Swaminathan J, Ehrhardt C (2012) Liposomal delivery of proteins and peptides. Expert Opin Drug Deliv 9(12):1489–1503
Takalo M, Salminen A, Soininen H, Hiltunen M, Haapasalo A (2013) Protein aggregation and degradation mechanisms in neurodegenerative diseases. Am J Neurodegener Dis 2(1):1–14
Tarnai I et al (2007) Effect of C111T polymorphism in exon 9 of the catalase gene on blood catalase activity in different types of diabetes mellitus. Free Radic Res 41(7):806–811

Tong P et al (2013) An engineered HIV-1 gp41 trimeric coiled coil with increased stability and anti-HIV-1 activity: implication for developing anti-HIV microbicides. J Antimicrob Chemother 68(11):2533–2544

Torosantucci R, Schoneich C, Jiskoot W (2014) Oxidation of therapeutic proteins and peptides: structural and biological consequences. Pharm Res 31(3):541–553

Tucker HM et al (2000) The plasmin system is induced by and degrades amyloid-beta aggregates. J Neurosci 20(11):3937–3946

Ulrich AS (2002) Biophysical aspects of using liposomes as delivery vehicles. Biosci Rep 22(2):129–150

Veronese FM, Mero A (2008) The impact of PEGylation on biological therapies. BioDrugs 22(5):315–329

Vigneri R, Squatrito S, Sciacca L (2010) Insulin and its analogs: actions via insulin and IGF receptors. Acta Diabetol 47(4):271–278

Vugmeyster Y, Xu X, Theil FP, Khawli LA, Leach MW (2012) Pharmacokinetics and toxicology of therapeutic proteins: advances and challenges. World J Biol Chem 3(4):73–92

Walsh G, Jefferis R (2006) Post-translational modifications in the context of therapeutic proteins. Nat Biotechnol 24(10):1241–1252

Wang T, Xia Y (2012) Inducible nitric oxide synthase aggresome formation is mediated by nitric oxide. Biochem Biophys Res Commun 426(3):386–389

Wang YS et al (2002) Structural and biological characterization of pegylated recombinant interferon alpha-2b and its therapeutic implications. Adv Drug Deliv Rev 54(4):547–570

Wang YJ et al (2010) PEGylation markedly enhances the in vivo potency of recombinant human non-glycosylated erythropoietin: a comparison with glycosylated erythropoietin. J Control Release 145(3):306–313

Wang XM, Lehky TJ, Brell JM, Dorsey SG (2012) Discovering cytokines as targets for chemotherapy-induced painful peripheral neuropathy. Cytokine 59(1):3–9

Weiner AL (1994) Liposomes for protein delivery: selecting manufacture and development processes. Immuno Methods 4(3):201–209

Wolf P et al (2000) Topical treatment with liposomes containing T4 endonuclease V protects human skin in vivo from ultraviolet-induced upregulation of interleukin-10 and tumor necrosis factor-alpha. J Invest Dermatol 114(1):149–156

Wondrak GT (2009) Redox-directed cancer therapeutics: molecular mechanisms and opportunities. Antioxid Redox Signal 11(12):3013–3069

Wyatt AR, Yerbury JJ, Ecroyd H, Wilson MR (2013) Extracellular chaperones and proteostasis. Annu Rev Biochem 82:295–322

Xiao G, Gan LS (2013) Receptor-mediated endocytosis and brain delivery of therapeutic biologics. Int J Cell Biol 2013:703545

Xu X, Vugmeyster Y (2012) Challenges and opportunities in absorption, distribution, metabolism, and excretion studies of therapeutic biologics. AAPS J 14(4):781–791

Xu J et al (2011) Gain of function of mutant p53 by coaggregation with multiple tumor suppressors. Nat Chem Biol 7(5):285–295

Yao Z, Dai W, Perry J, Brechbiel MW, Sung C (2004) Effect of albumin fusion on the biodistribution of interleukin-2. Cancer Immunol Immunother 53(5):404–410

Yerbury JJ, Rybchyn MS, Easterbrook-Smith SB, Henriques C, Wilson MR (2005) The acute phase protein haptoglobin is a mammalian extracellular chaperone with an action similar to clusterin. Biochemistry 44(32):10914–10925

Yin J et al (2005) Effects of excipients on the hydrogen peroxide-induced oxidation of methionine residues in granulocyte colony-stimulating factor. Pharm Res 22(1):141–147

Zheng K, Yarmarkovich M, Bantog C, Bayer R, Patapoff TW (2014) Influence of glycosylation pattern on the molecular properties of monoclonal antibodies. MAbs 6(3):649–658

Zhong X, Wright JF (2013) Biological insights into therapeutic protein modifications throughout trafficking and their biopharmaceutical applications. Int J Cell Biol 2013:273086

Zinman B (2013) Newer insulin analogs: advances in basal insulin replacement. Diabetes Obes Metab 15(Suppl 1):6–10

Antibody-Like Molecules Designed for Superior Targeting and Pharmacokinetics

Alexey A. Lugovskoy and Melissa L. Geddie

Introduction

Antibody therapeutics is a thriving field with currently over thirty treatments that are approved primarily for oncology and inflammatory disorders. Antibodies have many advantages as drugs, including high specificity to their targets, long half-lives, and generally highly favorable toxicity profiles. The majority of the currently approved antibody therapeutics are conventional immunoglobulin Gs directed at a single target. However, as the field of protein engineering has advanced, a significant number of new antibody-like formats have been developed that possess robust bioactivity and manufacturability profiles. A number of these molecules have demonstrated superiority to conventional monoclonal antibodies in preclinical settings and have entered clinical testing. This review will discuss engineering of antibody-like molecules that optimize efficacy by targeting multiple receptors or by incorporating additional mechanisms of action, including altered effector function against established therapeutic targets. These molecules are commonly termed biobetters, which, formally speaking, are biologic drugs that are developed against previously validated target antigens but have some properties that are superior compared to currently approved products for commercial use.

A.A. Lugovskoy (✉) • M.L. Geddie
Merrimack Pharmaceuticals, Inc.,
One Kendall Square Suite B7201, Cambridge, MA 02139, USA
e-mail: alugovskoy@merrimackpharma.com; mgeddie@merrimackpharma.com

A. Rosenberg, B. Demeule (eds.), *Biobetters*, AAPS Advances in the Pharmaceutical Sciences Series 19, DOI 10.1007/978-1-4939-2543-8_12

Engineering Antibody-Like Molecules for Success

While therapeutic antibodies have attracted a great deal of attention from the biopharmaceutical industry, their development continues to be far from straightforward. In fact, these molecules experience comparable attrition rates to small molecules through development stages, with the situation being particularly dire in cardiovascular disease and oncology (Hay et al. 2014). Most of these failures occur in Phase II studies, where the biological hypothesis is often being tested in the clinical setting for the first time. If this test and confirmatory studies are successful, the pathology of the disease becomes linked to the activity of the specific antigen (e.g. a growth factor receptor) and to its treatment by blockade of the antigen's activity by an antibody. It is tempting to think that this group of disease-linked antigens constitutes a special subset of targets that would be easier to interrogate. This belief, coupled with the commercial success of subsequent generation small molecules, (e.g., atorvastatin was the fifth drug in the statin class to be developed), strongly support the development of biobetter antibodies against validated targets. Biobetters can be engineered to have improved target affinity, enhanced crosslinking or degradation of target molecules, better engagement or recruitment of the immune system, longer circulation half-life, decreased immunogenicity, improved tolerability, and other potentially useful qualities. A classic example of a validated target is TNFα, where four antagonists have been approved by the United States Food and Drug Administration following infliximab (Tracey et al. 2008). While these molecules are considered to be largely similar efficacy-wise, they are differentiated by the convenience of dosing, tolerability and immunogenicity.

It is worth noting that while TNFα is a soluble molecule, a typical antibody target in the oncology setting is a cell-surface protein. Cancer cells may express differing amounts of the target antigen, may mutate in the face of selection pressure to express very low levels of such antigen, or develop other mechanisms to limit the activity of the mAb. For example, despite higher affinity for EGFR and lower immunogenicity, panitumumab has not been proven to be clinically superior to cetuximab (Price et al. 2013). Furthermore, biobetter development has proven to be challenging in the case of targeting the CD20 antigen, an established target for B cell malignancies and autoimmune disorders, for which rituximab was developed and approved (Cang et al. 2012). Second generation molecules focused on decreasing immunogenicity, by reducing the murine amino acid content in the antibody and/or increasing affinity to CD20, have failed to demonstrate clear advantages over rituximab. This led to the realization that immunogenicity may be a less important differentiator in therapeutic oncology candidates, but only when co-administration of cytotoxic drugs (Cunningham et al. 2013) depletes immune cells, thus minimizing the anti-antibody immune responses.

Surpassing the therapeutic efficacy of an innovator antibody by a biobetter through improvements in antibody pharmacokinetics and pharmacodynamics may be limited. If an antibody competes with the cognate interaction partner for its antigen or deposits on the cellular membrane in sufficient quantities to activate effector function and if its half-life is sufficient to optimally support these activities, the target mechanism may become saturated and further improvements in affinity or half-life may not lead to a significant increase in biological activity. Consequently, third generation biobetter mAb candidates of anti-CD20 molecules have focused on improving efficacy through diversifying the mechanism of action. One of these in particular, obinutuzumab, has shown a clear advantage over rituximab in a head-to-head clinical trial (Goede et al. 2014) and has been approved in combination with chlorambucil for patients with previously untreated chronic lymphocytic leukemia. In addition to being glycoengineered to induce more potent immune destruction of CD20 positive cells via ADCC (Golay et al. 2013), obinutuzumab targets an alternative CD20 epitope that allows it to induce direct cell death (Herter et al. 2013) thus complementing the additional immune effectors. The success of obinutuzumab underscores an important step in engineering a successful biobetter mAb: the analysis of the limitations of the innovator molecule thereby guiding engineering of additional mechanisms of action in the follow-on molecule. Confirming this observation, currently developed next-generation EGFR targeting agents, such as GA201, (Gerdes et al. 2013), CetuGEX (Reichert and Dhimolea 2012), Sym004 (Pedersen et al. 2010) and MM-151 (Fauvel and Yasri 2014) achieve superior activity by employing new mechanisms of action. They engage immune function more potently (GA201, CetuGEX and MM-151), induce removal of the receptor from the cancer cell surface by triggering internalization and degradation (Sym004 and MM-151), and act as EGFR superantagonists (MM-151). The engineering of new mechanisms of action into the biobetter often mandates modulating the effector function or topology of the classical immunoglobulin G (IgG) format.

Overview of Immunoglobulin G Structure

An immunoglobulin G is a symmetric homodimer that consists of two Fab arms with one Fc arm connected by a 'hinge', a cysteine-rich linker region (Fig. 1). The antigen binds to the Fab arms, so improvements in the affinity or stability are generally done through optimization of this region, by altering the complementary determining regions (CDRs) and framework regions (FR). The Fab region is a heterodimer comprising the VH and CH1 portion of the heavy chain and the entire light chain, which is composed of the VL and CL domains. The Fc region comprises the CH2 and CH3 domains of the heavy chain and is responsible for dimerization and for effector functions, i.e., the engagement of the immune system through the Fc receptor, complement binding, and maintaining a long half-life through interaction with the neonatal Fc receptor, FcRn. Distinct antibody isotypes have different affinities for Fc receptors and activities as mediators of complement function via binding to C1q. Thus, they vary in their ability to engage immune system components and in their circulation half-lives. (Alyanakian et al. 2003).

IgGs can be engineered in many ways, but they are primarily modified to improve affinity and stability through engineering of the Fab portion or to alter effector

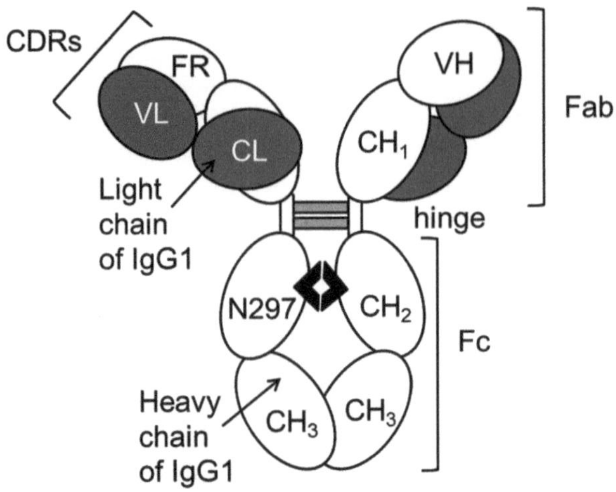

Fig. 1 Domain structure of an IgG1 antibody

function through engineering of the Fc domain. Although it was originally thought that the Fab arms and the Fc arm functioned independently of one another, there is increasing evidence that each can influence the function of the other. For example, when an antibody binds to a multivalent antigen or a cell-surface protein, the second Fab arm will have accelerated binding leading to an increase in the apparent affinity. This effect, termed cross-arm binding or avidity, is caused by increases in the local concentration of antibody in the vicinity of its target, triggered by the first binding event (Harms et al. 2014). Also, proteins with identical Fc regions, but different Fab arms, can have vastly different pharmacokinetic properties. This could be due to their different internalization capacities, or due to Fab-dependent modulation of FcRn binding (Wang et al. 2011). Further, antibodies with the same Fab fragment but different Fc regions can exhibit vastly different antigen cross-linking and internalization functions. This highlights the need for concordant optimization of Fab and Fc regions of an antibody-like molecule.

Effector Functions Are Mediated by the Constant Region

The constant region of an antibody mediates four primary different effector functions, which are often desirable in a therapeutic. An Fc containing antibody generally has a long half-life, typically weeks, which allows for greater exposure and less frequent dosing intervals. This is primarily due to the interaction of the Fc region with FcRn (Brambell et al. 1964) which promotes recycling of the antibody back into the circulation. Additionally, the constant region mediates interactions with FcγRs, which modulate antibody-dependent cellular toxicity (ADCC) and antibody-dependent cellular phagocytosis (ADCP). Furthermore, Fc-dependent interactions with complement proteins such as C1q trigger complement-dependent cellular cytotoxicity (CDC), which leads to damage of the cellular membrane of targeted cells. Retaining these effector functions has been shown to be critical for the efficacy of several marketed antibodies, including trastuzumab (Clynes et al. 2000) and rituximab (Weiner 2010). Given the importance of both effector function and half-life for antibody bioactivity, a large amount of engineering has been done to optimize the Fc to support these components of mechanism of action (Zalevsky et al. 2010). As a result, there are now well-characterized mutations within the Fc region that alter antibody effector function by modulating binding to Fc receptors and C1q (Presta 2006, 2008).

Engineering Effector Function to Enhance Activity of a Therapeutic Antibody

When considering the design of immune effector function for a therapeutic, it is important to determine the desired type and magnitude of effector function up-front. This level will depend both on the antigen characteristics and chosen molecular format, and also may need to be determined empirically for each new therapeutic to prevent life-threatening overactivation of the immune system ("Can super-antibody drugs be tamed?" 2006; Stebbings et al. 2007). For a biobetter, this determination is more straightforward as it is guided by the clinical data generated with the first generation molecule. There are three approaches that are commonly employed for altering immune effector function: isotype optimization, glycoengineering, and targeted mutagenesis within the Fc region.

Isotype Optimization

Selection of isotype can have a major impact on the properties of an antibody drug. While most therapeutic antibodies are IgG1s, IgG2s or IgG4s have been also used. In fact, the most straightforward way to reduce effector function is to select a non-IgG1 isotype. Both IgG2 and IgG4 show weaker binding to Fc receptors, particularly to the activating FcR, FcγRIIIA (Nimmerjahn and Ravetch 2008), with IgG4 also having low affinity for the complement protein C1q (Salfeld 2007). In addition, IgG4 has been shown to exchange Fab arms with endogenous human IgG4 due to its unique hinge and CH3 domain composition (Labrijn et al. 2009). Therefore, in human plasma, unmodified IgG4s should be considered to be functionally monovalent molecules that have poor antigen cross-linking capability. This property was shown to be present in natalizumab, an FDA approved IgG4 antibody targeting α-4-integrin which is used in the treatment of multiple sclerosis and Crohn's disease (Shapiro et al. 2011). Additional examples of IgG4 antibodies in the clinic include nivolumab and pembrolizumab, anti PD-1 antibodies that are being tested in Phase

III clinical trials in several oncology indications (Lu et al. 2014). These molecules contain IgG4 Fc domains to reduce the depletion of anti-tumor T cells, which are activated following inhibition of PD-1/PD-L1 interactions. However, due to their stabilized hinge regions via reversion to an IgG1 consensus motif, nivolumab and pembrolizumab do not scramble with serum IgG4 and are functionally bivalent.

Targeted Mutagenesis

Just as it is possible to change the antibody isotype, it is also possible to introduce mutations into the IgG1 Fc that reduce effector function. A single point mutation (K322A) in IgG1 has been found to reduce ADCC twofold while completely removing CDC, and a double mutation (L234A, L235A) removed both ADCC and CDC (Hezareh et al. 2001). The triple IgG1 mutant (L234F/L235A/P331S) has complete abolishment of both ADCC and CDC. Structural analysis of the mutant antibody revealed a very similar structure to unmodified antibodies (Oganesyan et al. 2008). Perhaps the most effectorless Fc variant described to date is an IgG2-based mutant, IgG2m4, which has four amino acid substitutions derived from IgG4 (H268Q, V309L, A330S, P331S). IgG2m4 has been shown to be completely devoid of ADCC, ADPC and CDC functions while maintaining stability (An et al. 2009). Antibody Fc mutants with reduced immune effector function are commonly used to treat immunological disorders. An example of an engineered clinical stage IgG1 antibody with reduced effector function is MPDL3280A, a therapeutic candidate in oncology which targets PD-L1, one of the PD-1 ligands important for regulation of immune checkpoint. Similar to anti-PD-1 antibodies, this is designed to minimize the depletion of PD-L1 positive anti-tumor immune cells.

Similar to reducing effector function, targeted mutagenesis can be implemented to enhance effector function. For example, triple mutations in the CH2 domain of IgG1 (S267E, H268F, S324T) can enhance complement activation approximately sevenfold (Moore et al. 2010). Similarly, there are IgG1 mutations that improve affinity to FcγRs (S239D, I332E), or are specific to activating receptors FcγRIIa and FcγRIIIa relative to the inhibitory receptor FcγRIIB (G236A/I332E) (Lazar et al. 2006; Richards et al. 2008). An extension of an antibody's half-life can decrease the frequency of dosing and/or increase its bioactivity. In that regard, there are mutations reported to increase the half-life of the antibody, primarily by promoting its binding to FcRn at endosomal pH. These include mutations in both the CH2 and CH3 regions (e.g., M252Y/S254T/T256E or M428L/N434S) (Hinton et al. 2004; Dall'Acqua et al. 2006; Richards et al. 2008; Zalevsky et al. 2010). Motavizumab, an antibody against RSV, has a serum half-life of 24 days in the clinic (Abarca et al. 2009) and requires monthly infusions. In contrast, the YTE IgG1 mutant (M252Y/S254T/T256E) showed a tenfold increase in the binding of FcRn at endosomal pH, which led to a fourfold increase in half-life in monkeys (Dall'Acqua et al. 2006). The reports of clinical Phase I data for this antibody demonstrate a marked improvement in half-life in humans extending it up to 100 days (Robbie et al. 2013).

The increased serum circulation time for engineered mAbs requires a more extensive evaluation of the stability of the antibody, as in vivo conditions of higher temperature (37 °C) and higher pH (7.4) are known to increase the rate of degradation reactions (such as oxidation or deamidation) that could lead to greater immunogenicity or loss of activity. Moreover, potential off target effects should also be rigorously evaluated.

A complementary technique to increasing half-life via FcRn is to promote antibody recycling (reviewed in Tesar and Björkman 2010) within the endosome through the engineering of pH-sensitive antigen binding. This is typically done by histidine scanning of the CDRs and variable regions to derive mutations that decrease target binding affinity at the endosomal pH, while maintaining it at neutral pH (Igawa et al. 2010). This technique was successfully used to improve the half-life of tocilizumab (Igawa et al. 2010) and a cholesterol lowering antibody targeting PCSK9 (Chaparro-Riggers et al. 2012).

Glycoengineering

An additional approach to alter effector function is to engineer the N-linked glycans that are present in position 297 in the CH2 domain of the heavy chain. Removing the glycan completely (by mutating the N297/S298/T299 glycosylation motif, typically N297Q or T299A) leads to a nearly complete ablation of FcγRIIIA binding. Reciprocally, one could also enhance immune effector function through the removal of core fucose. Afucosylated variants bind FcγRIIIA with greater affinity and show enhanced ADCC (Beck 2013; Listinsky et al. 2013). The first approved glycoengineered antibody, mogamulizumab, was produced in a Chinese Hamster Ovary (CHO) cell line with a genetic knock out of fucose transferase 8 (FUT8) (Beck and Reichert 2012). Alternatively, one can use a CHO cell line engineered to express *N*-acetylglucosaminyltransferase III (GnTIII), which catalyzes the addition of bisecting *N*-acetyl-D-glucosamine to block the addition of fucose. This method has been used to produce obinutuzumab, which was recently approved for the treatment of chronic myeloid leukemia (Mössner et al. 2010).

Improving Efficacy Through Bispecific Targeting

In addition to engineering of the constant region, the variable domain of antibodies can be optimized to improve activity by using altered antibody formats that allow the simultaneous targeting of different receptors. These multispecific antibodies have gained significant traction for treating diseases such as cancer, where cells commonly rely on different signaling pathways for proliferation and survival, and thus optimal treatment would consist of targeting both pathways (Kontermann 2012). Additionally, multispecific antibodies can be more potent than mixtures of separate antibodies, likely due to the increased avidity of multiple covalently-linked arms acting on proximal targets expressed on the same cell surface (Rudnick and Adams 2009; Harms et al. 2014). Currently there is only one bispecific antibody with an Fcbackbone approved, catumaxomab, an anti-EpCAM + anti CD3 molecule originally generated from a rodent quadroma (Linke et al. 2010). However, many more antibody-like inhibitors are now advancing through preclinical and clinical development. Multispecific antibodies have generally been engineered to optimize their activity within a certain disease-related pathway or to engage non-redundant pathways important for supporting the pathology of the disease.

Biobetter Antibody-Like Molecules in the Clinic

Multispecific antibody-like molecules can improve on existing therapies by targeting multiple receptors simultaneously, which can lead to greater efficacy or broader activity profile. They also could have greater potential for off-target effects. Many multispecific formats have been developed (Fig. 2) and several of them have entered clinical testing. One of them is DVD-Ig™, which contains two additional variable domains on each IgG monomer that are linked N-terminally to heavy and light

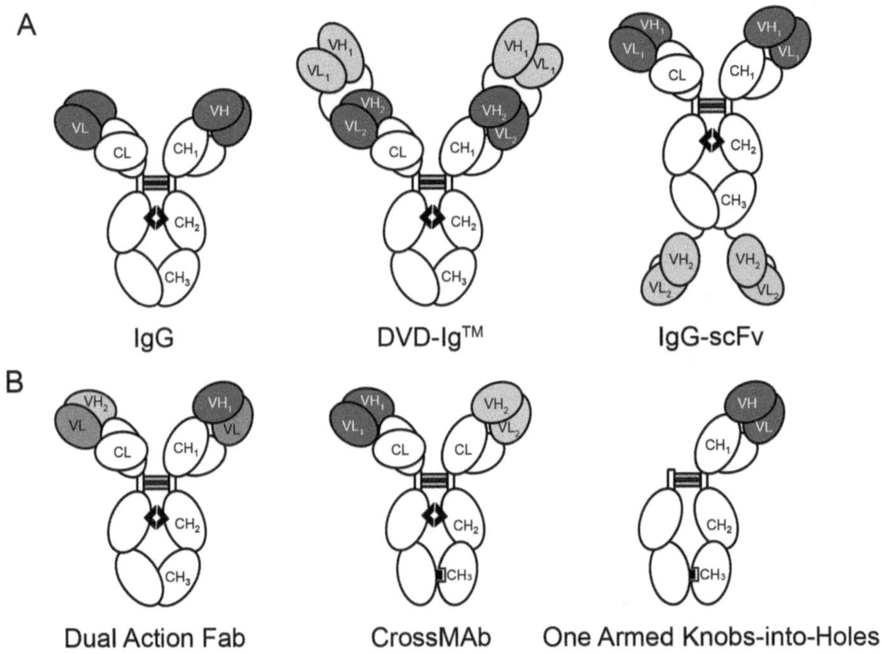

Fig. 2 Formats of bispecific antibody-like molecules in clinical trials. (**a**) Formats with symmetric pairing of the heavy chain and light chain, resulting in bivalent or tetravalent binding. (**b**) Formats that allow asymmetric binding of Fab arms that can lead to monovalent antigen binding

chains (Fig. 2a). The earliest reported DVD-Ig was designed as an anti-IL-12/anti IL-18 bispecific (Wu et al. 2007). This antibody was one of the first multispecific formats to show robust manufacturability, excellent pharmaceutical properties and good circulation half-life. Careful analysis of antigen recognition showed that this antibody-like molecule bound two molecules of each target, demonstrating that all four arms are simultaneously engaged. Because two variable regions are in tandem with each other, the inner variable domain has sometimes been found to have reduced binding affinity to its target. This reduced affinity can be mitigated by introducing longer linkers between the two domains, suggesting that shorter linkers may be sterically hindering the antigen access (Wu et al. 2007). More extensive linker optimization of the format revealed a complex relationship between length and composition; but overall, partially ordered longer linkers were more likely to allow antibodies in the inner domain to retain their affinity (Digiammarino et al. 2011). At least two of DVD-Ig molecules are currently in clinical trials, one targeting TNF and IL-17 and the other targeting IL-1 α and IL-β (Wu et al. 2009; DiGiammarino et al. 2012; Gu and Ghayur 2012).

Another example of a clinically-tested bispecific format with superior activity on validated targets is MEHD7945A, a Dual Action Fab (DAF) that targets EGFR and ErbB3 (Schaefer et al. 2011a) (Fig. 2b). This antibody retains the architecture of a conventional monoclonal antibody, but recognizes two antigens within its variable regions (Eigenbrot and Fuh 2013). DAFs are typically developed by taking an antibody that recognizes a first target and then randomizing either the light chain (Bostrom et al. 2009) or heavy chain CDRs (Lee et al. 2014) and selecting against a second target while introducing affinity retention screening against the first target. Preclinical work demonstrated that MEHD7945A had increased activity in multiple tumor models driven by EGFR when compared to a combination of two monospecific antibodies (Schaefer et al. 2011a). This antibody has now advanced to Phase II clinical trials in head and neck and colorectal cancer where it will be compared to cetuximab (NCT01577173, NCT01652482) on the backbone of standard of care chemotherapeutic regimen.

The DVD-Ig and DAF molecules are all symmetric due to the native pairing of the Fc fragment of their heavy chains, which renders the molecule bivalent or tetravalent. However, for a subset of targets that are activated through ligand-induced dimerization, such symmetric antibodies can activate agonistic, rather than the desired antagonistic, signaling due to crosslinking of antigens (Prat et al. 1998). This limitation triggered a significant interest in engineering monovalent antibodies through altering the constant regions to promote asymmetric pairing of heavy chains. Many of these antibodies build on "knobs-into-holes" technology (Ridgway et al. 1996), which relies on steric hindrance of asymmetrically mutated CH3 domains to impede symmetric pairing. This method has been used to engineer onartuzumab, a monovalent Hepatocyte Growth Factor Receptor (HGFR or c-Met) antagonist with one Fab arm and knobs-into-holes in the heavy chain (Fig. 2b); onartuzumab has advanced to Phase III clinical testing (Merchant et al. 2013). However, this original knobs-into-holes heterodimeric format does not address random association of the heavy and light chain chains in the Fab fragment, making it difficult to create a homogeneous bispecific with two different Fab specificities (Klein et al. 2012). One solution to this problem was provided by the CrossMab format (Schaefer et al. 2011b) that has one arm of the antibody dimer domain crossed over between VH-VL or the CH1-CL domains, allowing for specific light chain pairing (Fig. 2b). The heterodimerization of the Fc fragment is accomplished by introducing knobs-into-holes mutations. The CrossMab format has shown good manufacturability and demonstrated IgG-like circulation half-life. Recent preclinical work suggests that a bispecific CrossMab targeting Vascular Endothelial Growth Factor A (VEGF-A) and Angiopoietin-2 (Ang-2) (RO5520985, RG7221) may be a more effective anti-tumor agent (Kienast et al. 2013) when compared to VEGF-targeted therapies. Phase I trial results for this molecule have been recently reported with RO5520985 showing an acceptable safety and desired pharmacokinetic and pharmacodynamic profile (Lieu et al. 2014). An ongoing Phase II trial is evaluating the safety and efficacy of the RO5520985 compared to bevacizumab in combination with oxaliplatin/folinic acid/5-fluorouracil chemotherapy in metastatic colorectal cancer (NCT02141295).

Advances in antibody engineering described in this chapter have enabled the generation of many types of antibody-like molecules with adequate stability and manufacturability to support their clinical development. However, the choice of targets, formats, and modes of action that can yield superior clinical activity remain complicated by the redundancy and heterogeneity found in many human diseases. To address this challenge Merrimack Pharmaceuticals has developed a Network Biology approach that provides an analytical framework for highlighting disease-relevant cellular pathways for therapeutic intervention, guiding selection of actionable drug targets within these pathways, and identifying optimal inhibitor characteristics for these targets. In brief, this method involves building computational models of biochemical and cellular pathways, training these models using vast arrays of experimental data, conducting sensitivity analyses to identify the preferred intervention points and to reveal the desired inhibition profile, and optimizing properties of the hypothetical therapeutic relative to this profile (Schoeberl et al. 2009; Fitzgerald and Lugovskoy 2011; Harms et al. 2012, 2014).

Case Studies of Antibody-Like Molecules Designed for Superior Blockade of Growth Factor Induced Oncogenic Signaling

Growth-factor signaling pathways are activated to support host growth, control metabolic responses, and resist apoptotic stress. During malignant transformation, cancer cells upregulate these pathways to support their autonomous proliferation and to resist anti-tumor action of the immune system and cytotoxic therapies (Huang et al. 2010; Yano et al. 2012). Thus, it is not surprising that the strategy of blocking growth factor signaling pathways with mAbs has been an intense focus of development and yielded three classes of drugs: VEGF blockers (e.g., bevacizumab), EGFR blockers (e.g., cetuximab) and ErbB2 blockers (e.g., trastuzumab). However, tumors that are exquisitely sensitive to these agents are rare, as most of them use multiple growth factor signaling pathways to non-redundantly support their proliferation and survival. Even in a context of a dominant oncogene such as ErbB2, the merits of co-blocking multiple targets such as ErbB2 and ErbB3 (through antagonism of ErbB2/ErbB3 dimerization) have been demonstrated clinically, leading to the approval of pertuzumab, trastuzumab, and docetaxel combinations for the treatment of HER2-positive metastatic breast cancer (Swain et al. 2013). Using a Network Biology platform, Merrimack's bispecific antibody MM-111 (McDonagh et al. 2012) was designed to have superior activity on ErbB2/ErbB3 signaling by employing a novel mechanism of action. MM-111 directly targets ErbB2/ErbB3 heterodimers on the cell surface by binding to ErbB2 with high affinity and then to ErbB3 with high avidity due to ErbB2 anchoring. It promotes displacement of the ErbB3 ligand heregulin and sequesters ErbB2/ErbB3 heterodimers into inactive complexes. MM-111 does not trigger internalization of ErbB2/ErbB3 complexes, preserving the ability to co-target them with trastuzumab, which can elicit immune effector function. Preclinical studies have shown that MM-111 can be combined with trastuzumab and paclitaxel for increased antitumor activity. A randomized Phase II study of MM-111 with paclitaxel and trastuzumab in second line patients with ErbB2 positive carcinomas of the distal esophagus, gastroesophageal junction and stomach is currently on-going (NCT01774851).

The case study of MM-141, a dual antibody inhibitor of IGF-1R and ErbB3, provides another informative example of applying Network Biology to the design of next-generation antibody drugs against well-established targets that were suboptimally treated by the first-in-class molecules. It is well known that high levels of insulin-like growth factor 1 (IGF-1) are associated with poor prognosis in multiple cancers. In response to ligand binding, IGF-1 receptor (IGF-1R) activates pro-survival PI3K/AKT/mTOR signaling, rendering cancer cells resistant to therapy. Although antibody inhibition of IGF-1R was expected to eliminate this key resistance mechanism, clinical results to date have been disappointing (Pollak 2012). We have previously demonstrated that ErbB3 signaling, activated by its ligand heregulin, can limit the utility of IGF-1R inhibition. Specifically, we have shown that a majority of IGF-1 sensitive cell lines produce autocrine heregulin, which upregulates and activates ErbB3 in response to IGF-1R blockade. This compensation leads to rapid reactivation of PI3K/AKT/mTOR signaling and can result in the increase of oncogenic signaling rendering IGF-1R inhibitors ineffective. To facilitate the design of a dual inhibitor of IGF-1R and ErbB3, we constructed and trained a biochemical model of the PI3K/AKT/mTOR signaling network with IGF and ErbB family receptors, their ligands, and key intracellular proteins encoded (Fitzgerald et al. 2014). Modeling of the optimal antibody co-inhibitor of IGF-1R and ErbB3 (Fig. 3) demonstrated the need for two mechanisms of action: IGF-1, IGF-2 and heregulin ligand blockade through high affinity IGF-1R and ErbB3 binding and induction of IGF-1R and ErbB3 degradation. The simulation also revealed that high cross-target avidity was required, which led to the choice of a tetravalent bispecific format as opposed to a mixture of two immunoglobulins. We did not expect MM-141 to

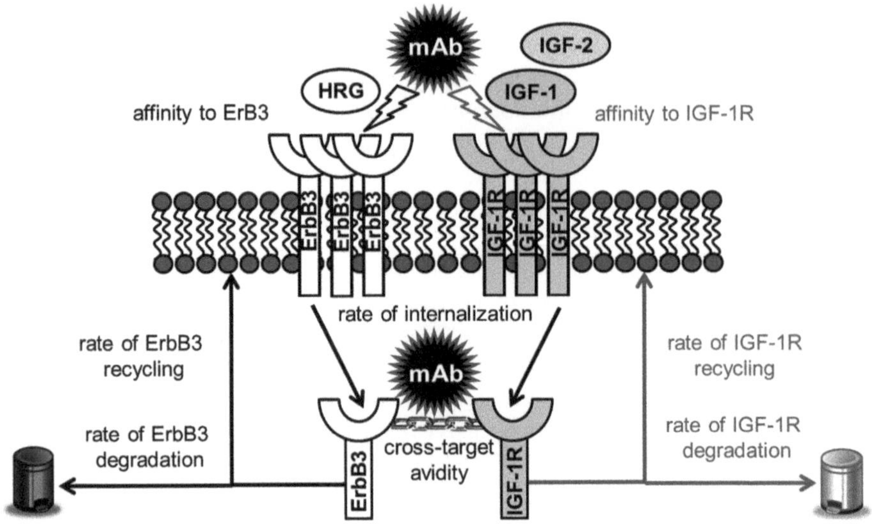

Fig. 3 Simulation guided design of the optimal dual inhibitor of IGF-1R and ErbB3. A sample antibody inhibitor was introduced into the trained model of IGF-1R and ErbB mediated PI3K/AKT/mTOR signaling with the following variable parameters: cross-target avidity, affinities to ErbB3 and IGF-1R, rates of internalization of antibody/receptor complexes and ErbB3 and IGF-1R recycling. High affinities to IGF-1R and ErbB3, cross-target avidity and the rapid receptor were identified as the most sensitive parameters for optimal PI3K/AKT/mTOR signaling inhibition

accumulate on the cell membrane in quantities sufficient to support ADCC, ADPC and CDC due to the low numbers of IGF-1R and ErbB3 expressed. Coupled with this expectation and with internalization being a major component of mechanism of action, our design was antibody isotype agnostic. We chose IgG1 Fc fragments due to its superior biophysical properties. As all highly engineered antibody-like molecules could be unstable, we conducted an extensive engineering campaign to derive lead molecules with robust pharmaceutical properties (Xu et al. 2013).

In preclinical studies MM-141 blocked IGF-1 and IGF-2 binding to IGF-1R as well as heregulin binding to ErbB3. MM-141 was also shown to trigger rapid degradation of receptor complexes containing IGF-1R and ErbB3, including their heterodimers with Insulin Receptor and ErbB2, showing superiority to monospecific antibodies targeting IGF-1R (Fitzgerald et al. 2014). By employing these novel components of mechanism of action, MM-141 was able to counteract pre-existing ligand-mediated resistance to several classes of cytotoxic therapies including antimetabolites (e.g., gemcitabine) and taxanes (e.g., nab-paclitaxel), as well as reverse the acquired resistance mediated by upregulation of IGF-1R and ErbB3 and their heterodimerization partners. MM-141 is currently undergoing Phase I development, where it is being tested as a monotherapy, in combination with everolimus, and in combination with nab-paclitaxel and gemcitabine (Isakoff et al. 2014). The Phase II study of MM-141 in combination with nab-paclitaxel and gemcitabine in metastatic pancreatic cancer is planned.

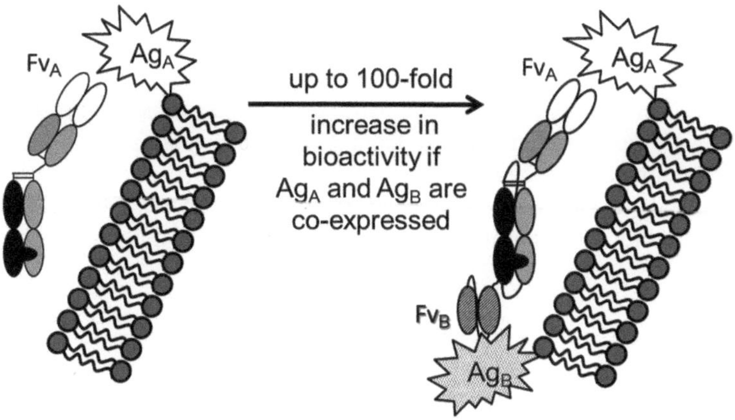

Fig. 4 Effect of avidity on the bioactivity of antibody-like molecules

Hepatocyte Growth Factor Receptor (HGFR or c-Met) is another important tyrosine kinase receptor in cancer biology. Normally involved in development and wound healing, c-Met and its ligand, hepatocyte growth factor (HGF), are often overexpressed in many cancers, including lung, colon, and colorectal. HGF binding triggers c-Met dimerization and activation. Conventional antibodies can also induce c-Met dimerization, therefore acting as agonists. One-armed antibody-like molecules, such as onartuzumab, do not cross-link c-Met and can inhibit its HGF-induced signaling (Feng and Ma 2011; Merchant et al. 2013). However, such monovalent antibodies can lose up to 100-fold in bioactivity due to the loss of avidity in their binding, making restoration of bivalent targeting beneficial (Fig. 4). To increase the therapeutic utility of a monovalent antibody c-Met antagonist, we have explored tumor antigens that could serve as anchors for bivalent cell surface binding (Harms et al. 2014). Network Biology simulations have suggested that the performance of the resulting bispecific biobetter would be dependent on the relative expression of the anchor target relative to the therapeutic target. These simulations have also led us to select an anchor antigen that is expressed at higher levels than c-Met. The resulting bispecific Fc-containing molecule, MM-131, has shown marked improvement in the inhibition of c-Met signaling in vitro and in vivo due to de novo engineered avid bivalent binding to the target cells. MM-131 is engineered to be manufacturable and is thus appropriate for development as a therapeutic for human diseases (Geddie et al., manuscript in preparation).

Concluding Remarks

Advances in protein engineering techniques have provided the means for precise manipulation of antibody structure and function. The development of these techniques has enabled the construction of many antibody-like therapeutic formats that possess unique properties while retaining the stability and manufacturability of conventional IgGs. These useful modifications have the ability to expand the utility of antibody therapies by broadening the spectrum of their activity and increasing efficacy against validated targets. However, such modifications by no means give a recipe for success, as the success of biobetter mAb molecules should be exquisitely tuned to a specific therapeutic task in hand. Comprehensive systems-level analysis of the limitations of first generation molecules, engineering modes of action that address these deficiencies, and prioritization of design objectives and success criteria are required for the continuous progress in the field of antibody-like biobetters.

References

(2006) Can super-antibody drugs be tamed? Nature 440:855–856

Abarca K, Jung E, Fernández P, Zhao L, Harris B, Connor EM, Losonsky GA (2009) Safety, tolerability, pharmacokinetics, and immunogenicity of motavizumab, a humanized, enhanced-potency monoclonal antibody for the prevention of respiratory syncytial virus infection in at-risk children. Pediatr Infect Dis J 28:267–272

Alyanakian M-A, Bernatowska E, Scherrmann J-M, Aucouturier P, Poplavsky J-L (2003) Pharmacokinetics of total immunoglobulin G and immunoglobulin G subclasses in patients undergoing replacement therapy for primary immunodeficiency syndromes. Vox Sang 84:188–192

An Z, Forrest G, Moore R, Cukan M, Haytko P, Huang L, Vitelli S, Zhao JZ, Lu P, Hua J, Gibson CR, Harvey BR, Montgomery D, Zaller D, Wang F, Strohl W (2009) IgG2m4, an engineered antibody isotype with reduced Fc function. MAbs 1:572–579

Beck A (ed) (2013) Glycosylation engineering of biopharmaceuticals. Springer, New York

Beck A, Reichert JM (2012) Marketing approval of mogamulizumab: a triumph for glyco-engineering. MAbs 4:419–425

Bostrom J, Yu S-F, Kan D, Appleton BA, Lee CV, Billeci K, Man W, Peale F, Ross S, Wiesmann C, Fuh G (2009) Variants of the antibody herceptin that interact with HER2 and VEGF at the antigen binding site. Science 323:1610–1614

Brambell F, Hemmings W, Morris I (1964) A theoretical model of gamma-globulin catabolism. Nature 26:1352–1354

Cang S, Mukhi N, Wang K, Liu D (2012) Novel CD20 monoclonal antibodies for lymphoma therapy. J Hematol Oncol 5:64

Chaparro-Riggers J, Liang H, DeVay RM, Bai L, Sutton JE, Chen W, Geng T, Lindquist K, Casas MG, Boustany LM, Brown CL, Chabot J, Gomes B, Garzone P, Rossi A, Strop P, Shelton D, Pons J, Rajpal A (2012) Increasing serum half-life and extending cholesterol lowering in vivo by engineering antibody with pH-sensitive binding to PCSK9. J Biol Chem 287: 11090–11097

Clynes RA, Towers TL, Presta LG, Ravetch JV (2000) Inhibitory Fc receptors modulate in vivo cytoxicity against tumor targets. Nat Med 6:443–446

Cunningham D, Hawkes EA, Jack A, Qian W, Smith P, Mouncey P, Pocock C, Ardeshna KM, Radford JA, McMillan A, Davies J, Turner D, Kruger A, Johnson P, Gambell J, Linch D (2013) Rituximab plus cyclophosphamide, doxorubicin, vincristine, and prednisolone in patients with newly diagnosed diffuse large B-cell non-Hodgkin lymphoma: a phase 3 comparison of dose intensification with 14-day versus 21-day cycles. Lancet 381:1817–1826

Dall'Acqua WF, Kiener PA, Wu H (2006) Properties of human IgG1s engineered for enhanced binding to the neonatal Fc receptor (FcRn). J Biol Chem 281:23514–23524

Digiammarino EL, Harlan JE, Walter KA, Ladror US, Edalji RP, Hutchins CW, Lake MR, Greischar AJ, Liu J, Ghayur T, Jakob CG (2011) Ligand association rates to the inner-variable-domain of a dual-variable-domain immunoglobulin are significantly impacted by linker design. MAbs 3:487–494

DiGiammarino E, Ghayur T, Liu J (2012) Design and generation of DVD-Ig™ molecules for dual-specific targeting. Methods Mol Biol 899:145–156

Eigenbrot C, Fuh G (2013) Two-in-One antibodies with dual action Fabs. Curr Opin Chem Biol 17:400–405

Fauvel B, Yasri A (2014) Antibodies directed against receptor tyrosine kinases: current and future strategies to fight cancer. MAbs 6:838–851

Feng Y, Ma PC (2011) Anti-MET targeted therapy has come of age: the first durable complete response with MetMAb in metastatic gastric cancer. Cancer Discov 1:550–554

Fitzgerald J, Lugovskoy A (2011) Rational engineering of antibody therapeutics targeting multiple oncogene pathways. MAbs 3:299–309

Fitzgerald JB, Johnson BW, Baum J, Adams S, Iadevaia S, Tang J, Rimkunas V, Xu L, Kohli N, Rennard R, Razlog M, Jiao Y, Harms BD, Olivier KJ, Schoeberl B, Nielsen UB, Lugovskoy AA (2014) MM-141, an IGF-IR- and ErbB3-directed bispecific antibody, overcomes network adaptations that limit activity of IGF-IR inhibitors. Mol Cancer Ther 13:410–425

Gerdes CA, Nicolini VG, Herter S, van Puijenbroek E, Lang S, Roemmele M, Moessner E, Freytag O, Friess T, Ries CH, Bossenmaier B, Mueller HJ, Umaña P (2013) GA201 (RG7160): a novel, humanized, glycoengineered anti-EGFR antibody with enhanced ADCC and superior in vivo efficacy compared with cetuximab. Clin Cancer Res 19:1126–1138

Goede V, Fischer K, Busch R, Engelke A, Eichhorst B, Wendtner CM, Chagorova T, de la Serna J, Dilhuydy M-S, Illmer T, Opat S, Owen CJ, Samoylova O, Kreuzer K-A, Stilgenbauer S, Döhner H, Langerak AW, Ritgen M, Kneba M, Asikanius E, Humphrey K, Wenger M, Hallek M (2014) Obinutuzumab plus chlorambucil in patients with CLL and coexisting conditions. N Engl J Med 370:1101–1110

Golay J, Da Roit F, Bologna L, Ferrara C, Leusen JH, Rambaldi A, Klein C, Introna M (2013) Glycoengineered CD20 antibody obinutuzumab activates neutrophils and mediates phagocytosis through CD16B more efficiently than rituximab. Blood 122:3482–3491

Gu J, Ghayur T (2012) Generation of dual-variable-domain immunoglobulin molecules for dual-specific targeting. Methods Enzymol 502:25–41

Harms BD, Kearns JD, Su SV, Kohli N, Nielsen UB, Schoeberl B (2012) Optimizing properties of antireceptor antibodies using kinetic computational models and experiments. Methods Enzymol 502:67–87

Harms BD, Kearns JD, Iadevaia S, Lugovskoy AA (2014) Understanding the role of cross-arm binding efficiency in the activity of monoclonal and multispecific therapeutic antibodies. Methods 65:95–104

Hay M, Thomas DW, Craighead JL, Economides C, Rosenthal J (2014) Clinical development success rates for investigational drugs. Nat Biotechnol 32:40–51

Herter S, Herting F, Mundigl O, Waldhauer I, Weinzierl T, Fauti T, Muth G, Ziegler-Landesberger D, Van Puijenbroek E, Lang S, Duong MN, Reslan L, Gerdes CA, Friess T, Baer U, Burtscher H, Weidner M, Dumontet C, Umana P, Niederfellner G, Bacac M, Klein C (2013) Preclinical activity of the type II CD20 antibody GA101 (obinutuzumab) compared with rituximab and ofatumumab in vitro and in xenograft models. Mol Cancer Ther 12:2031–2042

Hezareh M, Hessell AJ, Jensen RC, van de Winkel JG, Parren PW (2001) Effector function activities of a panel of mutants of a broadly neutralizing antibody against human immunodeficiency virus type 1. J Virol 75:12161–12168

Hinton PR, Johlfs MG, Xiong JM, Hanestad K, Ong KC, Bullock C, Keller S, Tang MT, Tso JY, Vásquez M, Tsurushita N (2004) Engineered human IgG antibodies with longer serum half-lives in primates. J Biol Chem 279:6213–6216

Huang GS, Brouwer-Visser J, Ramirez MJ, Kim CH, Hebert TM, Lin J, Arias-Pulido H, Qualls CR, Prossnitz ER, Goldberg GL, Smith HO, Horwitz SB (2010) Insulin-like growth factor 2 expression modulates Taxol resistance and is a candidate biomarker for reduced disease-free survival in ovarian cancer. Clin Cancer Res 16:2999–3010

Igawa T, Ishii S, Tachibana T, Maeda A, Higuchi Y, Shimaoka S, Moriyama C, Watanabe T, Takubo R, Doi Y, Wakabayashi T, Hayasaka A, Kadono S, Miyazaki T, Haraya K, Sekimori Y, Kojima T, Nabuchi Y, Aso Y, Kawabe Y, Hattori K (2010) Antibody recycling by engineered pH-dependent antigen binding improves the duration of antigen neutralization. Nat Biotechnol 28:1203–1207

Isakoff SJ, Lugovskoy A, Manoli S, Czibere A, LoRusso P, Arnedos M (2014) First-in-human study of MM-141: a novel tetravalent monoclonal antibody targeting IGF-1R and ErbB3 [abstract]. J Clin Oncol 32(Suppl 3):abstr 3068

Kienast Y, Klein C, Scheuer W, Raemsch R, Lorenzon E, Bernicke D, Herting F, Yu S, The HH, Martarello L, Gassner C, Stubenrauch K-G, Munro K, Augustin HG, Thomas M (2013) Ang-2-VEGF-A CrossMab, a novel bispecific human IgG1 antibody blocking VEGF-A and Ang-2 functions simultaneously, mediates potent antitumor, antiangiogenic, and antimetastatic efficacy. Clin Cancer Res 19:6730–6740

Klein C, Sustmann C, Thomas M, Stubenrauch K, Croasdale R, Schanzer J, Brinkmann U, Kettenberger H, Regula JT, Schaefer W (2012) Progress in overcoming the chain association issue in bispecific heterodimeric IgG antibodies. MAbs 4:653–663

Kontermann RE (2012) Dual targeting strategies with bispecific antibodies. MAbs 4:182–197

Labrijn AF, Buijsse AO, van den Bremer ETJ, Verwilligen AYW, Bleeker WK, Thorpe SJ, Killestein J, Polman CH, Aalberse RC, Schuurman J, van de Winkel JGJ, Parren PWHI (2009) Therapeutic IgG4 antibodies engage in Fab-arm exchange with endogenous human IgG4 in vivo. Nat Biotechnol 27:767–771

Lazar GA, Dang W, Karki S, Vafa O, Peng JS, Hyun L, Chan C, Chung HS, Eivazi A, Yoder SC, Vielmetter J, Carmichael DF, Hayes RJ, Dahiyat BI (2006) Engineered antibody Fc variants with enhanced effector function. Proc Natl Acad Sci U S A 103:4005–4010

Lee CV, Koenig P, Fuh G (2014) A two-in-one antibody engineered from a humanized interleukin 4 antibody through mutation in heavy chain complementarity-determining regions. MAbs 6:622–627

Lieu C, Harb WA, Beeram M, Power L, Kearns JD, Nering R, Moyo VM, Wolf BB, Adjei AA (2014) Phase I trial of MM-151, a novel oligoclonal anti-EGFR antibody combination in patients with refractory solid tumors [abstract]. J Clin Oncol 32:5s, abstr 2518

Linke R, Klein A, Seimetz D (2010) Catumaxomab: clinical development and future directions. MAbs 2:129–136

Listinsky JJ, Siegal GP, Listinsky CM (2013) Glycoengineering in cancer therapeutics: a review with fucose-depleted Trastuzumab as the model. Anticancer Drugs 24:219–227

Lu J, Lee-Gabel L, Nadeau MC, Ferencz TM, Soefje SA (2014) Clinical evaluation of compounds targeting PD-1/PD-L1 pathway for cancer immunotherapy. J Oncol Pharm Pract

McDonagh CF, Huhalov A, Harms BD, Adams S, Paragas V, Oyama S, Zhang B, Luus L, Overland R, Nguyen S, Gu J, Kohli N, Wallace M, Feldhaus MJ, Kudla AJ, Schoeberl B, Nielsen UB (2012) Antitumor activity of a novel bispecific antibody that targets the ErbB2/ErbB3 oncogenic unit and inhibits heregulin-induced activation of ErbB3. Mol Cancer Ther 11:582–593

Merchant M, Ma X, Maun HR, Zheng Z, Peng J, Romero M, Huang A, Yang N, Nishimura M, Greve J, Santell L, Zhang Y-W, Su Y, Kaufman DW, Billeci KL, Mai E, Moffat B, Lim A, Duenas ET, Phillips HS, Xiang H, Young JC, Vande Woude GF, Dennis MS, Reilly DE, Schwall RH, Starovasnik MA, Lazarus RA, Yansura DG (2013) Monovalent antibody design and mechanism of action of onartuzumab, a MET antagonist with anti-tumor activity as a therapeutic agent. Proc Natl Acad Sci U S A 110:E2987–E2996

Moore GL, Chen H, Karki S, Lazar GA (2010) Engineered Fc variant antibodies with enhanced ability to recruit complement and mediate effector functions. MAbs 2:181–189

Mössner E, Brünker P, Moser S, Püntener U, Schmidt C, Herter S, Grau R, Gerdes C, Nopora A, van Puijenbroek E, Ferrara C, Sondermann P, Jäger C, Strein P, Fertig G, Friess T, Schüll C, Bauer S, Dal Porto J, Del Nagro C, Dabbagh K, Dyer MJS, Poppema S, Klein C, Umaña P (2010) Increasing the efficacy of CD20 antibody therapy through the engineering of a new type II anti-CD20 antibody with enhanced direct and immune effector cell-mediated B-cell cytotoxicity. Blood 115:4393–4402

Nimmerjahn F, Ravetch JV (2008) Fcgamma receptors as regulators of immune responses. Nat Rev Immunol 8:34–47

Oganesyan V, Gao C, Shirinian L, Wu H, Dall'Acqua WF (2008) Structural characterization of a human Fc fragment engineered for lack of effector functions. Acta Crystallogr D Biol Crystallogr 64:700–704

Pollak M (2012) The insulin and insulin-like growth factor receptor family in neoplasia: an update. Nat Rev Cancer 12:159–169

Pedersen MW, Jacobsen HJ, Koefoed K, Hey A, Pyke C, Haurum JS, Krah M (2010) Sym004: a novel synergistic anti-epidermal growth factor receptor antibody mixture with superior anticancer efficacy. Cancer Res 70:588–97

Prat M, Crepaldi T, Pennacchietti S, Bussolino F, Comoglio PM (1998) Agonistic monoclonal antibodies against the Met receptor dissect the biological responses to HGF. J Cell Sci 111 (Pt 2):237–47

Presta LG (2006) Engineering of therapeutic antibodies to minimize immunogenicity and optimize function. Adv Drug Deliv Rev 58:640–656

Presta LG (2008) Molecular engineering and design of therapeutic antibodies. Curr Opin Immunol 20:460–470

Price TJ, Peeters M, Kim T, Li J, Cascinu S, Ruff P, Suresh A, Zhang K, Murugappan S, Sidhu R (2013) ASPECCT: a randomized, multicenter, open-label, phase 3 study of panitumumab (pmab) vs cetuximab (cmab) for previously treated wild-type (WT) KRAS metastatic colorectal cancer (mCRC). Abstract. ESMO, LBA18

Reichert JM, Dhimolea E (2012) The future of antibodies as cancer drugs. Drug Discov Today 17:954–963

Richards JO, Karki S, Lazar GA, Chen H, Dang W, Desjarlais JR (2008) Optimization of antibody binding to FcgammaRIIa enhances macrophage phagocytosis of tumor cells. Mol Cancer Ther 7:2517–2527

Ridgway JB, Presta LG, Carter P (1996) "Knobs-into-holes" engineering of antibody CH3 domains for heavy chain heterodimerization. Protein Eng 9:617–621

Robbie GJ, Criste R, Dall'acqua WF, Jensen K, Patel NK, Losonsky GA, Griffin MP (2013) A novel investigational Fc-modified humanized monoclonal antibody, motavizumab-YTE, has an extended half-life in healthy adults. Antimicrob Agents Chemother 57:6147–6153

Rudnick SI, Adams GP (2009) Affinity and avidity in antibody-based tumor targeting. Cancer Biother Radiopharm 24:155–161

Salfeld JG (2007) Isotype selection in antibody engineering. Nat Biotechnol 25:1369–1372

Schaefer G, Haber L, Crocker LM, Shia S, Shao L, Dowbenko D, Totpal K, Wong A, Lee CV, Stawicki S, Clark R, Fields C, Lewis Phillips GD, Prell RA, Danilenko DM, Franke Y, Stephan J-P, Hwang J, Wu Y, Bostrom J, Sliwkowski MX, Fuh G, Eigenbrot C (2011a) A two-in-one antibody against HER3 and EGFR has superior inhibitory activity compared with monospecific antibodies. Cancer Cell 20:472–486

Schaefer W, Regula JT, Bähner M, Schanzer J, Croasdale R, Dürr H, Gassner C, Georges G, Kettenberger H, Imhof-Jung S, Schwaiger M, Stubenrauch KG, Sustmann C, Thomas M, Scheuer W, Klein C (2011b) Immunoglobulin domain crossover as a generic approach for the production of bispecific IgG antibodies. Proc Natl Acad Sci U S A 108:11187–11192

Schoeberl B, Pace EA, Fitzgerald JB, Harms BD, Xu L, Nie L, Linggi B, Kalra A, Paragas V, Bukhalid R, Grantcharova V, Kohli N, West KA, Leszczyniecka M, Feldhaus MJ, Kudla AJ, Nielsen UB (2009) Therapeutically targeting ErbB3: a key node in ligand-induced activation of the ErbB receptor-PI3K axis. Sci Signal 2:ra31

Shapiro RI, Plavina T, Schlain BR, Pepinsky RB, Garber EA, Jarpe M, Hochman PS, Wehner NG, Bard F, Motter R, Yednock TA, Taylor FR (2011) Development and validation of immunoassays to quantify the half-antibody exchange of an IgG4 antibody, natalizumab (Tysabri®) with endogenous IgG4. J Pharm Biomed Anal 55:168–175

Stebbings R, Findlay L, Edwards C, Eastwood D, Bird C, North D, Mistry Y, Dilger P, Liefooghe E, Cludts I, Fox B, Tarrant G, Robinson J, Meager T, Dolman C, Thorpe SJ, Bristow A, Wadhwa M, Thorpe R, Poole S (2007) "Cytokine storm" in the phase I trial of monoclonal antibody TGN1412: better understanding the causes to improve preclinical testing of immunotherapeutics. J Immunol 179:3325–3331

Swain SM, Kim S-B, Cortés J, Ro J, Semiglazov V, Campone M, Ciruelos E, Ferrero J-M, Schneeweiss A, Knott A, Clark E, Ross G, Benyunes MC, Baselga J (2013) Pertuzumab, trastuzumab, and docetaxel for HER2-positive metastatic breast cancer (CLEOPATRA study): overall survival results from a randomised, double-blind, placebo-controlled, phase 3 study. Lancet Oncol 14:461–471

Tesar DB, Björkman PJ (2010) An intracellular traffic jam: Fc receptor-mediated transport of immunoglobulin G. Curr Opin Struct Biol 20:226–233

Tracey D, Klareskog L, Sasso EH, Salfeld JG, Tak PP (2008) Tumor necrosis factor antagonist mechanisms of action: a comprehensive review. Pharmacol Ther 117:244–279

Wang W, Lu P, Fang Y, Hamuro L, Pittman T, Carr B, Hochman J, Prueksaritanont T (2011) Monoclonal antibodies with identical Fc sequences can bind to FcRn differentially with pharmacokinetic consequences. Drug Metab Dispos 39:1469–1477

Weiner GJ (2010) Rituximab: mechanism of action. Semin Hematol 47:115–123

Wu C, Ying H, Grinnell C, Bryant S, Miller R, Clabbers A, Bose S, McCarthy D, Zhu R-R, Santora L, Davis-Taber R, Kunes Y, Fung E, Schwartz A, Sakorafas P, Gu J, Tarcsa E, Murtaza A, Ghayur T (2007) Simultaneous targeting of multiple disease mediators by a dual-variable-domain immunoglobulin. Nat Biotechnol 25:1290–1297

Wu C, Ying H, Bose S, Miller R, Medina L, Santora L, Ghayur T (2009) Molecular construction and optimization of anti-human IL-1alpha/beta dual variable domain immunoglobulin (DVD-Ig) molecules. MAbs 1:339–347

Xu L, Kohli N, Rennard R, Jiao Y, Razlog M, Zhang K, Baum J, Johnson B, Tang J, Schoeberl B, Fitzgerald J, Nielsen U, Lugovskoy A (2013) Rapid optimization and prototyping for therapeutic antibody-like molecules. MAbs 5:237–254

Yano S, Takeuchi S, Nakagawa T, Yamada T (2012) Ligand-triggered resistance to molecular targeted drugs in lung cancer: roles of hepatocyte growth factor and epidermal growth factor receptor ligands. Cancer Sci 103:1189–1194

Zalevsky J, Chamberlain AK, Horton HM, Karki S, Leung IWL, Sproule TJ, Lazar GA, Roopenian DC, Desjarlais JR (2010) Enhanced antibody half-life improves in vivo activity. Nat Biotechnol 28:157–159

Alternative Protein Scaffolds as Novel Biotherapeutics

Michaela Gebauer and Arne Skerra

Introduction: The Route to Antibody Alternatives as Biobetters

Proposed as a new type of medicine more than a century ago by Emil von Behring, initially as "antitoxins" and only later dubbed "antibodies" (Lindenmann 1984), monoclonal antibodies (MAbs) reached regulatory approval in 1986 and subsequently, in a humanized or human format, have become the most successful and fastest growing class of biopharmaceuticals with more than 30 marketed drugs today (Reichert 2012, 2014; Strohl and Strohl 2012). Antibody technology has gone through a long evolution, starting with animal sera for passive immunization against toxins. The first breakthrough with regard to developing monoclonal as opposed to polyclonal antibodies was the invention of hybridoma technology in 1975, thus allowing the preparation of monospecific antibodies obtained via immunization of rodents (Köhler and Milstein 1975). However, only one MAb derived from mouse hybridoma cell culture was successful in the clinic: Muromonab-CD3/Orthoclone (OKT3®), approved in 1986 (discontinued in 2010) as an immunosuppressant drug to reduce acute rejection in patients after organ transplantation (Strohl and Strohl 2012).

The recognition that MAbs from animal sources elicit an anti-drug immune response in patients prompted efforts to make them more alike their human counterparts. The first strategy was the construction of chimeric MAbs by combining the pair of murine variable domains, carrying the antigen-binding site, with the

M. Gebauer
XL-protein GmbH, 85354 Freising-Weihenstephan, Germany

A. Skerra (✉)
Lehrstuhl für Biologische Chemie, Technische Universität München,
Emil-Erlenmeyer-Forum 5, 85350 Freising-Weihenstephan, Germany
e-mail: skerra@tum.de

A. Rosenberg, B. Demeule (eds.), *Biobetters*, AAPS Advances
in the Pharmaceutical Sciences Series 19, DOI 10.1007/978-1-4939-2543-8_13

constant region of a human immunoglobulin (Ig) (Morrison et al. 1984). Since antigenic sites predominantly occur in the xenogenic constant regions, the resulting chimeric Ig proteins were well tolerated (although somewhat biased by their nature as immune suppressants), as first demonstrated with Rituximab (Rituxan®; anti-CD20) and Daclizumab (Zenapax®; anti-IL-2Rα), both approved by the FDA in 1997 for treating Non-Hodgkin's lymphoma (NHL) / rheumatoid arthritis (RA) and to prevent transplant rejection, respectively (Strohl and Strohl 2012).

Following this major breakthrough, the technology of CDR-grafting was invented (Jones et al. 1986), allowing more complete humanization of MAbs by further exchanging the framework regions within the variable domains with those from a human Ig. One of the first and most successful examples has been Trastuzumab (Herceptin®) (Carter et al. 1992), a humanized antibody directed against the HER2 cell surface receptor overexpressed on metastasizing breast cancer cells, which was approved in 1998. However, as the conformation of the six CDRs is also influenced by the framework region, this type of humanization is a tricky approach that usually necessitates introduction of additional affinity-preserving mutations via protein engineering (Carter et al. 1992). This caveat eventually stimulated the development of alternative methods for MAb humanization, for example by resurfacing (Strohl and Strohl 2012).

With the advent of phage display techniques, the age of so-called "human" MAbs arose, initially by selecting recombinant antibody fragments from naïve variable gene/cDNA libraries that were cloned from non-immunized human donors. The first MAb that emerged from this technology was Adalimumab (Humira®), which neutralizes tumor necrosis factor α (TNFα) and received approval for the treatment of RA and Crohn's disease in 2002 (Bain and Brazil 2003; Osbourn et al. 2005). Various other experimental strategies, both in vitro and in vivo, have followed, culminating in the breeding of transgenic animals that have their own Ig gene loci replaced by corresponding human chromosomal segments, thus directly allowing the biosynthesis of human MAbs—e.g. followed by cloning of the relevant V-genes from antibody producing B-cells (Lee et al. 2014; Lonberg 2005). However, it should be kept in mind that in spite of the commonly used term "human", each MAb developed by one of these technologies is still foreign to the body of an individual who receives corresponding treatment since it exhibits novel CDR sequences (Harding et al. 2010). Hence, immunogenicity in the patient cannot be fully prevented and, in fact, with regard to immunological tolerability, it is clear that the early chimeric antibody concept already had the strongest impact (Getts et al. 2010; Strohl and Strohl 2012).

Far from over, technology development in the antibody field continues apace (cf. the Chapters in Part II of this monograph). Apart from searching for innovative MAbs directed against new targets and/or novel epitopes, having greater affinities and specificities, or even agonistic activities, there are numerous ongoing approaches to improve the generic features and the Ig format as a whole (Strohl and Strohl 2012). In particular, these efforts aim at the following features: enhancing immune effector functions (usually associated with the Fc region) such as antibody-dependent cellular cytotoxicity (ADCC) and complement-dependent cytotoxicity (CDC);

prolonging (or adjusting) circulation half-life; achieving more homogeneous and functional glycosylation; arming with drugs or radioactive payloads; and creating bispecific MAbs, resulting in so-called second or third generation antibodies (Beck et al. 2010). Application of these strategies to marketed MAbs, while retaining the original validated epitope specificity, results in "biobetters" in the true sense. One recent example is the antibody Ado-Trastuzumab Emtansine (T-DM1, Kadcyla™), a toxic drug conjugate of Trastuzumab (Dirix et al. 2013).

Generally, the term biobetter refers to a therapeutic protein that is functionally similar (not identical) to an established biopharmaceutical but has improvements over the original in critical attributes and clinical performance (Beck 2011). However, in the case of antibodies this rather narrow definition deserves broader interpretation, especially if the focus is on the tunable antigen-binding function as dominating feature (Skerra 2003). In fact, the past two decades have given evidence that full size Igs with their characteristic Y-shaped molecular architecture are not the only suitable platform for the generation of protein reagents with novel ligand-binding properties for applications in biomedical research and human therapy. Once the technical challenges of engineering recombinant antibody fragments had been mastered, so-called alternative protein scaffolds were identified which, too, can provide specific binding sites with high affinities against a wide range of biochemical or cellular target structures (Skerra 2000a). Consequently, biobetters may be understood in this context also as advancements in the molecular format of Ig-based biologics as a class.

Functional Ig Fragments: from Fab to Domain Antibody

MAbs differ from classical biologics such as hormones or cytokines insofar as the generation of new molecular entities with differing target specificities and varying mode of action is possible on the basis of the same molecular format, the one of Igs. These proteins owe their remarkable ability for molecular recognition of almost any kind of antigen as well as their utility for diverse medical applications to their unique modular format (Bork et al. 1994; Padlan 1994; Strohl and Strohl 2012). On the first level, their molecular architecture comprises two types of domains which are either of constant or of variable character. Both rest upon the same basically conserved Ig fold which is characterized by a sandwich of two β-sheets with altogether 7 or 9 strands, respectively. Several constant domains assemble to form the Fc part at the stem of the Y-shaped protein, which is responsible for most effector functions, in particular binding and/or activation of cellular receptors or complement. The two arms of the Y are called antigen-binding fragments, Fabs, which are mutually identical in natural Igs and flexibly linked to the Fc stem via the hinge region. At each tip of the Y a pair of variable domains, VH from the heavy and VL from the light chain, forms the antigen-binding site.

Within these paired variable domains, there is another level of separation between structurally conserved and variable polypeptide segments, the so-called framework region(s) on the one hand and the hypervariable region(s) on the other. As regards the hypervariable regions there are in total six hypervariable loops, three per variable domain, also known as complementarity-determining regions (CDRs), which—upon tight association between VH and VL via their inner β-sheets—together form the combining site that intimately contacts the antigen during complex formation. Within each V-domain the three hypervariable loops are supported by the β-sandwich fold, which provides the structural framework. This discontinuous region shows high similarity in its backbone conformation, not only among the human antibody repertoire but also across species, a feature that provides the basis of the CDR-grafting technology for antibody humanization mentioned above.

This modular format explains the success of MAbs as almost universal protein tools for the tight and highly specific binding of a vast number of biomedically relevant molecular structures. Obviously, the β-sheet framework is sufficiently stable to structurally support a more or less unlimited number of CDR peptide sequences and to impose a defined three-dimensional conformation on them by providing the proper geometrical constraints. As a consequence, the number of possible CDR backbone conformations, called canonical structures, seems to be restricted (Al-Lazikani et al. 1997). However, the conformational stability of the Ig fold is not indefinite. In fact, it has been shown that the folding energy of the Ig framework statistically varies about some consensus sequence, which appears to constitute the thermodynamic minimum (Steipe et al. 1994), whereas the stability of the variable domain as a whole is clearly influenced by the individual sequences of the hypervariable loops (Jung and Plückthun 1997). Interestingly, these long known structural properties of Igs also match well with observations made for the alternative protein scaffolds discussed further below.

Nevertheless, with the ever increasing application of antibodies during the last two decades, several disadvantages have become apparent. For example, MAbs have a very large size and complicated composition, comprising four polypeptide chains, glycosylation of the heavy chains, at least, as well as multiple structurally relevant disulfide bonds. Thus, intact antibodies require manufacturing in eukaryotic expression systems, usually involving stably transfected mammalian cell lines, whose optimization and fermentation is laborious and costly (Steinmeyer and McCormick 2008; Werner 2004). Consequently, the exploration of alternative protein reagents that are capable of specifically recognizing and tightly binding 'antigens' has been stimulated, which led to the investigation of different types of antibody fragments and, eventually, even to the use of isolated Ig domains (Holliger and Hudson 2005; Skerra 1993).

Initially, the development of methods for the production and manipulation of functional Ig fragments in the bacterial host cell *Escherichia coli* (Better et al. 1988; Skerra and Plückthun 1988) enormously stimulated the field of antibody engineering (Humphreys 2003; Skerra 1993), not only with regard to research and discovery but also concerning biological drug development (Holliger and Hudson 2005; Nelson and Reichert 2009). Two major types of recombinant antibody fragments carrying the intact antigen-binding site of a MAb are commonly in use: first, the Fab fragment, which consists of the entire light chain and the N-terminal half of the heavy chain and has been traditionally prepared by selective proteolysis of full size Igs and, second, the Fv fragment; this smallest functional antibody fragment comprises just the two variable domains and is readily accessible only via recombinant DNA technology (Riechmann et al. 1988; Skerra and Plückthun 1988).

The conventional Fv fragment is made of the paired variable regions from both Ig chains which are held together just by non-covalent forces such that its stability is limited, in particular at elevated temperatures (like body temperature), low protein concentration and in the absence of antigen. Chain dissociation may be prevented by introducing Cys residues at appropriate locations into the framework of VH and VL, resulting in a disulfide crosslink (Glockshuber et al. 1990; Reiter et al. 1996) and yielding so-called disulfide-stabilized Fv fragments (dsFv). However, the gain in stability is often accompanied by poorer expression yields, as typically observed when introducing additional Cys residues and/or disulfide bridges into recombinant proteins.

A more widely applied approach is the construction of a fusion protein comprising the two variable domains artificially linked via a flexible peptide segment, yielding the so-called single chain Fv fragment (scFv) (Bird and Walker 1991). In this manner, VH and VL are not only covalently associated, but also the difficulty on the genetic level of simultaneously coexpressing two different polypeptide chains is avoided. Nevertheless, several disadvantages have been observed with this strategy, most notably the frequently reduced antigen-binding activity as a result of connecting one N-terminus—usually close to the combining site—with the peptide linker, which can lead to conformational changes or steric hindrance during antigen binding. This is especially the case for scFvs that are originally derived from full length MAbs and not directly selected from V-gene libraries (e.g. via phage display).

The scFv format also shows other undesired properties, such as low folding efficiency upon expression in *E. coli*, enhanced aggregation and, importantly, a pronounced tendency to form oligomers (Arndt et al. 1998; Demarest and Glaser 2008; Power et al. 2003), which can significantly hamper bioprocess development. Although numerous attempts were made to vary the length and amino acid sequence of the peptide linker and to alter the mutual order of the VH and VL domains within the fusion protein, these caveats have not been fundamentally alleviated and, therefore, laborious individual optimization is needed (Strohl and Strohl 2012). This may explain why, despite their flourishing in biomedical research for more than 25 years, no scFv has been approved as a biopharmaceutical to date (Nelson 2010). The currently most advanced scFv under clinical development (beside an immunotoxin fusion protein discussed below) is a PEG-linked conjugate with doxorubicin-charged liposomes targeting HER2, MM-302 (Merrimack, http://merrimackpharma.com), that has entered phase II/III studies for the treatment of breast cancer (Reichert 2015; Nielsen et al. 2002; Wickham et al. 2010).

Still, there were attempts to generate scFvs with elevated stability, for example using the so-called Immuna® screening platform (ESBAtech, http://www.esbatech.com) which involves the generation of stable humanized single-chain Fv fragments from rabbit monoclonal antibodies (Borras et al. 2010). A few scFv candidates, directed against TNFα (ESBA105) and VEGF were developed up to clinical phase

II for topical application on the eye (Strohl and Strohl 2012; Thiel et al. 2013). Further development was pursued in the form of the PENTRA®body technology (Delenex, http://www.delenex.com) for local application in dermatology, where the anti-TNFα scFv (DLX105) appeared to induce a clinical response in psoriasis patients after intradermal administration (Tsianakas et al. 2014).

In the area of antimicrobial therapy, Efungumab (Mycograb™) (Pachl et al. 2006), which binds to the heat shock protein 90 of *Candida albicans*, and Aurograb™ (NCT00217841), an scFv directed against a surface protein of methicillin resistant *Staphylococcus aureus*, were brought to advanced clinical stages before their development was discontinued. Efungumab formed aggregates in solution, was contaminated with host proteins, and administration was associated with cytokine release while no treatment benefit was seen (EMEA 2007). In fact, reevaluation of its in vitro activity indicated that the presumed antifungal potentiation of amphotericin B was a nonspecific effect that was also seen with a wide range of unrelated proteins (Richie et al. 2012). Aurograb was discontinued following a review of phase II data showing lack of efficacy (Nelson 2010).

Other scFvs were clinically exploited as part of fusion proteins, either with distinct functional modules or in bispecific form, as will be discussed in greater detail further below. For example, Darleukin (L19-IL2) is a fusion protein between the scFv L19 selected from a phage display library against the extra-domain B (ED-B) of oncofetal fibronectin (a splice form specific for neovasculature described below) and the cytokine interleukin-2 (IL-2) to trigger immunopotentiating and antineoplastic activities. This immunocytokine has been developed as combination therapy with dacarbazine for the treatment of melanoma in several clinical studies up to phase II (Philogen, http://www.philogen.com). The same L19 scFv moiety is also under clinical investigation in different formats and/or combination therapies, for example as TNFα fusion protein, as well as for radio-immunotherapy (RIT). While initially L19 was described as a high-affinity scFv having a remarkable K_D in the low picomolar range (Pini et al. 1998), a significantly diminished potency was measured when reformatted to a Fab (Gebauer et al. 2013). Indeed, detailed inspection of the scFv revealed its propensity to form a stable homodimer, and this form showed much enhanced ED-B binding activity (most likely due to an avidity effect) in a head to head comparison with both the isolated monomeric scFv and the inherently monovalent Fab.

In a similar fusion protein approach, an immunotoxin (VB4-845) has been developed by combining a humanized anti-EpCAM scFv with a truncated fragment of *Pseudomonas* exotoxin A (Premsukh et al. 2011). This biological drug candidate is in phase II clinical trials (Viventia, http://www.viventia.com) and has shown promising results in the treatment of loco-regional tumors such as non-invasive-bladder cancer (Vicinium™) as well as recurrent refractory cancer of the head and neck (Proxinium™). Notably, to allow repeated systemic dosing, VB4-845 was recently reformatted as a Fab fragment fused to a deimmunised version of the plant proteotoxin bouganin via a furin-cleavable linker (VB6-845) (Cizeau et al. 2009; Entwistle et al. 2012).

The currently most advanced recombinant immunotoxin utilizing an Fv moiety is Moxetumomab pasudotox (CAT-8015 or HA22), which is in phase III clinical trials (Reichert 2013). This fusion protein comprises PE38, a catalytically active fragment of *Pseudomonas* exotoxin A, fused to an anti-CD22 Fv. Originally, this program had started with an scFv fragment derived from the anti-CD22 MAb RFB4; however, to increase protein stability an interchain disulfide bond was introduced instead of the peptide linker, thus converting the Fv moiety into the dsFv format (Kreitman and Pastan 2011). In a phase I trial for refractory hairy-cell leukemia (HCL), Moxetumomab pasudotox showed a high response rate and a remarkable number of complete remissions, despite some residual immunogenicity (Kreitman et al. 2012).

A clearly more robust functional antibody fragment is the Fab, comprising the entire light chain and the variable domain and first constant domain of the heavy chain (also known as Fd fragment). Due to the additional tight packing between the pair of constant domains from both chains, compared with the association between VH and VL alone, and an interchain disulfide bond that is usually located close to the C-termini, the Fab is a structurally well-defined protein with high stability similar to a full size Ig. Thus, Fabs were considered early on for medical applications wherein either an Fc effector region is not needed or one of the usually beneficial features of intact MAbs is detrimental in the clinical context, for example the long circulation or the bivalency which may cause cellular signaling via receptor clustering. The first Fabs for clinical use were prepared according to the classical route by proteolysis from polyclonal antibodies elicited in immunized sheep and have served as antidotes to treat acute intoxications since decades (Holliger and Hudson 2005). Also, several Fabs were approved as diagnostic reagents for radio-immuno imaging of tumors via single-photon emission computed tomography (SPECT) in vivo: ^{99m}Tc-Sulesomab (LeukoScan) and ^{99m}Tc-Bectumomab (LymphoScan) (Mendler et al. 2014).

The first Fab that was approved for therapy by the FDA in 1994 was Abciximab (RheoPro®), which binds to the platelet glycoprotein IIb/IIIa (integrin α2b/β3) on blood platelets (Knight et al. 1995), thus inhibiting their aggregation in cardiovascular disease. Notably, this Fab is produced by papain cleavage from a full length chimeric MAb. In contrast, Ranibizumab (Lucentis®) was the first fully recombinant Fab produced in *E. coli* approved by the FDA in 2006 for treating wet age-related macular degeneration (AMD) by intravitreal injection (Rosenfeld et al. 2006). Ranibizumab (Y0317), which binds and neutralizes vascular endothelial growth factor (VEGF-A), was derived from the same parent mouse antibody as the humanized Ig Bevacizumab (Avastin®), previously developed for antiangiogenic cancer therapy, in a process involving in vitro affinity maturation via CDR mutagenesis and phage display (Chen et al. 1999).

The strictly monovalent nature and the lack of immunological effector functions are usually considered advantages of the Fab format, in particular if just the blocking of a biomolecular disease target or use as antidote is intended. However, the much shorter plasma half-life compared with full size Mabs, due to the smaller size and lack of endosomal recycling as explained in the next Chapter of this book, is a general disadvantage which hampers the broader clinical application of this protein class. One approach to circumvent this problem is PEGylation, as exemplified by

Certolizumab pegol (CDP-870, Cimzia®), a bacterially produced Fab directed against TNFα conjugated to a branched PEG chain of 40 kDa via a single Cys residue that was introduced into the hinge region (Melmed et al. 2008; Rose-John and Schooltink 2003). Cimzia is approved since 2008 for the treatment of Crohn's disease and RA (Blick and Curran 2007; Goel and Stephens 2010).

The number of marketed Fabs is still rather limited (Nelson 2010), despite previous optimistic expectations (Nelson and Reichert 2009). However, there is a robust clinical pipeline for this class of Ig fragments, often as PEGylated versions, with examples from major pharma and biotech companies: anti-TNFα Afelimomab (AZD9773, MAK195F, CytoFab™) for the treatment of severe sepsis; a PEGylated Fab directed against the type III secretion system of *Pseudomonas aeroginosa* (KB-001); anti-Factor D Lampalizumab (RG-7417) for treatment of dry AMD; anti-C5 Fab LFG316A against the same condition as well as other inflammatory eye diseases; the PEGylated humanised Fab CDP7657 developed as a monovalent CD40L antagonist for the treatment of systemic lupus erythematosus; and a humanized Fab (Idarucizumab, BI655075) destined as an antidote for dabigatran etexilate mesylate (Pradaxa®) to reverse its anticoagulation effect (Brennan 2014; Reichert 2015; Strohl and Strohl 2012).

Apart from scFv and Fab, another class of small Ig fragments known as domain antibodies (dAbs) or Nanobodies has attracted attention (Holliger and Hudson 2005). These artificial binding proteins comprise an isolated variable domain and exhibit a binding site made of just three CDRs. Consequently, such single-domain Ig fragments (sdIgs) cannot be derived from intact MAbs but must be generated by different methods. This approach is complicated by the fact that an unpaired variable Ig domain usually exposes a significant hydrophobic surface area to solvent, which is otherwise shielded by association either with the second variable domain from the other Ig chain or, as often seen for light chains, via formation of so-called Bence Jones homo-dimers (Stevens et al. 1991). Especially in the case of isolated heavy chain variable domains (VH) this fundamental property normally causes aggregation (Davies and Riechmann 1994; Glockshuber et al. 1990) and/or non-specific adsorption.

Natural antibodies that are devoid of light chains were initially identified as an Ig subclass in camels (Hamers-Casterman et al. 1993; Muyldermans et al. 2001; Wiecek 2010). In such "heavy-chain" antibodies, which are composed of a pair of heavy chains that lack the C_H1 domain and sometimes have an extended hinge region, the VH domains have apparently evolved to stay soluble without hetero- or homo-dimerization. These camelid VH domains (dubbed VHH), which are found in camels, dromedaries, and also in llamas, reveal elevated surface hydrophilicity in the region that faces the VL domain in ordinary Igs. In addition, they have a much longer CDR-H3, which often participates in an additional disulfide bridge within the VH domain, thus partially shielding this interface region from solvent (Conrath et al. 2005; Muyldermans 2013).

Engineered camelid VHH domains (detached from the constant region) are exploited for disease diagnosis and therapy (Siontorou 2013) as Nanobodies® (Ablynx, http://www.ablynx.com). As these sdIgs intrinsically have a very short plasma half-life, the corresponding drug candidates are usually developed as fusion proteins to enlarge size and retard kidney filtration. This involves combining either several copies of the same domain or domains with different specificities, including one that confers affinity to serum albumin in order to prolong circulation via endosomal recycling, a mechanism further explained in the following Chapter. Notably, the latter strategy also provides a general targeting activity in the case of cancer and inflammatory diseases as these tissues are known to actively take up albumin (Merlot et al. 2014; Tijink et al. 2008).

Two advanced clinical candidates are Caplacizumab and Ozoralizumab (Ablynx). Caplacizumab (ALX-0081 and ALX-0681, respectively, for i.v. and s.c. administration) is a bivalent Nanobody, with two domains fused in tandem, directed against the A1 domain of von Willebrand Factor (vWF) that has successfully completed phase II trials in hematology to treat the rare disorder thrombotic thrombocytopenic purpura (TTP) (Bartunek et al. 2013; Holz 2012). Ozoralizumab (ATN103 or PF-5230896) is a bivalent Nanobody comprising two anti-TNFα modules which are linked to a third Nanobody that binds to human serum albumin to extend plasma half-life. A phase IIa study demonstrated a statistically significant improvement of disease scores compared with placebo in patients with active RA (Comer et al. 2012). Similarly, a monovalent anti-IL-6 receptor (IL-6R) Nanobody (ALX-0061) comprising a bispecific fusion protein with a second Nanobody that confers binding activity towards albumin has completed phase IIa studies and is developed for the treatment of inflammatory diseases like RA (Van Beneden et al. 2014). Two additional Nanobodies are part of phase I programs: ALX-0171 is a trivalent Nanobody (three identical copies consecutively fused to each other) that neutralizes Respiratory Syncytial Virus (RSV) and offers potential for inhalational administration to treat pulmonary infections; ALX-0761 constitutes a bispecific half-life extended Nanobody that neutralizes both IL-17A and IL-17F—comprising an N-terminal IL-17F specific moiety, a C-terminal moiety that binds both IL-17A and F and a central module conferring albumin affinity—which has shown efficacy in a cynomolgus monkey model of RA (Vanheusden et al. 2013).

Beyond therapeutic applications, Nanobodies have attracted attention for in vivo imaging (Chakravarty et al. 2014; Vaneycken et al. 2011). In this setting, their small size and rapid renal elimination allows high tumor to blood contrast already shortly after administration. The lead optimized HER2-specific Nanobody 2Rs15d was prepared in current-good-manufacturing-practice grade, labeled with ^{68}Ga via a 1,4,7-triazacyclononane-1,4,7-triacetic acid (NOTA) derivative and assessed in biodistribution and positron emission tomography/computed tomography (PET/CT) studies using a murine SK-OV-3 tumor xenograft model, resulting in high-specific-contrast imaging of HER2-positive tumors with no observed toxicity (Xavier et al. 2013). ^{68}Ga-NOTA-2Rs15d is under development for first-in-human clinical trials.

Another type of sdIg has been derived from the human immune system as so-called domain antibodies, dAbs (Connelly 2005), by employing cloned repertoires of human V-genes together with stringent selection protocols to achieve high protein solubility, apart from target-binding activity (Holt et al. 2003). So far, two products have advanced to the clinical trial stage. An anti-TNFα dAb-Fc fusion protein

(Placulumab, ART-621, CEP-37247) (Gay et al. 2010) was tested in a phase II clinical study (NCT00928317) for the treatment of stable plaque psoriasis and RA (Strohl and Strohl 2012). Another dAb (dubbed AlbudAb™), which was selected to bind with affinities in the mid nM range to serum albumin from various species, was initially employed to create a fusion protein with the interleukin-1 receptor antagonist (IL-1Ra, also known as Anakinra), thus resulting in an immunomodulator for RA treatment that showed extended plasma half-life in a mouse model (Holt et al. 2008). AlbudAb was also combined with some bioactive peptides, in particular the orexigenic peptide YY (PYY) and Exendin-4, a clinically approved GLP-1 mimetic (Bao et al. 2013). A recombinant fusion protein of AlbudAb with Exendin-4 (GSK2374697) was investigated in a phase I study in normal and obese healthy volunteers (Hodge et al. 2013).

An alternative source of naturally occurring sdIgs are the so-called Ig novel antigen receptors (IgNARs) of shark (Kovaleva et al. 2014). Antigen-specific VNAR fragments of these receptors have similar structure and properties as VHH from camels and llamas (Stanfield et al. 2004; Streltsov et al. 2004), and they can be cloned as isolated proteins either from immunized animals or selected in vitro from combinatorial libraries. However, despite some efforts to exploit the shark VNAR format for therapeutic applications by biotech companies, e.g. AdAlta (http://adalta.com.au) and Haptogen/Wyeth (http://www.wyeth.co.za), clinical applications have not thus far been publicized (Zielonka et al. 2015).

The Concept of Alternative Protein Scaffolds

sdIgs offer a kind of conservative approach for generating MAb biobetters insofar as the methods for their selection (in vivo or in vitro) as well as their biophysical properties are related to conventionally engineered antibody fragments, except that they do not rely on the characteristic light/heavy chain association of natural human Igs. However, it is still not fully clear whether all sdIgs behave as truly monomeric proteins (Schiefner et al. 2011). At least in some cases, X-ray structural analysis has indicated a mode of dimerization in the crystal packing that resembles either the VH/VL pairing of classical Igs or the structurally analogous light chain homo-association that is typical for Bence-Jones proteins (George et al. 2014; Jespers et al. 2004b). Thus, like for scFvs, intrinsic oligomerization and aggregation propensities may affect bioprocess development and/or biological drug formulation and must be carefully assessed in each instance.

Interestingly, however, proteins with non-Ig fold can also be furnished with novel binding sites. This has been achieved using the methodology of combinatorial biotechnology, that is site-directed random mutagenesis in combination with phage display or other molecular selection techniques (Rothe et al. 2006; Skerra 2003). The resulting class of novel alternative binding reagents has also become known as engineered “protein scaffolds” (Skerra 2000a), illustrating the fact that a structurally rigid natural protein is utilized either to modify an existing binding site towards a prescribed target or to implement a new one. This development has triggered a paradigm shift since antibodies have no longer to be considered as a unique class of binding reagents in biomedical research or biotechnology (Sheridan 2007; Skerra 2003) and, eventually, it has enabled the concept of biobetters discussed in this Chapter.

Usually, a protein scaffold suitable for therapeutic applications may be derived from either of two sources: (1) a robust and small monomeric soluble protein (e.g. a lipocalin or a Kunitz protease inhibitor—for details on the latter, see (Fiedler and Skerra 2014; Nixon et al. 2014)) or (2) a stably folded domain excised from a cell surface (e.g. protein A or fibronectin) or cytoskeletal protein (e.g. ankyrin). Compared with antibodies or their recombinant fragments, such protein scaffolds promise practical advantages, in particular elevated stability and high production yield in microbial expression systems, together with an independent intellectual property situation (Fiedler and Skerra 2014).

As these novel binding proteins are obtained via biomolecular engineering in vitro, with the primary goal to achieve tight and highly specific target-binding activity, they can also be subjected to additional selection schemes in order to address other desired properties such as solubility, thermal stability and protease resistance (Jermutus et al. 2001; Jespers et al. 2004a; Jung et al. 1999; Sieber et al. 1998). Resulting protein reagents are expected to provide beneficial properties for subsequent drug development and manufacturing as a built-in feature. Since the effort to generate and/or develop such alternative binding proteins still is greater than that for preparation of a conventional antibody (or a recombinant Ig fragment), most of the ongoing activities in this area are directed towards therapeutic use—instead of aiming at research or in vitro diagnostic reagents—hence offering the chance of a high return on investment from a business perspective. This has led to a series of corresponding biobetter formats developed by biotech companies, some of which have already been adopted by the pharmaceutical industry, as will be described in detail in the following sections.

So far, more than 50 different protein scaffolds have been proposed during the past two decades and these approaches have been extensively summarized in several recent reviews (Binz et al. 2005; Gebauer and Skerra 2009; Gill and Damle 2006; Hey et al. 2005; Mintz and Crea 2013; Nuttall and Walsh 2008; Wurch et al. 2012). This field emerged in the late 1990s with research on three different protein scaffolds, initially: (1) the Z-domain of *staphylococcal* protein A, a three-helix bundle of 58 residues, providing an interface for protein binding on two of its α-helices, leading to the Affibodies® (Nord et al. 1995); (2) the 10th extracellular domain of human fibronectin III (^{10}FN3), which adopts an Ig-like β-sandwich fold (94 residues) with two to three exposed loops but lacks the central disulfide bridge, leading to the Monobodies (Koide et al. 1998), subsequently also dubbed Adnectins™; (3) the lipocalins, comprising a diverse family of soluble β-barrel proteins that naturally form binding sites for small ligands with four structurally variable loops, resulting in the Anticalins® (Beste et al. 1999).

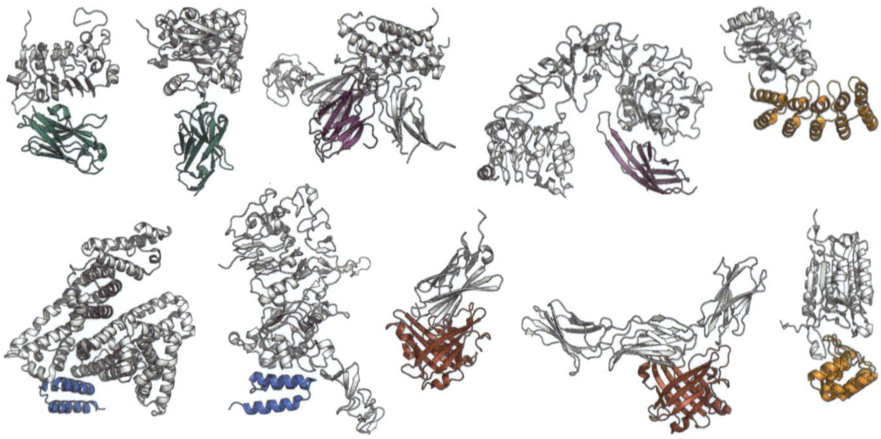

Fig. 1 Structural comparison between single domain Ig fragments and advanced alternative protein scaffolds (shown from left to right in *different colors*), each in complex with a biomedically relevant target protein (*light grey*): Nanobody (*green*) and EGFR fragment (PDB entry: 4KRL); Nanobody (*green*) and ricin (4LHQ); AdNectin (*violet*) and IL23 heterodimer (3QWR); AdNectin (*violet*) and EGFR extracellular region (3QWQ); DARPin (*orange*) and HER2 fragment (4HRL); Affibody (albumin binding domain; *blue*) and albumin (1TF0); Affibody (*blue*) and HER2 extracellular region (3MZW); Anticalin (*red*) and CTLA-4 extracellular region (3BX7); Anticalin (*red*) and ED-B as part of a three-domain fibronectin fragment (4GH7); DARPin (*orange*) and caspase 3 (2XZT)

Whereas the Affibodies were initially derived from a bacterial protein, Adnectins and Anticalins are both based on endogenous human proteins. In contrast, the designed ankyrin repeat proteins, DARPins, were more recently proposed as an alternative scaffold as result of a protein design effort. In this case an artificial consensus sequence was derived from a series of known ankyrin repeat proteins, which are present in virtually all species and provide a rigid interface arising from modules with typically two anti-parallel α-helices and a β-turn as the underlying principle (Binz et al. 2003). Together, these four scaffolds (Fig. 1) currently constitute the most advanced approaches in this field, each with multiple target specificities exemplified in basic research as well as preclinical studies and at least one drug candidate tested in the clinic (Gebauer and Skerra 2009).

Otherwise, the field of engineered protein scaffolds has undergone some consolidation and appears to focus on those few that may potentially yield products with commercial value. In particular, technical demands at the outset of biopharmaceutical drug development left behind many protein scaffolds that were once proposed in an academic setting, but never matured beyond an initial model case study (Gebauer and Skerra 2009). As result, only few scaffold technologies, in particular those mentioned above, have successfully expanded after first in vitro proof of concept and now see increasing application toward medical use, as discussed in the following sections.

Adnectins

The fibronectin type III domain (FN3), a small (10 kDa) autonomous folding unit found in the abundant extracellular matrix proteins fibronectin and tenascin, as well as in a large variety of other multidomain cell adhesion proteins (Hohenester and Engel 2002), was one of the first protein scaffolds employed for generating novel binding sites. The FN3 structure closely resembles the Ig domain fold, exhibiting three exposed loops, termed BC, DE, and FG, analogous to the CDRs described further above. In contrast, the FN3 domain lacks the central disulfide bond which normally connects the sandwich of β-sheets. Despite their conserved structure, individual FN3 domains show considerable sequence divergence, which hinted at their tolerance for mutations introduced in order to implement desired binding functions (Koide et al. 1998).

FN3 random libraries were based on the 10th type III (^{10}FN3) domain from human fibronectin to generate antibody (or, more precisely, sdIg) mimics initially by randomizing the BC and FG loops and, later, also the DE loop (Koide et al. 1998; Lipovsek et al. 2007; Xu et al. 2002). The first variant of this scaffold that had a significant, though still moderate affinity ($K_D \approx 5$ μM) for ubiquitin as a model target was selected from a phagemid display library and called "Monobody". In combination with mRNA display technology, this scaffold was subsequently commercialized as "Trinectin" and continued as "Adnectin™" (Lipovsek 2011).

The design of the ^{10}FN3-based random libraries was successively improved using insights from synthetic antibody libraries, including loop length variation and bias in amino acid composition within the binding site, which eventually allowed the generation of high-affinity binding proteins against various targets (with K_D values down to the pM range) (Hackel et al. 2008, 2010; Hackel and Wittrup 2010; Koide et al. 2007). Furthermore, the ^{10}FN3 scaffold was investigated for the potential to utilize all six loops—at both ends of the β-sandwich structure—to implement binding activities (Batori et al. 2002). No consensus has yet emerged as to which loop design strategy is most effective, although some studies indicate that the BC and FG loops, which are topologically equivalent to the CDRs H1 and H3 of a VH domain, are most important for target binding (Koide et al. 2007). Recently, an alternative recognition mode was described that utilizes a concave binding site on the surface of the β-sandwich (Gilbreth and Koide 2012; Ramamurthy et al. 2012). Hence, the wedge-shaped ^{10}FN3 scaffold offers the possibility to address either cleft-like epitopes, e.g. on interleukin 23 (IL-23), or those with a convex shape, e.g. on epidermal growth factor receptor (EGFR) (Fig. 1), depending on the interface.

In 2006 the first Adnectin, Pegdinetanib (CT-322, BMS-844203, Angiocept), which specifically inhibits the vascular endothelial growth factor receptor 2 (VEGF-R2), i.e. the primary receptor in mediating tumor angiogenesis, entered clinical trials (Bloom and Calabro 2009; Dineen et al. 2008; Mamluk et al. 2010). Pegdinetanib, whose penultimate Cys residue (no. 93) is modified with a doubly methoxy-PEG-derivatized (40 kDa) maleimide, binds VEGF-R2 in vitro with a K_D value of 11 nM (Mamluk et al. 2010). The first in human study confirmed the favorable pharmacokinetics of the PEGylated protein scaffold, showing slow plasma clearance with $\tau_{½}$ of 3–4 days (Tolcher et al. 2011). Notably, the majority of patients developed anti-drug antibodies (ADAs), which appeared to be specifically directed against the engineered binding loops, as no cross-reactivity with the original ^{10}FN3 scaffold was seen. However, no adverse events associated with the ADA response appeared in this phase I trial (Tolcher et al. 2011). Nevertheless, a novel design of the Adnectin libraries aiming at eliminating immunogenic hot spots was recently published (Davis et al. 2013). In this approach, key areas within the β-strand B, the BC loop and β-strand C were kept as wild-type sequence while modifications were introduced into other regions of the scaffold to achieve high affinity target binding. Pegdinetanib was subsequently evaluated in multiple phase II trials to assess its applicability in a range of cancers including non-small cell lung cancer, glioblastoma multiforme and metastatic colorectal cancer.

Recently, a phase I clinical study of a second Adnectin (BMS-986089), a myostatin inhibitor developed to enhance muscle growth in indications such as muscular dystrophy or sarcopenia in older people (Cload et al. 2014), was initiated (BMS, http://www.b-ms.co.uk/research/pipeline). Myostatin is a secreted protein and member of the TGF-β family that is produced primarily in skeletal muscle cells, preventing muscle cell growth and differentiation, thus being of great interest as a therapeutic target in myopathies (Dschietzig 2014).

Other Adnectins in preclinical development include a fusion protein between fibroblast growth factor 21 (FGF21) and an "Adnectin pharmacokinetic enhancer" (AdPKE) as a potential therapeutic agent in diabetes. In FGF21-AdPKE the Adnectin moiety confers binding activity towards human as well as monkey serum albumin, thus extending the very short circulation half-life of human FGF21 from 4 to 96 h in cynomolgus monkeys (Klöhn et al. 2013), potentially offering once weekly subcutaneous dosing in human patients (Mukherjee 2013). The second Adnectin in advanced preclinical stage, BMS-938790 (likely in a PEGylated form), is a potent inhibitor of the inflammatory cytokine IL-23, whose increased levels are associated with several autoimmune diseases such as RA, multiple sclerosis (MS) and Crohn's disease (Das Gupta 2014).

Apart from this ^{10}FN3 scaffold, similar domains derived from the same fibronectin or related extracellular matrix proteins were also employed to generate scaffold libraries. For example, so-called Pronectins™ were developed on the basis of the 14th type III domain of fibronectin utilizing a bioinformatics approach (Mintz and Crea 2013). The Pronectin phage display library (Cappuccilli et al. 2014) was designed after analysis of the natural loop diversity found in thousands of non-redundant human FN3 sequences and comprises a combination of three sublibraries, each having mutations in only two of the exposed BC, DE and FG loops (Protelica, http://www.protelica.com).

Another related concept was recently adopted during development of the Centyrins (Jacobs et al. 2012). In this case, random libraries were built on an artificial consensus sequence derived from the tenascin FN3 domains. Not only the CDR-like loops but also exposed side chains on some of the β-strands were randomized to generate novel binding proteins, e.g. against human hepatocyte growth factor receptor (HGFR, c-MET), murine IL-17A and rat TNFα (Diem et al. 2014). Centyrins also offer potential as bispecific reagents (for broader discussion see further below), for example c-METxEGFR, engineered as a fusion protein of two Centyrin modules that are additionally linked to an albumin-binding domain (Klein et al. 2013).

Affibodies

Affibody molecules are small (7 kDa) artificial binding proteins originally derived from protein Z (Nilsson et al. 1987), an isolated engineered IgG-binding domain of the "protein A" found on the bacterial cell surface of *Staphylococcus aureus*. Combinatorial phage display libraries were generated by targeting random mutagenesis at residues exposed on the first two α-helices of the three-helix bundle scaffold in a region that is naturally involved in binding to the Fc part of antibodies (Nord et al. 1995, 1997). To date, a large number of Affibodies have been described, not only as research or purification reagents, but also exerting binding activities towards various medically relevant protein targets, e.g. insulin, fibrinogen, transferrin, TNFα, IL-8, gp120, CD28, IgA, IgE, IgM, EGFR, HER2 (Löfblom et al. 2010) and, recently, ErbB3/HER3 (Kronqvist et al. 2011). One Affibody, $Z_{HER2:342}$, specific for the breast cancer target HER2, constitutes probably the most intensely studied protein of this class (>200 hits in PubMed; http://www.ncbi.nlm.nih.gov). This Affibody, which after extensive in vitro maturation shows extraordinary affinity for its target ($K_D = 22$ pM) (Orlova et al. 2006), will be discussed further below.

Affibodies have been predominantly exploited for use in diagnostic settings, especially molecular imaging, where they offer beneficial features. The small size of this scaffold ensures rapid tissue penetration, fast blood clearance and, moreover, allows production via solid-phase peptide synthesis, permitting the direct site-specific incorporation of various chemical functionalities such as fluorescent probes or chelating groups for radioactive metal ions. The generally high stability, independent of disulfide bonds, in concert with the high solubility of most Affibodies further enable these molecules to resist even harsh chemical labelling conditions while retaining binding activity. Nevertheless, some Affibodies may adopt molten globule structures in the absence of their target proteins, probably due to the extensive mutagenesis of secondary structural elements (Wahlberg et al. 2003). Also, one Affibody was described that surprisingly adapts to the hairpin conformation of its target, the Alzheimer amyloid-β peptide, by dimerizing and forming a four-stranded intermolecular β-sheet, thus giving rise to an extended groove (Hoyer et al. 2008).

Up to now, Affibodies were developed in the clinic mainly as tracers for HER2-specific molecular imaging. In 2010, the ^{68}Ga- or ^{111}In-radiolabeled chelator-functionalized Affibody ABY-002 (DOTA-$Z_{HER2:342\text{-}pep2}$) demonstrated for the first time that it is possible to localize via SPECT or PET/CT imaging even very small HER2-positive cancer lesions in patients with recurrent metastatic breast cancer and, notably, as early as 2 h after injection of the tracer (Baum et al. 2010). Imaging with ABY-002 was also successful in a patient that had already received Herceptin treatment. Indeed, ABY-002 binds to an epitope distinct from those of Trastuzumab (Herceptin®) and Pertuzumab (Perjeta®) (Eigenbrot et al. 2010) and, hence, this tracer may be useful for monitoring changes in HER2 expression during therapeutic intervention with these antibodies.

Compared to the current standard of tumor imaging, ^{18}F-FDG PET, which indicates increased glucose metabolism, the uptake of ABY-002 in some metastatic lesions was distinct, either due to differing levels of HER2 expression or to the limited sensitivity of ^{18}F-FDG PET/CT in early-stage cancer (Baum et al. 2010). However, this study also showed that the Affibody was mainly cleared via the liver, which obscured imaging of metastases in this organ. To address this issue, the scaffold of ABY-002 was reengineered by substituting surface-exposed side chains outside of the target interface to increase hydrophilicity, to achieve higher thermal and chemical stability, and to reduce unwanted binding to Ig as well as normal tissues, thus effecting elimination predominantly via the kidney (Feldwisch et al. 2010). The resulting anti-HER2 Affibody, $Z_{HER2:2891}$, had an optimized surface that, despite the considerable deviation from the parental scaffold, fully retained tumor targeting functionality in vitro and in vivo in a mouse xenograft model (Ahlgren et al. 2010).

The clinical safety and efficacy of $Z_{HER2:2891}$ for HER2 imaging in metastatic breast cancer was recently evaluated (Sörensen et al. 2014). In contrast to the parental molecule ABY-002 (DOTA-$Z_{HER2:342\text{-}pep2}$), the new lead candidate ABY-025 ([MMA-DOTA-Cys61]-$Z_{HER2:2891}$-Cys) can also be produced (apart from peptide synthesis) by recombinant expression in *E. coli* (Feldwisch et al. 2010) and, via a C-terminally introduced Cys residue ($Z_{HER2:2891}$-Cys), coupled to the chelator maleimido-1,4,7,10-tetraazacyclododecane-1,4,7,10-tetraacetic acid (DOTA). The resulting radiolabeled molecule, ^{111}In-ABY-025, allowed diagnosis of HER2 expression in seven patients while lacking the disadvantage of ABY-002, as the highest uptake of the new Affibody was seen in the kidneys (Sörensen et al. 2014). Probably as a benefit of the reengineered scaffold of ABY-025, and in good agreement with previous animal experiments (Ahlgren et al. 2010), no anti-ABY-025 antibodies were observed after single administration (100 μg). In one patient, molecular imaging with ABY-025 in conjunction with ^{18}F-FDG PET enabled detection of a converted receptor status on metastases despite a HER2-positive primary tumor, and, in another patient, a brain lesion was diagnosed that was not seen via previous ^{18}F-FDG PET alone (Sörensen et al. 2014).

Anti-HER2 Affibodies were also labeled with radionuclides suitable for RIT. Generally, the small size and rapid pharmacokinetics (Orlova et al. 2009), which is compatible with very short-lived positron emitting radionuclides like ^{68}Ga, is particularly advantageous for molecular imaging, especially if compared to imaging with ^{89}Zr-Trastuzumab, whose optimal time point to assess tumor accumulation is 4–5 days post injection (Dijkers et al. 2010). However, pronounced renal reabsorption of the Affibody radioconjugates and, thus, high levels of kidney exposure constitute a potential safety problem for therapeutic applications. Two approaches were evaluated to alleviate this problem for Affibodies: first, reducing renal uptake by modifying the pharmacokinetics via fusion to a second Affibody module that effects reversible binding to albumin (Orlova et al. 2013; Tolmachev et al. 2007, 2009b) and, second, decreasing renal retention of radioactivity by means of modified labelling chemistry (Orlova et al. 2010; Tolmachev et al. 2009a).

Notably, use of serum albumin as a drug carrier is often not only paired with longer circulation half-lives but also favors tumor targeting as albumin is known to preferentially localize to malignant versus normal tissue (Kratz 2008; Merlot et al. 2014). In the novel Affibody construct, ^{111}In-ABY-027, the same target-specific molecule used for ABY-025 was fused to a small, high-affinity albumin-binding domain, ABD_{035}, an engineered bacterial three-helix bundle protein with similar structure as the Z-domain (Jonsson et al. 2008), and site-specifically radiolabeled. In a mouse xenograft model ^{111}In-ABY-027 showed a prolonged kinetic profile and significantly lower renal accumulation of radioactivity together with an increase in the delivered dose to the tumor compared to previous anti-HER2 Affibody reagents (Orlova et al. 2013). The impact of the nature and position of the radionuclide chelator within the protein molecule on the biodistribution of Affibodies was demonstrated using $Z_{HER2:342}$ with different radiolabels, including mercaptoacetyl-containing peptide-based chelators for ^{99m}Tc-labelling. In this case, the use of more hydrophilic chelators resulted in a switch from predominantly hepatobiliary excretion to renal excretion, which significantly improved image contrast in the abdominal area (Tolmachev and Orlova 2009).

While most of the molecular imaging studies were performed with Affibodies specific for HER2, this approach was recently complemented by Affibodies targeting other important members of the transmembrane receptor tyrosine-kinase family such as EGFR (Malmberg et al. 2011; Tolmachev et al. 2010) and PDGF-Rβ (Strand et al. 2014; Tolmachev et al. 2014), which are both overexpressed in many malignancies.

Anticalins

Anticalins are antibody mimetics with designed ligand-binding properties derived from the natural scaffold of the lipocalins, a family of compact, soluble proteins that are abundant in human plasma and other body fluids and serve to transport or scavenge low molecular weight molecules. Lipocalins are found in many organisms including vertebrates, insects, plants and even bacteria (Åkerström et al. 2006). These functionally diverse small proteins, with a single polypeptide chain of 160–180 residues, generally form complexes with chemically sensitive biological compounds, in particular, vitamins, steroids, signaling molecules, lipids and other secondary metabolites. In humans, more than 12 different lipocalins have been identified thus far (Breustedt et al. 2006; Schiefner and Skerra 2015). Due to their high abundance in blood and their capacity for binding various physiologically active compounds, even some natural lipocalins have been considered for therapeutic use. In this regard, not only endogenous human lipocalins attracted attention, but also lipocalins from blood-sucking insects with medically relevant binding activities as well as low immunogenicity in humans.

For example, human neutrophil gelatinase-associated lipocalin (NGAL, also known as lipocalin 2, Lcn2, or as siderocalin), which tightly binds the catecholate-type siderophore Fe^{III} · enterochelin/enterobactin (Goetz et al. 2002) as well as other siderophores of mycobacteria, was proposed for anti-infective drug development (Holmes et al. 2005). Apart from its potential use as antibacterial agent, its therapeutic application for renal protection from ischemia-reperfusion injury was suggested—based on the notion that NGAL also mediates iron trafficking in kidney epithelial cells (Mori et al. 2005). Furthermore, two insect lipocalins have attracted interest for clinical use: the histamine-binding protein (HBP) from the saliva of the tick *Rhipicephalus appendiculatus* (Paesen et al. 1999) and the C5 complement inhibitor (OmCI) from the soft tick *Ornithodoros moubata* (Barratt-Due et al. 2011; Fredslund et al. 2008).

In vivo efficacy was demonstrated for recombinant HBP (rEV131) in preclinical studies, where it was found to inhibit murine allergic asthma (Couillin et al. 2004), and this lipocalin was further evaluated up to a phase II clinical trial (NCT00353964). Preclinical studies of recombinant OmCI, initially termed rEV576, showed effective inhibition of complement and significantly reduced weakness in vivo in two rat models of experimentally acquired myasthenia gravis (Soltys et al. 2009). Other preclinical studies of OmCI demonstrated a protective effect against neural injury in an in vitro mouse model of the Miller Fisher syndrome (Halstead et al. 2008) as well as reduction of excessive inflammatory reactions associated with severe forms of Influenza A virus infections in mice (Garcia et al. 2013). OmCI, meanwhile named Coversin (Varleigh Immuno Pharmaceuticals, http://www.vipimmunopharma.com), has been evaluated in a phase I clinical study (Weston-Davies et al. 2013).

The central folding motif of the lipocalins is a structurally highly conserved β-barrel: eight antiparallel β-strands wind around a central axis and form an almost cylindrical cup-shaped structure. The bottom of this β-barrel is closed by short loops, and a hydrophobic core is formed there by densely packed amino acid side chains. At the open end, a set of four loops forms the entrance to the ligand pocket. While in general the amino acid sequence homology among the lipocalins is extremely low (Flower 1996), these four loops additionally exhibit large differences in their lengths and backbone conformations, in line with the diverse ligand specificities seen for the different family members (Skerra 2000b; Schiefner and Skerra 2015). Thus, in principle, the ligand-binding sites of lipocalins with their four structurally hypervariable loops resemble those of Igs with their set of six CDRs. However, a major advantage of the lipocalin scaffold is its simple composition of just one rather short polypeptide chain. Thus, biotechnological production and purification of lipocalins and their engineered variants is much easier and more cost-effective than antibody production, especially in connection with an *E. coli* expression system.

The peculiar ligand-binding site of the lipocalins with its (outside the Ig superfamily) unprecedented structural plasticity has prompted the engineering of novel antigen-binding proteins with antibody-like properties that were named "Anticalins" (Beste et al. 1999; Skerra 2001). Employing the methods of combinatorial protein design to alter the four hypervariable loops at the entrance to the ligand pocket as well as adjoining β-strand regions in a directed manner, artificial lipocalins with novel ligand specificities for prescribed targets can be generated (Gebauer and Skerra 2012). Meanwhile, several different natural lipocalin scaffolds for which three-dimensional structures became available have been utilized to generate Anticalins (Richter et al. 2014): e.g. the bilin-binding protein (BBP) from the butterfly *Pieris brassicae*, the human apolipoprotein D (ApoD), human lipocalin 1 (Lcn1, also known as tear lipocalin, Tlc) and human lipocalin 2 (Lcn2, NGAL).

In the case of BBP, a blue pigment protein that naturally protects insects from oxidative stress (Schmidt and Skerra 1994), the first Anticalins with novel specificities were selected via phage display from a semi-synthetic random library to bind the dye fluorescein (Beste et al. 1999), the plant steroid digoxigenin (Schlehuber et al. 2000) and a phthalic ester plasticiser (Mercader and Skerra 2002). These findings demonstrated the high tolerance of the lipocalin scaffold for introduction of artificial side chains into the binding site on a wider scale and, with the help of site-directed mutagenesis studies and crystal structure analyses (Korndörfer et al. 2003a, b; Vopel et al. 2005), confirmed the fundamental similarity with the combining site of antibodies.

Building on this knowledge, Anticalins were subsequently designed on the basis of human lipocalins, in particular Lcn1 and Lcn2, with the goal to ensure high tolerability to patients upon therapeutic application (Mendler and Skerra 2013). Again, these Anticalins were generated via combinatorial protein engineering in an in vitro process that in principle resembles the humoral immune response against an antigen (Gebauer and Skerra 2012). One notable aspect of this technology is that due to the large differences between natural members of the lipocalin family each scaffold involves an individually optimized random library design. This pertains to choosing the most suitable set of residues to be targeted for mutagenesis, whereby the total number of randomized positions is generally limited according to theoretical considerations (Richter et al. 2014). In case of the Lcn2 scaffold it was nicely demonstrated that the library design can be optimized in a few iterative cycles, also taking advantage of X-ray structural information (Gebauer et al. 2013). The third generation Lcn2-based Anticalin library resulting from this endeavor is highly potent in yielding binding proteins against a wide range of biomolecular targets, showing high affinities in the nM to pM range directly after phage display selection from the naive library (Richter et al. 2014) while still offering the potential for further improvement via in vitro affinity maturation.

Anticalins have been selected against a series of disease-relevant protein antigens. An early example was the extracellular domain of cytotoxic T-lymphocyte-associated antigen 4 (CTLA-4, CD152), which has attracted considerable interest as target of immune-checkpoint inhibitors (Pardoll 2012). In agreement with its function as a negative regulatory T-cell co-receptor, the lead Anticalin showed potent blocking activity and, consequently, an immunostimulatory effect on the T-cell response in an animal model of infectious disease (Schönfeld et al. 2009).

Furthermore, this Anticalin has shown potential for the immunotherapy of cancer in preclinical studies (Pieris, http://www.pieris-ag.com). Another protein-specific Anticalin recognizes a disease-specific splice variant of the extracellular matrix protein fibronectin (Fn) exhibiting the extra-domain B (ED-B) (Gebauer et al. 2013). Since in adults ED-B-positive, so-called oncofetal Fn is almost exclusively expressed during tumor angiogenesis, apart from wound healing and some other pathophysiological states that involve neovascularization, it has emerged as a promising marker for various cancers and constitutes a validated target for in vivo imaging (Kaspar et al. 2006), as already mentioned further above. Cognate Anticalins show specific staining of ED-B positive cells in immunofluorescence microscopy and, notably, allow the detection of primary glioblastoma multiforme in human patients (Albrecht et al. submitted).

Beside Lcn2-based Anticalins, other human lipocalin scaffolds have been successfully applied to generate highly specific and functionally active binding proteins against medically relevant targets. The human Lcn1/Tlc scaffold served to develop an Anticalin that effectively blocks VEGF-A (Chekhonin et al. 2013) and constitutes a drug candidate for the treatment of solid cancers. This Anticalin, dubbed Angiocal® (PRS-050; conjugated with 40 kDa branched PEG), has been investigated as inhibitor of tumor angiogenesis in a first-in-human phase I trial, demonstrating safety and high tolerability as well as a terminal half-life ($\tau_{½}$) ranging from 5.5 to 7.0 days (Mross et al. 2013). Importantly, PRS-050 appeared to lack immunogenicity, based on the absence of an ADA response in 24 patients with post-baseline samples available, including six patients who received biweekly dosing and one who had received altogether 17 doses. Based on the encouraging data from this repeat dose escalating study, Angiocal shows promise for phase II clinical trials.

Recently, another Anticalin based on the Lcn1/Tlc scaffold (PRS-110; conjugated with 40 kDa branched PEG) was shown to act as a highly potent and specific c-Met antagonist with both ligand-dependent and ligand-independent activity in mouse xenograft models (Olwill et al. 2013). Moreover, a radiolabeled version of this PEGylated Anticalin, ^{89}Zr-PRS-110, allowed visualization of c-Met tumor expression via in vivo imaging in mice (Terwisscha van Scheltinga et al. 2014).

Like the natural lipocalins and as mentioned above for the BBP platform, Anticalins derived from a human scaffold, in particular Lcn2, can also tightly bind low molecular weight substances, i.e. hapten-like targets. This has been shown for an Anticalin developed with pM affinity against a metal-chelate complex comprising a lanthanide ion, e.g. Y^{III}, and a derivative of diethylenetriamine pentaacetic acid (DTPA) (Eggenstein et al. 2013; Kim et al. 2009), which offers a promising reagent for in vivo pretargeting radioimmuno therapy and diagnostics. Hapten-specific Anticalins based on the BBP scaffold have been investigated for therapeutic applications as well. The fluorescein-specific Anticalin FluA was used, after genetic fusion to an scFv fragment recognizing the Fn extra-domain A as angiogenesis marker, for pretargeted payload delivery in a mouse tumor xenograft model (Steiner et al. 2013). Also, an affinity-improved digoxigenin-binding Anticalin (DigiCal) showed promising results in a rat study as antidote to treat digitalis intoxication (Eyer et al. 2012).

High affinity Anticalins with blocking activity were also selected against peptides, for example the Aβ peptide (Rauth et al. in preparation), which plays a crucial role in the pathophysiology of Alzheimer's disease, and hepcidin (Hohlbaum et al. 2011), a negative regulator of iron homeostasis relevant in anemia (Ruchala and Nemeth 2014). The development of Anticalins against the hepcidin peptide has been funded by the European Commission via the EUROCALIN Consortium (EUROCALIN, http://www.eurocalin-fp7.eu) to promote clinical investigation. PRS-080, an anti-hepcidin Anticalin designed to treat anemia, has entered a phase I clinical trial in 2014 (Pieris). Generally, the lipocalin scaffold appears to offer the most versatile platform to engineer binding proteins against a wide range of target molecules with different sizes, shapes and biomolecular properties (Fig. 1), including the potential to create multispecific or multifunctional fusion proteins (e.g. so-called Duocalins). Furthermore, it is the only type of scaffold for which natural representatives have been subjected to biopharmaceutical development.

DARPins

The so-called designed ankyrin repeat protein (DARPin) scaffold emerged from a consensus design approach (Binz et al. 2004; Forrer et al. 2004) based on the highly abundant natural ankyrin repeat proteins, ARPs (Grove et al. 2008; Mosavi et al. 2002). ARPs are found, for example, as cytoskeletal proteins, transcriptional initiators, cell cycle regulators and signal transducers mostly in the cellular cytoplasm, where they mediate protein-protein interactions. This protein class exhibits a rigid and also modular fold comprising characteristic repeat units of 33 amino acids, each with two anti-parallel α-helices that are mutually connected by a short loop and are linked to the next unit by an extended β-turn. Natural ARPs usually contain four to six of these repeats (while this number can rise up to $\geq$24, e.g. for ankyrin itself), which are stacked on each other to form a compact, overall rod-shaped domain, also called linear solenoid (Fig. 1). To generate DARPins, random amino acid substitutions are introduced into the β-turn and the first α-helix of each module, usually employing three repeat units (depending on the DARPin library), together with N- and C-terminal capping units. Typically, six potential target interaction positions per internal repeat are mutated in the artificial ARP consensus sequence.

From such combinatorial libraries DARPins with high affinities have been selected, normally via ribosome display, against a variety of proteins, including several medically relevant targets: for example, HIV gp120 (Mann et al. 2013; Schweizer et al. 2008), EpCAM (Stefan et al. 2011), Alzheimer amyloid-β peptide (Aβ) (Hanenberg et al. 2014) and, as will be described in greater detail below, VEGF-A (Stahl et al. 2013). Since the advantageous biophysical properties of the parental ankyrin scaffold such as high-level expression, solubility, and stability are often retained in these DARPins, they are also well suited for the generation of multispecific constructs (Molecular Partners, http://www.molecularpartners.com). Indeed, the therapeutic potential of DARPins in the area of cancer therapy is supported by the development of bispecific and/or biparatopic DARPins (Jost et al. 2013) as well as toxin fusions (Simon et al. 2014).

Furthermore, methods were established to extend the intrinsically very short plasma half-life of DARPins (Zahnd et al. 2010), which appears to be only few minutes in cynomolgus monkeys (Klöhn et al. 2013). To this end, DARPins were site-specifically functionalized using unique Cys residues or, alternatively, through incorporation of non-natural amino acids via metabolic pressure in order to allow selective conjugation with PEG, serum albumin and, optionally, with cytotoxins (Moody et al. 2014; Simon et al. 2013, 2014). An interesting concept of DARPins directed against tumor-specific cellular surface receptors such as HER2, EGFR and EpCAM aims at gene therapy: different DARPin-targeted viruses were shown to efficiently discriminate between tissues, potentially offering cell type-specific in vivo gene delivery (Friedrich et al. 2013; Munch et al. 2013). Notably, viral envelope proteins fused with DARPins showed superior incorporation to corresponding fusion proteins with scFvs, which are typically used for this kind of approach (Munch et al. 2011), probably owing to the more robust nature of this non-Ig scaffold.

The lead clinical candidate DARPin, Abicipar pegol (initially developed as MP0112, after reformulation named AGN-150998), is a potent antagonist of VEGF-A and inhibits all relevant subtypes with $IC_{50} \leq 10$ pM (Binz et al. 2014; Klöhn et al. 2013; Stahl et al. 2013). Abicipar pegol is currently tested in two human eye diseases related to neovascularization: age-related macular degeneration (AMD), a painless eye condition that leads to the gradual loss of central vision, and diabetic macular edema (DME) (Campochiaro et al. 2013). In two phase I/II clinical trials, MP0112 was shown to be well tolerated and to efficiently neutralize VEGF-A after a single intravitreal injection. However, application of the DARPin was also associated with adverse events such as ocular inflammation in most of the patients, probably caused by impurities from the bacterially produced drug. MP0112 was subsequently reformulated into AGN-150998 (EudraCT 2011). Although the sponsor has not yet publicized which modifications were made, it appears that the formulation of AGN-150998 comprises a DARPin covalently linked to a single molecule of methoxy maleimide polyethylene glycol 20 (mPEG20), possibly even containing a modified Fc part of human Ig (CROSS 2011; INN 2012).

Safety and efficacy of AGN-150998 were recently tested in a phase II clinical trial (REACH study), where the reformulated drug proved to be at least as effective as monthly dosed Ranibizumab (Lucentis), the standard of care for wet AMD, while showing longer duration of action (NCT01397409; Reuters 2014). However, in contrast to Ranibizumab, a humanized high affinity anti-VEGF Fab produced in *E. coli* (Rosenfeld et al. 2006) mentioned already further above, patients treated with AGN-150998 still experienced ocular inflammation, although to a lesser extent than in phase I (Campochiaro et al. 2013). Improvements to the manufacturing process for AGN-150998 are expected before commencing phase III studies (Bloomberg 2013).

Apart from this monospecific protein drug, the dual-antagonistic DARPin MP0260, which is directed both against VEGF-A and platelet-derived growth factor B (PDGF-B), is under development for the treatment of wet AMD and related conditions (Molecular Partners). Considering that another anti-VEGF/anti-PDGF combination therapy of wet AMD, i.e. Fovista® with Ranibizumab, had already shown superior efficacy in a phase II trial compared to Lucentis monotherapy (Boyer 2013), a bispecific "all-in-one" drug such as MP0260 might prove beneficial. However, in light of the potential complications and burden for the patient associated with repeated intravitreal drug injections for the treatment of chronic ocular diseases, an alternative route of administration would be highly desirable. Yet, only preclinical data were published so far with regard to the potential application of DARPins as eye-drops (Stahl et al. 2013).

Bispecific Constructs

Engineered protein scaffolds are no longer regarded as alternatives to conventional MAbs just for the sake of intellectual property or other commercial aspects; rather, there is increasing recognition that they, together with robust Ig fragments, may enable unique therapeutic modes of action more easily owing to their small size and facile manipulation, possibly extending to bioprocess development. One area of great interest (Garber 2014) is the preparation of bispecific binding proteins, in particular by way of the modular fusion approaches mentioned above. In fact, despite the long history of bispecific MAbs (Milstein 2000)—starting with hybrid hybridoma (quadroma) technology—this classical format still poses a challenge with regard to biomanufacturing, even though several strategies were recently developed to drive heterodimerization of the two halves of the Y-shaped Ig molecule more efficiently (Strohl and Strohl 2012).

One successful strategy that makes use of scFv fragments is the bispecific T-cell engager format, BiTE® (Bäuerle and Reinhardt 2009), directing the patient's T-cell immune response to cancer cells by creating a physical link. This concept arose from the notion that bispecific MAbs which redirect non-cognate T-lymphocytes to tumor cells were found to be generally more efficient in triggering cell killing or elimination than conventional MAbs of the same target specificity (Riethmüller 2012; Weiner and Hillstrom 1991). To provide a simpler molecular format, BiTEs are composed of two flexibly linked scFvs, fused in tandem, one binding to CD3 as part of the T-cell receptor (TCR) complex and the other one to a predefined surface antigen on the tumor cell, for example CD19 on B-cell lymphomas. The small size of BiTEs allows bringing target and effector cells in close proximity, thereby enabling the formation of a cytolytic synapse and triggering tumor cell killing—a process that is normally seen only between T-cells and antigen presenting cells (Dreier et al. 2003; Hoffmann et al. 2005; Offner et al. 2006).

In the BiTE approach, systemic cytokine release and toxicities are low or negligible, probably as a result of the monovalent format, which does not activate the CD3 receptor on non-engaged T-cells in the absence of target cells (Brischwein et al. 2007). So far, BiTEs have been generated against several tumor-associated target molecules such as CD19, CD33, EpCAM, HER2, CEA, ephrin A2 (EphA2), and melanoma-associated chondroitin sulfate proteoglycan (MCSP) (Bäuerle et al. 2009).

Blinatumomab (AMG103, formerly MT103 or MEDI-538) (Löffler et al. 2000) was the first genetically engineered bispecific antibody that entered clinical trials and constitutes the most advanced BiTE (Garber 2014). Besides CD3, it targets the B-cell surface marker CD19 which is ubiquitously expressed throughout all stages of B-cell differentiation and, in particular, present on virtually all tested malignant B-lineage cells in acute lymphoblastic leukemia (ALL) (Raponi et al. 2011). This biological drug candidate has been tested in combination with chemotherapy in a phase III trial in ALL patients who were not cured by chemotherapy alone and were given a poor prognosis (NCT02013167). A preceding large phase II study had confirmed antileukemic activity of Blinatumomab alone in a difficult-to-treat population with relapsed/refractory ALL (Topp et al. 2014). Considering these encouraging results, the FDA designated Blinatumomab in July 2014 as breakthrough therapy for ALL and, furthermore, this compound received orphan drug designation for the treatment of non-Hodgkin's lymphoma (NHL) (Viardot et al. 2013). Blinatumomab was approved by the FDA in the end of 2014 as Blincyto™ for treating a rare form of B-cell ALL (Sheridan 2015).

Another BiTE (AMG330) targets CD33, a cell surface marker of acute myeloid leukemia (AML), i.e. the most frequent form of acute leukemia in adults. Ex vivo studies of AMG330 with primary AML cells revealed potent cytotoxicity, showing efficient T-cell activation and expansion even in samples with low CD33 expression. Moreover, AMG330 did not modulate surface expression of CD33, suggesting that long-term exposure to the BiTE, as in the clinical application of Blinatumomab, should not diminish expression of the receptor (Krupka et al. 2014; Laszlo et al. 2014). Given the fact that all previous attempts to introduce novel targeted therapies for AML have failed to date—ultimately including the anti-CD33 antibody-drug conjugate (ADC) Mylotarg®, which had been approved in 2000 but was subsequently withdrawn from the market due to safety issues and limited benefit (Strohl and Strohl 2012)—it will be interesting to see if this BiTE may offer a more potent treatment.

The success of this bispecific format triggered the design of other scFv-based constructs with similar size and valence. A competing technology is the dual-affinity re-targeting (DART®) molecule (Johnson et al. 2010), a heterodimer composed of two polypeptide chains, each comprising a VH domain from one Fv fused to a VL domain from another one, thus resembling the classical Diabody format (Holliger et al. 1993). However, DARTs are covalently linked by a C-terminal interchain disulfide bond, which makes them independent of the stability of the natural VL/VH association and avoids domain exchange, which is known to affect conventional Diabodies (Johnson et al. 2010).

In head-to-head in vitro experiments comparing the DART and BiTE formats with the same Fv sequences, the CD19xCD3 DART was 4–60 fold more potent in redirecting cell killing (Moore et al. 2011). This was surprising insofar as the DART protein displayed only a twofold higher affinity for each target molecule. Consequently, this platform has been developed for different targets (MacroGenics, http://www.macrogenics.com). The first DART that entered phase I clinical study is MGD006 which binds the tumor target CD123 and redirects, via its CD3 specificity, T-cells against corresponding leukemic cells, resulting in the clearance of AML blasts in vitro and in vivo (Al Hussaini et al. 2013). Furthermore, the DART

MGD007, directed against gpA33, is under development for the treatment of colorectal cancer and other gastrointestinal tumors (MacroGenics).

The relatively short biological half-life of BiTEs and DARTs, which is only a few hours in humans (Hoffman and Gore 2014), requires an appropriate dosing schedule and/or formulation to ensure lasting activation of T-cells against target cells. Blinatumomab, for instance, is administered as continuous intravenous infusion over four weeks to achieve sufficient steady-state plasma levels of the drug.

In comparison, a more favorable pharmacokinetic profile has been described for TandAb®s (Kipriyanov 2009; Kipriyanov et al. 1999; Klöhn et al. 2013). In this format, two pairs of scFvs with different specificities are genetically fused in tandem to self-assemble in a head-to-tail fashion, thus forming a tandem Diabody structure (Kontermann 2012). By individually engineering the lengths and compositions of the linker sequences that join the four Fv moieties, the individual V regions can be forced to specifically pair with their counterparts in the second polypeptide chain, thus stabilizing the bispecific protein (Le Gall et al. 2004). With their higher molecular weight (110 kDa) the molecular size of TandAbs is slightly above the renal threshold for first-pass kidney clearance and their half-life in non-human primates is between 3 and 23 h, depending on the dosage (Klöhn et al. 2013). Exhibiting binding activity for both the T-cell effector CD3 and the disease target CD19, the TandAb AFM11 was demonstrated to induce killing of tumor cells in preclinical studies, without any cytotoxicity toward target-negative cells (Zhukovsky et al. 2012b, 2014). Also, compared to similar BiTE and DART reagents, the tetravalent AFM11 showed increased avidity for both CD19 and CD3 as well as higher potency in preclinical experiments (McAleese and Eser 2012; Zhukovsky et al. 2014).

As in classical MAbs, bivalent binding activity is expected to lead to higher avidity and also efficacy, but this must be evaluated for each target with great care. Interestingly, application of elevated concentrations of AFM11 to T-cells in the absence of target-positive cells caused down-modulation of the CD3/TCR-complex following an initial T-cell boost (Zhukovsky et al. 2012a, b), a phenomenon that has not been reported for monovalent BiTEs and DARTs. A phase I clinical study with AFM11 in patients with CD19-positive B-cell malignancies was started in 2014 (NCT02106091). The most clinically advanced TandAb AFM13 (CD30xCD16A), which recruits host natural killer cells and macrophages via CD16A (FcγRIIIa) to kill CD30-positive cells in both B- and T-cell malignancies (Rajkovic et al. 2012), has been scheduled for a phase II study after it was demonstrated to be well tolerated and had shown anti-tumor activity in advanced Hodgkin lymphoma patients (Affimed, http://www.affimed.com). Like BiTEs and DARTs, TandAbs are also under development for the treatment of solid tumors (e.g. AFM21; EGFRvIIIxCD3).

In the context of bispecific biobetters, beyond the Ig fragments and alternative scaffolds discussed so far, there are also some miscellaneous approaches that deserve mention, especially as these have already reached the clinic or are close to it. For example, Fcab™s are engineered (homo-dimeric) Fc fragments that carry mutated loop sequences at the C-terminal tip of the C_H3 domain, thus conferring target specificity while retaining FcRn binding as well as immunological effector functions (Wozniak-Knopp et al. 2010). A HER2-specific drug candidate, FS102

(Batey et al. 2013), is expected to reach a phase I investigation in the near future (F-star, http://www.f-star.com). Notably, this technology forms a link between antibody fragments and alternative scaffolds insofar as the binding site in the Fc fragment is generated via combinatorial design in a manner typical for non-Ig proteins. In an extension of the Fcab technology the mAb^{2}™ format comprises a full-length MAb in which the natural Fc region has been replaced by an Fcab (Woisetschläger et al. 2013). Such bispecific MAbs can bind (bivalently) two different antigens at the same time, with corresponding drug candidates in the discovery to preclinical stage in immuno-oncology.

Furthermore, Fynomers® are non-Ig scaffolds based on the SH3 domain of the human intracellular Fyn tyrosine kinase, which carries two flexible loops suitable for randomization (Grabulovski et al. 2007). The FynomAb® technology platform involves fusion of several Fynomers with an Ig Fc region, thus creating a hybrid homodimeric construct. The lead product COVA322, a bispecific anti-TNFα/anti-IL17A FynomAb (Covagen, http://covagen.com), is in a phase Ib/IIa study for psoriasis (NCT02243787).

Conclusions and Outlook

Both recombinant antibody fragments and engineered binding proteins based on non-Ig scaffolds have emerged as biobetters which complement and, in the future, may even supersede MAbs for various applications in medical therapy (Table 1). With regard to Ig-derived binding proteins, either as recombinant fragments of existing MAbs or directly selected from cloned V-gene libraries, the Fab format is currently most successful, with several established biopharmaceuticals and a growing number of drug candidates in the clinical pipeline (Nelson and Reichert 2009). While their stably monovalent behavior provides a clear mechanistic benefit in a range of therapeutic settings, the predominant disadvantage of Fabs is their short circulation in the blood, which they share with all other platforms discussed in this Chapter. With Cimzia there is already a clinically approved version that shows an in vivo half-life comparable to MAbs, achieved with the help of PEGylation. Further conjugates of this kind are under development, both for Fabs and for most of the alternative scaffolds. However, PEGylation, with its potential burden of a non-biodegradable synthetic polymer, will likely be replaced by more advanced technologies, for example by albumin-binding peptides (Dennis et al. 2002) or by PASylation (Schlapschy et al. 2013), as will be discussed in the next Chapter.

Despite its long history and broad use in basic science, the scFv format is less likely to become a successful therapeutic drug class by itself, especially with regard to systemic application. The reason lies in the generally poor biochemical behavior and inferior stability at the level of protein folding and solubility (Demarest and Glaser 2008). The scFv format appears more promising as part of fusion proteins such as bispecific antibody reagents, e.g. the BITEs, or if linked to cytotoxic enzymes as recombinant immunotoxins. Nevertheless, even in these settings the

Table 1 Engineered protein scaffolds in clinical development

Scaffold, generic name	Target(s)	Molecule	Sponsor	Clinical trials identifier	Name of trial	Phase, start, status
Adnectin: CT-322, BMS-844203, Pegdinetanib, Angiocept	Vascular endothelial growth factor receptor 2 (VEGF-R2)	PEGylated Adnectin; engineered human ^{10}FN3 domain, coupled via its penultimate C-terminal cysteine residue to a branched 40 kDa PEG molecule	Adnexus/ Bristol-Myers Squibb	NCT00374179	CT-322 in treating patients with advanced solid tumors and non-Hodgkin's lymphoma	Phase 1, Sep 2006, completed
				NCT00768911	CT-322 in combination with radiation therapy and temozolomide to treat newly diagnosed glioblastoma multiforme	Phase 1, Oct 2008, closed
				NCT00562419	CT-322 in treating patients with recurrent glioblastoma multiforme and combination therapy with irinotecan	Phase 2, Nov 2007, closed
			Bristol-Myers Squibb	NCT00851045	Ph II trial of a novel anti-angiogenic agent in combination with chemotherapy for the second-line treatment of metastatic colorectal cancer	Phase 2, Feb 2009, completed
				NCT00850577	Ph II of a novel anti-angiogenic agent in combination with chemotherapy for the treatment of non-small cell lung cancer	Phase 2, Feb 2009, terminated
BMS-962476	Proprotein convertase subtilisin/kexin type 9 (PCSK9)	PEGylated Adnectin; engineered human ^{10}FN3 domain, coupled via its penultimate C-terminal cysteine residue to a branched 40 kDa PEG molecule	Bristol-Myers Squibb	NCT01587365	Single ascending dose safety study of BMS-962476 in healthy subjects and patients with elevated cholesterol on statins	Phase 1, May 2012, completed

(continued)

Table 1 (continued)

Scaffold, generic name	Target(s)	Molecule	Sponsor	Clinical trials identifier	Name of trial	Phase, start, status
BMS-986089	Myostatin	Engineered human ^{10}FN3 domain	Bristol-Myers Squibb	NCT02145234	Single and multiple ascending-dose study to evaluate the safety, tolerability, immunogenicity, pharmacokinetics and pharmacodynamics in healthy adult and elderly subjects	Phase 1, May 2014, ongoing
Affibody: ABY-025	Her2	Radioconjugate composed of the affibody $Z_{HER2:2891}$-Cys, conjugated via its C-terminal cysteine to maleimide-DOTA and labeled with radioisotope ^{111}In or ^{68}Ga	Biomedical Radiation Sciences, Swedish Cancer Society	NCT01216033	Exploratory study of breast cancer with ABY025 (ABY0125); drug: ^{111}In-ABY-025	Phase 1, Oct 2010, completed
				NCT01858116	PET study of breast cancer patients using [68Ga]ABY-025	Phase 1, Apr 2013, ongoing
			Dorte Nielsen, Herlev Hospital, Denmark	NCT02095210	HER2 PET imaging in breast cancer patients using [68Ga]ABY-025	Phase 1, Apr 2013, ongoing
Anticalin: PRS-050, Angiocal	Vascular endothelial growth factor (VEGF-A)	Engineered human tear lipocalin (Tlc), conjugated to 40 kDa PEG	Pieris	NCT01141257	Study of Angiocal® in patients with solid tumors, investigating safety, tolerability, blood concentration of study drug	Phase 1, May 2010, completed
PRS-080	Hepcidin	Engineered human neutrophil gelatinase-associated lipocalin (NGAL), conjugated to 30 kDa PEG	Pieris		Anemia of chronic disease (ACD)	Phase 1, Dec 2014, ongoing

BiTE: AMG103, MT103, MEDI-538, Blinatumomab, Blincyto™	CD19xCD3	Bispecific immunoglobulin format comprising a single-chain fusion of an anti-human CD19 mouse monoclonal scFv fragment with an anti-human CD3 mouse monoclonal scFv fragment	National Cancer Institute (NCI)	NCT02101853	Blinatumomab in treating younger patients with relapsed B-cell acute lymphoblastic leukemia	Phase 3, Aug 2014, ongoing
			Amgen	NCT02013167	Ph 3 trial of blinatumomab versus investigator's choice of chemotherapy in patients with relapsed or refractory ALL	Phase 3, Dec 2013, ongoing
			National Cancer Institute (NCI)	NCT02003222	Combination chemotherapy with or without blinatumomab in treating patients with newly diagnosed BCR-ABL-negative B lineage acute lymphoblastic leukemia	Phase 3, Dec 2013, ongoing
DART: MGD006	CD123xCD3	Bispecific and disulfide-stabilized immunoglobulin format, comprising a humanized Fv heterodimer made of the chains $V_{L2}+V_{H1}$ and $V_{L1}+V_{H2}$	MacroGenics	NCT02152956	Safety study of MGD006 in relapsed/refractory AML	Phase 1, May 2014, ongoing
DARPin: MP0112, AGN-150998	Vascular endothelial growth factor (VEGF-A)	Alternative scaffold domain based on ankyrin repeats, conjugated via cysteine 135 to a single linear 20 kDa PEG	Molecular Partners/Allergan	NCT01086761	Study of MP0112 intravitreal injection in patients with wet age related macular degeneration	Phase 1, Mar 2010, terminated
				NCT01042678	Study of MP0112 intravitreal injection in patients with diabetic macula edema	Phase 1, Feb 2010, terminated
			Allergan	NCT01397409	Evaluation of AGN-150998 in exudative age-related macular degeneration (AMD)	Phase 2, Sep 2011, completed

(continued)

Table 1 (continued)

Scaffold, generic name	Target(s)	Molecule	Sponsor	Clinical trials identifier	Name of trial	Phase, start, status
Nanobody: ALX-0081, (ALX-0681), anti-vWF Nanobody, Caplacizumab	human von Willebrand factor (A1 domain; vWF)	Biparatopic immunoglobulin format, comprising a single-chain VHH-linker-VHH fusion of humanized *Lama glama* monoclonal variable heavy chain (VHH) fragments	Ablynx	NCT01020383	Comparative study of ALX-0081 versus GPIIb/IIIa inhibitor in high risk percutaneous coronary intervention (PCI) patients	Phase 2, Sep 2009, completed
			Ablynx	NCT01151423	Study to assess efficacy and safety of Anti-von Willebrand Factor Nanobody in patients with acquired thrombotic thrombocytopenic purpura (TTP)	Phase 2, Sep 2010, completed
			Ablynx	NCT01020383	Comparative study of ALX-0081 versus GPIIb/IIIa inhibitor in high risk percutaneous coronary intervention (PCI) patients	Phase 2, Sept 2009, completed
ALX-0061	Human IL-6R and human serum albumin (HSA)	Bispecific immunoglobulin format, comprising a single-chain VHH_1-linker-VHH_2 fusion of an anti-human IL-6R humanized *Lama glama* monoclonal VHH fragment with an anti-HSA humanized *Lama glama* VHH fragment	Ablynx	NCT01284569	Study to assess safety and efficacy of anti-interleukin 6-receptor (IL6R) Nanobody in Rheumatoid Arthritis (RA) patients	Phase 1/2, Mar 2011, completed
Ozoralizumab ATN-103, PF-5230896	Human TNFα (cachectin) and human serum albumin (HSA)	Bispecific, trivalent immunoglobulin format, comprising a single-chain VHH_{TNF}-linker-VHH_{HSA}-linker-VHH_{TNF} fusion of two anti-human TNFα and a single anti-HSA humanized *Lama glama* monoclonal VHH fragment	Ablynx	NCT01063803	Study evaluating long-term safety of ATN-103 in subjects with Rheumatoid Arthritis	Phase 2, Feb 2010, completed

ALX-0171	Anti-human respiratory syncytial virus (RSV)	Trivalent immunoglobulin format, comprising a single-chain VHH-linker-VHH-linker-VHH fusion of humanized *Lama glama* monoclonal VHH fragments	Ablynx	NCT01483911	ALX-0171 phase I study, evaluating single ascending dose and multiple dose in healthy male volunteers	Phase 1, Dec 2011, completed
			Ablynx	NCT01875926	ALX-0171 phase I pharmacokinetic study in healthy male volunteers	Phase 1, Jun 2013, completed
			Ablynx	NCT01909843	ALX-0171 safety study in adults with hyperresponsive airways	Phase 1, Aug 2013, completed
TandAb: AFM11	CD19xCD3	Bispecific, tetravalent immunoglobulin format, comprising a humanized Fv homodimer made of the chains V_{H1}-linker-V_{L2}-linker-V_{H2}-linker-V_{L1}	Affimed Therapeutics	NCT02106091	Safety study to assess AFM11 in patients with relapsed and/or refractory CD19 positive B-cell NHL or B-precursor ALL	Phase 1, Apr 2014, ongoing
AFM13	CD30xCD16A	Bispecific, tetravalent immunoglobulin format, comprising a humanized Fv homodimer made of the chains V_{H1}-linker-V_{L2}-linker-V_{H2}-linker-V_{L1}	Affimed Therapeutics	NCT01221571	A study to assess AFM13 in patients with Hodgkin lymphoma	Phase 1, Oct 2010, completed

labile VH/VL pairing and the presence of two crucial disulfide bonds often cause problems during biosynthesis and bioprocess development. In this context the advantages of truly single-domain binding proteins, including sdIgs and alternative scaffolds, are brought to light, which likely will replace scFv fragments as target-specific fusion partners in the long run.

Currently, Nanobodies are the most broadly developed class of sdIgs with regard to molecules in the clinic and disease-relevant target specificities. In contrast to dAbs, Nanobodies are generated from immunized animals and, although it is claimed that their immunogenic potential in patients is low, still a certain risk prevails due to the set of conserved mutations that are necessary to maintain solubility and folding stability of the camelid VHH format but that are distinct from the human germline V-gene sequences. Also, it is interesting to note that, at least for therapeutic applications, Nanobodies are not employed as single domain proteins per se but are usually fused with other Nanobody modules. Apart from achieving higher avidity for the biomedical target in case of the biparatopic constructs, the resulting larger size generally slows down kidney filtration, even if not incorporating an albumin-specific Nanobody module to further extend plasma half-life.

In some sense, sdIgs of this kind bridge conventional antibody technology, including their antigen-binding fragments, to the alternative scaffolds with their fully synthetic binding sites. The non-Ig scaffold that is structurally most closely related to dAbs or Nanobodies is the Adnectin, whose fold still remotely belongs to the Ig superfamily but shows important deviations, in particular the missing central disulfide bond. Nevertheless, Adnectins exhibit three CDR-like loops in homologous regions at one end of the wedge-like β-sandwich, which can be randomized in order to create Adnectin libraries suitable for in vitro selection against prescribed targets. Consequently, Adnectins are likely to show a similar preference for pocket-like epitopes as has been discussed for Nanobodies (Gebauer and Skerra 2009). While the number of publicized development programs has remained low, there is still considerable academic research on this protein scaffold in progress (Gilbreth and Koide 2012; Hackel and Wittrup 2010). Of note, the Adnectin scaffold constitutes an isolated domain from the abundant extracellular matrix protein fibronectin and, thus, it does not occur as such in the body. Consequently, there has been concern about the exposure of immunological neoepitopes in Adnectins. However, in clinical trials conducted so far, ADA responses, though detected, were mostly directed against the engineered loops (Tolcher et al. 2011).

Like the FN3 format, the scaffold for Affibodies is an isolated domain excised from the larger extracellular region of a cell surface protein, in this case the Z domain of protein A from *Staphylococcus aureus*. As this small three-helix bundle does not offer a set of spatially contiguous loops, exposed side chains in two neighboring α-helices were chosen for random mutagenesis. This leads to a rather flat interface for complex formation, fixed by the underlying secondary structure. A problem inherited by this scaffold is its bacterial origin, which poses a risk with regard to ADA formation. Even though a deimmunized version has been developed recently (Feldwisch et al. 2010), it is probably wise to apply this class of alternative binding proteins preferentially either to in vivo imaging, where only minute protein amounts are administered and multiple doses are not required, or to highly immune-suppressed patients.

DARPins are based on an engineered consensus sequence from the ankyrin repeat proteins and, albeit the fold differs, this scaffold is characterized by a conformationally rigid secondary structure not unlike Affibodies. This may be considered an advantage and explains the stability and beneficial biochemical behavior of many DARPins. Generally, it appears that this scaffold with its rather flat binding site is optimally suited to recognize protein targets (Binz et al. 2004; Boersma and Plückthun 2011); in fact, similar to Nanobodies, DARPins have found use as reagents for co-crystallisation of sensitive proteins to promote their X-ray structural analysis (Bukowska and Grütter 2013). To make DARPins also amenable to small molecule or peptide targets, so-called loop-DARPins were recently designed (Schilling et al. 2014). However, it remains to be seen if this concept is useful for wider application, especially with regard to biopharmaceutical development. Notably, so far, DARPins have only been studied in phase II clinical trials for local application in the eye, which is a relatively immune-privileged organ.

Anticalins distinguish themselves from the other non-Ig scaffolds discussed here as they are derived from natural soluble plasma proteins abundant in the human body. This is not the only feature they share with Igs. In addition, their natural binding site, which is made of four structurally hypervariable loops mounted on a sandwich-like, almost circular arrangement of β-strands, shows fundamental similarity to the CDRs of MAbs. However, there are also two key differences: first, natural lipocalins prefer small molecules as ligands and, second, they are not subject to genetic mechanisms of somatic recombination and hypermutation like antibodies. Nevertheless, the methodology of combinatorial protein design has closed this gap, and Anticalins have convincingly demonstrated the ability to address the full spectrum of conceivable medically relevant targets, i.e. proteins (both soluble and as part of cell surface receptors), peptides and, naturally, haptens, similar to their endogenous counterparts.

The preparation of Fc fusion proteins, which is discussed more broadly in the next Chapter of this book, has become popular for most of the non-Ig scaffolds and even for scFv fragments, in this case yielding so-called small modular immunopharmaceuticals (SMIPs) (Hayden-Ledbetter et al. 2009). This concept has been carried further by creating fusions between engineered binding proteins specific for one target and a full length MAb that carries its own specificity for another target, as exemplified by the mAb^2 platform (Woisetschläger et al. 2013). These approaches form another link between the more conservative antibody engineering field and the presumably more innovative area of alternative scaffolds, indicating that at least from the industry perspective, both territories are on their way to optimize novel biopharmaceuticals by incorporating the best properties of each platform. However, with increasing complexity of the resulting constructs (Kontermann 2012), which also creates challenges for bioprocess development and quality control, the rationale behind the molecular design is not always clear, especially if compared to the classical bispecific antibody format (Milstein 2000).

In fact, it has to be questioned whether the diverse constructs that are currently in discussion really provide for biobetters in a functional sense. From the protein design perspective, it may be worthwhile examining more critically the features as well as benefits and drawbacks of the individual concepts to generate binding proteins before combining them in a hasty manner. A fast and successful biological drug development program is most likely based on a simple molecular design. This is probably easiest to realize for the construction of antagonists which just block a cell surface receptor or a signaling molecule, and there are already numerous examples of that in the alternative scaffold field. Another area with future potential for biobetters could be their combination with toxic drugs, as an alternative to the currently flourishing ADCs. Here, the non-Ig scaffolds may provide decisive advantages, including (1) the capability for site-directed conjugation with the drug in defined stoichiometry while retaining high target-binding activity, (2) possibly better endosomal escape into the cytoplasm of the target cell, enhancing cytotoxic effects, and (3) lower liver toxicity due to distinct routes of clearance.

References

Ahlgren S, Orlova A, Wållberg H, Hansson M, Sandström M, Lewsley R, Wennborg A, Abrahmsen L, Tolmachev V, Feldwisch J (2010) Targeting of HER2-expressing tumors using ^{111}In-ABY-025, a second-generation affibody molecule with a fundamentally reengineered scaffold. J Nucl Med 51:1131–1138

Åkerström B, Borregaard N, Flower DA, Salier J-S (2006) Lipocalins. Landes Bioscience, Georgetown

AL Hussaini MH, Ritchey J, Rettig MP, Eissenberg L, Uy GL, Chichili G, Moore PA, Johnson S, Collins L, Bonvini E, DiPersio JF (2013) Targeting CD123 in leukemic stem cells using dual affinity re-targeting molecules (DARTs®). Blood 122:360

Albrecht V, Richter A, Pfeiffer S, Gebauer M, Lindner S, Gieser E, Schüller U, Schichor C, Gildehaus FJ, Bartenstein P, Tonn JC, Skerra A, Glass R. Anticalins directed against the fibronectin extra domain B (ED-B) as diagnostic tracers for glioblastomas. Submitted

Al-Lazikani B, Lesk AM, Chothia C (1997) Standard conformations for the canonical structures of immunoglobulins. J Mol Biol 273:927–948

Arndt KM, Müller KM, Plückthun A (1998) Factors influencing the dimer to monomer transition of an antibody single-chain Fv fragment. Biochemistry 37:12918–12926

Bain B, Brazil M (2003) Adalimumab. Nat Rev Drug Discov 2:693–694

Bao W, Holt LJ, Prince RD, Jones GX, Aravindhan K, Szapacs M, Barbour AM, Jolivette LJ, Lepore JJ, Willette RN, DeAngelis E, Jucker BM (2013) Novel fusion of GLP-1 with a domain antibody to serum albumin prolongs protection against myocardial ischemia/reperfusion injury in the rat. Cardiovasc Diabetol 12:148

Barratt-Due A, Thorgersen EB, Lindstad JK, Pharo A, Lissina O, Lambris JD, Nunn MA, Mollnes TE (2011) *Ornithodoros moubata* complement inhibitor is an equally effective C5 inhibitor in pigs and humans. J Immunol 187:4913–4919

Bartunek J, Barbato E, Heyndrickx G, Vanderheyden M, Wijns W, Holz JB (2013) Novel antiplatelet agents: ALX-0081, a Nanobody directed towards von Willebrand factor. J Cardiovasc Transl Res 6:355–363

Batey S, Leung K, Rowlands R, Isaac S, Carvalho J, Weller S, Wydro M, Gaspar M, Medcalf M, Pegram R, Drewett V, Tuna M, Haurum J, Sun HH (2013) Preclinical evaluation of FS102: a HER2-specific Fcab with a novel mechanism of action. Mol Cancer Ther 12:B123

Batori V, Koide A, Koide S (2002) Exploring the potential of the monobody scaffold: effects of loop elongation on the stability of a fibronectin type III domain. Protein Eng 15:1015–1020

Bäuerle PA, Reinhardt C (2009) Bispecific T-cell engaging antibodies for cancer therapy. Cancer Res 69:4941–4944

Bäuerle PA, Kufer P, Bargou R (2009) BiTE: teaching antibodies to engage T-cells for cancer therapy. Curr Opin Mol Ther 11:22–30

Baum RP, Prasad V, Muller D, Schuchardt C, Orlova A, Wennborg A, Tolmachev V, Feldwisch J (2010) Molecular imaging of HER2-expressing malignant tumors in breast cancer patients using synthetic ^{111}In- or ^{68}Ga-labeled affibody molecules. J Nucl Med 51:892–897

Beck A (2011) Biosimilar, biobetter and next generation therapeutic antibodies. mAbs 3:107–110

Beck A, Wurch T, Bailly C, Corvaia N (2010) Strategies and challenges for the next generation of therapeutic antibodies. Nat Rev Immunol 10:345–352

Beste G, Schmidt FS, Stibora T, Skerra A (1999) Small antibody-like proteins with prescribed ligand specificities derived from the lipocalin fold. Proc Natl Acad Sci U S A 96:1898–1903

Better M, Chang CP, Robinson RR, Horwitz AH (1988) *Escherichia coli* secretion of an active chimeric antibody fragment. Science 240:1041–1043

Binz HK, Stumpp MT, Forrer P, Amstutz P, Plückthun A (2003) Designing repeat proteins: well-expressed, soluble and stable proteins from combinatorial libraries of consensus ankyrin repeat proteins. J Mol Biol 332:489–503

Binz HK, Amstutz P, Kohl A, Stumpp MT, Briand C, Forrer P, Grütter MG, Plückthun A (2004) High-affinity binders selected from designed ankyrin repeat protein libraries. Nat Biotechnol 22:575–582

Binz HK, Amstutz P, Plückthun A (2005) Engineering novel binding proteins from nonimmunoglobulin domains. Nat Biotechnol 23:1257–1268

Binz HK, Forrer P, Stumpp MT (2014) Modified binding proteins inhibiting the VEGF-A receptor interaction. Patent publication US20140221295A1

Bird RE, Walker BW (1991) Single chain antibody variable regions. Trends Biotechnol 9:132–137

Blick SK, Curran MP (2007) Certolizumab pegol: in Crohn's disease. BioDrugs 21:195–201

Bloom L, Calabro V (2009) FN3: a new protein scaffold reaches the clinic. Drug Discov Today 14:949–955

Bloomberg (2013) Allergan shares fall after CEO says two studies delayed. http://www.bloomberg.com/news/2013-05-01/allergan-shares-fall-after-ceo-says-drug-trial-will-be-delayed.html

Boersma YL, Plückthun A (2011) DARPins and other repeat protein scaffolds: advances in engineering and applications. Curr Opin Biotechnol 22:849–857

Bork P, Holm L, Sander C (1994) The immunoglobulin fold. Structural classification, sequence patterns and common core. J Mol Biol 242:309–320

Borras L, Gunde T, Tietz J, Bauer U, Hulmann-Cottier V, Grimshaw JP, Urech DM (2010) Generic approach for the generation of stable humanized single-chain Fv fragments from rabbit monoclonal antibodies. J Biol Chem 285:9054–9066

Boyer D (2013) A phase 2b study of Fovista™, a platelet derived growth factor (PDGF) inhibitor in combination with a vascular endothelial growth factor (VEGF) inhibitor for neovascular age-related macular degeneration (AMD). Invest Ophthalmol Vis Sci 54:2175

Brennan FR (2014) Monoclonal antibodies in phase 1 and 2 studies for immunological disorders. In: Dübel S, Reichert JM (eds) Handbook of therapeutic antibodies, 2nd edn. Wiley, Weinheim

Breustedt DA, Schönfeld DL, Skerra A (2006) Comparative ligand-binding analysis of ten human lipocalins. Biochim Biophys Acta 1764:161–173

Brischwein K, Parr L, Pflanz S, Volkland J, Lumsden J, Klinger M, Locher M, Hammond SA, Kiener P, Kufer P, Schlereth B, Baeuerle PA (2007) Strictly target cell-dependent activation of T cells by bispecific single-chain antibody constructs of the BiTE class. J Immunother 30:798–807

Bukowska MA, Grütter MG (2013) New concepts and aids to facilitate crystallization. Curr Opin Struct Biol 23:409–416

Campochiaro PA, Channa R, Berger BB, Heier JS, Brown DM, Fiedler U, Hepp J, Stumpp MT (2013) Treatment of diabetic macular edema with a designed ankyrin repeat protein that binds vascular endothelial growth factor: a phase I/II study. Am J Ophthalmol 155:697–704

Cappuccilli G, Crea R, Shen R, Hokanson CA, Kirk PB, Liston DR (2014) Universal fibronectin type III binding-domain libraries. Patent publication US8680019B2

Carter P, Presta L, Gorman CM, Ridgway JB, Henner D, Wong WL, Rowland AM, Kotts C, Carver ME, Shepard HM (1992) Humanization of an anti-p185HER2 antibody for human cancer therapy. Proc Natl Acad Sci U S A 89:4285–4289

Chakravarty R, Goel S, Cai W (2014) Nanobody: the "magic bullet" for molecular imaging? Theranostics 4:386–398

Chekhonin VP, Shein SA, Korchagina AA, Gurina OI (2013) VEGF in tumor progression and targeted therapy. Curr Cancer Drug Targets 13:423–443

Chen Y, Wiesmann C, Fuh G, Li B, Christinger HW, McKay P, de Vos AM, Lowman HB (1999) Selection and analysis of an optimized anti-VEGF antibody: crystal structure of an affinity-matured Fab in complex with antigen. J Mol Biol 293:865–881

Cizeau J, Grenkow DM, Brown JG, Entwistle J, MacDonald GC (2009) Engineering and biological characterization of VB6-845, an anti-EpCAM immunotoxin containing a T-cell epitope-depleted variant of the plant toxin bouganin. J Immunother 32:574–584

Cload S, Engle L, Lipovsek D, Madireddi M, Rakestraw GC, Swain J, Zhao W (2014) Fibronectin based scaffold domain proteins that bind to myostatin. Patent publication US20140105896A1

Comer G, Hegen M, Sharma A, Shields K (2012) Methods of treating immune disorders with single domain antibodies against TNF-alpha. Patent publication WO2012131053A1

Connelly R (2005) Fully human domain antibody therapeutics: the best of both worlds. Innov Pharm Technol 2005:42–45

Conrath K, Vincke C, Stijlemans B, Schymkowitz J, Decanniere K, Wyns L, Muyldermans S, Loris R (2005) Antigen binding and solubility effects upon the veneering of a camel VHH in framework-2 to mimic a VH. J Mol Biol 350:112–125

Couillin I, Maillet I, Vargaftig BB, Jacobs M, Paesen GC, Nuttall PA, Lefort J, Moser R, Weston-Davies W, Ryffel B (2004) Arthropod-derived histamine-binding protein prevents murine allergic asthma. J Immunol 173:3281–3286

CROSS (2011) N189084: the tariff classification of AGN-150998 from Germany. http://rulings.cbp.gov

Das Gupta R (2014) Preclinical development of an anti-IL-23 Adnectin and advancement into the clinic. Abstract; IBC's 9th annual next generation protein therapeutics Summit, June 4–6. San Francisco, CA

Davies J, Riechmann L (1994) 'Camelising' human antibody fragments: NMR studies on VH domains. FEBS Lett 339:285–290

Davis J, Lipovsek D, Camphausen R (2013) Fibronectin binding domains with reduced immunogenicity. Patent publication WO2013067029A2

Demarest SJ, Glaser SM (2008) Antibody therapeutics, antibody engineering, and the merits of protein stability. Curr Opin Drug Discov Devel 11:675–687

Dennis MS, Zhang M, Meng YG, Kadkhodayan M, Kirchhofer D, Combs D, Damico LA (2002) Albumin binding as a general strategy for improving the pharmacokinetics of proteins. J Biol Chem 277:35035–35043

Diem MD, Hyun L, Yi F, Hippensteel R, Kuhar E, Lowenstein C, Swift EJ, O'Neil KT, Jacobs SA (2014) Selection of high-affinity Centyrin FN3 domains from a simple library diversified at a combination of strand and loop positions. Protein Eng Des Sel 27:419–429

Dijkers EC, Oude Munnink TH, Kosterink JG, Brouwers AH, Jager PL, de Jong JR, van Dongen GA, Schroder CP, Lub-de Hooge MN, de Vries EG (2010) Biodistribution of ^{89}Zr-trastuzumab and PET imaging of HER2-positive lesions in patients with metastatic breast cancer. Clin Pharmacol Ther 87:586–592

Dineen SP, Sullivan LA, Beck AW, Miller AF, Carbon JG, Mamluk R, Wong H, Brekken RA (2008) The Adnectin CT-322 is a novel VEGF receptor 2 inhibitor that decreases tumor burden in an orthotopic mouse model of pancreatic cancer. BMC Cancer 8:352

Dirix LY, Rutten A, Huget P, Dirix M (2013) Trastuzumab emtansine in breast cancer. Expert Opin Biol Ther 13:607–614

Dreier T, Baeuerle PA, Fichtner I, Grün M, Schlereth B, Lorenczewski G, Kufer P, Lutterbüse R, Riethmüller G, Gjorstrup P, Bargou RC (2003) T cell costimulus-independent and very efficacious inhibition of tumor growth in mice bearing subcutaneous or leukemic human B cell lymphoma xenografts by a CD19-/CD3-bispecific single-chain antibody construct. J Immunol 170:4397–4402

Dschietzig TB (2014) Myostatin – from the mighty mouse to cardiovascular disease and cachexia. Clin Chim Acta 433:216–224

Eggenstein E, Eichinger A, Kim HJ, Skerra A (2013) Structure-guided engineering of Anticalins with improved binding behavior and biochemical characteristics for application in radio-immuno imaging and/or therapy. J Struct Biol 185:203–214

Eigenbrot C, Ultsch M, Dubnovitsky A, Abrahmsen L, Hard T (2010) Structural basis for high-affinity HER2 receptor binding by an engineered protein. Proc Natl Acad Sci U S A 107:15039–15044

EMEA (2007) Refusal CHMP assessment report for Mycograb. http://www.ema.europa.eu/docs/en_GB/document_library/EPAR_-_Public_assessment_report/human/000658/WC500070523.pdf

Entwistle J, Brown JG, Chooniedass S, Cizeau J, MacDonald GC (2012) Preclinical evaluation of VB6-845: an anti-EpCAM immunotoxin with reduced immunogenic potential. Cancer Biother Radiopharm 27:582–592

EudraCT (2011) No 2011-002526-43: Single and repeat dose study of the safety and efficacy of AGN-150998 in patients with exudative age-related macular degeneration. http://www.clinicaltrialsregister.eu/ctr-search/trial/2011-002526-43/DE

Eyer F, Steimer W, Nitzsche T, Jung N, Neuberger H, Müller C, Schlapschy M, Zilker T, Skerra A (2012) Intravenous application of an anticalin dramatically lowers plasma digoxin levels and reduces its toxic effects in rats. Toxicol Appl Pharmacol 263:352–359

Feldwisch J, Tolmachev V, Lendel C, Herne N, Sjoberg A, Larsson B, Rosik D, Lindqvist E, Fant G, Hoiden-Guthenberg I, Galli J, Jonasson P, Abrahmsen L (2010) Design of an optimized scaffold for affibody molecules. J Mol Biol 398:232–247

Fiedler M, Skerra A (2014) Non-antibody scaffolds as alternative therapeutic agents. In: Dübel S, Reichert JM (eds) Handbook of therapeutic antibodies, 2nd edn. Wiley, Weinheim

Flower DR (1996) The lipocalin protein family: structure and function. Biochem J 318:1–14

Forrer P, Binz HK, Stumpp MT, Plückthun A (2004) Consensus design of repeat proteins. ChemBioChem 5:183–189

Fredslund F, Laursen NS, Roversi P, Jenner L, Oliveira CL, Pedersen JS, Nunn MA, Lea SM, Discipio R, Sottrup-Jensen L, Andersen GR (2008) Structure of and influence of a tick complement inhibitor on human complement component 5. Nat Immunol 9:753–760

Friedrich K, Hanauer JR, Prüfer S, Münch RC, Völker I, Filippis C, Jost C, Hanschmann KM, Cattaneo R, Peng KW, Plückthun A, Buchholz CJ, Cichutek K, Mühlebach MD (2013) DARPin-targeting of measles virus: unique bispecificity, effective oncolysis, and enhanced safety. Mol Ther 21:849–859

Garber K (2014) Bispecific antibodies rise again. Nat Rev Drug Discov 13:799–801

Garcia CC, Weston-Davies W, Russo RC, Tavares LP, Rachid MA, Alves-Filho JC, Machado AV, Ryffel B, Nunn MA, Teixeira MM (2013) Complement C5 activation during influenza A infection in mice contributes to neutrophil recruitment and lung injury. PLoS One 8:e64443

Gay RD, Clarke AW, Elgundi Z, Domagala T, Simpson RJ, Le NB, Doyle AG, Jennings PA (2010) Anti-TNFα domain antibody construct CEP-37247: full antibody functionality at half the size. mAbs 2:625–638

Gebauer M, Skerra A (2009) Engineered protein scaffolds as next-generation antibody therapeutics. Curr Opin Chem Biol 13:245–255

Gebauer M, Skerra A (2012) Anticalins: small engineered binding proteins based on the lipocalin scaffold. Methods Enzymol 503:157–188

Gebauer M, Schiefner A, Matschiner G, Skerra A (2013) Combinatorial design of an Anticalin directed against the extra-domain B for the specific targeting of oncofetal fibronectin. J Mol Biol 425:780–802
George J, Compton JR, Leary DH, Olson MA, Legler PM (2014) Structural and mutational analysis of a monomeric and dimeric form of a single domain antibody with implications for protein misfolding. Proteins 82:3101–3116
Getts DR, Getts MT, McCarthy DP, Chastain EM, Miller SD (2010) Have we overestimated the benefit of human(ized) antibodies? mAbs 2:682–694
Gilbreth RN, Koide S (2012) Structural insights for engineering binding proteins based on non-antibody scaffolds. Curr Opin Struct Biol 22:413–420
Gill DS, Damle NK (2006) Biopharmaceutical drug discovery using novel protein scaffolds. Curr Opin Biotechnol 17:653–658
Glockshuber R, Malia M, Pfitzinger I, Plückthun A (1990) A comparison of strategies to stabilize immunoglobulin F_v-fragments. Biochemistry 29:1362–1367
Goel N, Stephens S (2010) Certolizumab pegol. mAbs 2:137–147
Goetz DH, Holmes MA, Borregaard N, Bluhm ME, Raymond KN, Strong RK (2002) The neutrophil lipocalin NGAL is a bacteriostatic agent that interferes with siderophore-mediated iron acquisition. Mol Cell 10:1033–1043
Grabulovski D, Kaspar M, Neri D (2007) A novel, non-immunogenic Fyn SH3-derived binding protein with tumor vascular targeting properties. J Biol Chem 282:3196–3204
Grove TZ, Cortajarena AL, Regan L (2008) Ligand binding by repeat proteins: natural and designed. Curr Opin Struct Biol 18:507–515
Hackel BJ, Wittrup KD (2010) The full amino acid repertoire is superior to serine/tyrosine for selection of high affinity immunoglobulin G binders from the fibronectin scaffold. Protein Eng Des Sel 23:211–219
Hackel BJ, Kapila A, Wittrup KD (2008) Picomolar affinity fibronectin domains engineered utilizing loop length diversity, recursive mutagenesis, and loop shuffling. J Mol Biol 381:1238–1252
Hackel BJ, Ackerman ME, Howland SW, Wittrup KD (2010) Stability and CDR composition biases enrich binder functionality landscapes. J Mol Biol 401:84–96
Halstead SK, Humphreys PD, Zitman FM, Hamer J, Plomp JJ, Willison HJ (2008) C5 inhibitor rEV576 protects against neural injury in an in vitro mouse model of Miller Fisher syndrome. J Peripher Nerv Syst 13:228–235
Hamers-Casterman C, Atarhouch T, Muyldermans S, Robinson G, Hamers C, Songa EB, Bendahman N, Hamers R (1993) Naturally occurring antibodies devoid of light chains. Nature 363:446–448
Hanenberg M, McAfoose J, Kulic L, Welt T, Wirth F, Parizek P, Strobel L, Cattepoel S, Spani C, Derungs R, Maier M, Plückthun A, Nitsch RM (2014) Amyloid-β peptide-specific DARPins as a novel class of potential therapeutics for Alzheimer disease. J Biol Chem 289:27080–27089
Harding FA, Stickler MM, Razo J, DuBridge RB (2010) The immunogenicity of humanized and fully human antibodies: residual immunogenicity resides in the CDR regions. mAbs 2:256–265
Hayden-Ledbetter MS, Cerveny CG, Espling E, Brady WA, Grosmaire LS, Tan P, Bader R, Slater S, Nilsson CA, Barone DS, Simon A, Bradley C, Thompson PA, Wahl AF, Ledbetter JA (2009) CD20-directed small modular immunopharmaceutical, TRU-015, depletes normal and malignant B cells. Clin Cancer Res 15:2739–2746
Hey T, Fiedler E, Rudolph R, Fiedler M (2005) Artificial, non-antibody binding proteins for pharmaceutical and industrial applications. Trends Biotechnol 23:514–522
Hodge RJ, O'Connor-Semmes RL, Lin J, Chism JP, Andrews SM, Gaddy JR, Nunez DJ (2013) GSK2374697, a long-acting GLP-1 mimetic: first use of an AlbudAb™ in humans – pharmacokinetics, pharmacodynamics, safety, and tolerability in healthy volunteers. Abstract No. 60-LB; American Diabetes Association 73rd scientific sessions, June 21–25. Chicago, IL
Hoffman LM, Gore L (2014) Blinatumomab, a bi-specific anti-CD19/CD3 BiTE® antibody for the treatment of acute lymphoblastic leukemia: perspectives and current pediatric applications. Front Oncol 4:63

Hoffmann P, Hofmeister R, Brischwein K, Brandl C, Crommer S, Bargou R, Itin C, Prang N, Baeuerle PA (2005) Serial killing of tumor cells by cytotoxic T cells redirected with a CD19-/CD3-bispecific single-chain antibody construct. Int J Cancer 115:98–104

Hohenester E, Engel J (2002) Domain structure and organisation in extracellular matrix proteins. Matrix Biol 21:115–128

Hohlbaum AM, Trentman S, Gille H, Allersdorfer A, Belaiba RS, Huelsmeyer M, Christian J, Sandal T, Matschiner G, Jensen K, Skerra A, Audoly LP (2011) Discovery and preclinical characterization of a novel hepcidin antagonist with tunable PK/PD properties for the treatment of anemia in different patient populations. Blood 118: Abstract 687

Holliger P, Hudson PJ (2005) Engineered antibody fragments and the rise of single domains. Nat Biotechnol 23:1126–1136

Holliger P, Prospero T, Winter G (1993) "Diabodies": small bivalent and bispecific antibody fragments. Proc Natl Acad Sci U S A 90:6444–6448

Holmes MA, Paulsene W, Jide X, Ratledge C, Strong RK (2005) Siderocalin (Lcn 2) also binds carboxymycobactins, potentially defending against mycobacterial infections through iron sequestration. Structure 13:29–41

Holt LJ, Herring C, Jespers LS, Woolven BP, Tomlinson IM (2003) Domain antibodies: proteins for therapy. Trends Biotechnol 21:484–490

Holt LJ, Basran A, Jones K, Chorlton J, Jespers LS, Brewis ND, Tomlinson IM (2008) Anti-serum albumin domain antibodies for extending the half-lives of short lived drugs. Protein Eng Des Sel 21:283–288

Holz JB (2012) The TITAN trial – assessing the efficacy and safety of an anti-von Willebrand factor Nanobody in patients with acquired thrombotic thrombocytopenic purpura. Transfus Apher Sci 46:343–346

Hoyer W, Grönwall C, Jonsson A, Ståhl S, Härd T (2008) Stabilization of a β-hairpin in monomeric Alzheimer's amyloid-β peptide inhibits amyloid formation. Proc Natl Acad Sci U S A 105:5099–5104

Humphreys DP (2003) Production of antibodies and antibody fragments in *Escherichia coli* and a comparison of their functions, uses and modification. Curr Opin Drug Discov Devel 6:188–196

INN (2012) Proposed International Nonproprietary Names for Pharmaceutical Substances (INN): List 108. WHO Drug Inf 26:401–471

Jacobs SA, Diem MD, Luo J, Teplyakov A, Obmolova G, Malia T, Gilliland GL, O'Neil KT (2012) Design of novel FN3 domains with high stability by a consensus sequence approach. Protein Eng Des Sel 25:107–117

Jermutus L, Honegger A, Schwesinger F, Hanes J, Plückthun A (2001) Tailoring in vitro evolution for protein affinity or stability. Proc Natl Acad Sci U S A 98:75–80

Jespers L, Schon O, Famm K, Winter G (2004a) Aggregation-resistant domain antibodies selected on phage by heat denaturation. Nat Biotechnol 22:1161–1165

Jespers L, Schon O, James LC, Veprintsev D, Winter G (2004b) Crystal structure of HEL4, a soluble, refoldable human V_H single domain with a germ-line scaffold. J Mol Biol 337:893–903

Johnson S, Burke S, Huang L, Gorlatov S, Li H, Wang W, Zhang W, Tuaillon N, Rainey J, Barat B, Yang Y, Jin L, Ciccarone V, Moore PA, Koenig S, Bonvini E (2010) Effector cell recruitment with novel Fv-based dual-affinity re-targeting protein leads to potent tumor cytolysis and in vivo B-cell depletion. J Mol Biol 399:436–449

Jones PT, Dear PH, Foote J, Neuberger MS, Winter G (1986) Replacing the complementarity-determining regions in a human antibody with those from a mouse. Nature 321:522–525

Jonsson A, Dogan J, Herne N, Abrahmsén L, Nygren P-Å (2008) Engineering of a femtomolar affinity binding protein to human serum albumin. Protein Eng Des Sel 21:515–527

Jost C, Schilling J, Tamaskovic R, Schwill M, Honegger A, Plückthun A (2013) Structural basis for eliciting a cytotoxic effect in HER2-overexpressing cancer cells via binding to the extracellular domain of HER2. Structure 21:1979–1991

Jung S, Plückthun A (1997) Improving in vivo folding and stability of a single-chain Fv antibody fragment by loop grafting. Protein Eng 10:959–966

Jung S, Honegger A, Plückthun A (1999) Selection for improved protein stability by phage display. J Mol Biol 294:163–180

Kaspar M, Zardi L, Neri D (2006) Fibronectin as target for tumor therapy. Int J Cancer 118:1331–1339

Kim HJ, Eichinger A, Skerra A (2009) High-affinity recognition of lanthanide(III) chelate complexes by a reprogrammed human lipocalin 2. J Am Chem Soc 131:3565–3576

Kipriyanov SM (2009) Generation of bispecific and tandem diabodies. Methods Mol Biol 562:177–193

Kipriyanov SM, Moldenhauer G, Schuhmacher J, Cochlovius B, Von der Lieth CW, Matys ER, Little M (1999) Bispecific tandem diabody for tumor therapy with improved antigen binding and pharmacokinetics. J Mol Biol 293:41–56

Klein D, Jacobs S, Sheri M, Anderson M, Attar R, Barnakov A, Brosnan K, Bushey B, Chevalier K, Chin D, Cornejo C, Diem M, Hyun L, Kuhar E, McCabe F, Picha K, Spinka-Doms T, Swift E, O'Neil K (2013) Bispecific Centyrin simultaneously targeting EGFR and c-Met demonstrates improved activity compared to the mixture of single agents. Cancer Res 73(8 Suppl): Abstract nr LB-312

Klöhn PC, Wuellner U, Zizlsperger N, Zhou Y, Tavares D, Berger S, Zettlitz KA, Proetzel G, Yong M, Begent RH, Reichert JM (2013) IBC's 23rd annual antibody engineering, 10th annual antibody therapeutics international conferences and the 2012 annual meeting of the antibody society, San Diego, 3–6 Dec 2012. mAbs 5, 178–201

Knight DM, Wagner C, Jordan R, McAleer MF, DeRita R, Fass DN, Coller BS, Weisman HF, Ghrayeb J (1995) The immunogenicity of the 7E3 murine monoclonal Fab antibody fragment variable region is dramatically reduced in humans by substitution of human for murine constant regions. Mol Immunol 32:1271–1281

Köhler G, Milstein C (1975) Continuous cultures of fused cells secreting antibody of predefined specificity. Nature 256:495–497

Koide A, Bailey CW, Huang X, Koide S (1998) The fibronectin type III domain as a scaffold for novel binding proteins. J Mol Biol 284:1141–1151

Koide A, Gilbreth RN, Esaki K, Tereshko V, Koide S (2007) High-affinity single-domain binding proteins with a binary-code interface. Proc Natl Acad Sci U S A 104:6632–6637

Kontermann R (2012) Dual targeting strategies with bispecific antibodies. mAbs 4:182–197

Korndörfer IP, Beste G, Skerra A (2003a) Crystallographic analysis of an "anticalin" with tailored specificity for fluorescein reveals high structural plasticity of the lipocalin loop region. Proteins 53:121–129

Korndörfer IP, Schlehuber S, Skerra A (2003b) Structural mechanism of specific ligand recognition by a lipocalin tailored for the complexation of digoxigenin. J Mol Biol 330:385–396

Kovaleva M, Ferguson L, Steven J, Porter A, Barelle C (2014) Shark variable new antigen receptor biologics – a novel technology platform for therapeutic drug development. Expert Opin Biol Ther 14:1527–1539

Kratz F (2008) Albumin as a drug carrier: design of prodrugs, drug conjugates and nanoparticles. J Control Release 132:171–183

Kreitman RJ, Pastan I (2011) Antibody fusion proteins: anti-CD22 recombinant immunotoxin moxetumomab pasudotox. Clin Cancer Res 17:6398–6405

Kreitman RJ, Tallman MS, Robak T, Coutre S, Wilson WH, Stetler-Stevenson M, Fitzgerald DJ, Lechleider R, Pastan I (2012) Phase I trial of anti-CD22 recombinant immunotoxin moxetumomab pasudotox (CAT-8015 or HA22) in patients with hairy cell leukemia. J Clin Oncol 30:1822–1828

Kronqvist N, Malm M, Göstring L, Gunneriusson E, Nilsson M, Hoiden Guthenberg I, Gedda L, Frejd FY, Ståhl S, Löfblom J (2011) Combining phage and *staphylococcal* surface display for generation of ErbB3-specific Affibody molecules. Protein Eng Des Sel 24:385–396

Krupka C, Kufer P, Kischel R, Zugmaier G, Bögeholz J, Köhnke T, Lichtenegger FS, Schneider S, Metzeler KH, Fiegl M, Spiekermann K, Baeuerle PA, Hiddemann W, Riethmüller G, Subklewe M (2014) CD33 target validation and sustained depletion of AML blasts in long-term cultures by the bispecific T-cell-engaging antibody AMG 330. Blood 123:356–365

Laszlo GS, Gudgeon CJ, Harrington KH, Dell'Aringa J, Newhall KJ, Means GD, Sinclair AM, Kischel R, Frankel SR, Walter RB (2014) Cellular determinants for preclinical activity of a novel CD33/CD3 bispecific T-cell engager (BiTE) antibody, AMG 330, against human AML. Blood 123:554–561

Le Gall F, Reusch U, Little M, Kipriyanov SM (2004) Effect of linker sequences between the antibody variable domains on the formation, stability and biological activity of a bispecific tandem diabody. Protein Eng Des Sel 17:357–366

Lee EC, Liang Q, Ali H, Bayliss L, Beasley A, Bloomfield-Gerdes T, Bonoli L, Brown R, Campbell J, Carpenter A, Chalk S, Davis A, England N, Fane-Dremucheva A, Franz B, Germaschewski V, Holmes H, Holmes S, Kirby I, Kosmac M, Legent A, Lui H, Manin A, O'Leary S, Paterson J, Sciarrillo R, Speak A, Spensberger D, Tuffery L, Waddell N, Wang W, Wells S, Wong V, Wood A, Owen MJ, Friedrich GA, Bradley A (2014) Complete humanization of the mouse immunoglobulin loci enables efficient therapeutic antibody discovery. Nat Biotechnol 32:356–363

Lindenmann J (1984) Origin of the terms 'antibody' and 'antigen'. Scand J Immunol 19:281–285

Lipovsek D (2011) Adnectins: engineered target-binding protein therapeutics. Protein Eng Des Sel 24:3–9

Lipovsek D, Lippow SM, Hackel BJ, Gregson MW, Cheng P, Kapila A, Wittrup KD (2007) Evolution of an interloop disulfide bond in high-affinity antibody mimics based on fibronectin type III domain and selected by yeast surface display: molecular convergence with single-domain camelid and shark antibodies. J Mol Biol 368:1024–1041

Löfblom J, Feldwisch J, Tolmachev V, Carlsson J, Ståhl S, Frejd FY (2010) Affibody molecules: engineered proteins for therapeutic, diagnostic and biotechnological applications. FEBS Lett 584:2670–2680

Löffler A, Kufer P, Lutterbüse R, Zettl F, Daniel PT, Schwenkenbecher JM, Riethmüller G, Dörken B, Bargou RC (2000) A recombinant bispecific single-chain antibody, CD19 x CD3, induces rapid and high lymphoma-directed cytotoxicity by unstimulated T lymphocytes. Blood 95:2098–2103

Lonberg N (2005) Human antibodies from transgenic animals. Nat Biotechnol 23:1117–1125

Malmberg J, Tolmachev V, Orlova A (2011) Imaging agents for in vivo molecular profiling of disseminated prostate cancer – targeting EGFR receptors in prostate cancer: comparison of cellular processing of [111In]-labeled affibody molecule $Z_{EGFR:2377}$ and cetuximab. Int J Oncol 38:1137–1143

Mamluk R, Carvajal IM, Morse BA, Wong H, Abramowitz J, Aslanian S, Lim AC, Gokemeijer J, Storek MJ, Lee J, Gosselin M, Wright MC, Camphausen RT, Wang J, Chen Y, Miller K, Sanders K, Short S, Sperinde J, Prasad G, Williams S, Kerbel R, Ebos J, Mutsaers A, Mendlein JD, Harris AS, Furfine ES (2010) Anti-tumor effect of CT-322 as an adnectin inhibitor of vascular endothelial growth factor receptor-2. mAbs 2:199–208

Mann A, Friedrich N, Krarup A, Weber J, Stiegeler E, Dreier B, Pugach P, Robbiani M, Riedel T, Moehle K, Robinson JA, Rusert P, Plückthun A, Trkola A (2013) Conformation-dependent recognition of HIV gp120 by designed ankyrin repeat proteins provides access to novel HIV entry inhibitors. J Virol 87:5868–5881

McAleese F, Eser M (2012) RECRUIT-TandAbs: harnessing the immune system to kill cancer cells. Future Oncol 8:687–695

Melmed GY, Targan SR, Yasothan U, Hanicq D, Kirkpatrick P (2008) Certolizumab pegol. Nat Rev Drug Discov 7:641–642

Mendler CT, Skerra A (2013) Anticalins: an emerging class of novel biologics to treat cancer and other severe diseases. Drug Future 38:169–179

Mendler CT, Friedrich L, Schlapschy M, Schwaiger M, Wester H-J, Skerra A (2015) High contrast tumor imaging with radio-labelled antibody Fab fragments tailored for optimized pharmacokinetics via PASylation. mAbs 7:96–109

Mercader JV, Skerra A (2002) Generation of anticalins with specificity for a nonsymmetric phthalic acid ester. Anal Biochem 308:269–277

Merlot AM, Kalinowski DS, Richardson DR (2014) Unraveling the mysteries of serum albumin – more than just a serum protein. Front Physiol 5:299

Milstein C (2000) With the benefit of hindsight. Immunol Today 21:359–364

Mintz CS, Crea R (2013) Protein scaffolds – the next generation of protein therapeutics? BioProcess Int 11:40–48

Moody P, Chudasama V, Nathani RI, Maruani A, Martin S, Smith ME, Caddick S (2014) A rapid, site-selective and efficient route to the dual modification of DARPins. Chem Commun (Camb) 50:4898–4900

Moore PA, Zhang W, Rainey GJ, Burke S, Li H, Huang L, Gorlatov S, Veri MC, Aggarwal S, Yang Y, Shah K, Jin L, Zhang S, He L, Zhang T, Ciccarone V, Koenig S, Bonvini E, Johnson S (2011) Application of dual affinity retargeting molecules to achieve optimal redirected T-cell killing of B-cell lymphoma. Blood 117:4542–4551

Mori K, Lee HT, Rapoport D, Drexler IR, Foster K, Yang J, Schmidt-Ott KM, Chen X, Li JY, Weiss S, Mishra J, Cheema FH, Markowitz G, Suganami T, Sawai K, Mukoyama M, Kunis C, D'Agati V, Devarajan P, Barasch J (2005) Endocytic delivery of lipocalin-siderophore-iron complex rescues the kidney from ischemia-reperfusion injury. J Clin Invest 115:610–621

Morrison SL, Johnson MJ, Herzenberg LA, Oi VT (1984) Chimeric human antibody molecules: mouse antigen-binding domains with human constant region domains. Proc Natl Acad Sci U S A 81:6851–6855

Mosavi LK, Minor DL Jr, Peng ZY (2002) Consensus-derived structural determinants of the ankyrin repeat motif. Proc Natl Acad Sci U S A 99:16029–16034

Mross K, Richly H, Fischer R, Scharr D, Buchert M, Stern A, Gille H, Audoly LP, Scheulen ME (2013) First-in-human phase I study of PRS-050 (Angiocal), an Anticalin targeting and antagonizing VEGF-A, in patients with advanced solid tumors. PLoS One 8:e83232

Mukherjee R (2013) FGF21-Adnectin-Pharmacokinetic Enhancer: a modified FGF21 protein with uniquely extended pharmacokinetic profile for the treatment of metabolic diseases. Abstract; Ramanbhai Foundation 6th international symposium, February 4–6. Ahmedabad, India

Munch RC, Muhlebach MD, Schaser T, Kneissl S, Jost C, Plückthun A, Cichutek K, Buchholz CJ (2011) DARPins: an efficient targeting domain for lentiviral vectors. Mol Ther 19:686–693

Munch RC, Janicki H, Volker I, Rasbach A, Hallek M, Buning H, Buchholz CJ (2013) Displaying high-affinity ligands on adeno-associated viral vectors enables tumor cell-specific and safe gene transfer. Mol Ther 21:109–118

Muyldermans S (2013) Nanobodies: natural single-domain antibodies. Annu Rev Biochem 82:775–797

Muyldermans S, Cambillau C, Wyns L (2001) Recognition of antigens by single-domain antibody fragments: the superfluous luxury of paired domains. Trends Biochem Sci 26:230–235

NCT00217841 Aurograb and Vancomycin in MRSA Infection. http://clinicaltrials.gov

NCT00353964 Safety and efficacy study of rEV131 in the treatment of ocular inflammation after cataract surgery. http://clinicaltrials.gov

NCT00928317 Dose ranging study of ART621 in subjects diagnosed with rheumatoid arthritis taking methotrexate. http://clinicaltrials.gov

NCT01397409 Evaluation of AGN-150998 in exudative age-related macular degeneration (AMD). http://clinicaltrials.gov

NCT02013167 Ph 3 trial of Blinatumomab vs investigator's choice of chemotherapy in patients with relapsed or refractory ALL. http://clinicaltrials.gov

NCT02106091 Safety study to assess AFM11 in patients with relapsed and/or refractory CD19 positive B-cell NHL or B-precursor ALL. http://clinicaltrials.gov

NCT02243787 Safety and tolerability study of COVA322 in patients with stable chronic moderate-to-severe plaque psoriasis. http://clinicaltrials.gov

Nelson AL (2010) Antibody fragments: hope and hype. mAbs 2:77–83

Nelson AL, Reichert JM (2009) Development trends for therapeutic antibody fragments. Nat Biotechnol 27:331–337

Nielsen UB, Kirpotin DB, Pickering EM, Hong K, Park JW, Shalaby MR, Shao Y, Benz CC, Marks JD (2002) Therapeutic efficacy of anti-ErbB2 immunoliposomes targeted by a phage antibody selected for cellular endocytosis. Biochim Biophys Acta 1591:109–118

Nilsson B, Moks T, Jansson B, Abrahmsén L, Elmblad A, Holmgren E, Henrichson C, Jones TA, Uhlén M (1987) A synthetic IgG-binding domain based on *staphylococcal* protein A. Protein Eng 1:107–113

Nixon AE, Sexton DJ, Ladner RC (2014) Drugs derived from phage display – from candidate identification to clinical practice. mAbs 6:73–85

Nord K, Nilsson J, Nilsson B, Uhlén M, Nygren P-Å (1995) A combinatorial library of an α-helical bacterial receptor domain. Protein Eng 8:601–608

Nord K, Gunneriusson E, Ringdahl J, Ståhl S, Uhlén M, Nygren P-Å (1997) Binding proteins selected from combinatorial libraries of an α-helical bacterial receptor domain. Nat Biotechnol 15:772–777

Nuttall SD, Walsh RB (2008) Display scaffolds: protein engineering for novel therapeutics. Curr Opin Pharmacol 8:609–615

Offner S, Hofmeister R, Romaniuk A, Kufer P, Baeuerle PA (2006) Induction of regular cytolytic T cell synapses by bispecific single-chain antibody constructs on MHC class I-negative tumor cells. Mol Immunol 43:763–771

Olwill SA, Joffroy C, Gille H, Vigna E, Matschiner G, Allersdorfer A, Lunde BM, Jaworski J, Burrows JF, Chiriaco C, Christian HJ, Hülsmeyer M, Trentmann S, Jensen K, Hohlbaum AM, Audoly L (2013) A highly potent and specific MET therapeutic protein antagonist with both ligand-dependent and ligand-independent activity. Mol Cancer Ther 12:2459–2471

Orlova A, Magnusson M, Eriksson TL, Nilsson M, Larsson B, Hoiden-Guthenberg I, Widstrom C, Carlsson J, Tolmachev V, Ståhl S, Nilsson FY (2006) Tumor imaging using a picomolar affinity HER2 binding affibody molecule. Cancer Res 66:4339–4348

Orlova A, Wallberg H, Stone-Elander S, Tolmachev V (2009) On the selection of a tracer for PET imaging of HER2-expressing tumors: direct comparison of a ^{124}I-labeled affibody molecule and trastuzumab in a murine xenograft model. J Nucl Med 50:417–425

Orlova A, Tran TA, Ekblad T, Karlström AE, Tolmachev V (2010) ^{186}Re-maSGS-$Z_{HER2:342}$, a potential Affibody conjugate for systemic therapy of HER2-expressing tumours. Eur J Nucl Med Mol Imaging 37:260–269

Orlova A, Jonsson A, Rosik D, Lundqvist H, Lindborg M, Abrahmsen L, Ekblad C, Frejd FY, Tolmachev V (2013) Site-specific radiometal labeling and improved biodistribution using ABY-027, a novel HER2-targeting affibody molecule-albumin-binding domain fusion protein. J Nucl Med 54:961–968

Osbourn J, Groves M, Vaughan T (2005) From rodent reagents to human therapeutics using antibody guided selection. Methods 36:61–68

Pachl J, Svoboda P, Jacobs F, Vandewoude K, van der Hoven B, Spronk P, Masterson G, Malbrain M, Aoun M, Garbino J, Takala J, Drgona L, Burnie J, Matthews R, Mycograb Invasive Candidiasis Study Group (2006) A randomized, blinded, multicenter trial of lipid-associated amphotericin B alone versus in combination with an antibody-based inhibitor of heat shock protein 90 in patients with invasive candidiasis. Clin Infect Dis 42:1404–1413

Padlan EA (1994) Anatomy of the antibody molecule. Mol Immunol 31:169–217

Paesen GC, Adams PL, Harlos K, Nuttall PA, Stuart DI (1999) Tick histamine-binding proteins: isolation, cloning, and three-dimensional structure. Mol Cell 3:661–671

Pardoll DM (2012) The blockade of immune checkpoints in cancer immunotherapy. Nat Rev Cancer 12:252–264

Pini A, Viti F, Santucci A, Carnemolla B, Zardi L, Neri P, Neri D (1998) Design and use of a phage display library. Human antibodies with subnanomolar affinity against a marker of angiogenesis eluted from a two-dimensional gel. J Biol Chem 273:21769–21776

Power BE, Doughty L, Shapira DR, Burns JE, Bayly AM, Caine JM, Liu Z, Scott AM, Hudson PJ, Kortt AA (2003) Noncovalent scFv multimers of tumor-targeting anti-Lewis^{y} hu3S193 humanized antibody. Protein Sci 12:734–747

Premsukh A, Lavoie JM, Cizeau J, Entwistle J, MacDonald GC (2011) Development of a GMP phase III purification process for VB4-845, an immunotoxin expressed in *E. coli* using high cell density fermentation. Protein Expr Purif 78:27–37

Rajkovic E, Knackmuss S, Reusch U, Rothe A, Topp M, Younes A, Ravic M, Hucke C, Zhukvosky E, Little M (2012) RECRUIT-TandAb AFM13: overcoming limitations of monoclonal antibodies in Hodgkin lymphoma. Cancer Res 72(8 Suppl):Abstract nr 3521

Ramamurthy V, Krystek SR Jr, Bush A, Wei A, Emanuel SL, Das Gupta R, Janjua A, Cheng L, Murdock M, Abramczyk B, Cohen D, Lin Z, Morin P, Davis JH, Dabritz M, McLaughlin DC, Russo KA, Chao G, Wright MC, Jenny VA, Engle LJ, Furfine E, Sheriff S (2012) Structures of adnectin/protein complexes reveal an expanded binding footprint. Structure 20:259–269

Raponi S, De Propris MS, Intoppa S, Milani ML, Vitale A, Elia L, Perbellini O, Pizzolo G, Foa R, Guarini A (2011) Flow cytometric study of potential target antigens (CD19, CD20, CD22, CD33) for antibody-based immunotherapy in acute lymphoblastic leukemia: analysis of 552 cases. Leuk Lymphoma 52:1098–1107

Rauth S, Hinz D, Eichinger A, Schneider M, Uhrig M, Mayhaus M, Riemenschneider M, Skerra A. High-affinity Anticalins with aggregation-blocking activity directed against the Alzheimer β-amyloid peptide. In preparation

Reichert JM (2012) Marketed therapeutic antibodies compendium. mAbs 4:413–415

Reichert JM (2013) Antibodies to watch in 2013: mid-year update. mAbs 5:513–517

Reichert JM (2014) Antibodies to watch in 2014: mid-year update. mAbs 6:799–802

Reichert JM (2015) Antibodies to watch in 2015. mAbs 7:1–8

Reiter Y, Brinkmann U, Lee B, Pastan I (1996) Engineering antibody Fv fragments for cancer detection and therapy: disulfide-stabilized Fv fragments. Nat Biotechnol 14:1239–1245

Reuters (2014) Stage 3 phase 2 study of DARPin abicipar pegol (previously MP0112) supports progressing to phase III development program. http://uk.reuters.com/article/2014/07/01/molecular-partners-idUKnBw015546a+100+BSW20140701

Richie DL, Ghannoum MA, Isham N, Thompson KV, Ryder NS (2012) Nonspecific effect of Mycograb on amphotericin B MIC. Antimicrob Agents Chemother 56:3963–3964

Richter A, Eggenstein E, Skerra A (2014) Anticalins: exploiting a non-Ig scaffold with hypervariable loops for the engineering of binding proteins. FEBS Lett 588:213–218

Riechmann L, Foote J, Winter G (1988) Expression of an antibody Fv fragment in myeloma cells. J Mol Biol 203:825–828

Riethmüller G (2012) Symmetry breaking: bispecific antibodies, the beginnings, and 50 years on. Cancer Immun 12:12

Rose-John S, Schooltink H (2003) CDP-870 – Celltech/Pfizer. Curr Opin Investig Drugs 4:588–592

Rosenfeld PJ, Brown DM, Heier JS, Boyer DS, Kaiser PK, Chung CY, Kim RY, MARINA Study Group (2006) Ranibizumab for neovascular age-related macular degeneration. N Engl J Med 355:1419–1431

Rothe A, Hosse RJ, Power BE (2006) In vitro display technologies reveal novel biopharmaceutics. FASEB J 20:1599–1610

Ruchala P, Nemeth E (2014) The pathophysiology and pharmacology of hepcidin. Trends Pharmacol Sci 35:155–161

Schiefner A, Skerra A (2015) The menagerie of human lipocalins: a natural protein scaffold for molecular recognition of physiological compounds. Acc Chem Res 48:976–985

Schiefner A, Chatwell L, Korner J, Neumaier I, Colby DW, Volkmer R, Wittrup KD, Skerra A (2011) A disulfide-free single-domain V_L intrabody with blocking activity towards Huntingtin reveals a novel mode of epitope recognition. J Mol Biol 414:337–355

Schilling J, Schöppe J, Plückthun A (2014) From DARPins to LoopDARPins: novel LoopDARPin design allows the selection of low picomolar binders in a single round of ribosome display. J Mol Biol 426:691–721

Schlapschy M, Binder U, Börger C, Theobald I, Wachinger K, Kisling S, Haller D, Skerra A (2013) PASylation: a biological alternative to PEGylation for extending the plasma half-life of pharmaceutically active proteins. Protein Eng Des Sel 26:489–501

Schlehuber S, Beste G, Skerra A (2000) A novel type of receptor protein, based on the lipocalin scaffold, with specificity for digoxigenin. J Mol Biol 297:1105–1120

Schmidt FS, Skerra A (1994) The bilin-binding protein of *Pieris brassicae*. cDNA sequence and regulation of expression reveal distinct features of this insect pigment protein. Eur J Biochem 219:855–863

Schönfeld D, Matschiner G, Chatwell L, Trentmann S, Gille H, Hülsmeyer M, Brown N, Kaye PM, Schlehuber S, Hohlbaum AM, Skerra A (2009) An engineered lipocalin specific for CTLA-4 reveals a combining site with structural and conformational features similar to antibodies. Proc Natl Acad Sci U S A 106:8198–8203

Schweizer A, Rusert P, Berlinger L, Ruprecht CR, Mann A, Corthesy S, Turville SG, Aravantinou M, Fischer M, Robbiani M, Amstutz P, Trkola A (2008) CD4-specific designed ankyrin repeat proteins are novel potent HIV entry inhibitors with unique characteristics. PLoS Pathog 4:e1000109

Sheridan C (2007) Pharma consolidates its grip on post-antibody landscape. Nat Biotechnol 25:365–366

Sheridan C (2015) Amgen's bispecific antibody puffs across finish line. Nat Biotechnol 33:219–221

Sieber V, Plückthun A, Schmid FX (1998) Selecting proteins with improved stability by a phage-based method. Nat Biotechnol 16:955–960

Simon M, Frey R, Zangemeister-Wittke U, Plückthun A (2013) Orthogonal assembly of a designed ankyrin repeat protein-cytotoxin conjugate with a clickable serum albumin module for half-life extension. Bioconjug Chem 24:1955–1966

Simon M, Stefan N, Borsig L, Plückthun A, Zangemeister-Wittke U (2014) Increasing the antitumor effect of an EpCAM-targeting fusion toxin by facile click PEGylation. Mol Cancer Ther 13:375–385

Siontorou CG (2013) Nanobodies as novel agents for disease diagnosis and therapy. Int J Nanomedicine 8:4215–4227

Skerra A (1993) Bacterial expression of immunoglobulin fragments. Curr Opin Immunol 5:256–262

Skerra A (2000a) Engineered protein scaffolds for molecular recognition. J Mol Recognit 13:167–187

Skerra A (2000b) Lipocalins as a scaffold. Biochim Biophys Acta 1482:337–350

Skerra A (2001) 'Anticalins': a new class of engineered ligand-binding proteins with antibody-like properties. J Biotechnol 74:257–275

Skerra A (2003) Imitating the humoral immune response. Curr Opin Chem Biol 7:683–693

Skerra A, Plückthun A (1988) Assembly of a functional immunoglobulin Fv fragment in *Escherichia coli*. Science 240:1038–1041

Soltys J, Kusner LL, Young A, Richmonds C, Hatala D, Gong B, Shanmugavel V, Kaminski HJ (2009) Novel complement inhibitor limits severity of experimentally myasthenia gravis. Ann Neurol 65:67–75

Sörensen J, Sandberg D, Sandström M, Wennborg A, Feldwisch J, Tolmachev V, Aström G, Lubberink M, Garske-Román U, Carlsson J, Lindman H (2014) First-in-human molecular imaging of HER2 expression in breast cancer metastases using the ^{111}In-ABY-025 affibody molecule. J Nucl Med 55:730–735

Stahl A, Stumpp MT, Schlegel A, Ekawardhani S, Lehrling C, Martin G, Gulotti-Georgieva M, Villemagne D, Forrer P, Agostini HT, Binz HK (2013) Highly potent VEGF-A-antagonistic DARPins as anti-angiogenic agents for topical and intravitreal applications. Angiogenesis 16:101–111

Stanfield RL, Dooley H, Flajnik MF, Wilson IA (2004) Crystal structure of a shark single-domain antibody V region in complex with lysozyme. Science 305:1770–1773

Stefan N, Martin-Killias P, Wyss-Stoeckle S, Honegger A, Zangemeister-Wittke U, Plückthun A (2011) DARPins recognizing the tumor-associated antigen EpCAM selected by phage and ribosome display and engineered for multivalency. J Mol Biol 413:826–843

Steiner M, Gutbrodt K, Krall N, Neri D (2013) Tumor-targeting antibody-anticalin fusion proteins for in vivo pretargeting applications. Bioconjug Chem 24:234–241

Steinmeyer DE, McCormick EL (2008) The art of antibody process development. Drug Discov Today 13:613–618

Steipe B, Schiller B, Plückthun A, Steinbacher S (1994) Sequence statistics reliably predict stabilizing mutations in a protein domain. J Mol Biol 240:188–192

Stevens FJ, Solomon A, Schiffer M (1991) Bence Jones proteins: a powerful tool for the fundamental study of protein chemistry and pathophysiology. Biochemistry 30:6803–6805

Strand J, Varasteh Z, Eriksson O, Abrahmsen L, Orlova A, Tolmachev V (2014) Gallium-68-labeled Affibody molecule for PET imaging of PDGFRβ expression in vivo. Mol Pharm 11:3957–3964

Streltsov VA, Varghese JN, Carmichael JA, Irving RA, Hudson PJ, Nuttall SD (2004) Structural evidence for evolution of shark Ig new antigen receptor variable domain antibodies from a cell-surface receptor. Proc Natl Acad Sci U S A 101:12444–12449

Strohl WR, Strohl LM (2012) Therapeutic antibody engineering: current and future advances driving the strongest growth area in the pharmaceutical industry. Woodhead, Cambridge, UK

Terwisscha van Scheltinga AG, Lub-de Hooge MN, Hinner MJ, Verheijen RB, Allersdorfer A, Hülsmeyer M, Nagengast WB, Schröder CP, Kosterink JG, de Vries EG, Audoly L, Olwill SA (2014) In vivo visualization of MET tumor expression and anticalin biodistribution with the MET-specific anticalin ^{89}Zr-PRS-110 PET tracer. J Nucl Med 55:665–671

Thiel MA, Wild A, Schmid MK, Job O, Bochmann F, Loukopoulos V, Alcantara W, Schmidt A, Lichtlen P, Escher D (2013) Penetration of a topically administered anti-tumor necrosis factor alpha antibody fragment into the anterior chamber of the human eye. Ophthalmology 120:1403–1408

Tijink BM, Laeremans T, Budde M, Stigter-van Walsum M, Dreier T, de Haard HJ, Leemans CR, van Dongen GA (2008) Improved tumor targeting of anti-epidermal growth factor receptor Nanobodies through albumin binding: taking advantage of modular Nanobody technology. Mol Cancer Ther 7:2288–2297

Tolcher AW, Sweeney CJ, Papadopoulos K, Patnaik A, Chiorean EG, Mita AC, Sankhala K, Furfine E, Gokemeijer J, Iacono L, Eaton C, Silver BA, Mita M (2011) Phase I and pharmacokinetic study of CT-322 (BMS-844203), a targeted Adnectin inhibitor of VEGFR-2 based on a domain of human fibronectin. Clin Cancer Res 17:363–371

Tolmachev V, Orlova A (2009) Update on Affibody molecules for in vivo imaging of targets for cancer therapy. Minerva Biotecnol 21:21–30

Tolmachev V, Orlova A, Pehrson R, Galli J, Baastrup B, Andersson K, Sandström M, Rosik D, Carlsson J, Lundqvist H, Wennborg A, Nilsson FY (2007) Radionuclide therapy of HER2-positive microxenografts using a ^{177}Lu-labeled HER2-specific Affibody molecule. Cancer Res 67:2773–2782

Tolmachev V, Mume E, Sjoberg S, Frejd FY, Orlova A (2009a) Influence of valency and labelling chemistry on in vivo targeting using radioiodinated HER2-binding Affibody molecules. Eur J Nucl Med Mol Imaging 36:692–701

Tolmachev V, Wållberg H, Andersson K, Wennborg A, Lundqvist H, Orlova A (2009b) The influence of Bz-DOTA and CHX-A″-DTPA on the biodistribution of ABD-fused anti-HER2 Affibody molecules: implications for ^{114m}In-mediated targeting therapy. Eur J Nucl Med Mol Imaging 36:1460–1468

Tolmachev V, Rosik D, Wallberg H, Sjoberg A, Sandstrom M, Hansson M, Wennborg A, Orlova A (2010) Imaging of EGFR expression in murine xenografts using site-specifically labelled anti-EGFR ^{111}In-DOTA-$Z_{EGFR:2377}$ Affibody molecule: aspect of the injected tracer amount. Eur J Nucl Med Mol Imaging 37:613–622

Tolmachev V, Varasteh Z, Honarvar H, Hosseinimehr SJ, Eriksson O, Jonasson P, Frejd FY, Abrahmsen L, Orlova A (2014) Imaging of platelet-derived growth factor receptor β expression in glioblastoma xenografts using Affibody molecule ^{111}In-DOTA-Z09591. J Nucl Med 55:294–300

Topp MS, Goekbuget N, Stein AS, Bargou RC, Dombret H, Fielding AK, Ribera JM, Foà R, Zugmaier G, Holland C, Maniar T, Huber B, Nagorsen D, Kantarjian HM (2014) Confirmatory open-label, single-arm, multicenter phase 2 study of the BiTE antibody blinatumomab in patients (pts) with relapsed/refractory B-precursor acute lymphoblastic leukemia (r/r ALL). J Clin Oncol 32:5s (suppl; abstr 7005)

Tsianakas A, Brunner P, Ghoreschi K, Berger C, Loser K, Röcken M, Stingl G, Luger T, Jung T (2014) Topical administration of the single-chain anti-TNFα antibody DLX105 suppresses TNFα and Th17 cytokines in psoriatic skin. Abstract No. LB011; 44th annual meeting of the European Society for Dermatological Research (ESDR), Sept 10–13. Copenhagen, Denmark

Van Beneden K, Verschueren K, Willems W, Wouters H, D'Artois J, De Swert K, Arold G, De Bruyn S (2014) Impact of clinical remission on physical function in patients with rheumatoid arthritis treated with ALX-0061: post-hoc analysis of phase I/II data. Ann Rheum Dis 73:506

Vaneycken I, D'Huyvetter M, Hernot S, De Vos J, Xavier C, Devoogdt N, Caveliers V, Lahoutte T (2011) Immuno-imaging using nanobodies. Curr Opin Biotechnol 22:877–881

Vanheusden K, Detalle L, Hemeryck A, Vicari A, Grenningloh R, Poelmans S, Wouters H, Stöhr T (2013) Pre-clinical proof-of-concept of ALX-0761, a Nanobody neutralising both IL-17A and IL-17F in a cynomolgus monkey collagen induced arthritis model. Abstract No. 1287; annual meeting of the American College of Rheumatology (ACR), Oct 26–30. San Diego, CA

Viardot A, Goebeler M, Pfreundschuh M, Adrian N, Libicher M, Degenhard E, Stieglmaier J, Zhang A, Nagorsen D, Bargou RC (2013) Open-label phase 2 study of the bispecific T-cell engager (BiTE®) Blinatumomab in patients with relapsed/refractory diffuse large B-cell lymphoma. Blood 122:1811

Vopel S, Mühlbach H, Skerra A (2005) Rational engineering of a fluorescein-binding anticalin for improved ligand affinity. Biol Chem 386:1097–1104

Wahlberg E, Lendel C, Helgstrand M, Allard P, Dincbas-Renqvist V, Hedqvist A, Berglund H, Nygren P-Å, Härd T (2003) An affibody in complex with a target protein: structure and coupled folding. Proc Natl Acad Sci U S A 100:3185–3190

Weiner GJ, Hillstrom JR (1991) Bispecific anti-idiotype/anti-CD3 antibody therapy of murine B cell lymphoma. J Immunol 147:4035–4044

Werner RG (2004) Economic aspects of commercial manufacture of biopharmaceuticals. J Biotechnol 113:171–182

Weston-Davies W, Westwood JP, Nunn M (2013) Phase 1 clinical trial of novel complement C5 inhibitor coversin. Mol Immunol 56:264

Wickham TJ, Reynolds J, Drummond DC, Kirpotin DB, Lahdenranta J, Leonard S, Geretti E, Lee H, Klinz S, Hendriks BS, Olivier K, Eckelhofer I, Park JW, Benz CC, Moyo VM, Niyikiza C, Nielsen UB (2010) Preclinical safety and activity of MM-302, a HER2-targeted liposomal doxorubicin designed to have an improved safety and efficacy profile over approved anthracyclines. Cancer Res 70(24 Suppl):Abstract nr P3-14-09.

Wiecek AS (2010) Nanobodies: going single-domain. BioTechniques. http://www.biotechniques.com/news/Nanobodies-Going-single-domain/biotechniques-257771.html

Woisetschläger M, Rüker F, Mudde GC, Wozniak-Knopp G, Bauer A, Himmler G (2013) Modular antibody engineering: antigen binding immunoglobulin Fc CH3 domains as building blocks for bispecific antibodies (mAb²). In: Schmidt SR (ed) Fusion protein technologies for biopharmaceuticals (p. 583–589). Wiley, Chichester

Wozniak-Knopp G, Bartl S, Bauer A, Mostageer M, Woisetschläger M, Antes B, Ettl K, Kainer M, Weberhofer G, Wiederkum S, Himmler G, Mudde GC, Rüker F (2010) Introducing antigen-binding sites in structural loops of immunoglobulin constant domains: Fc fragments with engineered HER2/neu-binding sites and antibody properties. Protein Eng Des Sel 23:289–297

Wurch T, Pierré A, Depil S (2012) Novel protein scaffolds as emerging therapeutic proteins: from discovery to clinical proof-of-concept. Trends Biotechnol 30:575–582

Xavier C, Vaneycken I, D'Huyvetter M, Heemskerk J, Keyaerts M, Vincke C, Devoogdt N, Muyldermans S, Lahoutte T, Caveliers V (2013) Synthesis, preclinical validation, dosimetry, and toxicity of ^{68}Ga-NOTA-anti-HER2 Nanobodies for iPET imaging of HER2 receptor expression in cancer. J Nucl Med 54:776–784

Xu L, Aha P, Gu K, Kuimelis RG, Kurz M, Lam T, Lim AC, Liu H, Lohse PA, Sun L, Weng S, Wagner RW, Lipovsek D (2002) Directed evolution of high-affinity antibody mimics using mRNA display. Chem Biol 9:933–942

Zahnd C, Kawe M, Stumpp MT, de Pasquale C, Tamaskovic R, Nagy-Davidescu G, Dreier B, Schibli R, Binz HK, Waibel R, Plückthun A (2010) Efficient tumor targeting with high-affinity designed ankyrin repeat proteins: effects of affinity and molecular size. Cancer Res 70:1595–1605

Zhukovsky E, Reusch U, Burkhardt C, Knackmuss S, Fucek I, Eser M, McAleese F, Ellwanger K (2012a) A T cell-engaging CD3 Recruit-Tandab potently kills CD19+ tumor B cells. Blood 120:3721

Zhukovsky E, Reusch U, Burkhardt C, Knackmuss S, Fucek I, Eser M, McAleese F, Ellwanger K, Little M (2012b) High affinity CD3 RECRUIT TandAb for T cell-mediated lysis of CD19+ tumor B cells. J Clin Oncol 30:8059

Zhukovsky E, Knackmuss S, Reusch U, Wall C, Ellwanger K, Fucek I, Burkhardt C (2014) Preclinical development, primary and secondary pharmacodynamics, of the CD19/CD3 Tandab (AFM11). J Clin Oncol 32:e19546

Zielonka S, Empting M, Grzeschik J, Könning D, Barelle CJ, Kolmar H (2015) Structural insights and biomedical potential of IgNAR scaffolds from sharks. mAbs 7:15–25

Current Strategies for Pharmacokinetic Optimization

Uli Binder and Arne Skerra

Introduction: The Generation of Biobetters via Half-Life Extension

One of the most straightforward strategies to design biobetters from existing biologics is to prolong their in vivo life-time. Looking at the published pharmacokinetic (PK) data of the first generation of biotechnological drugs, which includes blockbuster biopharmaceuticals such as erythropoietin (EPO), granulocyte colony stimulating factor (G-CSF), growth hormone (GH) and various interferons (IFNs), their plasma half-lives in human patients are disappointingly short, typically just a few hours (Tang et al. 2004). The quick elimination from circulation provokes excessive dosing at short intervals to reach a therapeutic effect. However, the alternating peak/trough-like shape of the resulting drug concentration profile in the body makes it difficult for the physician to optimally adjust the therapeutic window, whereas administration of overt high bolus amounts favors side effects and may lead to adverse reactions at the injection site.

The only class of conventional biopharmaceuticals that does not suffer from this trait comprises human(ized) monoclonal antibodies (MAbs)—also known as immunoglobulins (Igs)—which possess a long natural half-life of typically 1–2 weeks (Lobo et al. 2004), certainly one of the reasons for their current clinical and commercial success (Elvin et al. 2013). While glycosylated IgGs with their large size of

U. Binder
XL-protein GmbH, 85354 Freising, Germany

A. Skerra (✉)
Lehrstuhl für Biologische Chemie, Technische Universität München,
Emil-Erlenmeyer-Forum 5, 85350 Freising-Weihenstephan, Germany
e-mail: skerra@tum.de

A. Rosenberg, B. Demeule (eds.), *Biobetters*, AAPS Advances
in the Pharmaceutical Sciences Series 19, DOI 10.1007/978-1-4939-2543-8_14

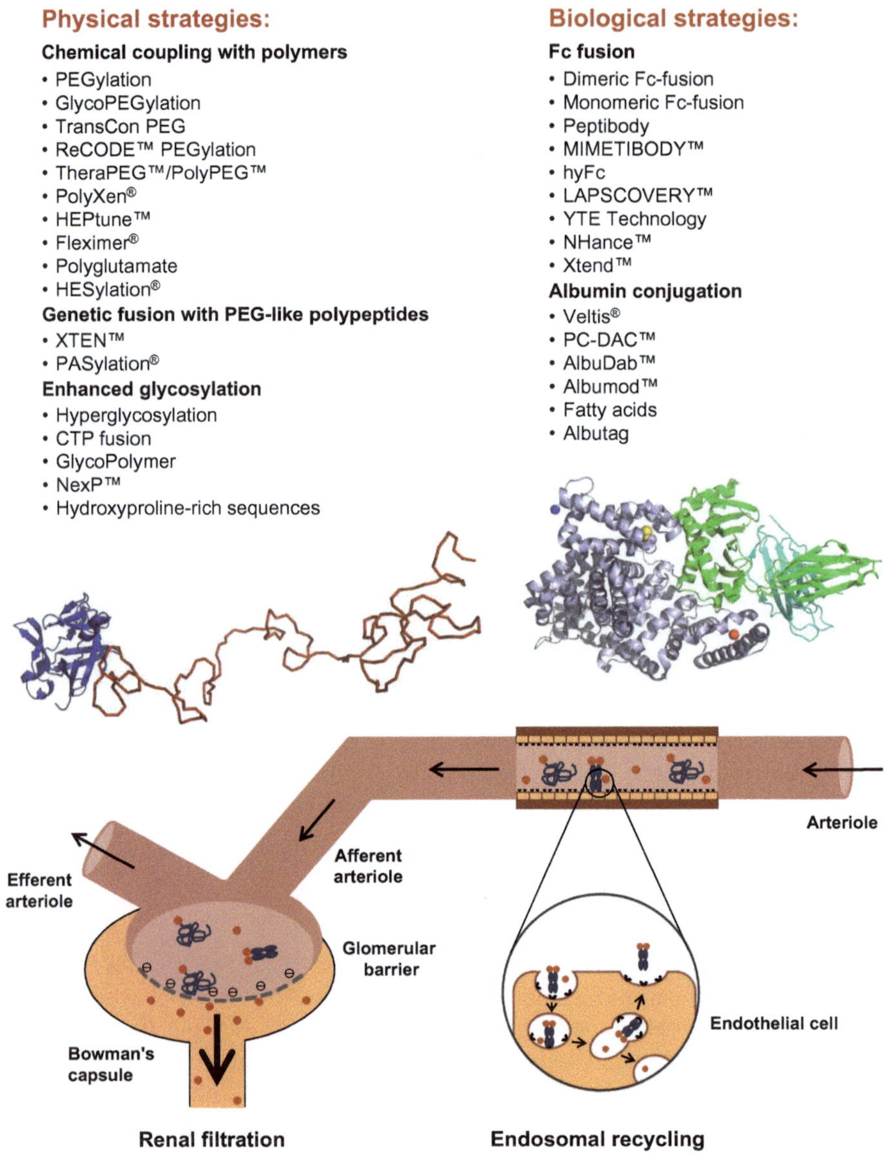

Fig. 1 Strategies for plasma half-life extension. Available technologies prolong circulation of biologics either by increasing the hydrodynamic molecular volume beyond the kidney pore size and, thus, retarding kidney filtration via a biophysical effect (*left*) or by utilising FcRn-mediated endosomal recycling via biomolecular complex formation mediated by an Ig Fc fragment or HSA (*right*). Both strategies are explained in greater detail in the text. The molecular model to the *right* shows the crystal structure (PDB entry 4N0F) of the complex between HSA (*light blue*) and FcRn (*cyan/green*). In HSA the N- and C-termini, which are available for genetic fusion, are depicted as *blue* and *red spheres*, respectively, whereas the free thiol side chain of Cys 34, which can be used for chemical coupling, is highlighted as *yellow sphere*. The molecular model to the *left* shows the crystal structure of IL1-Ra (PDB entry 1IRA; *blue*) with a PAS chain of 200 residues (*red*) fused to its C-terminus and depicted in a simulated exemplary random coil conformation to illustrate its fluctuating structure

approximately 150 kDa are mostly eliminated in receptor-dependent processes, including antigen- and FcγR-mediated catabolism (Tabrizi et al. 2006) as well as carbohydrate receptor-mediated clearance in the liver (Wright et al. 2000), conventional biologics with typical sizes between 10 and 30 kDa are predominantly subject to plasma filtration in the kidneys. There, the narrow pores of the renal slit diaphragm, 35–40 Å wide, and the mesh-like glomerular basement membrane allow rapid permeation into the primary urine in a size-dependent manner, followed by reabsorption and degradation in the proximal tubular cells (Haraldsson et al. 2008). Apart from these differences in molecular physiology, MAbs can take advantage of a peculiar mechanism called endosomal recycling, mediated by the neonatal Fc receptor (FcRn) (Kuo and Aveson 2011; Roopenian and Akilesh 2007), which further prolongs their circulation.

Accordingly, already in the 1990s first attempts were made to modify non-Ig biopharmaceuticals in order to boost their in vivo half-life. To this end, two different mechanisms were exploited: (1) increasing the apparent molecular size to slow down renal filtration and (2) conferring binding activity towards FcRn to effect endosomal recycling (Fig. 1). Meanwhile, several advanced technologies have emerged to accomplish these goals, which will be described in the following sections: Ig Fc fusion, albumin conjugation, conjugation with synthetic polymers, genetic fusion with disordered polypeptides and, finally, enhanced glycosylation (for an overview, see Table 1).

Fc Fusion Strategies

Beside their large molecular size, the main reason for the atypically long plasma half-life of IgGs in humans is endosomal recycling via interaction with FcRn. FcRn is present on several cells types, in particular vascular endothelial cells (Borvak et al. 1998), hematopoietic cells (Zhu et al. 2001), podocytes (Akilesh et al. 2008) and liver cells (Telleman and Junghans 2000). Binding of an antibody via its Fc portion to FcRn occurs under low pH conditions but not at the quasi-neutral pH 7.4 in the blood. Thus, taken up into endosomes by fluid phase pinocytosis, MAbs are able to transiently associate with FcRn under the acidic condition in early endosomes. Whereas most common plasma proteins are subsequently degraded in the lysosome, FcRn-bound IgG is recycled via exocytosis to the cell surface and there gets released back into circulation (Rath et al. 2013).

Capon et al. (1989) first demonstrated that this mechanism can be transferred to non-Ig molecules by fusing the Fc fragment of IgG1 with a biologically active protein. The first Fc hybrid was a so-called CD4 immunoadhesin, which blocked in vitro cell killing of CD4-positive T-cells by a HIV-1 T-lymphotrophic isolate with the same potency as soluble recombinant CD4, but showed much increased plasma half-life in rabbits, from 0.25 to 48.0 h. Since then, nine approved Fc fusion proteins have appeared on the market, among those the highly successful drug Enbrel® (Etanercept) (Kerensky et al. 2012), originally designed by Bruce A. Beutler and coworkers (Peppel et al. 1991).

Table 1 Overview of current PK extension strategies

Technology	Company	Biobetters[a]	Website or Reference
Fc-fusion			
Heterodimeric Fc-fusion	Regeneron	IL1 receptor[b] (Arcalyst®)	www.regeneron.com
Homodimeric Fc-fusion	Amgen	TNF receptor (Enbrel®)	www.amgen.com
Homodimeric Fc-fusion	Astellas Pharma	LFA3[b] (Amevive®)	www.astellas.com
Homodimeric Fc-fusion	Bristol-Myers Squibb	CTLA-4[b] (Nulojix®), CTLA-4 mutant[b] (Orencia®)	www.bms.com
Monomeric Fc-fusion	BiogenIdec/SOBI	Factor VIII (Elocate™), Factor IX (Alprolix™)	www.biogenidec.de/www.sobi.com
Peptibody	Amgen	Thrombopoietin receptor agonist[b] (NPlate®)	www.amgen.com
Peptibody	EliLilly	GLP-1 analog (Dulaglutide)	www.lilly.com
MIMETIBODY™	Centocor (Janssen Biotech)	GLP-1 analog (CNTO0736)	www.janssenbiotech.com
hyFc	Genexine	Epo (GX-E2), hGH (GX-H9), G-CSF (GX-G3)	www.genexine.com
LAPSCOVERY™	Hamni Pharmaceuticals	Insulin (HM12470), hGH (HM10560A), G-CSF (HM10460A), CA-Exendin-4 (HM11260C)	www.hanmipharm.com
NHance™	arGen-X	Anti-IL-20/IL-22 antibody[b] (ARGX-112)	www.argen-x.com
Xtend™	Xencor	Anti-TNF antibody (Xtend-TNF)	www.xencor.com
YTE-Technology	Medimmune	Anti-RSV antibody (Motovizumab)	www.medimmune.com
Albumin conjugation			
Veltis®	Novozymes Biopharma	GLP-1 analog	www.biopharma.novozymes.com
AlbuDab™	GlaxoSmithKline	IL1-Ra, IFNα2b	www.gsk.com
Albumod™	Affibody	Anti C5[b] (SOBI002)	www.affibody.com
PC-DAC™	ConjuChem	Exendin-4 (CJC-1134-PC), PC-DAC™: Insulin	www.conjuchem.com
Fatty acids	Novo Nordisk	GLP-1 analog (Victoza®), Insulin (Tresiba®, Levemir®)	www.novonordisk.de
Albumin-binding nanobodies	Ablynx	Anti-TNFα domain antibody (ATN-103)	www.ablynx.com
Albutag	–	Anti-ED-A scFv[b]	Trussel et al. (2009)
PEGylation			
PEGylation	Enzon Pharmaceuticals	IFNα2b (PEG-Intron®/MSD), αTNFα (Cimzia®/UCB)	www.enzon.com

PEGylation	Nektar Therapeutics	FVIII (BAX885/Baxter), G-CSF (Neulasta®/Amgen), IFNα2a (Pegasys®/Roche)	www.nektar.com
GlycoPEGylation	Novo Nordisk	Factor IX (NN7999), Factor VIII (NN7088), Factor VIIa (NP7-GP)	www.novonordisk.de
GlycoPEGylation	TEVA	G-CSF (Lonquex®)	www.tevapharm.com
TransCon PEG	Ascendis Pharma	hGH (ACP-001)	www.ascendispharma.com
ReCODE™	Ambrx	hGH (ARX201), Leptin (ARX328)	www.ambrx.com
TheraPEG™	Polytherics	L-Asparaginase, IFNα-2b	www.polytherics.com
Degradable polymer conjugation			
PolyXen®	Xenetic Biosciences	Epo (ErepoXen®), Insulin (SuliXen®), rFVIII	www.xeneticbio.com
HEPtune™	Caisson Biotech	Insulin	www.caissonbiotech.com
Polyglutamate	CTI Biopharma	Taxol® (Paclitaxel poliglumex)	www.ctibiopharma.com
Fleximer®	Mersana Therapeutics	Camptothecin (XMT-1001)	www.mersana.com
HESylation®	Fresenius Kabi	IL1-Ra	www.fresenius-kabi.de
Genetic fusion with PEG-like polypeptides			
PASylation®	XL-protein	hGH (xl020), Exendin-4 (xl110), Leptin (xl100), IFNα (xl080), IFN superagonist (xl082)	www.xl-protein.com
XTEN™	Amunix Operating	GLP-2 (AMX256), hGH (VRS317), Exendin-4 (VRS859), FVIII	www.amunix.com
Enhanced glycosylation			
Hyperglycosylation	Amgen and others	Epo (Aranesp®)	www.amgen.com
CTP-fusion	Opko Health	hGH, Factor VIIa, oxyntomodulin	www.opko.com
CTP-fusion	Merck Sharp & Dohme	FSH (Elonva®)	www.msd.de
GlycoPolymer	Aequus Biopharma	G-CSF (AQB-101)	www.aequusbiopharma.com
NexP™	Alteogen	–	www.alteogen.com
Hydroxyproline-rich proteins	–	hGH, IFNα2b	Xu et al. (2010); Xu et al. (2007)

[a]Only some representative examples are given

[b]Innovative biologic

Beyond the half-life extending effect, Fc fusion also facilitates purification of hybrid proteins via protein A affinity chromatography, a widely applied method for full size antibodies (Shukla et al. 2007). The majority of the approved Fc fusion proteins so far are homodimeric receptor antagonists derived from the extracellular domains of membrane proteins such as the TNF receptor 1 (Enbrel®, Etanercept) (Goldenberg 1999), cytotoxic T lymphocyte-associated antigen 4 (Orencia®, Abatacept and Nulojix®, Belatacept, respectively) (Grinyo et al. 2013; Keating 2013) or the lymphocyte associated antigen 3 (Amevive®, Alefacept) (Ortonne and Prinz 2004). In contrast, Arcalyst® (Rilonacept) (Stahl et al. 2009) and Eylea® (Aflibercept) (Jin et al. 2010) comprise combinations of different receptor domains fused in line to each Fc chain: the IL-1 accessory protein followed by the ligand-binding domain of IL1 receptor in case of Rilonacept, the second domain of VEGF receptor 1 and the third domain of VEGF receptor 2 in case of Aflibercept.

Recently, Aprolix®, a long-acting coagulation factor IX (rFIX-Fc) (Shapiro et al. 2012), and Eloctate™, an engineered FVIII with deleted B-domain (rFVIII-Fc) (Peters et al. 2013), were approved, thus extending the original concept of receptor-Fc fusions to enzymes as cargoes and, furthermore, demonstrating that biobetters of conventional first generation biotech products can be created in this way. Notably, the molecular design of these latter products deviates insofar as only one chain of the dimeric Fc domain is fused to one blood clotting factor. Initially developed for pulmonary delivery, such monomeric Fc fusions show improved properties compared to their homodimeric counterparts like enhanced transepithelial delivery and, apparently, even greater extension of the circulating half-life. In addition, this format leads to higher bioactivity in vitro and in vivo as demonstrated e.g. for IFNβ-Fc, IFNα-Fc and EPO-Fc (Dumont et al. 2006). In phase I/IIa clinical studies, rFIX-Fc and rFVIII-Fc reached 3.0- to 4.0-fold and 1.5- to 1.7-fold longer elimination half-lives, respectively, than the natural blood clotting factors, thus offering a once-weekly and an every 5 day dosing regimen, respectively (Mancuso and Mannucci 2014).

Some innovative antibody constructs may be considered as other examples for monomeric Fc fusion proteins, e.g. MetMAb (Genentech) and the Unibody™ format (Genmab). MetMAb blocks the hepatocyte growth factor receptor—also known as c-Met—a receptor tyrosine kinase, which promotes tumor survival, growth, angiogenesis and metastasis (Teng and Lu 2013). This novel engineered one-armed antibody comprises one Fab fragment linked to an Fc part, a molecular design that prevents unwanted tumor cell activation due to receptor crosslinking and also allows production in *E. coli* (Martens et al. 2006). Unibodies™ consist of only one pair of heavy and light chains based on IgG4 (instead of the conventional IgG hetero-tetramer), which avoids activation of the immune system. The deletion of the IgG4 hinge region abolishes heavy chain homodimerization, thus resulting in antigen-specific monovalent light/heavy chain heterodimers while retaining an Fc region to provide stability in vivo. Again, the monovalency precludes cross-linking of cell surface antigens, thus avoiding unwanted receptor activation (Nelson 2010).

Recently, some new Fc fusion platforms were developed with the goal of improving safety and activity profiles. For instance, the hybrid Fc (hyFc) technology (Genexine) involves the fusion of a therapeutic protein, such as erythropoietin

(EPO), to the Fc part of human IgG4 linked via the flexible hinge region of human IgD (Yang et al. 2012). This highly flexible natural spacer minimizes the loss of bioactivity upon fusion. As the IgG4 Fc region does not bind to phagocytes or to the complement factor C1q, typical immunological effector functions such as antibody-dependent cellular cytotoxicity (ADCC) and complement-dependent cytotoxicity (CDC) are avoided, which results in better safety. Nevertheless, interaction with FcRn guarantees a long plasma half-life as shown for GX-E2, a (homodimeric) EPO-hyFc fusion currently tested in clinical phase II. The terminal plasma half-life of EPO-hyFc after i.v. injection in Sprague–Dawley rats was slightly longer than that of conventional EPO-IgG1-Fc (29.1 vs. 24.7 h). Furthermore, the bioactivity of GX-E2 measured in a cell culture assay showed an ED_{50} of 27.6 pM, which was approximately fourfold better than EPO-IgG1-Fc (ED_{50} = 111.5 pM) (Im et al. 2011). Other hyFc biobetters with prolonged half-life and improved efficacy, e.g., GX-H9 (hGH-hyFc) (Kim et al. 2013b) or GX-G3 (G-CSF-hyFc), are currently under evaluation in phase I trials (cf. Table 1).

The LAPSCOVERY™ (Long Acting Protein/Peptide Discovery) technology (Hanmi Pharmaceuticals) presents a different solution to reducing steric hindrance upon Fc conjugation. By chemical coupling of a pharmacologically active compound via a synthetic PEG linker to an aglycosylated IgG Fc region, bioactivity is largely retained while certain advantages over genetic fusion are offered by this strategy. For example, the linker is resistant to proteases, the optimal site for linker attachment can be chosen, a monovalent Fc conjugate can be easily prepared and production costs are lower since the Fc part can be produced in *E. coli* (Kim et al. 2012)—even though there is additional downstream effort for chemical coupling and purification of the conjugate. In this manner, long-acting biobetters of follicle-stimulating hormone (FSH) (Jung et al. 2014) and insulin (Park et al. 2012) were prepared. In PK studies, s.c. injected LAPS-insulin showed significantly extended elimination half-lives of 15 and 32 h in normal rats and dogs, respectively, whereas native insulin is normally eliminated with a short terminal half-life of just ~20 min in rats and of 50–90 min in beagle dogs (Plum et al. 2000).

LAPS-GLP/GCG, a monovalent long-acting oxyntomodulin analog, was developed by site-specific conjugation of the peptide analog to an aglycosylated Fc moiety of human IgG4 linked via a PEG chain. LAPS-GLP/GCG acts both as GLP-1 (glucagon like peptide) and glucagon agonist, leading to synergistic effects in obesity. In high fat diet induced obese mice, a once weekly s.c. 5 nmol/kg dose of LAPS-GLP/GCG reduced body weight by ~31 % after 2 weeks, whereas a daily 100 nmol/kg dose of the gold standard liraglutide only led to ~17 % body weight loss (Jung et al. 2013). Further biobetters such as LAPS-hGH, LAPS-G-CSF and LAPS-CA-Exendin-4 are currently tested in clinical phase II (cf. Table 1).

Beyond linker optimization, several studies have demonstrated a correlation between affinity of IgG to FcRn and plasma half-life, which prompted engineering of the Fc region to further prolong circulation. For instance, Fc variants with increased binding to FcRn at pH 6.0 have been reported (Xencor) and, in particular, one mutant with the substitutions M428L and N434S provided an 11-fold improvement in FcRn affinity (Zalevsky et al. 2010). The correspondingly mutated anti-

VEGF antibody Bevacizumab showed a 3.2-fold improvement in plasma half-life in monkeys ($\tau_{1/2}$=31.1 days) compared with the native Fc region ($\tau_{1/2}$=9.7 days). In addition, this so-called Xtend™ Bevacizumab variant showed a statistically greater level of xenograft tumor reduction compared to the native IgG1 version in an hFcRn/Rag1$^{-/-}$ mouse model (Zalevsky et al. 2010).

Another example is the anti-respiratory syncytial virus (RSV) monoclonal antibody Motavizumab (Medimmune). Introduction of the three point mutations M252Y/S254T/T256E (YTE) led to a tenfold increase in affinity toward human FcRn at pH 6.0. In healthy adults, these amino acid replacements resulted in an 2–4-fold increase in plasma half-life, from 19 to 34 days for the original Motavizumab to 70–100 days for Motavizumab(YTE) (Robbie et al. 2013). The NHance™ technology (arGEN-X) uses a different engineered IgG1 Fc region, in this case carrying the amino acid substitutions H433K and N434F, which resulted in increased FcRn affinity at acidic pH and prolonged plasma half-life of a correspondingly modified anti-lysozyme antibody (Ward Ober 2012). So far, Fc engineering has only been applied to create biobetters of full-length MAbs. On the other hand, it is well known that Fc fusion proteins exhibit a three to fivefold weaker binding to FcRn, in line with their generally shorter half-life (Suzuki et al. 2010). Consequently, Fc engineering should also be an option to improve immunoadhesins.

Furthermore, several biotech companies have used Fc fragments as carriers to modulate the PK properties of peptides. The most advanced technology is the "peptibody" platform (Amgen). Expressed at high levels as inclusion bodies in *E. coli*, manufacturing of peptibodies is inexpensive, leading to unglycosylated homogenous products after refolding and purification. In contrast to the rapid elimination of peptides from the human body within just a few minutes, the plasma half-life of the peptibody format reaches 3–8 days due to the large size increase and endosomal recycling as described above (Wu and Sun 2014). Peptibodies are constructed by fusion either to the N-terminus of IgG1 Fc (Furie et al. 2014; Glaesner et al. 2010) or to its C-terminus (Cines et al. 2008). Alternatively, peptides can be inserted into the Fc framework within the CH3 domain, which resulted in improved refolding efficiency, yield and in vitro stability of the bacterially produced Fc chimeric proteins compared with the N/C-terminal fusions (Hall et al. 2010; Shimamoto et al. 2012). In addition, improved exposure may be achieved in the latter case as the Fc context protects the peptide from proteolytic degradation.

The functional bivalency of peptibodies can lead to increased receptor binding activity via an avidity effect. In some cases, avidity can be further enhanced by employing peptide concatamers spaced by linkers, as demonstrated for NPlate® (Romiplostim), which was approved by the U.S. Food and Drug Administration (FDA) in 2008 for the treatment of thrombocytopenia. This peptibody was constructed from the human IgG1 Fc via fusion with two copies of a 14 amino acid thrombopoietin receptor binding peptide on each chain. Another example is Trebananib, an angiopoietin antagonist currently investigated in a phase III clinical trial (Monk et al. 2014; Shimamoto et al. 2012), which likewise carries two peptides per Fc chain and has shown a tenfold increased potency in a cell culture assay compared with the single peptide fusion. Other peptibody products such as Dulaglutide

(LY2189265; Eli Lilly), a GLP-1 analog N-terminally fused to IgG4 Fc (Umpierrez et al. 2014), or Blisibimod (Anthera), an antagonist of the B-cell activating factor (BAFF, also known as B-lymphocyte stimulator or BLyS) (Furie et al. 2014), are currently undergoing phase III clinical studies.

At first glance, peptide fusion to an antibody Fc fragment seems a simple approach; however, several challenges have to be considered. In most cases, the fused peptide dramatically loses activity due to loss of conformational flexibility and/or steric hindrance (Glaesner et al. 2010). Furthermore, unwanted effector functions and cytotoxicity may cause issues in clinical trials (Czajkowsky et al. 2012). Therefore, each peptide fusion necessitates extensive protein engineering efforts to restore bioactivity, to ensure product homogeneity and stability and to control Fc functionality. For example, in the case of Dulaglutide, conjugation of a dipeptidyl dipeptidase IV (DPP4)-protected GLP-1 variant (see below) to the hinge region of IgG1 Fc initially resulted in a >95 % loss of potency in vitro. Optimization of the linker sequence between the peptide and the IgG hinge region was required to rescue bioactivity. In addition, to avoid ADCC, an IgG4 Fc was chosen and Phe234 as well as Leu235 were both replaced by Ala to reduce interaction with high affinity Fc receptors. Moreover, Ser228 was mutated to Pro to stabilize the Fc homodimer, and its C-terminal Lys residue was deleted to reduce product heterogeneity (Glaesner et al. 2010).

The MIMETIBODY™ platform (Centocor) uses specially designed amino acid sequences as linkers to fuse peptides to an Fc fragment, claiming improved bioactivity (Bugelski et al. 2008). MIMETIBODIES can be produced as functional proteins in mammalian cell culture but, like peptibodies, they require protein engineering to individually optimize activity, stability and PK (Picha et al. 2014). The MIMETIBODY platform has been applied to various bioactive peptides, among those a GLP-1 agonist (CNTO0736) (Picha et al. 2008) and an EPO-mimetic peptide (CNTO528). The latter exhibited a terminal half-life of ~5.9 days in a phase I clinical study (Perez-Ruixo et al. 2009). Apart from genetic fusion, peptides can also be chemically linked to the antibody scaffold, which is of special interest for synthetic peptides with non-natural modifications.

Half-Life Extension by Albumin Conjugation

Human serum albumin (HSA) is the most abundant plasma protein in the body, found at concentrations of up to 50 g/l (Yousefpour and Chilkoti 2014). Similar to Igs, HSA exhibits a very long plasma half-life of ca. 19 days (Peters 1996) owing to several mechanisms. First, HSA is a large, acidic protein with dimensions of ca. $80 \times 80 \times 30$ Å^3 (Sugio et al. 1999) and an isoelectric point (pI) near 5. Both of these properties hamper permeation through the narrow pores of the kidney slit diaphragm and the likewise negatively charged glomerular basement membrane (Haraldsson et al. 2008; Tojo and Kinugasa 2012). Second, HSA interacts with FcRn, which protects this plasma protein—like Igs—from premature intracellular degradation and further prolongs its circulation (Andersen et al. 2012).

As albumin has a complex fold, comprising three domains and 17 disulfide bridges, initial attempts to produce this protein in *E. coli* failed (Yousefpour and Chilkoti 2014) whereas efficient yeast-based expression systems were developed in the early 90s (Fleer et al. 1991; Sleep et al. 1990, 1991). Shortly after these microbial secretion systems were established, the first albumin fusion protein was published, consisting of the two N-terminal extracellular domains of CD4 fused to the C-terminus of HSA. The CD4 moiety retained HIV gp160-binding as well as antiviral activity in vitro and, notably, the plasma half-life in rabbits was 140-fold increased compared to the soluble extracellular fragment of CD4 comprising domains D1–D4 (Yeh et al. 1992).

Since then, a variety of pharmaceutically relevant proteins have been fused to albumin in attempts to create biobetters, including IFNα2b (Joulferon®, Zalbin™, Albuferon), hGH (Albutropin®), G-CSF (Neugranin®, Albugranin), GLP-1 agonists (Tanzeum™ (US) / Eperzan® (EU), Albiglutide), insulin (Albulin) or the blood clotting factors VIIa (CSL689) and IX (CSL654). However, although CD4-HSA was developed more than two decades ago, the first albumin fusion protein, Tanzeum, received FDA approval only recently in 2014 for the treatment of type 2 diabetes (Poole and Nowlan 2014). Tanzeum (GlaxoSmithKline) was constructed by fusing two tandem copies of a modified human GLP-1 fragment (His7–Gly36) to HSA. The GLP-1 sequence was modified to achieve resistance against proteolysis by DPP4 via replacement of Ala8 by Gly (Bush et al. 2009). While the natural, non DPP4-resistant GLP-1 exhibits a terminal half-life of just 5 min in humans, the plasma half-life of Tanzeum is 5 days (Matthews et al. 2008).

Typical half-lives of albumin fusions in humans range between 1 and 6 days, thus offering no more than a once weekly dosing frequency (Sleep 2015). Beside impairment of the interaction with FcRn, enhanced receptor-mediated clearance and/or proteolysis mediated by the biologically active fusion partner are reasons for the substantially shorter half-life of albumin fusions compared to HSA itself. To pave the way for bi-weekly or even monthly dosing, the Veltis® technology (Novozyme) has been developed. This approach involves engineered albumins with altered binding to FcRn, thus leading to prolonged circulation (Andersen et al. 2012, 2014). For example, the single amino acid substitution Lys573Pro resulted in a 60 % increase in plasma half-life. With an albumin variant carrying several substitutions even a 2.5-fold longer circulation time was achieved in a non-human primate model (Sleep 2015).

Other albumin fusion proteins like Albugranin (Halpern et al. 2002) and Albuferon (Subramanian et al. 2007) raised high expectations but did not obtain regulatory approval. Albuferon, developed to treat hepatitis C, was constructed by fusing human IFNα2b to the C-terminus of HSA, which led to prolongation of the plasma half-life from 4 to 141 h in humans (Rustgi 2009). These data supported biweekly dosing, compared to the once weekly dosing of Pegasys, the current gold standard PEG-IFNα2a (described further below) with a published terminal half-life of 72 h (Garcia-Garcia et al. 2010). In two phase III clinical studies, ACHIEVE 1 in patients with chronic hepatitis C genotype 1 (Zeuzem et al. 2010) and ACHIEVE 2/3 in patients with genotypes 2 and 3 (Nelson et al. 2010), the biweekly injected Albuferon in combination with ribavirin met the primary efficacy endpoint of non-inferiority of sustained viral response rate (SVR), but achieved a numerically lower

SVR than the once weekly administered Pegasys. Adverse events of a 900 μg dose were similar to Pegasys, but the 1,200 μg dose arm showed a higher rate of serious pulmonary adverse events and necessitated treatment discontinuation. Following concerns by the FDA and the European Medicines Agency (EMA) regarding the risk-benefit assessment, the marketing application was withdrawn (GEN 2010).

As HSA fusion proteins are complex molecules, special caution is required with regard to immunogenicity. The molecular complexity can result in poor solution stability and aggregation, which may stimulate an immune response (Mitragotri et al. 2014). In addition, the free cysteine of HSA can cause disulfide-mediated aggregation associated with increased immunogenicity as shown for IFNα2b-HSA (Zhao et al. 2009). Furthermore, the junction in the fusion protein between albumin and a therapeutic protein or peptide may generate novel T-cell epitopes. Resulting anti-drug antibodies could be problematic in the light of potential cross-reactivity with endogenous albumin.

Apart from genetic fusion, drugs can also be chemically coupled to HSA, especially by utilizing the single unpaired thiol side chain of the Cys residue at position 34 (Sleep et al. 2013). This strategy was exemplified with Albuviritide, an albumin-conjugated HIV entry inhibitor related to the approved antiviral peptide Fuzeon® (T20, Enfurvitide) (Chong et al. 2012), which suffers from a poor plasma half-life of just 4 h and, therefore, requires twice daily injections (Patel et al. 2005). The synthetic peptide that forms part of the long-acting Albuviritide differs in three amino acid residues from the natural (so-called C34) peptide sequence within the helical region of the C-terminal heptad repeat 2 domain of HIV glycoprotein 41 (gp41). Whereas the Met2 to Glu and Ser17 to Glu substitutions were introduced into the C34 peptide to improve stability, solubility and antiviral activity, the Ser residue at position 13 was replaced by Lys to introduce a 3-maleimidopropionic acid, allowing subsequent chemical coupling to the free Cys residue of recombinant HSA. The plasma half-life of Albuviritide, which is currently subject to phase III clinical studies (Frontier Biotechnologies), has been reported to be 10–13 days in humans, while the HIV RNA level was suppressed for 6–10 days after a single injection, indicating suitability for once-weekly dosing (Chong et al. 2012).

Another example of a chemical HSA conjugate, developed using the PC-DAC™ (Preformed Conjugate-Drug Affinity Complex) technology (ConjuChem), is CJC-131, an Exendin-4 peptide with an additional C-terminal 2-[2-(2-maleimidopropion-amido) ethoxy]ethoxy-acetamide-modified Lys covalently coupled to the free Cys of albumin (Elsadek and Kratz 2012). This substance was well tolerated as an adjunct therapy to metformin in a phase I/II trial, achieving up to 2.0 kg reduction in body weight for patients as well as significantly diminished HbA1c levels (Wang et al. 2009).

Although in principle albumin conjugation appears as a promising technology to design monovalent biobetters with extended plasma half-lives, it should be kept in mind that efficient production of albumin fusion proteins is restricted to yeast or mammalian expression systems. In addition, both fusion and chemical coupling with the 66.5 kDa HSA molecule often leads to steric hindrance of its fusion partner, resulting in lower affinity for the biomedical target and, hence, decreased bioactivity. For instance, in a cell-culture assay using Madin-Darby bovine kidney cells infected with vesicular stomatitis virus (VSV), Albuferon was clearly less potent on a molar basis than native IFNα while still being approximately equivalent to PEG-IFNα2a (Subramanian et al. 2007). In the case of an albumin fusion with factor IX, bioactivity was rescued by inserting a cleavable linker peptide sequence containing one of the FIX activation sites between both moieties, thus allowing release of FIX after cleavage by FXIa or FVIIa/tissue factor (Metzner et al. 2013).

Alternatively, it is possible to take advantage of HSA as a well known carrier protein in the blood which transports physiological ligands such as thyroid hormones, bilirubin, vitamin A and fatty acids as well as many pharmacological substances (Elsadek and Kratz 2012; Kratz 2014). This feature led to the development of a novel insulin analog for diabetes treatment by coupling myristic acid to its Lys residue at position B29, thus effecting non-covalent complex formation with endogenous albumin in human plasma (NovoNordisk). This molecule, insulin detemir, approved in 2004 in EU and 2005 in US and marketed as Levemir®, shows an extended plasma half-life of 5–7 h in humans and is used as a basal bolus. The second generation product Tresiba® uses palmitic acid linked via a γ-L-glutamyl spacer to LysB29 of insulin, resulting in a threefold greater half-life of 17–21 h, and was launched in 2013 (Kratz 2014).

In a similar manner, Liraglutide ((γ-L-glutamoyl(*N*-α-hexadecanoyl)-Lys^{26}Arg^{34}-GLP-1(7–37)), a once-daily dosed GLP-1 agonist marketed under the tradename Victoza®, was generated. The palmitic acid side chain of Victoza® leads to formation of heptamers by self-association, allowing slow absorption via the subcutaneous route, while subsequent albumin binding retards glomerular filtration and also likely hinders DPP4 access (Kratz 2014). Together, this results in a functional plasma half-life of ~13 h (Deacon 2009; Sjöholm 2010), that is more than 150-fold longer compared to the natural peptide (1–5 min) (Hui et al. 2002). Alternatively, the synthetic compound 2-(3-maleimidopropanamido)-6-(4-(4-iodophenyl)-butanamido)-hexanoate, termed Albu-tag, can be employed to promote non-covalent HSA association—possibly due to its similarity with thyroid hormones, which are natural albumin ligands (Bartalena and Robbins 1993). Chemically conjugated to an scFv antibody fragment that recognizes the extra domain A (ED-A) of oncofetal fibronectin, the plasma half-life was extended from 20–30 min in mice to approximately 1,000 min (Trussel et al. 2009). In addition, the uptake into murine F9 tumors increased by about tenfold.

Association with HSA can also be mediated by genetic fusion with small albumin-specific binding peptides or proteins. For example, the naturally occurring albumin-binding domain (ABD) of *streptococcal* protein G supports virulence of pathogenic bacteria by coating the pathogen surface with plasma proteins (Kraulis et al. 1996). The three-helical ABD comprises just 46 amino acid residues and has been used as fusion partner to prolong plasma half-life of tumor targeting agents such as Fab fragments (Schlapschy et al. 2007), single chain diabodies (scDb) (Stork et al. 2007) and Affibodies (Andersen et al. 2011; Tolmachev et al. 2007) (see previous Chapter). In a comparative study of an scDb fused to ABD mutants having low ($K_D = 634$ nM), medium (21.4 nM), and high affinity (1.8 nM) for murine serum albumin, respectively, the terminal plasma half lives in mice were 28.4 h (low affinity variant), 36.4 h (medium affinity variant) and 47.5 h (high affinity variant).

Notably, this gain in half-life was less than expected, considering the more than 350-fold variation in affinity (Hopp et al. 2010). This may be explained by the high concentration of albumin in plasma, which promotes biomolecular complex formation according to the Law of Mass Action even for low affinity interactions, such that a tunable plasma half-life is hard to achieve (Andersen et al. 2011).

Nevertheless, the ABD was subjected to an affinity maturation toward human albumin using phage display, and the variant ABD034 with an extremely low dissociation constant in the femtomolar range was selected (Affibody). ABD034 showed cross-reactivity with rat, mouse and cynomolgus serum albumin, which facilitates preclinical development. Subsequently, ABD034 was deimmunized, resulting in ABD094 (Frejd 2012), which forms the basis of the Albumod™ half-life extension platform. ABD094 was applied to generate a long-acting GLP-1 agonist (Lindgren et al. 2014) and is currently being tested in further development programs (Nilvebrant and Hober 2013). In principle, due to a binding constant in the fM range and the high albumin concentration in blood plasma, ABD094 fusion proteins should predominantly exist in the albumin-bound state and, consequently, biodistribution should be similar to albumin. This may be an advantage for radionuclide tumor therapy, as it is well known that albumin accumulates in the malignant tissue due to a defective vascular architecture and the enhanced permeability and retention (EPR) effect (Kratz 2010; Merlot et al. 2014). Reciprocally, biodistribution and extravasation into healthy tissue may be hampered by this albumin-binding approach.

Meanwhile, several other engineered albumin-binding proteins have been generated by immunization and/or phage display, including peptides (Dennis et al. 2002) and proteins as explained in the previous Chapter. Among those are small proteins based on the single Ig domain format, e.g., human domain antibodies (dAb) (Bao et al. 2013; Walker et al. 2010), the new antigen receptor domains from shark (VNAR) (Muller et al. 2012) and camelid heavy chain antibody fragments, so-called Nanobodies (Tijink et al. 2008). In addition, albumin-binding proteins based on non-Ig scaffolds such as Adnectins (Gosselin et al. 2011) and DARPins (Binz et al. 2012) were discovered.

However, it remains to be seen if these artificially designed binding proteins elicit an immune response upon repeated dosing in humans and, thus, require additional deimmunization efforts—as described for the ABD above and as discussed for some of the other binding proteins in the preceding Chapter. Another concern relates to the complex and multiple biological interactions of serum albumin. HSA not only binds FcRn, but also a wide range of cellular receptors including albondin, gp18, gp30, calreticulin, cubilin and megalin (Merlot et al. 2014). Some of these receptors, in particular gp18 and gp30, are scavengers for chemically and/or conformationally modified albumins (Sleep 2015) that may be responsible for accelerated elimination of drug-modified albumins. Generally, binding to these and other receptors could result in off-target effects if a biopharmaceutical is linked to albumin itself or via a binding domain. Also, HSA binds a wide variety of endogenous and exogenous ligands including fatty acids, thyroxine, bilirubin and many small molecule drugs such as warfarin or diazepam (Ghuman et al. 2005); these may undergo preferential receptor targeting along with an albumin-tagged protein or peptide.

Finally, from volume of distribution measurements in PK studies it appears that albumin association confines a biopharmaceutical predominantly to the vascular compartment (Schlapschy et al. 2007). Even though albumin is also abundant in the extravascular space, the mutual exchange seems to be slow.

Nevertheless, several biologics have been fused with AlbudAb™, a human domain antibody directed against HSA (GSK), to study the effect of extended circulation. These include interleukin 1 receptor antagonist (IL1-Ra) (Holt et al. 2008) and IFNα2b (Walker et al. 2010). After s.c. injection into rats, the plasma half-life of the AlbudAb-IFN was 28.3 h, more than 18-fold longer than that of the unfused IFN (1.5 h). Interestingly, direct fusion with HSA resulted in a shorter plasma half-life of just 19.7 h while the bioavailability and in vivo efficacy of the AlbudAb fusion was greater (Walker et al. 2010), probably due to steric hindrance and impairment of IFN folding upon direct fusion with albumin (as in the case of Albuferon). Exendin-4 fused to an AlbudAb (GSK2374697) has been investigated in a phase I study (Hodge et al. 2013). Apart from that, Ozoralizumab (ATN-103), a trivalent Nanobody® (Ablynx) consisting of two anti-TNF-α modules and one albumin-binding Nanobody, is currently under investigation in phase II clinical trials, evaluating once-monthly dosing (Elsadek and Kratz 2012; see also the preceding Chapter).

PEGylation and Chemical Coupling to Other Polymers

PEGylation, the conjugation of a drug with the synthetic polymer polyethylene glycol (PEG), currently constitutes one of the most successful strategies to develop biobetters with extended PK. The chemically inert, uncharged and highly water soluble PEG adopts a disordered structure in aqueous solution, and each ethylene glycol unit can interact via its polar ether oxygen as a hydrogen bond acceptor with up to two water molecules. As a consequence, the hydrodynamic molecular volume of the linked biopharmaceutical is increased above the glomerular pore size of the kidneys, which effectively retards excretion (Veronese and Pasut 2008).

Initially dubbed "pegnology" (Davis 2002), this strategy was devised to suppress the formation of antibodies against xenogenic therapeutic proteins (Abuchowski et al. 1977). Indeed, bovine adenosine deaminase coupled with multiple PEG chains of 5 kDa molecular weight (Davis et al. 1981) showed reduced immunogenicity and slower clearance in mice. Accordingly, Adagen® (Pegademase bovine; Enzon) was the first PEGylated protein that received approval by the FDA in 1990 for the treatment of severe combined immunodeficiency disease (SCID) (Booth and Gaspar 2009). This was quickly followed by likewise multi-PEGylated L-asparaginase (Abuchowski et al. 1984; Park et al. 1981) from *E. coli* (Pegaspargase, Oncaspar®) for treating acute lymphoblastic leukemia (ALL) in patients who are hypersensitive to the native unmodified form of the bacterial enzyme (Patel and Benfield 1996). While multiple attachments of short chain PEG molecules to a non-human protein seem to have a shielding effect on the immune system and to suppress formation of anti-drug antibodies (ADA) (Singh 2011), this has not been reported for products that were later developed utilizing a single modification with a long, sometimes branched PEG chain. Moreover, PEG itself appears to be immunogenic as will be discussed further below. Hence, the retarding effect of PEGylation on kidney elimination has become the focus of current applications.

Remarkably, this strategy has led to the approval of more than ten PEGylated proteins (Kolate et al. 2014). Half of them are biobetters of well-known first generation biologics, most notably interferons such as PEG-Intron®, Pegasys® and the recently approved PLEGRIDY™, where the longer duration of action and less frequent dosing provide clearly better convenience for patients. Moreover, a *meta*-analysis revealed a lower probability of serious adverse events for PLEGRIDY compared with other therapies (Tolley et al. 2014). PEGylated biologics are conjugated with either multiple short (5–10 kDa) PEG chains (Krystexxa™, Adagen®, Oncaspar®, Somavert®), single longer chains (12–30 kDa; Lonquex®, Neulasta™, PLEGRIDY™, PEG-Intron®, Mircera®) or branched 40 kDa PEG derivatives (Pegasys®, Cimzia®) using different coupling chemistries (Alconcel et al. 2011). Branched PEG derivatives are more expensive but exhibit lower viscosity and sometimes lead to longer plasma half-life than the linear polymer (Kling 2013).

Pegasys, for instance, is produced by random coupling of chemically activated branched 40 kDa PEG to Lys side chains of IFNα2a (Reddy et al. 2002), which is accompanied by a significantly reduced bioactivity (see below). On the other hand, Cimzia, a PEGylated Fab fragment of an anti-TNFα antibody, represents the second approved biological equipped with a branched 40 kDa PEG, in this case site-specifically coupled to a free Cys residue, resulting in a well-defined conjugate without loss in antigen affinity (Blick and Curran 2007). Cimzia has been approved for Crohn's disease, rheumatoid arthritis and psoriatic arthritis, offering the advantage of monthly dosing (Pasut 2014a). In this context, PEG attachment provides an additional benefit because the large hydrodynamic volume leads to accumulation in inflamed tissue due to the EPR effect (Veronese and Mero 2008). The missing CDC and ADCC in the absence of an Fc moiety result in an improved safety profile (Pasut 2014a).

Several further PEGylated products are currently in clinical trials, e.g., blood clotting factors such as PEGylated rFVIII (BAX855; Baxter) (Ehrlich et al. 2013), developed using a proprietary PEGylation technology (Nektar Therapeutics), and PEGylated B-domain-deleted recombinant factor VIII (BAY94-9027; Bayer) (Ivens et al. 2013), as well as several PEGylated versions of human growth hormone (hGH), including NNC126-0083 (Novo Nordisk) (de Schepper et al. 2011), ARX201 (Ambrx) and ACP001 (Ascendis). However, despite the success of PEGylation technology so far, drawbacks have emerged and the pharmaceutical industry has started to search for PEG alternatives. Generally, PEGylation is associated with high production costs, as clinical grade PEG is very expensive and chemical coupling requires additional processing and purification steps. Furthermore, the inherent polydispersity of the chemical polymer (Bagal et al. 2008), together with the additional product heterogeneity that arises from nonspecific chemical coupling, e.g., to Lys side chains (Foser et al. 2003b), hampers downstream processing, analytics and affects product performance (Kling 2013).

Notably, a substantial proportion of patients develop antibodies against the PEG moiety. In the case of KRYSTEXXA, a PEGylated uricase approved in 2010, 41 % of treated subjects developed anti-PEG antibodies in a phase III clinical trial (EMA 2013b). Also, for PEG-asparaginase (Armstrong et al. 2007) and PEGylated phenylalanine ammonia lyase (Kling 2013) the formation of anti-PEG antibodies has been reported. Recently, it was found that anti-PEG antibodies are detectable in 20–25 % of all healthy blood donors, which is probably caused by the increased exposure to PEG from cosmetics, pharmaceuticals and processed food products (Garay et al. 2012). Anti-PEG antibodies may adversely affect drug efficacy, not only in the course of primary treatment but also for other PEGylated drugs that may be administered to the same individual later on (Armstrong 2009; Verhoef et al. 2014). Based on the notion that the terminal methoxy group constitutes the major immunogen, hydroxy-PEG has been developed (Mountain View Therapeutics), which is expected to lead to a reduced ADA response (Saifer et al. 2014).

Another concern regarding PEGylated products is the decrease in binding activity that is usually observed for cytokines, hormones or other binding proteins, including small antibody fragments or alternative scaffolds. Generally, in vivo biological activity of PEGylated proteins depends on the PEG attachment site within the protein (Cho et al. 2011) as well as on the number, length and structure of conjugated PEG chains (Pfister and Morbidelli 2014). In the case of Pegasys, for example, only 7 % of the antiviral activity of the unmodified interferon was retained (Bailon et al. 2001). Even though the loss in pharmaceutical potency is compensated by the increased overall systemic exposure due to the prolonged circulation, indicated by the much enlarged area under the curve (AUC) in PK studies, substantial efforts have been made during the last 20 years to reduce adverse effects of PEG on bioactivity.

Several previously approved PEGylated products, including Pegasys and Mircera, are randomly modified with one long PEG chain, and the resulting mono-PEGylated products are purified and formulated as a mixture of positional isomers (Foser et al. 2003a). Nowadays, however, there is a clear trend toward site-specific PEGylation, which promises a more homogeneous product, less reduction in bioactivity and higher yield during purification. First attempts to couple PEG site-specifically to a biologic were made during the development of PEG-Intron. This conjugate is prepared by coupling of 12 kDa PEG-succinimidyl carbonate to IFNα2b at low pH, resulting predominantly (47 %) in a carbamate linkage to His34. This unstable conjugate avoids loss of activity by releasing native IFNα2b in the plasma over time (Alconcel et al. 2011). Compared to Pegasys, PEG-Intron shows a ca. fourfold higher in vitro activity (Veronese and Mero 2008). Today, site-specific conjugation is often accomplished by selective PEGylation at the N-terminal amino group of the polypeptide via reductive alkylation with PEG-aldehyde. For instance, Neulasta was prepared in this way using a 20 kDa linear PEG chain, resulting in a ca. 12-fold longer plasma half-life (42 h) compared to the first generation G-CSF, Neupogen (3.5 h) (Molineux 2004).

Alternatively, site-specific PEGylation can be achieved by modifying a single unpaired thiol side chain using maleimide coupling chemistry. However, as most therapeutic proteins carry structural disulfide bonds, the introduction of an additional Cys residue usually results in lower expression yields and less homogeneous product, due to formation of non-physiological disulfide isomers and/or adducts with thiol-containing metabolites from the host cell or culture medium. Thus, the identification of a viable drug candidate with retained binding activity as well as efficient folding properties requires considerable positional optimization, as demonstrated for G-CSF (Doherty et al. 2005; Rosendahl et al. 2005) or thyroid stimulating hormone (Qiu et al. 2013), for example.

A potentially more elegant approach for site specific PEGylation is the use of an exposed natural disulfide bridge. In the TheraPEG™ conjugation procedure (PolyTherics), a structurally accessible disulfide bond is first reduced, and the two contributing Cys side chains are then reconnected by bis-alkylation via a three-carbon bridge to which the PEG chain is attached in the middle. This mimics the geometry of a disulfide cross-link, thus maintaining the fold and function of the protein, as demonstrated for L-asparaginase and IFNα2b (Balan et al. 2007; Brocchini et al. 2008). Another technology, HiPEG™, allows site-specific PEGylation via an N- or C-terminal His_6-tag using PEG sulfone (Cong et al. 2012).

In a different approach, *N*- or *O*-glycosylation sites within a protein are targeted via glycoPEGylation. As natural sugar side chains often show low interference with target binding or folding stability of a protein, they provide useful conjugation sites. Recently, the first glycoPEGylated product was approved in Europe: Lonquex (TEVA) is based on recombinant G-CSF produced in *E. coli*, which is selectively glycosylated by in vitro treatment with a recombinant *O*-GalNAc-transferase followed by conjugation with a 20 kDa PEG-sialic acid derivative employing a sialyltransferase (DeFrees et al. 2006; Scheckermann et al. 2013). Lonquex possesses an approximately tenfold prolonged plasma half-life (28–34 h) compared to Neupogen. GlycoPEGylated factors VIIa (Ljung et al. 2013; Stennicke et al. 2008), VIII (NN7088) (Stennicke et al. 2013) and IX (NN7999) (Ostergaard et al. 2011) were developed using a slightly modified procedure (Novo Nordisk). First, the glycosylated blood clotting factors were produced in mammalian cell culture; then, the terminal sialic acid groups were removed using a sialidase in order to expose terminal galactose (Gal) sugars, to which a PEG-modified sialic acid was enzymatically attached.

The ReCODE™ technology (Ambrx) uses synthetic biotechnology with an expanded genetic code to introduce non-natural amino acids having "bioorthogonal" chemical reactivity—e.g. p-acetylphenylalanine (pAcF)—for subsequent site-specific PEGylation of proteins. To this end, a UAG (amber) stop codon is introduced at a predefined site within the structural gene. The non-natural amino acid is then precisely inserted at this position in the course of protein biosynthesis in a cell that co-expresses an engineered aminoacyl-tRNA synthetase with the required substrate specificity together with a cognate amber suppressor tRNA (Kim et al. 2013a). The purified recombinant gene product carrying the exposed keto side chain efficiently reacts with an aminooxy derivative of PEG by forming a covalent oxime. In this way, a mono-PEGylated hGH (ARX201), conjugated at residue 35 with a linear 30 kDa PEG chain, has been prepared. In GH-deficient adults, ARX201 demonstrated efficacy and safety comparable to therapy with native hGH while showing a plasma half-life of 89–102 h (Cho et al. 2011), which is ca. 30-fold longer than the one of the unmodified recombinant hGH, Genotropin® (~3.0 h).

An alternative to permanent PEGylation is TransCon (Ascendis), a transient PEG conjugation technology that takes advantage of hydrolyzable linkers when coupling PEG to the protein or peptide of interest. Depending on temperature and pH, both precisely controlled in the human body fluids, the linker undergoes autohydrolysis and slowly releases the unmodified, fully active protein (Hersel et al. 2009). A correspondingly modified hGH is under current investigation in several phase II clinical studies.

Beyond coupling chemistry, advances have been made with regard to novel shapes and lengths of the PEG molecule. The first approved PEG products Adagen and Oncaspar were randomly coupled via multiple Lys residues (modifying 60–70 % of all free side chains) (Davis et al. 1981; Park et al. 1981) to an activated derivative of linear low molecular weight (5 kDa) PEG. As the main goal initially was the reduction of immunogenicity, the greater shielding effect of multiple PEG chains outweighed a potential steric hindrance, which also played a lesser role for these enzymes that act on highly diffusible small molecule substrates. Later products used single linear PEGs of greater lengths or a 40 kDa branched PEG as discussed above, resulting in even more pronounced half-life extension. Generally, an increase in PEG length prolongs plasma half-life (Jevševar et al. 2010), but at the same time decreases binding activity for receptors or other macromolecular interaction partners (Caserman et al. 2009). Branched PEG can lead to longer circulation compared with linear PEG of the same size (Veronese et al. 1997) while, similarly, the loss of binding activity often is stronger. On the other hand, viscosity of branched PEG is much lower (Sim et al. 2012), which is of particular interest if a protein has to be administered at high concentration using thin needles, e.g., via s.c. injection. In a related approach to overcome viscosity issues, PolyPEG™ was designed as a multi-branched macromolecule comprising PEG 'teeth' stitched along a poly(methacrylate) backbone to form a comb-shaped structure (Kling 2013).

Novel PEGylation technologies primarily address the loss of bioactivity, the heterogeneity of the protein conjugate or its high viscosity; however, the main drawbacks of PEG, namely its immunogenicity and, most importantly, its lack of biodegradability, are not solved. Especially during chronic treatment and/or for drugs applied at high doses, the synthetic PEG polymer can accumulate in various organs. In a number of animal studies, PEG-rich vacuoles were observed in the liver or kidney (Bendele et al. 1998), in macrophages (Young et al. 2007) or even in ependymal cells of the choroid plexus in the brain (EMA 2010, 2012). Recently, the EMA sent a reminder on elevated precaution regarding the use of PEGylated drug products in the paediatric population (EMA 2012).

To tackle this problem, several biodegradable PEG substitutes have been proposed. Semi-synthetic polyacetals, e.g. poly(1-hydroxymethylethylene hydroxymethylformal), also called PHF or Fleximer® (Papisov et al. 2005), are biochemically inert, hydrophilic polymers and, thus, may provide an alternative to PEG. PHF is degradable due to the pH-sensitive acetal groups which are stable under physiological conditions, at pH 7–7.5, but are hydrolysed under the acidic conditions faced after endocytosis or phagocytosis (Yurkovetskiy et al. 2005). So far, the Fleximer technology (Mersana Therapeutics) has been mainly applied to small molecule oncology drugs such as the topoisomerase 1 inhibitor camptothecin (CTP) (XMT-1001) (Yurkovetskiy and Fram 2009) or an anti-angiogenic fumagillin analog (XMT-1107). In these instances, the drug is coupled to the Fleximer backbone via a customized linker and, as result of its chemical design, a slow and sustained release of the active compound is achieved, allowing higher drug levels to be dosed while minimizing off-target side effects (Yurkovetskiy and Fram 2009).

Fleximer conjugation not only improves solubility of the drug and decreases its toxicity, but also prolongs plasma half-life as shown in a phase I clinical study. Indeed, a CTP analog conjugated to a 60 kDa Fleximer, XMT-1001, showed an extended half-life of 4.8–12.3 h in humans (Sausville et al. 2009). In principle, the Fleximer can also be coupled to biologics such as antibody fragments or full-length antibodies, e.g. to develop antibody-drug conjugates (ADCs) that carry multiple toxin cargoes (Carlson 2012).

Currently emerging biobetters that are coupled to biodegradable polymers are based on polysaccharides, in particular polysialic acid (PSA), hydroxyethyl starch (HES) and heparosan. The conjugation of biologics to the naturally occurring, biodegradable $\alpha(2 \rightarrow 8)$ linked PSA to prolong circulation was proposed more than 20 years ago (Gregoriadis et al. 1993) and is marketed today as PolyXen® technology (Xenetic Biosciences). PSA is a linear, non-immunogenic polymer of *N*-acetylneuraminic acid naturally produced by mammalian and bacterial cells and present on cell surfaces as well as glycoproteins. In fact, bacteria decorated with PSA are protected against host phagocytosis and complement activation (Gregoriadis et al. 2005). However, one bottleneck is the pronounced polydispersity of PSA, which results from its complex bacterial biosynthesis, depending on several factors such as pH, growth rate and temperature (Zheng et al. 2013).

Nevertheless, Polysialylation has been successfully applied to a number of biologics to generate biobetters with prolonged plasma half-lives: α1-antitrypsin (A1AT) (Lindhout et al. 2011), oxyntomodulin (Vorobiev et al. 2013), L-asparaginase (Fernandes and Gregoriadis 2001), insulin (Jain et al. 2003; Zhang et al. 2010), G-CSF, IFNα2b, rFVIII, DNAse I and EPO (Bader and Wardwell 2014). L-Asparaginase from *Erwinia carotovora* chemically conjugated via its Lys side chains with multiple 10 kDa PSA chains showed a three- to fourfold greater terminal plasma half-life in mice compared to the native enzyme, while the modification reduced its antigenicity (Fernandes and Gregoriadis 2001). PSA is safe and well tolerated in humans as shown in a clinical phase I trial for SuliXen®, an insulin with a 14 kDa PSA attached to the α-amino group of the B chain (Zhang et al. 2010). ErepoXen®, a polysialylated EPO, is currently being tested in a phase II/III clinical trial (Xenetic Biosciences).

In addition to chemical coupling, PSA can be enzymatically attached in vitro to preexisting *N*-linked glycans on glycoproteins by first using *Campylobacter jejuni* Cst-II α2,8-sialyltransferase to form a primer, followed by polysialylation using the *Neisseria meningitidis* α2,8-polysialyltransferase (Lindhout et al. 2011). The range of the PSA chain length can be controlled by variation of the acceptor protein concentration in the polysialylation reaction. Using this approach, A1AT was modified with a PSA polymer of ca. 20 kDa, resulting in a terminal half-life of 27 h compared to 5 h for the unmodified protein and an 18-fold greater AUC (Lindhout et al. 2011).

HESylation®, the covalent coupling of a protein with the semi-synthetic, biodegradable and highly water soluble polymer hydroxyethyl starch (HES), is another strategy to increase the molecular size of a drug above the renal clearance threshold. The half-life of the HES polymer can be controlled by its rate of enzymatic degradation in plasma, depending on the extent of hydroxyethylation of the plant starch backbone (Jungheinrich and Neff 2005). So far, there are only a few published studies on proteins modified with HESylation technology (Fresenius Kabi). For instance, the interleukin-1 receptor antagonist (IL-1Ra) Kineret® was conjugated predominantly at its N-terminus with 85 kDa propionaldehyde-HES via reductive amination (Liebner et al. 2014). In surface plasmon resonance measurements binding activity towards the IL1 receptor was decreased from 50 to 320 pM, mainly due to a more than eightfold slower association rate compared to the native IL-1Ra. This is a common phenomenon, also seen for other polymers such as PEG, most likely due to a shielding effect and the lower diffusion coefficient of the conjugate with its enlarged molecular dimensions (Kubetzko et al. 2005). In PK studies in rats, the HESylated IL-1Ra showed a ca. 6.5-fold increase in plasma half-life and a 45-fold AUC compared to Kineret (Liebner et al. 2014).

To avoid potential detrimental effects on the protein upon chemical coupling, an enzymatic HES conjugation strategy using microbial transglutaminase has been developed as an alternative (Besheer et al. 2009). Although HES was initially considered to be non-toxic and safe in its primary use as a plasma volume expander (Brecher et al. 1997), this assumption has been challenged as clinical trials revealed an increased risk of death and renal impairment in septic patients treated with HES (Ghijselings and Rex 2014; Perel and Roberts 2007; Zarychanski et al. 2013). As a consequence, the Pharmacovigilance Risk Assessment Committee (PRAC) of the EMA recommended suspending marketing authorizations for infusion solutions containing hydroxyethyl starch (EMA 2013a) and the FDA also communicated a serious warning with respect to the use of HES as plasma expander (FDA 2013). Whether these safety concerns are also relevant for the HESylation technology in the area of biobetters, where the polymer is applied at a much lower concentration, remains to be seen.

Finally, another sugar-based biodegradable polymer is heparosan ($[GlcUA\text{-}1,4\text{-}GlcNAc\text{-}1,4]_n$), the unbranched, unsulfated, negatively charged polysaccharide precursor of heparin and heparan sulfate, two polysaccharides ubiquitously found in mammalian cells. Several pathogenic bacteria, e.g. *E. coli* K5 or *Pasteurella multocida* Type D, are decorated by heparosan to evade the host immune system (Chavaroche et al. 2013). Therefore, heparosan has been presumed to be non-immunogenic. Also, this polymer likely is biodegradable, as intracellular glucuronidase and hexosaminidase enzymes can efficiently hydrolyze heparosan into non-toxic monosaccharides. The lack of *O*-sulfation prevents both cleavage in plasma by the mammalian heparanase and clearance via binding to the hyaluronan receptor, which would lead to endocytosis. In contrast to heparin, heparosan does not affect clotting of human plasma (DeAngelis 2013). Intramuscular (i.m.) injection of a radioiodinated 100 kDa heparosan polymer in rats showed a plasma half-life of around 3 days and no signs of organ accumulation.

Heparosan is produced in a chemoenzymatic process from UDP-GlcUA and UDP-GlcNAc monomer precursors using PmHS1, a bacterial polymerizing enzyme called heparosan synthase. Since UDP sugars required for this synthesis are quite expensive, production costs are high (Chavaroche et al. 2013). The size of the polymer can be controlled by the concentration of the acceptor, a mixture of heparosan tetrasaccharide and hexasaccharide, and in this manner polymers up to 800 kDa with a polydispersity in the range of 1.06–1.18 can be produced (Sismey-Ragatz et al. 2007). Functionalized heparosan can be coupled to amino-, carbonyl-, or sulfhydryl groups of biologics. Although data on heparosan-conjugated biologics are not yet published, the corresponding HEPtune™ technology (Caisson) may evolve to become a biodegradable alternative for PEGylation in the future.

Apart from polysaccharides, poly-L-amino acids provide another class of quasi-natural and biodegradable polymers. A synthetic polymer of L-glutamic acid has been used to enhance solubility, to decrease toxicity and to achieve tumor targeting of the cytotoxic drug paclitaxel (Northfelt et al. 2014). Due to the large hydrodynamic volume of the polyanionic polymer conjugate, the drug cannot extravasate from blood vessels in normal tissue; however, it accumulates in tumors due to the EPR effect (Iyer et al. 2006). In human clinical studies, conjugation of paclitaxel to polyglutamate led to a 3- to 14-fold prolonged plasma half-life compared to the small molecule itself (Singer 2005). Paclitaxel poliglumex (Opaxio™, formerly known as XYOTAX™) is currently tested for various cancer types, including ovarian cancer, glioblastoma and head and neck cancer (Cell Therapeutics).

This technology was also proposed for the development of biobetters with less frequent dosing. For instance, G-CSF was genetically fused at its N-terminus with 175 Glu residues and IFNα2b was C-terminally fused with 84 Glu residues (see also next section). Both fusion proteins were successfully produced in a soluble state in *E. coli* and showed bioactivity in cell culture assays (Leung et al. 2002). However, no further data have been published.

Genetic Fusion with Recombinant Biopolymers

In the past few years, several approaches for the generation of biobetters were published that involve the genetic fusion of therapeutic proteins or peptides with specifically designed polypeptides, i.e., polymers of L-amino acids with defined sequences. This developing area is considered as one of the most important breakthroughs in biological drug development (Pasut 2014b), offering multiple advantages over the chemical coupling with synthetic polymers. In particular, two distinct technologies have emerged, PASylation® (Schlapschy et al. 2013) and XTEN™ (Schellenberger et al. 2009), which both employ recombinant polypeptides to mimic the size-enlarging effect of PEG for plasma half-life extension.

XTEN (Amunix) comprises unstructured, non-repetitive, charged and hydrophilic amino acid sequences (Schellenberger et al. 2009). These sequences were derived from an extensive screening process for a soluble, chemically stable and predominantly unstructured polypeptide that provides a maximal hydrodynamic radius to shield its protein payload. In this process, hydrophobic amino acids, which are prone to aggregation and may contribute to HLA/MHC-II mediated immune responses, were excluded. In addition, crosslinking Cys residues and positively charged amino acids, which may interact with cell membranes, as well as amide-containing amino acids, which are considered unstable during long-term storage, were avoided. Consequently, the remaining six amino acids P, E, S, T, A and G were chosen to create a library of random, non-repetitive amino acid segments.

Interestingly, these amino acids are quite similar to a deimmunized version of the natural sequence repeat found in the secreted *trans*-sialidase of *Trypanosoma cruzi* comprising the amino acids P, S, T, A, D. This parasitic protozoan appears to use repeat sequence extensions in order to gain prolonged life-time for its virulence factors in the host organism. In fact, such sequences were demonstrated to extend the plasma half-life if fused to the catalytic domain of trypanosomal *trans*-sialidase or to a rat tyrosine aminotransferase by a factor 4.5–6 in mice (Alvarez et al. 2004).

After initial screening of the XTEN amino acid library, highly expressed sequences were iteratively ligated and rescreened, finally yielding several different XTEN sequences, which are all characterized by a substantial amount of Glu residues, thus resulting in a high overall negative charge (Schellenberger et al. 2009). Meanwhile, a series of biologics were fused to XTEN, including exenatide (Schellenberger et al. 2009), glucagon (Geething et al. 2010), glucagon-like peptides 1 and 2 (Alters et al. 2012), annexin A5 (Haeckel et al. 2014) and hGH (Cleland et al. 2012).

The mammalian glucagon peptide sequence was C-terminally fused with XTEN sequences of 288, 144, 72, or 36 residues, and the observed terminal half-lives were found to correlate with the sequence length. Glucagon fused with the 288 residue XTEN showed the longest circulation, with 9 h in cynomolgus monkeys. As XTENylated glucagon was intended for the treatment of nocturnal hypoglycemia, the fusion with 144 residues having a shorter plasma half-life, termed Gcg-XTEN, better matched clinical requirements. Notably, in a cell culture assay, *E. coli* produced Gcg-XTEN only retained 15 % bioactivity compared to the unmodified peptide. Nevertheless, due to the optimized PK, Gcg-XTEN was effective in preventing hypoglycemia in fasted dogs without associated hyperglycemia as it would have been expected for glucagon (Geething et al. 2010).

The most advanced XTEN program is a long-acting hGH (VRS-317) fused both at its N-terminus with a 83.3 kDa XTEN sequence and at its C-terminus with a shorter 13.3 kDa XTEN peptide. The polyanionic XTEN changed the pI of the protein from 5.2 to 3 and led to a 12-fold lower in vitro potency in a cell culture assay. However, this turned out as an advantage since the lower affinity retarded receptor-mediated clearance and potentially minimized certain side effects such as lipoatrophy. In PK studies in cynomolgus monkeys plasma half-life after s.c. injection was extended to 110 h. Only a mild antibody response directed against hGH and not against the XTEN moiety was detected (Cleland et al. 2012).

Efficacy and safety of VRS-317 in pediatric patients with growth hormone deficiency was tested in a clinical phase II trial (Versartis). Results at the 6 month time point indicate that monthly administration of VRS-317 at a concentration of 5 mg/kg (daily rhGH mass equivalent of 30 μg/kg/day), achieved a similar height velocity compared to daily injections of conventional rhGH administered at 33 μg/kg/day. There were no signs of nodules or lipoatrophy and only mild adverse events. In addition, IGF-I levels were significantly increased and maintained during the dosing interval (Bright et al. 2014). XTEN polypeptides can also be applied for chemical conjugation to drugs, as recently exemplified with the antiviral peptide T20 (Ding et al. 2014) and the glucagon-like peptide 2 (Podust et al. 2013).

PASylation® is the conjugation of a drug with conformationally disordered, repetitive and uncharged polypeptides having defined sequences composed of the small L-amino acids Pro, Ala, and/or Ser (Schlapschy et al. 2013). This technology was developed based on previous studies with Gly-rich homo-amino acid polymers (HAP), intended to mimic the physical and hydrodynamic properties of PEG as closely as possible with a polymer made of simple uncharged amino acids (Schlapschy et al. 2007). In this earlier investigation, repeat sequences of the type $(Gly_4Ser)_n$ were found to form a stable random coil in solution, conferring an increased hydrodynamic volume to a recombinant Fab fragment upon fusion to the C-terminus of its light chain. However, such a Gly-rich sequence with up to 200 residues only led to moderate half-life extension while also showing an aggregation tendency under certain buffer conditions—in line with historical knowledge of the limited solubility of Gly-based poly-amino acids (Schlapschy et al. 2007).

Consequently, the more advanced PAS polypeptides were rationally designed to avoid any aggregation propensity and to adopt a maximally expanded hydrodynamic volume—proportional to the length of the PAS chain—while exhibiting an unstructured, uncharged and highly hydrophilic character. This was managed successfully by combination of the small amino acids Pro, Ala and Ser (Schlapschy et al. 2013). Whereas a homopolymer of each one of these amino acids is known to adopt a defined secondary structure, an appropriate mixture breaks regular hydrogen-bonding patterns and results in a stable random coil polypeptide with expanded molecular dimensions under physiological conditions. The well known *cis/trans* isomerization propensity at the N-terminal peptide bond of the interspersed Pro residues appears to further promote this behavior.

Similar to PEG, these biophysical features retard permeation through the filtration slits of the glomerular basement membrane in the kidney, thus prolonging circulation of a conjugated biopharmaceutical. In contrast to PEGylation, genetic fusion allows cheaper single-step production, obviating in vitro coupling or modification steps while expensive clinical-grade PEG material and additional bioprocessing procedures to separate unconjugated species are not required. Furthermore, PAS polypeptides are stable in the blood plasma, but are degraded by intracellular proteases into non-toxic, metabolizable L-amino acids, thus avoiding tissue accumulation.

PAS sequences are typically designed to be repetitive at the amino acid sequence level, e.g. comprising 12mer, 20mer or 24mer repeats. The repetitive nature of these PAS amino acid sequences guarantees uniform biophysical properties over the entire length of the polymer, which can range from 100 to 1,000 residues and beyond.

The use of biopolymers composed of Pro, Ala and/or Ser has several advantages for the design of biobetters with regard to bioactivity and manufacturing. PAS is highly hydrophilic but does not influence the isoelectric point (pI) of the fusion partner due to its complete lack of charged side chains. Consequently, diffusion into the negatively charged extracellular matrix, where target receptors of most biopharmaceuticals are located, is not hampered by electrostatic repulsion. The uncharged nature of these polypeptides also allows efficient translocation across biological membranes, both in bacteria, where secretion into the oxidizing milieu of the periplasm facilitates disulfide bond formation, and out of mammalian cells, resulting in high titers in the culture medium. Furthermore, endogenous biosynthesis of the small amino acids is feasible in most biotechnological production organisms and does not limit protein expression. Another advantage with regard to bioprocess development is the resistance to posttranslational modification, as well as to chemical degradation, due to the lack of reactive amino acid side chains, thus yielding uniform protein preparations usually characterized by a single peak in mass-spectrometric analysis (Schlapschy et al. 2013).

So far, PAS sequences with lengths up to 1,000 residues have been successfully fused to the N- or C-terminus, or to both termini, of more than 20 biologics, including hGH, GM-CSF, leptin, interferons (including engineered interferon superagonists and antagonists), diagnostic antibody fragments (suitable for in vivo imaging) or peptides such as a GLP-1 analog (Harari et al. 2014; Mendler et al. 2015; Morath et al., 2015; Schlapschy et al. 2013). Even bispecific molecules, where the PAS sequence serves as a linker between two functional modules, have been generated. In each case, the PASylated protein was directly produced in a soluble, fully active state (without refolding), showed a dramatically increased hydrodynamic volume and a much prolonged plasma half-life in animals while retaining high biological activity. Remarkably, in all animal studies so far, no antibody response against the PAS moiety was detected, even after active immunization (Schlapschy et al. 2013).

For instance, a PASylated IFN superagonist carrying 600 PAS residues to treat MOG_{35-55} peptide-induced experimental autoimmune encephalomyelitis (EAE), a mouse model of human multiple sclerosis (MS), did not induce antibodies upon repeated administration, and its biological efficacy remained unchanged after 21 days of treatment. The tenfold extended pharmacodynamic half-life led to improved disease protection compared to IFNβ, the standard of care drug, despite being injected with a fourfold less frequency and at an overall 16-fold lower dosage (Harari et al. 2014).

In the case of hGH, N-terminal fusion with a PAS sequence comprising 600 amino acids prolonged plasma half-life after i.v. injection in mice by 95-fold, from 2.8 min for the unmodified hormone to 265 min for PAS(600)-hGH. A similar terminal half-life and excellent bioavailability were evident after s.c. injection. In vitro receptor-binding activity measured via surface plasmon resonance was only marginally influenced by PAS fusion, with 780 versus 300 pM for the unmodified protein. The small difference in affinity was a consequence of a lower k_{on}, most likely due to a slight shielding effect of the polymer and/or slower diffusion of the enlarged protein, as already described for PEGylation further above. Pharmacodynamics of

PAS(600)-hGH was investigated in so-called "little" mice (Bellini and Bartolini 1993), an established mouse model having a lower level of endogenous growth hormone and responding to the human ortholog. S.c. injected PAS(600)-hGH led to a threefold increased net weight gain compared with unmodified hGH, even if applied at double the dosing interval (Schlapschy et al. 2013).

Multiple PK studies in mice, rats, dogs, swine and monkeys have demonstrated that PASylation is suited for intramuscular, subcutaneous and intravenous injection routes, showing fast and high bioavailability in each case. Furthermore, plasma half-life can be tuned by variation of the PAS length. Whereas for in vivo imaging Fab fragments with PAS sequences around 200 amino acids result in optimal tumor contrast (Mendler et al. 2015), PAS polypeptides with 600 to 1,000 residues are suitable for long-lasting biobetters with projected once or bi-weekly dosing in humans.

PASylation was also successfully applied to leptin, a satiety hormone which acts on receptors in the hypothalamus. From recent studies with PASylated leptin (as well as a PASylated leptin antagonist), it seems that PASylation does not hamper crossing of the blood–brain barrier via leptin-specific receptors (Morath et al., 2015). After a single i.p. injection of PAS(600)-leptin into C57BL/6 J mice, strongly increased and sustained hypothalamic STAT3 phosphorylation (13-fold AUC) was observed compared with the unmodified hormone. Furthermore, a reduction in food intake by 60 % and a loss in body weight of more than 10 %, which lasted for more than 5 days, were achieved, whereas the unmodified leptin (applied at the same molar dose) resulted in minute detectable effects for only 1 day after administration.

PASylated proteins can be produced at high yields in established expression systems including mammalian cells (CHO, HEK, COS), yeast (*P. pastoris*, *K. lactis*) and *E. coli*. Bacterial production of PASylated proteins is compatible with both cytoplasmic and periplasmic routes and even with secretion into the bacterial culture medium as demonstrated using ESETEC® technology (Wacker Biotech). Using this *E. coli* production system, high density fermentations of PASylated hGH and of a PASylated Fab fragment directed against an immunological disease target yielded multiple g/L of soluble, fully functional protein (Di Cesare et al. 2013).

An alternative to genetic fusion is chemical coupling with isolated PAS polymers. These substances can be produced in high yields in *E. coli*, and the only reactive N-terminal amino group can be specifically activated for regioselective coupling to small molecules, DNA/RNA aptamers, cyclic- or modified peptides and even to proteins. Available preclinical data and multiple in vivo proof of concept studies indicate that PASylation has the potential to become an ideal tool to generate biobetters with extended plasma half-lives and potentially less side effects than PEGylation while offering surprisingly similar biophysical properties. Several PASylated biobetters are currently under development (XL-protein).

Compared with PASylation and XTEN, the genetic fusion with Elastin-Like Polypeptides (ELPs) represents a different approach to prolonging the plasma half-life of biologics. ELPylation technology (Phasebio) uses artificially designed, repetitive amino acid sequences derived from a five residue motif found in tropoelastin, the natural precursor of elastin. Proteins or peptides are fused with repeating ELP building blocks, e.g., $(VPGXG)_n$, wherein the "guest residue" X may be any amino acid except Pro. The resulting fusion protein can be expressed in a soluble form in the cytoplasm of *E. coli* (Floss et al. 2010) or in plants (Phan and Conrad 2011). ELPs undergo reversible phase transition at a specific temperature, which is dependent on the type of guest residue, the length of the ELP moiety, the cargo protein and salt concentration. Above the transition temperature, the ELP collapses from an extended chain to a highly ordered hydrogen-bonded structure that reversibly forms microaggregates. These insoluble ELP conjugates can be precipitated by centrifugation, thereby facilitating purification (Floss et al. 2010).

Importantly, ELPs can be designed to remain soluble at room temperature but to form microaggregates at body temperature. This leads to the formation of a gel-like depot after s.c. injection, which significantly prolongs the in vivo half-life of the fused biologic (Betre et al. 2006; Shamji et al. 2007). For instance, in a phase IIa clinical study, a 636 amino acid polypeptide comprising GLP-1 genetically fused to numerous ELP repeats significantly reduced mean average glucose levels in humans over a 7 day period with minimal loss of glycemic control (PB1023; Phasebio). PB1023 was well tolerated and only one patient developed low titer non-neutralizing antibodies. These data suggest a once weekly treatment for hyperglycemia in patients with type 2 diabetes (Christiansen et al. 2012). Other biologics such as Vasomera™ (del Rio et al. 2013), a novel long-acting VPAC2-selective, vasoactive intestinal peptide agonist, or PE0139, a basal insulin, are currently being evaluated in clinical phase I.

Bolstering Protein Glycosylation

Apart from the in vitro conjugation with polysaccharides mentioned further above, it is well known that natural glycosylation of therapeutic proteins can have a stabilizing effect in plasma by reducing susceptibility to proteolysis and increasing the hydrodynamic volume and/or negative charge, thus prolonging in vivo half-life (Sinclair and Elliott 2005). Consequently, boosting the number or size of *N*- or *O*-linked sugar side chains in recombinant proteins produced in eukaryotic host cells—which provide the necessary biochemical glycosylation machinery, if necessary, after appropriate genetic engineering (Lepenies and Seeberger 2014)—has emerged as another means to create biobetters.

This technology reached the market in 2010 with the FDA approval of the "hyperglycosylated" EPO derivative Aranesp® (Amgen), which was created by exchanging five amino acid side chains leading to two additional (in total five) *N*-glycosylation sites. Whereas the additional glycosylation decreased receptor affinity 4.3-fold, probably as a result of steric shielding, in vivo activity was overall increased due to the longer circulation (Egrie and Browne 2001, 2002). In human clinical trials the mean terminal half-life was 26.3 h, approximately threefold longer than for the conventional biologic Epoetin alfa (8.5 h), thus allowing once weekly dosing (Macdougall et al. 1999).

Nevertheless, hyperglycosylation is a complex process, as carbohydrate addition not only depends on the presence of a glycosylation consensus sequence (the so-called sequon), but also on protein folding and local conformation as well as accessibility to glycosyltransferases during biosynthesis. Both the introduction of changes into the amino acid sequence and the attachment of new oligosaccharide side chains (with their high steric demand and, potentially, charge) can adversely affect receptor binding, recombinant protein folding or immunological tolerance. Therefore, an extensive screening for mutants with the desired properties is required.

NexP™ technology (Alteogen) offers an alternative and faster approach to direct hyperglycosylation of biologics. By fusion of a therapeutic protein or peptide to an already hyperglycosylated long-acting α1-antitrypsin mutant, the plasma half-life of the fusion partner can be significantly extended (Chung et al. 2010). However, an even more elegant way is the use of a small unstructured glycosylated peptide as fusion partner, instead of a large globular glycoprotein that may influence folding efficiency or binding activity of the pharmaceutically active component. In this regard, an arabinoglycan module was proposed to create biobetters with prolonged circulation. Such glycomodules are found in hydroxyproline (Hyp)-rich glycoproteins (HRGPs) of plants and green algae which contain Hyp-rich clusters that are highly *O*-glycosylated. An X-Hyp-X-Hyp (X = Ser or Ala) repeat motif has been identified that promotes addition of long *O*-linked arabinogalactan chains to the hydroxyl group of Hyp if expressed in tobacco cells (Tan et al. 2003).

When genetically fused with a glycomodule consisting of multiple Ser-Hyp repeats, therapeutic proteins such as human IFNα2b (Xu et al. 2007) and hGH (Xu et al. 2010) not only showed higher yield in tobacco cell culture, but also a prolonged plasma half-life in mice. For instance, secretion of hGH C-terminally fused with 10 Ser-Hyp repeats into the medium yielded a 500-fold higher level than wild type hGH. hGH-$(\text{Ser-Hyp})_{10}$ comprised relatively few (~8) glycan compositional isomers, with an arabinogalactan carrying in total 25 sugar residues *O*-linked to Hyp as the predominant species. Plasma half-life in mice reached 2.5 h in this case, which is six times longer than for unmodified hGH (roughly 25 min) (Xu et al. 2010). Unfortunately, this technology is restricted to plant expression systems as Hyp incorporation into the recombinant polypeptide is a prerequisite for this type of glycosylation. Furthermore, caution has to be taken with regard to potential immunogenicity of the non-mammalian oligosaccharide, also considering that β-arabinosyl-Hyp is present on pollen allergens (Leonard et al. 2005).

GlycoPolymer technology (Aequus Biopharma) involves the genetic fusion of a therapeutic protein or peptide with repeating units of the human *N*-glycosylation sequon Asn-Asn-Thr (NNT) or Asn-Asn-Ser (NNS), which results in multiple additional *N*-linked glycans on the Asn side chains upon expression in mammalian cell culture. The bulky, solvated carbohydrate structures increase the hydrodynamic volume of the biologic, thus leading to prolonged PK. The lead product, AQB-101, a human G-CSF C-terminally fused with 51 NNT repeats, has been successfully produced in CHO cells. In preclinical studies the plasma half-life of this protein was similar to PEGylated recombinant hG-CSF, Neulasta. When a single s.c. injection of AQB-101 was compared to an equimolar amount of Neulasta a similar increase in neutrophil counts was observed, although the duration of the AQB-101-induced response was slightly shorter.

The first biobetter that makes use of fusion with a natural human glycosylation moiety and has received market approval in Europe in 2010 is Elonva®, a long-acting follicle stimulating hormone (FSH) for the treatment of female infertility. Elonva was generated by fusing FSH at the C-terminus of its β-subunit with the 28 amino acid C-terminal peptide (CTP) of the human chorionic gonadotropin (hCG) β-subunit (Fares et al. 1992), which carries four *O*-linked glycosylation sites (Sugahara et al. 1996). Peptide extension with the additionally negatively charged, sialylated oligosaccharides prolonged the elimination half-life of this chimeric FSH to 60–75 h in healthy pituitary-suppressed females, which is approximately twice as long as that of rFSH (Duijkers et al. 2002). A formulation allowing once weekly dosing of this biobetter led to an equally effective ovarian response in terms of the number of growing follicles and number of eggs compared to conventional daily FSH injection (de Lartigue 2011; Ledger et al. 2011).

Further CTP fusions with thyrotropin (Joshi et al. 1995), hGH (Fares et al. 2010) and EPO (Fares et al. 2011) demonstrated that the attachment of this module does not affect secretion in mammalian cells, receptor affinity or bioactivity in vitro. To increase the moderate plasma half-life extending effect of a single module, multiple CTPs have also been used. In this manner, a long-acting EPO was created by fusing one and two CTPs, respectively, to the N- and C-termini of the cytokine. The circulatory half-life of this EPO-$(CTP)_3$ after i.v. injection was 13.1 h, which is roughly threefold longer than the one of rhEPO and even slightly longer than that of the hyperglycosylated EPO derivative Aranesp (10.8 h). Moreover, EPO-$(CTP)_3$ was more effective than both rhEPO and Aranesp in increasing reticulocyte number in mouse blood (Fares et al. 2011). Clinical development is underway for several CTP fusion proteins, including factor VIIa-CTP in the area of hemophilia, oxyntomodulin-CTP (MOD-6030) for the treatment of obesity and a once-weekly dosed hGH-CTP for use in growth hormone deficiency, which is currently subject to a phase III trial (OPKO Health).

Conclusions and Outlook

During the last two decades numerous biopharmaceuticals have reached the market, among those a growing number of biobetters with extended plasma half-life. These long-lasting biologics have led to a new generation of drugs, promising improved efficacy as well as patient compliance. Consequently, implementation of prolonged action in vivo is no longer considered just an opportunity for the life cycle management of existing biologics but has become a crucial aspect also for the development of innovative biopharmaceuticals (Leader et al. 2008). Following increasing demand by the pharmaceutical industry, multiple half-life extension technologies have been devised, mostly by biotech companies or in an academic environment. Their underlying principle is either to expand the hydrodynamic molecular volume of a biological drug, to form a reversible depot, or to make use of endosomal recycling via FcRn interaction.

For the latter strategy, the most successful approach is the Fc fusion technology, with nine approved products so far. Notably, among these, only two long-acting blood clotting factors actually represent improvements of previously approved therapeutic proteins, whereas all other Fc fusions received regulatory approval as new biological entities (mostly derived from cell surface receptor domains). One explanation for this situation may be the dimeric nature of the Fc moiety, which is compatible with the structure of homo-dimeric receptors but interferes with the typical monomeric state of conventional therapeutic proteins such as cytokines, hormones or growth factors. Indeed, the bivalency may cause unwanted side effects due to target receptor clustering, leading to agonistic signalling. Consequently, a new generation of monomeric Fc fusions has entered the market with the approval of Eloctate and Aprolix, the two blood-clotting factor fusion proteins mentioned above. However, the biotechnological production of these novel monomeric Fc fusion proteins is clearly more complex.

Other concerns may be unwanted immunological side effects caused by the Fc moiety, which constitutes the effector region of Igs, and the frequent loss in bioactivity of the pharmaceutically active component as part of an Fc fusion protein. Novel technologies such as LAPSCOVERY or hyFc may partially solve this problem. Another aspect is that plasma half-life extension via Fc fusion may just be too long for certain medical indications—with the potential to cause adverse events. One should keep in mind that strong PK prolongation for short acting factors is highly unphysiologic and may have unknown long term consequences. Just as one example, hGH has a pulsatile secretion naturally and it cannot be fully excluded that off-target effects might be accrued from long acting hGH with consequent induction of IGFs which may be linked to cancer (Yu and Rohan 2000).

Nevertheless, adjusting—or even further extending—the in vivo life-time by altering affinity to FcRn is possible, as exemplified with the Xtend technology (although merely applied to full-length antibodies up to now) and as also demonstrated with the engineered albumin fusion technology Veltis. On the other hand, approaches like CTP fusion or hyperglycosylation rather achieve moderate prolongation of circulation per se. In comparison, the most promising strategy for a precisely tuned plasma half-life is based on polymer conjugation, in particular PEGylation technology, but also its innovative biopolymer alternatives PASylation or XTEN (Binder and Skerra 2012). In these cases, the in vivo life-time of a drug can be tailored in a wide range by adjusting the length of the polymer.

All glycosylation technologies and most of the Fc fusion technologies (except for LAPSCOVERY and PEPTIBODIES) are limited to mammalian expression systems. Likewise, albumin with its 17 disulfide bridges necessitates yeast or mammalian expression hosts while Hyp-rich protein fusions even require expression in plant cells. In contrast, polymer conjugation technologies are not restricted to a specific expression system. For PEGylation, HESylation, Sialylation or HepTune, the protein or peptide of interest is first produced in the host cell of choice and is then conjugated to the polymer in a second step. Of course, this in vitro modification causes additional effort for the chemical or enzymatic coupling procedure itself and for subsequent purification. However, this can be avoided by using genetic fusion with XTEN or PAS polypeptides. For instance, PASylated proteins can be produced efficiently in one step in *E. coli*, yeast, or typical mammalian cells such as CHO; this offers freedom to choose the optimal host cell according to the requirements of the pharmaceutically active protein with regard to disulfide bridge formation or other kinds of posttranslational modifications. As PAS sequences are uncharged, they are compatible with high yield secretion across cellular membranes into the bacterial periplasm or into the culture medium, which is a clear benefit compared to other technologies. Genetic fusion with unstructured polypeptide sequences also has advantages for GMP production and product analytics as the corresponding fusion proteins are usually monodisperse. This is in clear contrast to all synthetic polymers, in particular PEG or polysaccharide-based polymers like HES, heparosan or polysialic acid, which are inherently polydisperse.

At present, PEGylation still represents the gold standard for plasma half-life extension with more than ten approved products, including several drugs with annual sales of more than 1 billion US dollars, which provides a successful track record. Novel developments in this area such as site-specific coupling technologies, glycoPEGylation, branched PEG derivatives or transient PEGylation will yield future biobetters. However, these advanced PEGylation strategies do not solve a fundamental problem of this synthetic polymer, which is its lack of biodegradability. To address this issue, novel biopolymers based either on polysaccharides (HES, polysialic acid, heparosan) or on polypeptides (XTEN, PAS) have been developed and should allow, in principle, to profit from the beneficial macromolecular features of PEG while avoiding tissue and organ accumulation. To the extent data are available so far, PASylation appears to be the closest mimic of PEGylation with regard to biophysical characteristics, since representing an uncharged, highly solvated and flexible random chain polymer.

For comparison, many other conjugation technologies, based e.g. on albumin, heparosan, polyglutamate, XTEN or the Fleximer polymers, involve multiple anionic charges and significantly lower the pI of the biologic. While this electrostatic effect may additionally retard renal clearance due to repulsion from the likewise negatively charged glomerular basement membrane, endothelial surfaces of blood vessels as well as many cell surfaces are negatively charged, too, which could hamper efficient extravasation and receptor targeting.

Taken together, facing the looming patent cliff for first generation biologics and the need for more effective and convenient biopharmaceuticals, biobetters with extended plasma half-life offer a lucrative market opportunity with a much lower attrition rate compared with the original development of new biological entities. Beyond the two classical approaches, Fc fusion and PEGylation, multiple new technologies have become available. Although these are still novel platforms, their promise of cost effectiveness, enhanced bioactivity and high patient compliance render them of great interest to the pharmaceutical industry.

References

Abuchowski A, van Es T, Palczuk NC, Davis FF (1977) Alteration of immunological properties of bovine serum albumin by covalent attachment of polyethylene glycol. J Biol Chem 252:3578–3581

Abuchowski A, Kazo GM, Verhoest CR Jr, Van Es T, Kafkewitz D, Nucci ML, Viau AT, Davis FF (1984) Cancer therapy with chemically modified enzymes. I. Antitumor properties of polyethylene glycol-asparaginase conjugates. Cancer Biochem Biophys 7:175–186

Akilesh S, Huber TB, Wu H, Wang G, Hartleben B, Kopp JB, Miner JH, Roopenian DC, Unanue ER, Shaw AS (2008) Podocytes use FcRn to clear IgG from the glomerular basement membrane. Proc Natl Acad Sci U S A 105:967–972

Alconcel SNS, Baas AS, Maynard HD (2011) FDA-approved poly(ethylene glycol)–protein conjugate drugs. Polym Chem 2:1142–1148

Alters SE, McLaughlin B, Spink B, Lachinyan T, Wang CW, Podust V, Schellenberger V, Stemmer WP (2012) GLP2-2G-XTEN: a pharmaceutical protein with improved serum half-life and efficacy in a rat Crohn's disease model. PLoS One 7:e50630

Alvarez P, Buscaglia CA, Campetella O (2004) Improving protein pharmacokinetics by genetic fusion to simple amino acid sequences. J Biol Chem 279:3375–3381

Andersen JT, Pehrson R, Tolmachev V, Daba MB, Abrahmsen L, Ekblad C (2011) Extending half-life by indirect targeting of the neonatal Fc receptor (FcRn) using a minimal albumin binding domain. J Biol Chem 286:5234–5241

Andersen JT, Dalhus B, Cameron J, Daba MB, Plumridge A, Evans L, Brennan SO, Gunnarsen KS, Bjoras M, Sleep D, Sandlie I (2012) Structure-based mutagenesis reveals the albumin-binding site of the neonatal Fc receptor. Nat Commun 3:610

Andersen JT, Dalhus B, Viuff D, Ravn BT, Gunnarsen KS, Plumridge A, Bunting K, Antunes F, Williamson R, Athwal S, Allan E, Evans L, Bjoras M, Kjaerulff S, Sleep D, Sandlie I, Cameron J (2014) Extending serum half-life of albumin by engineering neonatal Fc receptor (FcRn) binding. J Biol Chem 289:13492–13502

Armstrong JK (2009) The occurrence, induction, specificity and potential effect of antibodies against poly(ethylene glycol). In: Veronese FM (ed) PEGylated protein drugs: basic science and clinical applications. Birkhäuser, Basel

Armstrong JK, Hempel G, Koling S, Chan LS, Fisher T, Meiselman HJ, Garratty G (2007) Antibody against poly(ethylene glycol) adversely affects PEG-asparaginase therapy in acute lymphoblastic leukemia patients. Cancer 110:103–111

Bader RA, Wardwell PR (2014) Polysialic acid: overcoming the hurdles of drug delivery. Ther Deliv 5:235–237

Bagal D, Zhang H, Schnier PD (2008) Gas-phase proton-transfer chemistry coupled with TOF mass spectrometry and ion mobility-MS for the facile analysis of poly(ethylene glycols) and PEGylated polypeptide conjugates. Anal Chem 80:2408–2418

Bailon P, Palleroni A, Schaffer CA, Spence CL, Fung WJ, Porter JE, Ehrlich GK, Pan W, Xu ZX, Modi MW, Farid A, Berthold W, Graves M (2001) Rational design of a potent, long-lasting form of interferon: a 40 kDa branched polyethylene glycol-conjugated interferon alpha-2a for the treatment of hepatitis C. Bioconjug Chem 12:195–202

Balan S, Choi JW, Godwin A, Teo I, Laborde CM, Heidelberger S, Zloh M, Shaunak S, Brocchini S (2007) Site-specific PEGylation of protein disulfide bonds using a three-carbon bridge. Bioconjug Chem 18:61–76

Bao W, Holt LJ, Prince RD, Jones GX, Aravindhan K, Szapacs M, Barbour AM, Jolivette LJ, Lepore JJ, Willette RN, DeAngelis E, Jucker BM (2013) Novel fusion of GLP-1 with a domain antibody to serum albumin prolongs protection against myocardial ischemia/reperfusion injury in the rat. Cardiovasc Diabetol 12:148

Bartalena L, Robbins J (1993) Thyroid hormone transport proteins. Clin Lab Med 13:583–598

Bellini MH, Bartolini P (1993) In vivo bioassay for the potency determination of human growth hormone in dwarf "little" mice. Endocrinology 132:2051–2055

Bendele A, Seely J, Richey C, Sennello G, Shopp G (1998) Short communication: renal tubular vacuolation in animals treated with polyethylene-glycol-conjugated proteins. Toxicol Sci 42:152–157

Besheer A, Hertel TC, Kressler J, Mader K, Pietzsch M (2009) Enzymatically catalyzed HES conjugation using microbial transglutaminase: proof of feasibility. J Pharm Sci 98:4420–4428

Betre H, Liu W, Zalutsky MR, Chilkoti A, Kraus VB, Setton LA (2006) A thermally responsive biopolymer for intra-articular drug delivery. J Control Release 115:175–182

Binder U, Skerra A (2012) Half-life extension of therapeutic proteins via genetic fusion to recombinant PEG mimetics. In: Kontermann R (ed) Therapeutic proteins – strategies to modulate their plasma half-lives. Wiley-VCH, Weinheim

Binz HK, Gulotti-Georgieva M, Merz FW, Phillips D, Sonderegger I, Steiner D (2012) Designed repeat proteins binding to serum albumin. Patent Publication WO2012069654A1

Blick SK, Curran MP (2007) Certolizumab pegol: in Crohn's disease. BioDrugs 21:195–201

Booth C, Gaspar HB (2009) Pegademase bovine (PEG-ADA) for the treatment of infants and children with severe combined immunodeficiency (SCID). Biologics 3:349–358

Borvak J, Richardson J, Medesan C, Antohe F, Radu C, Simionescu M, Ghetie V, Ward ES (1998) Functional expression of the MHC class I-related receptor, FcRn, in endothelial cells of mice. Int Immunol 10:1289–1298

Brecher ME, Owen HG, Bandarenko N (1997) Alternatives to albumin: starch replacement for plasma exchange. J Clin Apher 12:146–153

Bright GM, Moore WV, Nguyen HQ, Kletter GB, Miller BS, Rogers DG, Humphriss E, Cleland JL (2014) Safety and efficacy results of a 6 month, randomized, multi-center trial of a novel long-acting rhGH (VRS-317) in naive to treatment, pre-pubertal children with growth hormone deficiency (GHD). Abstract No. MON-0147; 16th international congress of endocrinology, June 21–24. Chicago, IL

Brocchini S, Godwin A, Balan S, Choi JW, Zloh M, Shaunak S (2008) Disulfide bridge based PEGylation of proteins. Adv Drug Deliv Rev 60:3–12

Bugelski PJ, Capocasale RJ, Makropoulos D, Marshall D, Fisher PW, Lu J, Achuthanandam R, Spinka-Doms T, Kwok D, Graden D, Volk A, Nesspor T, James IE, Huang C (2008) CNTO 530: molecular pharmacology in human UT-7EPO cells and pharmacokinetics and pharmacodynamics in mice. J Biotechnol 134:171–180

Bush MA, Matthews JE, De Boever EH, Dobbins RL, Hodge RJ, Walker SE, Holland MC, Gutierrez M, Stewart MW (2009) Safety, tolerability, pharmacodynamics and pharmacokinetics of albiglutide, a long-acting glucagon-like peptide-1 mimetic, in healthy subjects. Diabetes Obes Metab 11:498–505

Capon DJ, Chamow SM, Mordenti J, Marsters SA, Gregory T, Mitsuya H, Byrn RA, Lucas C, Wurm FM, Groopman JE, Broder S, Smith DH (1989) Designing CD4 immunoadhesins for AIDS therapy. Nature 337:525–531

Carlson B (2012) Antibody-drug conjugates: were the action is: ADCs – the new frontier. Biotechnol Healthc 9:28–31

Caserman S, Kusterle M, Kunstelj M, Milunovic T, Schiefermeier M, Jevševar S, Porekar VG (2009) Correlations between in vitro potency of polyethylene glycol-protein conjugates and their chromatographic behavior. Anal Biochem 389:27–31

Chavaroche AA, van den Broek LA, Eggink G (2013) Production methods for heparosan, a precursor of heparin and heparan sulfate. Carbohydr Polym 93:38–47

Cho H, Daniel T, Buechler YJ, Litzinger DC, Maio Z, Putnam AM, Kraynov VS, Sim BC, Bussell S, Javahishvili T, Kaphle S, Viramontes G, Ong M, Chu S, Becky GC, Lieu R, Knudsen N, Castiglioni P, Norman TC, Axelrod DW, Hoffman AR, Schultz PG, DiMarchi RD, Kimmel BE (2011) Optimized clinical performance of growth hormone with an expanded genetic code. Proc Natl Acad Sci U S A 108:9060–9065

Chong H, Yao X, Zhang C, Cai L, Cui S, Wang Y, He Y (2012) Biophysical property and broad anti-HIV activity of albuvirtide, a 3-maleimimidopropionic acid-modified peptide fusion inhibitor. PLoS One 7:e32599

Christiansen M, Matson M, Braz R, Georgopoulos L, Arnold S, Kramer W, Shi L, Strange P (2012) Weekly subcutaneous doses of glymera (PB1023) a novel GLP-1 analogue reduce glucose exposure dose-dependently. Abstract No. 946-P; 72nd scientific sessions of American Diabetes Association, June 8–12. Philadelpia, PA

Chung HS, Yoo SB, Lee SM (2010) In vivo half life increased fusion protein or peptide maintained by sustained in vivo release, and method for increasing in vivo half-life using same. Patent Publication WO2010123290A2

Cines DB, Yasothan U, Kirkpatrick P (2008) Romiplostim. Nat Rev Drug Discov 7:887–888

Cleland JL, Geething NC, Moore JA, Rogers BC, Spink BJ, Wang CW, Alters SE, Stemmer WP, Schellenberger V (2012) A novel long-acting human growth hormone fusion protein (VRS-317): enhanced in vivo potency and half-life. J Pharm Sci 101:2744–2754

Cong Y, Pawlisz E, Bryant P, Balan S, Laurine E, Tommasi R, Singh R, Dubey S, Peciak K, Bird M, Sivasankar A, Swierkosz J, Muroni M, Heidelberger S, Farys M, Khayrzad F, Edwards J, Badescu G, Hodgson I, Heise C, Somavarapu S, Liddell J, Powell K, Zloh M, Choi JW, Godwin A, Brocchini S (2012) Site-specific PEGylation at histidine tags. Bioconjug Chem 23:248–263

Czajkowsky DM, Hu J, Shao Z, Pleass RJ (2012) Fc-fusion proteins: new developments and future perspectives. EMBO Mol Med 4:1015–1028

Davis FF (2002) The origin of pegnology. Adv Drug Deliv Rev 54:457–458

Davis S, Abuchowski A, Park YK, Davis FF (1981) Alteration of the circulating life and antigenic properties of bovine adenosine deaminase in mice by attachment of polyethylene glycol. Clin Exp Immunol 46:649–652

de Lartigue J (2011) Corifollitropin alfa: a new option to treat female infertility. Drugs Today 47:583–590

de Schepper J, Rasmussen MH, Gucev Z, Eliakim A, Battelino T (2011) Long-acting pegylated human GH in children with GH deficiency: a single-dose, dose-escalation trial investigating safety, tolerability, pharmacokinetics and pharmacodynamics. Eur J Endocrinol 165:401–409

Deacon CF (2009) Potential of liraglutide in the treatment of patients with type 2 diabetes. Vasc Health Risk Manag 5:199–211

DeAngelis PL (2013) HEPtune: a process of conjugating a naturally occurring sugar molecule, heparosan, to a drug for enhanced drug delivery. Drug Dev Deliv 13:34–38

DeFrees S, Wang ZG, Xing R, Scott AE, Wang J, Zopf D, Gouty DL, Sjoberg ER, Panneerselvam K, Brinkman-Van der Linden EC, Bayer RJ, Tarp MA, Clausen H (2006) GlycoPEGylation of recombinant therapeutic proteins produced in *Escherichia coli*. Glycobiology 16:833–843

del Rio CL, George R, Kloepfer P, Ueyama Y, Youngblood B, Georgopoulos L, Arnold S, Hamlin RL (2013) Vasomera™, a novel VPAC 2-selective vasoactive intestinal peptide agonist, enhances contractility and decreases myocardial demand in dogs with both normal hearts and with pacing-induced dilated cardiomyopathy. J Am Coll Cardiol 61:E645

Dennis MS, Zhang M, Meng YG, Kadkhodayan M, Kirchhofer D, Combs D, Damico LA (2002) Albumin binding as a general strategy for improving the pharmacokinetics of proteins. J Biol Chem 277:35035–35043

Di Cesare S, Binder U, Maier T, Skerra A (2013) High-yield production of PASylated human growth hormone using secretory *E. coli* technology. Bioprocess Int 11:30–38

Ding S, Song M, Sim BC, Gu C, Podust VN, Wang CW, McLaughlin B, Shah TP, Lax R, Gast R, Sharan R, Vasek A, Hartman MA, Deniston C, Srinivas P, Schellenberger V (2014) Multivalent antiviral XTEN-peptide conjugates with long in vivo half-life and enhanced solubility. Bioconjug Chem 25:1351–1359

Doherty DH, Rosendahl MS, Smith DJ, Hughes JM, Chlipala EA, Cox GN (2005) Site-specific PEGylation of engineered cysteine analogues of recombinant human granulocyte-macrophage colony-stimulating factor. Bioconjug Chem 16:1291–1298

Duijkers IJ, Klipping C, Boerrigter PJ, Machielsen CS, De Bie JJ, Voortman G (2002) Single dose pharmacokinetics and effects on follicular growth and serum hormones of a long-acting recombinant FSH preparation (FSH-CTP) in healthy pituitary-suppressed females. Hum Reprod 17:1987–1993

Dumont JA, Low SC, Peters RT, Bitonti AJ (2006) Monomeric Fc fusions: impact on pharmacokinetic and biological activity of protein therapeutics. BioDrugs 20:151–160

Egrie JC, Browne JK (2001) Development and characterization of novel erythropoiesis stimulating protein (NESP). Br J Cancer 84:3–10

Egrie JC, Browne JK (2002) Development and characterization of darbepoetin alfa. Oncology 16:13–22

Ehrlich HJ, Wong WY, Ewenstein BM, Dockal M, Turecek PL, Gringeri A, Chehadeh H, Low-Baselli A, Scheiflinger F, Reininger AJ (2013) Development of novel treatment options for patients with haemophilia. Hamostaseologie 33(Suppl 1):S36–S38

Elsadek B, Kratz F (2012) Impact of albumin on drug delivery – new applications on the horizon. J Control Release 157:4–28

Elvin JG, Couston RG, van der Walle CF (2013) Therapeutic antibodies: market considerations, disease targets and bioprocessing. Int J Pharm 440:83–98

EMA (2010) European Public Assessment Report (EPAR) Cimzia®. http://www.ema.europa.eu/docs/en_GB/document_library/EPAR_-_Product_Information/human/001037/WC500069763.pdf

EMA (2012) CHMP safety working party's response to the PDCO regarding the use of PEGylated drug products in the paediatric population. http://www.ema.europa.eu/docs/en_GB/document_library/Scientific_guideline/2012/11/WC500135123.pdf

EMA (2013a) PRAC recommends suspending marketing authorisations for infusion solutions containing hydroxyethyl starch. http://www.ema.europa.eu/ema/index.jsp?curl=pages/news_and_events/news/2013/06/news_detail_001814.jsp&mid=WC0b01ac058004d5c1

EMA (2013b) Krystexxa: summary of product characteristics. http://www.ema.europa.eu/docs/en_GB/document_library/EPAR_-_Product_Information/human/002208/WC500138318.pdf

Fares FA, Suganuma N, Nishimori K, LaPolt PS, Hsueh AJ, Boime I (1992) Design of a long-acting follitropin agonist by fusing the C-terminal sequence of the chorionic gonadotropin beta subunit to the follitropin beta subunit. Proc Natl Acad Sci U S A 89:4304–4308

Fares F, Guy R, Bar-Ilan A, Felikman Y, Fima E (2010) Designing a long-acting human growth hormone (hGH) by fusing the carboxyl-terminal peptide of human chorionic gonadotropin beta-subunit to the coding sequence of hGH. Endocrinology 151:4410–4417

Fares F, Havron A, Fima E (2011) Designing a long acting erythropoietin by fusing three carboxyl-terminal peptides of human chorionic gonadotropin beta subunit to the N-terminal and C-terminal coding sequence. Int J Cell Biol 2011:275063

FDA (2013) FDA safety communication: boxed warning on increased mortality and severe renal injury, and additional warning on risk of bleeding, for use of hydroxyethyl starch solutions in some settings. http://www.fda.gov/biologicsbloodvaccines/safetyavailability/ucm358271.htm

Fernandes AI, Gregoriadis G (2001) The effect of polysialylation on the immunogenicity and antigenicity of asparaginase: implication in its pharmacokinetics. Int J Pharm 217:215–224

Fleer R, Yeh P, Amellal N, Maury I, Fournier A, Bacchetta F, Baduel P, Jung G, L'Hote H, Becquart J, Fukuhara H, Mayaux JF (1991) Stable multicopy vectors for high-level secretion of recombinant human serum albumin by *Kluyveromyces* yeasts. Biotechnology 9:968–975

Floss DM, Schallau K, Rose-John S, Conrad U, Scheller J (2010) Elastin-like polypeptides revolutionize recombinant protein expression and their biomedical application. Trends Biotechnol 28:37–45

Foser S, Schacher A, Weyer KA, Brugger D, Dietel E, Marti S, Schreitmüller T (2003a) Isolation, structural characterization, and antiviral activity of positional isomers of monopegylated interferon α-2a (PEGASYS). Protein Expr Purif 30:78–87

Foser S, Weyer K, Huber W, Certa U (2003b) Improved biological and transcriptional activity of monopegylated interferon-α-2a isomers. Pharmacogenomics J 3:312–319

Frejd F (2012) Half-life extension by binding to albumin through an albumin binding domain. In: Kontermann R (ed) Therapeutic proteins: strategies to modulate their plasma half-lives. Wiley-VCH, Weinheim

Furie RA, Leon G, Thomas M, Petri MA, Chu AD, Hislop C, Martin RS, Scheinberg MA, For the PEARL-SC Study (2014) A phase 2, randomised, placebo-controlled clinical trial of blisibimod,

an inhibitor of B cell activating factor, in patients with moderate-to-severe systemic lupus erythematosus, the PEARL-SC study. Ann Rheum Dis. doi:10.1136/annrheumdis-2013-205144
Garay RP, El-Gewely R, Armstrong JK, Garratty G, Richette P (2012) Antibodies against polyethylene glycol in healthy subjects and in patients treated with PEG-conjugated agents. Expert Opin Drug Deliv 9:1319–1323
Garcia-Garcia I, Gonzalez-Delgado CA, Valenzuela-Silva CM, Diaz-Machado A, Cruz-Diaz M, Nodarse-Cuni H, Perez-Perez O, Bermudez-Badell CH, Ferrero-Bibilonia J, Paez-Meireles R, Bello-Rivero I, Castro-Odio FR, Lopez-Saura PA, FarmaPEG Study Group (2010) Pharmacokinetic and pharmacodynamic comparison of two "pegylated" interferon alpha-2 formulations in healthy male volunteers: a randomized, crossover, double-blind study. BMC Pharmacol 10:15
Geething NC, To W, Spink BJ, Scholle MD, Wang CW, Yin Y, Yao Y, Schellenberger V, Cleland JL, Stemmer WP, Silverman J (2010) Gcg-XTEN: an improved glucagon capable of preventing hypoglycemia without increasing baseline blood glucose. PLoS One 5:e10175
GEN News (2010) GEN News highlight: novartis and HGS ditch HCV drug as EMEA and FDA express risk-benefit concern. http://www.genengnews.com/gen-news-highlights/novartis-and-hgs-ditch-hcv-drug-as-emea-and-fda-express-risk-benefit-concerns/81244022
Ghijselings I, Rex S (2014) Hydroxyethyl starches in the perioperative period. A review on the efficacy and safety of starch solutions. Acta Anaesthesiol Belg 65:9–22
Ghuman J, Zunszain PA, Petitpas I, Bhattacharya AA, Otagiri M, Curry S (2005) Structural basis of the drug-binding specificity of human serum albumin. J Mol Biol 353:38–52
Glaesner W, Vick AM, Millican R, Ellis B, Tschang SH, Tian Y, Bokvist K, Brenner M, Koester A, Porksen N, Etgen G, Bumol T (2010) Engineering and characterization of the long-acting glucagon-like peptide-1 analogue LY2189265, an Fc fusion protein. Diabetes Metab Res Rev 26:287–296
Goldenberg MM (1999) Etanercept, a novel drug for the treatment of patients with severe, active rheumatoid arthritis. Clin Ther 21:75–87
Gosselin ML, Fabrizio D, Swain JF, Mitchell TS, Camphausen R, Cload ST, Furfine E, Morin PE, Mukherjee R, Taylor SI (2011) Serum albumin binding molecules. Patent Publication WO2011140086A2
Gregoriadis G, McCormack B, Wang Z, Lifely R (1993) Polysialic acids: potential in drug delivery. FEBS Lett 315:271–276
Gregoriadis G, Jain S, Papaioannou I, Laing P (2005) Improving the therapeutic efficacy of peptides and proteins: a role for polysialic acids. Int J Pharm 300:125–130
Grinyo JM, Budde K, Citterio F, Charpentier B (2013) Belatacept utilization recommendations: an expert position. Expert Opin Drug Saf 12:111–122
Haeckel A, Appler F, Figge L, Kratz H, Lukas M, Michel R, Schnorr J, Zille M, Hamm B, Schellenberger E (2014) XTEN-annexin A5: XTEN allows complete expression of long-circulating protein-based imaging probes as recombinant alternative to PEGylation. J Nucl Med 55:508–514
Hall MP, Gegg C, Walker K, Spahr C, Ortiz R, Patel V, Yu S, Zhang L, Lu H, DeSilva B, Lee JW (2010) Ligand-binding mass spectrometry to study biotransformation of fusion protein drugs and guide immunoassay development: strategic approach and application to peptibodies targeting the thrombopoietin receptor. AAPS J 12:576–585
Halpern W, Riccobene TA, Agostini H, Baker K, Stolow D, Gu ML, Hirsch J, Mahoney A, Carrell J, Boyd E, Grzegorzewski KJ (2002) Albugranin, a recombinant human granulocyte colony stimulating factor (G-CSF) genetically fused to recombinant human albumin induces prolonged myelopoietic effects in mice and monkeys. Pharm Res 19:1720–1729
Haraldsson B, Nystrom J, Deen WM (2008) Properties of the glomerular barrier and mechanisms of proteinuria. Physiol Rev 88:451–487
Harari D, Kuhn N, Abramovich R, Sasson K, Zozulya A, Smith P, Schlapschy M, Aharoni R, Köster M, Eliam R, Skerra A, Schreiber G (2014) Enhanced in vivo efficacy of a type I interferon superagonist with extended plasma half-life in a mouse model of multiple sclerosis. J Biol Chem 289:29014–29029

Hersel U, Kreuzer M, Rau H, Vetter D, Wegge T (2009) Tailor made pharmacokinetics of peptides by transient PEGylation. Adv Exp Med Biol 611:277–278

Hodge RJ, O'Connor-Semmes RL, Lin J, Chism JP, Andrews SM, Gaddy JR, Nunez DJ (2013) GSK2374697, a long-acting GLP-1 mimetic: first use of an AlbudAb™ in humans – pharmacokinetics, pharmacodynamics, safety, and tolerability in healthy volunteers. Abstract No. 60-LB; American Diabetes Association 73rd Scientific Sessions, June 21–25. Chicago, IL

Holt LJ, Basran A, Jones K, Chorlton J, Jespers LS, Brewis ND, Tomlinson IM (2008) Anti-serum albumin domain antibodies for extending the half-lives of short lived drugs. Protein Eng Des Sel 21:283–288

Hopp J, Hornig N, Zettlitz KA, Schwarz A, Fuss N, Müller D, Kontermann RE (2010) The effects of affinity and valency of an albumin-binding domain (ABD) on the half-life of a single-chain diabody-ABD fusion protein. Protein Eng Des Sel 23:827–834

Hui H, Farilla L, Merkel P, Perfetti R (2002) The short half-life of glucagon-like peptide-1 in plasma does not reflect its long-lasting beneficial effects. Eur J Endocrinol 146:863–869

Im SJ, Yang SI, Yang SH, Choi DH, Choi SY, Kim HS, Jang do S, Jin KS, Chung YK, Kim SH, Paik SH, Park YC, Chung MK, Kim YB, Han KH, Choi KY, Sung YC (2011) Natural form of noncytolytic flexible human Fc as a long-acting carrier of agonistic ligand, erythropoietin. PLoS One 6:e24574

Ivens IA, Baumann A, McDonald TA, Humphries TJ, Michaels LA, Mathew P (2013) PEGylated therapeutic proteins for haemophilia treatment: a review for haemophilia caregivers. Haemophilia 19:11–20

Iyer AK, Khaled G, Fang J, Maeda H (2006) Exploiting the enhanced permeability and retention effect for tumor targeting. Drug Discov Today 11:812–818

Jain S, Hreczuk-Hirst DH, McCormack B, Mital M, Epenetos A, Laing P, Gregoriadis G (2003) Polysialylated insulin: synthesis, characterization and biological activity in vivo. Biochim Biophys Acta 1622:42–49

Jevševar S, Kunstelj M, Porekar VG (2010) PEGylation of therapeutic proteins. Biotechnol J 5:113–128

Jin K, Shen Y, He K, Xu Z, Li G, Teng L (2010) Aflibercept (VEGF Trap): one more double-edged sword of anti-VEGF therapy for cancer? Clin Transl Oncol 12:526–532

Joshi L, Murata Y, Wondisford FE, Szkudlinski MW, Desai R, Weintraub BD (1995) Recombinant thyrotropin containing a beta-subunit chimera with the human chorionic gonadotropin-beta carboxy-terminus is biologically active, with a prolonged plasma half-life: role of carbohydrate in bioactivity and metabolic clearance. Endocrinology 136:3839–3848

Jung SY, Kim JS, Choi IY, Lee GH, Kim YH, Kang JH, Kwon SC (2013) The novel long acting GLP-1/Glucagon dual agonist HM12525A reduces body weight and improves glycemic control in rodent models. Diabetes 62:A268

Jung S, Park Y, Kim Y, Kim YY, Choi HJ, Son WC, Kwon S (2014) LAPS-FSH: a new and effective long-acting follicle-stimulating hormone analogue for the treatment of infertility. Reprod Fertil Dev 26:1142–1153

Jungheinrich C, Neff TA (2005) Pharmacokinetics of hydroxyethyl starch. Clin Pharmacokinet 44:681–699

Keating GM (2013) Abatacept: a review of its use in the management of rheumatoid arthritis. Drugs 73:1095–1119

Kerensky TA, Gottlieb AB, Yaniv S, Au SC (2012) Etanercept: efficacy and safety for approved indications. Expert Opin Drug Saf 11:121–139

Kim YM, Bae SM, Kim DJ, Song DH, Lim CK, Kwon SC, Lee GS (2012) Immunoglobulin Fc fragment modified by non-peptide polymer and pharmaceutical composition comprising the same. Patent Publication US8124094B2

Kim CH, Axup JY, Schultz PG (2013a) Protein conjugation with genetically encoded unnatural amino acids. Curr Opin Chem Biol 17:412–419

Kim ES, Jang do S, Yang SY, Lee MN, Jin KS, Cha HJ, Kim JK, Sung YC, Choi KY (2013b) Controlled release of human growth hormone fused with a human hybrid Fc fragment through a nanoporous polymer membrane. Nanoscale 5:4262–4269

Kling J (2013) PEGylation of biologics. Bioprocess Int 11:34–43

Kolate A, Baradia D, Patil S, Vhora I, Kore G, Misra A (2014) PEG – a versatile conjugating ligand for drugs and drug delivery systems. J Control Release 192:67–81

Kratz F (2010) Albumin, a versatile carrier in oncology. Int J Clin Pharmacol Ther 48:453–455

Kratz F (2014) A clinical update of using albumin as a drug vehicle – a commentary. J Control Release 190:331–336

Kraulis PJ, Jonasson P, Nygren PÅ, Uhlén M, Jendeberg L, Nilsson B, Kördel J (1996) The serum albumin-binding domain of streptococcal protein G is a three-helical bundle: a heteronuclear NMR study. FEBS Lett 378:190–194

Kubetzko S, Sarkar CA, Plückthun A (2005) Protein PEGylation decreases observed target association rates via a dual blocking mechanism. Mol Pharmacol 68:1439–1454

Kuo TT, Aveson VG (2011) Neonatal Fc receptor and IgG-based therapeutics. mAbs 3:422–430

Leader B, Baca QJ, Golan DE (2008) Protein therapeutics: a summary and pharmacological classification. Nat Rev Drug Discov 7:21–39

Ledger WL, Fauser BC, Devroey P, Zandvliet AS, Mannaerts BM (2011) Corifollitropin alfa doses based on body weight: clinical overview of drug exposure and ovarian response. Reprod Biomed Online 23:150–159

Leonard R, Petersen BO, Himly M, Kaar W, Wopfner N, Kolarich D, van Ree R, Ebner C, Duus JO, Ferreira F, Altmann F (2005) Two novel types of *O*-glycans on the mugwort pollen allergen Art v 1 and their role in antibody binding. J Biol Chem 280:7932–7940

Lepenies B, Seeberger PH (2014) Simply better glycoproteins. Nat Biotechnol 32:443–445

Leung D, Bergman P, Lofquist A, Pietz G, Tompkins C, Waggoner D (2002) Recombinant production of polyanionic polymers, and uses thereof. Patent Publication WO2002077036A3

Liebner R, Mathaes R, Meyer M, Hey T, Winter G, Besheer A (2014) Protein HESylation for half-life extension: synthesis, characterization and pharmacokinetics of HESylated anakinra. Eur J Pharm Biopharm 87:378–385

Lindgren J, Refai E, Zaitsev SV, Abrahmsén L, Berggren PO, Karlström AE (2014) A GLP-1 receptor agonist conjugated to an albumin-binding domain for extended half-life. Biopolymers 102:252–259

Lindhout T, Iqbal U, Willis LM, Reid AN, Li J, Liu X, Moreno M, Wakarchuk WW (2011) Site-specific enzymatic polysialylation of therapeutic proteins using bacterial enzymes. Proc Natl Acad Sci U S A 108:7397–7402

Ljung R, Karim FA, Saxena K, Suzuki T, Arkhammar P, Rosholm A, Giangrande P, Pioneer I (2013) 40K glycoPEGylated, recombinant FVIIa: 3-month, double-blind, randomized trial of safety, pharmacokinetics and preliminary efficacy in hemophilia patients with inhibitors. J Thromb Haemost 11:1260–1268

Lobo ED, Hansen RJ, Balthasar JP (2004) Antibody pharmacokinetics and pharmacodynamics. J Pharm Sci 93:2645–2668

Macdougall IC, Gray SJ, Elston O, Breen C, Jenkins B, Browne J, Egrie J (1999) Pharmacokinetics of novel erythropoiesis stimulating protein compared with epoetin alfa in dialysis patients. J Am Soc Nephrol 10:2392–2395

Mancuso ME, Mannucci PM (2014) Fc-fusion technology and recombinant FVIII and FIX in the management of the hemophilias. Drug Des Devel Ther 8:365–371

Martens T, Schmidt NO, Eckerich C, Fillbrandt R, Merchant M, Schwall R, Westphal M, Lamszus K (2006) A novel one-armed anti-c-Met antibody inhibits glioblastoma growth in vivo. Clin Cancer Res 12:6144–6152

Matthews JE, Stewart MW, De Boever EH, Dobbins RL, Hodge RJ, Walker SE, Holland MC, Bush MA, Albiglutide Study Group (2008) Pharmacodynamics, pharmacokinetics, safety, and tolerability of albiglutide, a long-acting glucagon-like peptide-1 mimetic, in patients with type 2 diabetes. J Clin Endocrinol Metab 93:4810–4817

Mendler CT, Friedrich L, Schlapschy M, Schwaiger M, Wester H-J, Skerra A (2015) High contrast tumor imaging with radio-labelled antibody Fab fragments tailored for optimized pharmacokinetics via PASylation. mAbs 7:96–109

Merlot AM, Kalinowski DS, Richardson DR (2014) Unraveling the mysteries of serum albumin – more than just a serum protein. Front Physiol 5:299

Metzner HJ, Pipe SW, Weimer T, Schulte S (2013) Extending the pharmacokinetic half-life of coagulation factors by fusion to recombinant albumin. Thromb Haemost 110:931–939

Mitragotri S, Burke PA, Langer R (2014) Overcoming the challenges in administering biopharmaceuticals: formulation and delivery strategies. Nat Rev Drug Discov 13:655–672

Molineux G (2004) The design and development of pegfilgrastim (PEG-rmetHuG-CSF, Neulasta). Curr Pharm Des 10:1235–1244

Monk BJ, Poveda A, Vergote I, Raspagliesi F, Fujiwara K, Bae DS, Oaknin A, Ray-Coquard I, Provencher DM, Karlan BY, Lhomme C, Richardson G, Rincon DG, Coleman RL, Herzog TJ, Marth C, Brize A, Fabbro M, Redondo A, Bamias A, Tassoudji M, Navale L, Warner DJ, Oza AM (2014) Anti-angiopoietin therapy with trebananib for recurrent ovarian cancer (TRINOVA-1): a randomised, multicentre, double-blind, placebo-controlled phase 3 trial. Lancet Oncol 15:799–808

Morath V, Bolze F, Schlapschy M, Schneider S, Sedlmayer F, Seyfarth K, Klingenspor M, Skerra A (2015) PASylation of murine leptin leads to extended plasma half-life and enhanced in vivo efficacy. Mol Pharm. 12:1431–1442

Müller MR, Saunders K, Grace C, Jin M, Piche-Nicholas N, Steven J, O'Dwyer R, Wu L, Khetemenee L, Vugmeyster Y, Hickling TP, Tchistiakova L, Olland S, Gill D, Jensen A, Barelle CJ (2012) Improving the pharmacokinetic properties of biologics by fusion to an anti-HSA shark VNAR domain. mAbs 4:673–685

Nelson AL (2010) Antibody fragments: hope and hype. mAbs 2:77–83

Nelson DR, Benhamou Y, Chuang WL, Lawitz EJ, Rodriguez-Torres M, Flisiak R, Rasenack JW, Kryczka W, Lee CM, Bain VG, Pianko S, Patel K, Cronin PW, Pulkstenis E, Subramanian GM, McHutchison JG, ACHIEVE-2/3 Study Team (2010) Albinterferon Alfa-2b was not inferior to pegylated interferon-α in a randomized trial of patients with chronic hepatitis C virus genotype 2 or 3. Gastroenterology 139:1267–1276

Nilvebrant J, Hober S (2013) The albumin-binding domain as a scaffold for protein engineering. Comput Struct Biotechnol J 6:e201303009

Northfelt DW, Allred JB, Liu H, Hobday TJ, Rodacker MW, Lyss AP, Fitch TR, Perez EA, North Central Cancer Treatment G (2014) Phase 2 trial of Paclitaxel polyglumex with capecitabine for metastatic breast cancer. Am J Clin Oncol 37:167–171

Ortonne JP, Prinz JC (2004) Alefacept: a novel and selective biologic agent for the treatment of chronic plaque psoriasis. Eur J Dermatol 14:41–45

Østergaard H, Bjelke JR, Hansen L, Petersen LC, Pedersen AA, Elm T, Møller F, Hermit MB, Holm PK, Krogh TN, Petersen JM, Ezban M, Sørensen BB, Andersen MD, Agersø H, Ahmadian H, Balling KW, Christiansen ML, Knobe K, Nichols TC, Bjørn SE, Tranholm M (2011) Prolonged half-life and preserved enzymatic properties of factor IX selectively PEGylated on native N-glycans in the activation peptide. Blood 118:2333–2341

Papisov MI, Hiller A, Yurkovetskiy A, Yin M, Barzana M, Hillier S, Fischman AJ (2005) Semisynthetic hydrophilic polyals. Biomacromolecules 6:2659–2670

Park YK, Abuchowski A, Davis S, Davis F (1981) Pharmacology of Escherichia coli-L-asparaginase polyethylene glycol adduct. Anticancer Res 1:373–376

Park YJ, Lim CK, Lim KJ, Kim D, Hyun JS, Lee GH, Kim YI, Kim YH, Oh HH, Choi CS, Kwon SC (2012) Pharmacokinetics and pharmacodynamics of ultra-long acting insulin (LAPS-Insulin) in animal models. Diabetes 61:A234

Pasut G (2014a) Pegylation of biological molecules and potential benefits: pharmacological properties of certolizumab pegol. BioDrugs 28(Suppl 1):15–23

Pasut G (2014b) Polymers for protein conjugation. Polymers 6:160–178

Patel SS, Benfield P (1996) Pegaspargase (polyethylene glycol-L-asparaginase). Clin Immunother 5:492–496

Patel IH, Zhang X, Nieforth K, Salgo M, Buss N (2005) Pharmacokinetics, pharmacodynamics and drug interaction potential of enfuvirtide. Clin Pharmacokinet 44:175–186

Peppel K, Crawford D, Beutler B (1991) A tumor necrosis factor (TNF) receptor-IgG heavy chain chimeric protein as a bivalent antagonist of TNF activity. J Exp Med 174:1483–1489

Perel P, Roberts I (2007) Colloids versus crystalloids for fluid resuscitation in critically ill patients. Cochrane Database Syst Rev CD000567

Perez-Ruixo JJ, Krzyzanski W, Bouman-Thio E, Miller B, Jang H, Bai SA, Zhou H, Yohrling J, Cohen A, Burggraaf J, Franson K, Davis HM (2009) Pharmacokinetics and pharmacodynamics of the erythropoietin Mimetibody™ construct CNTO 528 in healthy subjects. Clin Pharmacokinet 48:601–613

Peters TJ (1996) All about albumin: biochemistry, genetics, and medical applications. Academic, San Diego

Peters RT, Toby G, Lu Q, Liu T, Kulman JD, Low SC, Bitonti AJ, Pierce GF (2013) Biochemical and functional characterization of a recombinant monomeric factor VIII-Fc fusion protein. J Thromb Haemost 11:132–141

Pfister D, Morbidelli M (2014) Process for protein PEGylation. J Control Release 180:134–149

Phan HT, Conrad U (2011) Membrane-based inverse transition cycling: an improved means for purifying plant-derived recombinant protein-elastin-like polypeptide fusions. Int J Mol Sci 12:2808–2821

Picha KM, Cunningham MR, Drucker DJ, Mathur A, Ort T, Scully M, Soderman A, Spinka-Doms T, Stojanovic-Susulic V, Thomas BA, O'Neil KT (2008) Protein engineering strategies for sustained glucagon-like peptide-1 receptor-dependent control of glucose homeostasis. Diabetes 57:1926–1934

Picha K, Huang C, Bugelski P, O'Neil K (2014) Engineering peptide therapeutics using MIMETIBODY technology. Methods Mol Biol 1088:125–145

Plum A, Agerso H, Andersen L (2000) Pharmacokinetics of the rapid-acting insulin analog, insulin aspart, in rats, dogs, and pigs, and pharmacodynamics of insulin aspart in pigs. Drug Metab Dispos 28:155–160

Podust VN, Sim BC, Kothari D, Henthorn L, Gu C, Wang CW, McLaughlin B, Schellenberger V (2013) Extension of in vivo half-life of biologically active peptides via chemical conjugation to XTEN protein polymer. Protein Eng Des Sel 26:743–753

Poole RM, Nowlan ML (2014) Albiglutide: first global approval. Drugs 74:929–938

Qiu H, Boudanova E, Park A, Bird JJ, Honey DM, Zarazinski C, Greene B, Kingsbury JS, Boucher S, Pollock J, McPherson JM, Pan CQ (2013) Site-specific PEGylation of human thyroid stimulating hormone to prolong duration of action. Bioconjug Chem 24:408–418

Rath T, Baker K, Dumont JA, Peters RT, Jiang H, Qiao SW, Lencer WI, Pierce GF, Blumberg RS (2013) Fc-fusion proteins and FcRn: structural insights for longer-lasting and more effective therapeutics. Crit Rev Biotechnol 1–20

Reddy KR, Modi MW, Pedder S (2002) Use of peginterferon alfa-2a (40 KD) (Pegasys®) for the treatment of hepatitis C. Adv Drug Deliv Rev 54:571–586

Robbie GJ, Criste R, Dall'acqua WF, Jensen K, Patel NK, Losonsky GA, Griffin MP (2013) A novel investigational Fc-modified humanized monoclonal antibody, motavizumab-YTE, has an extended half-life in healthy adults. Antimicrob Agents Chemother 57:6147–6153

Roopenian DC, Akilesh S (2007) FcRn: the neonatal Fc receptor comes of age. Nat Rev Immunol 7:715–725

Rosendahl MS, Doherty DH, Smith DJ, Bendele AM, Cox GN (2005) Site-specific protein PEGylation – application to cysteine analogs of recombinant human granulocyte colony-stimulating factor. Bioprocess Int 3:52–60

Rustgi VK (2009) Albinterferon alfa-2b, a novel fusion protein of human albumin and human interferon alfa-2b, for chronic hepatitis C. Curr Med Res Opin 25:991–1002

Saifer MG, Williams LD, Sobczyk MA, Michaels SJ, Sherman MR (2014) Selectivity of binding of PEGs and PEG-like oligomers to anti-PEG antibodies induced by methoxyPEG-proteins. Mol Immunol 57:236–246

Sausville EA, Garbo L, Weiss GJ, Shkolny D, Yurkovetskiy AV, Bethune C, Ramanathan RK, Fram RJ (2009) Phase 1 study of XMT-1001, a novel water soluble camptothecin conjugate, given as an intravenous infusion once every three weeks to patients with advanced solid tumors. Abstract No. B52; AACR-NCI-EORTC international conference: molecular targets and cancer therapeutics, November 15–19. Boston, MA

Scheckermann C, Schmidt K, Abdolzade-Bavil A, Allgaier H, Mueller UW, Shen WD, Liu P (2013) Lipegfilgrastim: a long-acting, once-per-cycle, glycopegylated recombinant human filgrastim. J Clin Oncol 31:abstr e13548

Schellenberger V, Wang CW, Geething NC, Spink BJ, Campbell A, To W, Scholle MD, Yin Y, Yao Y, Bogin O, Cleland JL, Silverman J, Stemmer WP (2009) A recombinant polypeptide extends the in vivo half-life of peptides and proteins in a tunable manner. Nat Biotechnol 27:1186–1190

Schlapschy M, Theobald I, Mack H, Schottelius M, Wester HJ, Skerra A (2007) Fusion of a recombinant antibody fragment with a homo-amino-acid polymer: effects on biophysical properties and prolonged plasma half-life. Protein Eng Des Sel 20:273–284

Schlapschy M, Binder U, Börger C, Theobald I, Wachinger K, Kisling S, Haller D, Skerra A (2013) PASylation: a biological alternative to PEGylation for extending the plasma half-life of pharmaceutically active proteins. Protein Eng Des Sel 26:489–501

Shamji MF, Betre H, Kraus VB, Chen J, Chilkoti A, Pichika R, Masuda K, Setton LA (2007) Development and characterization of a fusion protein between thermally responsive elastin-like polypeptide and interleukin-1 receptor antagonist: sustained release of a local antiinflammatory therapeutic. Arthritis Rheum 56:3650–3661

Shapiro AD, Ragni MV, Valentino LA, Key NS, Josephson NC, Powell JS, Cheng G, Thompson AR, Goyal J, Tubridy KL, Peters RT, Dumont JA, Euwart D, Li L, Hallen B, Gozzi P, Bitonti AJ, Jiang H, Luk A, Pierce GF (2012) Recombinant factor IX-Fc fusion protein (rFIXFc) demonstrates safety and prolonged activity in a phase 1/2a study in hemophilia B patients. Blood 119:666–672

Shimamoto G, Gegg C, Boone T, Queva C (2012) Peptibodies: a flexible alternative format to antibodies. mAbs 4:586–591

Shukla AA, Hubbard B, Tressel T, Guhan S, Low D (2007) Downstream processing of monoclonal antibodies – application of platform approaches. J Chromatogr B Analyt Technol Biomed Life Sci 848:28–39

Sim SL, He T, Tscheliessnig A, Mueller M, Tan RB, Jungbauer A (2012) Branched polyethylene glycol for protein precipitation. Biotechnol Bioeng 109:736–746

Sinclair AM, Elliott S (2005) Glycoengineering: the effect of glycosylation on the properties of therapeutic proteins. J Pharm Sci 94:1626–1635

Singer JW (2005) Paclitaxel poliglumex (XYOTAX, CT-2103): a macromolecular taxane. J Control Release 109:120–126

Singh SK (2011) Impact of product-related factors on immunogenicity of biotherapeutics. J Pharm Sci 100:354–387

Sismey-Ragatz AE, Green DE, Otto NJ, Rejzek M, Field RA, DeAngelis PL (2007) Chemoenzymatic synthesis with distinct *Pasteurella* heparosan synthases: monodisperse polymers and unnatural structures. J Biol Chem 282:28321–28327

Sjöholm A (2010) Liraglutide therapy for type 2 diabetes: overcoming unmet needs. Pharmaceuticals 3:764–781

Sleep D, Belfield GP, Goodey AR (1990) The secretion of human serum albumin from the yeast *Saccharomyces cerevisiae* using five different leader sequences. Biotechnology 8:42–46

Sleep D, Belfield GP, Ballance DJ, Steven J, Jones S, Evans LR, Moir PD, Goodey AR (1991) *Saccharomyces cerevisiae* strains that overexpress heterologous proteins. Biotechnology 9:183–187

Sleep D, Cameron J, Evans LR (2013) Albumin as a versatile platform for drug half-life extension. Biochim Biophys Acta 1830:5526–5534

Sleep D (2015) Albumin and its application in drug delivery. Expert Opin Drug Deliv 12:793–812

Stahl N, Radin A, Mellis S (2009) Rilonacept-CAPS and beyond. Ann N Y Acad Sci 1182:124–134

Stennicke HR, Østergaard H, Bayer RJ, Kalo MS, Kinealy K, Holm PK, Sørensen BB, Zopf D, Bjørn SE (2008) Generation and biochemical characterization of glycoPEGylated factor VIIa derivatives. Thromb Haemost 100:920–928

Stennicke HR, Kjalke M, Karpf DM, Balling KW, Johansen PB, Elm T, Øvlisen K, Möller F, Holmberg HL, Gudme CN, Persson E, Hilden I, Pelzer H, Rahbek-Nielsen H, Jespersgaard C, Bogsnes A, Pedersen AA, Kristensen AK, Peschke B, Kappers W, Rode F, Thim L, Tranholm M, Ezban M, Olsen EH, Bjørn SE (2013) A novel B-domain *O*-glycoPEGylated FVIII (N8-GP) demonstrates full efficacy and prolonged effect in hemophilic mice models. Blood 121:2108–2116

Stork R, Müller D, Kontermann RE (2007) A novel tri-functional antibody fusion protein with improved pharmacokinetic properties generated by fusing a bispecific single-chain diabody with an albumin-binding domain from streptococcal protein G. Protein Eng Des Sel 20:569–576

Subramanian GM, Fiscella M, Lamouse-Smith A, Zeuzem S, McHutchison JG (2007) Albinterferon α-2b: a genetic fusion protein for the treatment of chronic hepatitis C. Nat Biotechnol 25:1411–1419

Sugahara T, Pixley MR, Fares F, Boime I (1996) Characterization of the *O*-glycosylation sites in the chorionic gonadotropin β subunit in vivo using site-directed mutagenesis and gene transfer. J Biol Chem 271:20797–20804

Sugio S, Kashima A, Mochizuki S, Noda M, Kobayashi K (1999) Crystal structure of human serum albumin at 2.5 A resolution. Protein Eng 12:439–446

Suzuki T, Ishii-Watabe A, Tada M, Kobayashi T, Kanayasu-Toyoda T, Kawanishi T, Yamaguchi T (2010) Importance of neonatal FcR in regulating the serum half-life of therapeutic proteins containing the Fc domain of human IgG1: a comparative study of the affinity of monoclonal antibodies and Fc-fusion proteins to human neonatal FcR. J Immunol 184:1968–1976

Tabrizi MA, Tseng CM, Roskos LK (2006) Elimination mechanisms of therapeutic monoclonal antibodies. Drug Discov Today 11:81–88

Tan L, Leykam JF, Kieliszewski MJ (2003) Glycosylation motifs that direct arabinogalactan addition to arabinogalactan-proteins. Plant Physiol 132:1362–1369

Tang L, Persky AM, Hochhaus G, Meibohm B (2004) Pharmacokinetic aspects of biotechnology products. J Pharm Sci 93:2184–2204

Telleman P, Junghans RP (2000) The role of the Brambell receptor (FcRB) in liver: protection of endocytosed immunoglobulin G (IgG) from catabolism in hepatocytes rather than transport of IgG to bile. Immunology 100:245–251

Teng L, Lu J (2013) cMET as a potential therapeutic target in gastric cancer. Int J Mol Med 32:1247–1254

Tijink BM, Laeremans T, Budde M, Stigter-van Walsum M, Dreier T, de Haard HJ, Leemans CR, van Dongen GA (2008) Improved tumor targeting of anti-epidermal growth factor receptor nanobodies through albumin binding: taking advantage of modular nanobody technology. Mol Cancer Ther 7:2288–2297

Tojo A, Kinugasa S (2012) Mechanisms of glomerular albumin filtration and tubular reabsorption. Int J Nephrol 2012:481520

Tolley K, Hutchinson M, Pachner A, Kinter E, Sperling B, You X, Wang P, Taneja A, Siddiqui MK (2014) Systematic literature review and network meta-analysis of pegylated interferon beta-1a and disease modifying therapies for relapsing-remitting multiple sclerosis. Neurology 82:Supplement P7.234

Tolmachev V, Orlova A, Pehrson R, Galli J, Baastrup B, Andersson K, Sandström M, Rosik D, Carlsson J, Lundqvist H, Wennborg A, Nilsson FY (2007) Radionuclide therapy of HER2-positive microxenografts using a ^{177}Lu-labeled HER2-specific Affibody molecule. Cancer Res 67:2773–2782

Trussel S, Dumelin C, Frey K, Villa A, Buller F, Neri D (2009) New strategy for the extension of the serum half-life of antibody fragments. Bioconjug Chem 20:2286–2292

Umpierrez G, Povedano ST, Manghi FP, Shurzinske L, Pechtner V (2014) Efficacy and safety of dulaglutide monotherapy versus metformin in type 2 diabetes in a randomized controlled trial (AWARD-3). Diabetes Care 37:2168–2176

Verhoef JJ, Carpenter JF, Anchordoquy TJ, Schellekens H (2014) Potential induction of anti-PEG antibodies and complement activation toward PEGylated therapeutics. Drug Discov Today 19:1945–1952
Veronese FM, Mero A (2008) The impact of PEGylation on biological therapies. BioDrugs 22:315–329
Veronese FM, Pasut G (2008) PEGylation: posttranslational bioengineering of protein biotherapeutics. Drug Discov Today Technol 5:e57–e64
Veronese FM, Caliceti P, Schiavon O (1997) Branched and linear poly(ethylene glycol): influence of the polymer structure on enzymological, pharmacokinetic and immunological properties of protein conjugates. J Bioact Compat Polym 12:196–207
Vorobiev I, Matskevich V, Kovnir S, Orlova N, Knorre V, Jain S, Genkin D, Gabibov A, Miroshnikov A (2013) Chemical polysialylation: design of conjugated human oxyntomodulin with a prolonged anorexic effect in vivo. Biochimie 95:264–270
Walker A, Dunlevy G, Rycroft D, Topley P, Holt LJ, Herbert T, Davies M, Cook F, Holmes S, Jespers L, Herring C (2010) Anti-serum albumin domain antibodies in the development of highly potent, efficacious and long-acting interferon. Protein Eng Des Sel 23:271–278
Wang M, Matheson S, Picard J, Pezzullo J, Ulich T (2009) PC-DAC™: exendin-4 (CJC-1134-PC) significantly reduces HbA1c and body weight as an adjunct therapy to metformin: two randomized, double-blind, placebo-controlled, 12 week, phase II studies in patients with type 2 diabetes mellitus. Abstract No. 553-P; 69th Scientific Sessions of the American Diabetes Association meeting, June 5–9. New Orleans, LA
Ward Ober ES (2012) Immunoglobulin molecules with improved characteristics. Patent Publication US8163881B2
Wright A, Sato Y, Okada T, Chang K, Endo T, Morrison S (2000) In vivo trafficking and catabolism of IgG1 antibodies with Fc associated carbohydrates of differing structure. Glycobiology 10:1347–1355
Wu B, Sun YN (2014) Pharmacokinetics of peptide-Fc fusion proteins. J Pharm Sci 103:53–64
Xu J, Tan L, Goodrum KJ, Kieliszewski MJ (2007) High-yields and extended serum half-life of human interferon α2b expressed in tobacco cells as arabinogalactan-protein fusions. Biotechnol Bioeng 97:997–1008
Xu J, Okada S, Tan L, Goodrum KJ, Kopchick JJ, Kieliszewski MJ (2010) Human growth hormone expressed in tobacco cells as an arabinogalactan-protein fusion glycoprotein has a prolonged serum life. Transgenic Res 19:849–867
Yang SH, Yang SI, Chung YK (2012) A long-acting erythropoietin fused with noncytolytic human Fc for the treatment of anemia. Arch Pharm Res 35:757–759
Yeh P, Landais D, Lemaitre M, Maury I, Crenne JY, Becquart J, Murry-Brelier A, Boucher F, Montay G, Fleer R, Hirel P-H, Mayaux J-F, Klatzmann D (1992) Design of yeast-secreted albumin derivatives for human therapy: biological and antiviral properties of a serum albumin-CD4 genetic conjugate. Proc Natl Acad Sci U S A 89:1904–1908
Young MA, Malavalli A, Winslow N, Vandegriff KD, Winslow RM (2007) Toxicity and hemodynamic effects after single dose administration of MalPEG-hemoglobin (MP4) in rhesus monkeys. Transl Res 149:333–342
Yousefpour P, Chilkoti A (2014) Co-opting biology to deliver drugs. Biotechnol Bioeng 111:1699–1716
Yu H, Rohan T (2000) Role of the insulin-like growth factor family in cancer development and progression. J Natl Cancer Inst 92:1472–1489
Yurkovetskiy AV, Fram RJ (2009) XMT-1001, a novel polymeric camptothecin pro-drug in clinical development for patients with advanced cancer. Adv Drug Deliv Rev 61:1193–1202
Yurkovetskiy A, Choi S, Hiller A, Yin M, McCusker C, Syed S, Fischman AJ, Papisov MI (2005) Fully degradable hydrophilic polyals for protein modification. Biomacromolecules 6:2648–2658
Zalevsky J, Chamberlain AK, Horton HM, Karki S, Leung IW, Sproule TJ, Lazar GA, Roopenian DC, Desjarlais JR (2010) Enhanced antibody half-life improves in vivo activity. Nat Biotechnol 28:157–159

Zarychanski R, Abou-Setta AM, Turgeon AF, Houston BL, McIntyre L, Marshall JC, Fergusson DA (2013) Association of hydroxyethyl starch administration with mortality and acute kidney injury in critically ill patients requiring volume resuscitation: a systematic review and meta-analysis. J Am Med Assoc 309:678–688

Zeuzem S, Sulkowski MS, Lawitz EJ, Rustgi VK, Rodriguez-Torres M, Bacon BR, Grigorescu M, Tice AD, Lurie Y, Cianciara J, Muir AJ, Cronin PW, Pulkstenis E, Subramanian GM, McHutchison JG, ACHIEVE-1 Study Team (2010) Albinterferon Alfa-2b was not inferior to pegylated interferon-α in a randomized trial of patients with chronic hepatitis C virus genotype 1. Gastroenterology 139:1257–1266

Zhang R, Jain S, Rowland M, Hussain N, Agarwal M, Gregoriadis G (2010) Development and testing of solid dose formulations containing polysialic acid insulin conjugate: next generation of long-acting insulin. J Diabetes Sci Technol 4:532–539

Zhao HL, Xue C, Wang Y, Sun B, Yao XQ, Liu ZM (2009) Elimination of the free sulfhydryl group in the human serum albumin (HSA) moiety of human interferon-α2b and HSA fusion protein increases its stability against mechanical and thermal stresses. Eur J Pharm Biopharm 72:405–411

Zheng ZY, Wang SZ, Li GS, Zhan XB, Lin CC, Wu JR, Zhu L (2013) A new polysialic acid production process based on dual-stage pH control and fed-batch fermentation for higher yield and resulting high molecular weight product. Appl Microbiol Biotechnol 97:2405–2412

Zhu X, Meng G, Dickinson BL, Li X, Mizoguchi E, Miao L, Wang Y, Robert C, Wu B, Smith PD, Lencer WI, Blumberg RS (2001) MHC class I-related neonatal Fc receptor for IgG is functionally expressed in monocytes, intestinal macrophages, and dendritic cells. J Immunol 166:3266–3276

Part IV
Economic and Regulatory Considerations

Biosimilar and Biobetter Scenarios for the US and Europe: What Should We Expect?

Ernst R. Berndt and Mark R. Trusheim

Introduction and Background

FDA-approved biologics have been marketed in the US for more than 30 years (human insulin was the first recombinant therapeutic approved by the FDA in 1982).[1] In spite of this considerable history, the pathway to follow-on biologics (dubbed "biosimilars" by some observers) subsequent to originator loss of patent protection or other exclusivity has only recently begun to evolve, albeit at a more rapid pace in Europe than in the U.S. In this manuscript we discuss how biosimilar development, regulatory approval, manufacturing, branding and marketing, distribution, utilization, pricing and cost savings will differ from the historical experiences of generic small molecules, chemically synthesized therapies, both in the U.S. and Europe. We will also address how and why the European biosimilar diffusion process will likely differ from that in the U.S.

We begin by overviewing the stylized facts and existing literature concerning the evolution of U.S. and European generic small molecule markets, and then review the smaller but rapidly growing body of literature concerning U.S. and European markets for biosimilars and other specialty drugs. Finally, we explore the financial incentives that may drive development of biobetter vs. biosimilar therapeutics.

Ernst R. Berndt is the Louis E. Seley Professor in Applied Economics, and Mark R. Trusheim is Visiting Scientist, both at the Massachusetts Institute of Technology, Alfred P. Sloan School of Management, Cambridge, Massachusetts, USA.

[1] Trusheim et al. (2010).

E.R. Berndt (✉) • M.R. Trusheim
Alfred P. Sloan School of Management, Massachusetts Institute of Technology, 100 Main Street, E62-518, Cambridge, MA 02142, USA
e-mail: eberndt@mit.edu

A. Rosenberg, B. Demeule (eds.), *Biobetters*, AAPS Advances in the Pharmaceutical Sciences Series 19, DOI 10.1007/978-1-4939-2543-8_15

Stylized Facts of U.S. Generic Small Molecule Market Evolution

The seminal legislation governing generic entry of small molecules into the U.S. market is the 1984 Hatch-Waxman Act. Rather than requiring duplicative clinical studies demonstrating safety and efficacy, this legislation facilitated more rapid and less costly generic entry by only requiring the abbreviated New Drug Application (ANDA) applicant to demonstrate bioequivalence to the innovator drug, as well as compliance with current good manufacturing practices (cGMP). If the innovator's patents are not successfully challenged, typically relatively unfettered and massive generic entry now occurs on the day of loss of exclusivity (LOE), with the number of entrants depending in large part on the dollar or prescription volume of pre-LOE sales, and/or on the complexity the manufacturing or use of the drug entails.[2] Although generally increasing over time since passage of the Hatch-Waxman legislation, the generic efficiency rate (for molecules having multisource entry following LOE, the proportion of brand plus generic scripts dispensed as generics) has exceeded 90 % in the U.S. retail market in recent years, approaching its theoretical limit of 100 %, and has done so increasingly rapidly—on average over all small molecules, within 3–4 months of LOE the generic penetration rate (the proportion of all prescriptions dispensed as generic) now approaches 80–90 %.[3] Though not utilized initially for more than a decade, the legislation also provided incentives for generics to challenge innovators' patents as being invalid or not infringed by the ANDA applicant; the successful first-to-file ANDA Paragraph IV challenger is rewarded by being given 180 days of exclusivity, during which time no other ANDA holder can market the drug formulation/strength. However, brands can enter under their original New Drug Application (NDA) during the 180-day exclusivity period, launching their authorized generic (AG) to compete in a triopoly setting with the

[2] Grabowski et al. (2011) define a complex small molecule as one meeting two or more of the following criteria: black box warning, narrow therapeutic index, prescribed by specialists, oncology products, or manufacturing technology that is available to only a limited number of firms. The more complex the drug, *ceteris paribus*, the fewer the number of generic entrants at the time of LOE. Olson and Wendling (2013) find that entry both during and after the 180-day exclusivity depends not only on pre-LOE market size, but also is greater if the drug at issue was originally designated a New Chemical Entity (NCE) by the Food and Drug Administration during the NDA approval process.

[3] IMS Institute for Healthcare Informatics (2011a), p. 21, and Aitken et al. (2013). Earlier studies include those by Hurwitz and Caves (1988), Ellison et al. (1997), Cook (1998), Reiffen and Ward (2005), Saha et al. (2006), Aitken et al. (2008), Aitken and Berndt (2011), Berndt and Aitken (2011) and Berndt and Newhouse (2012). Regarding complexity, as discussed in the previous endnote, Grabowski et al. (2011, pp. 540–541) report that "On average, drugs with two or more characteristics faced 2.5 geneic entrants 1 year following initial generic entry, while drugs with one or no complexity characteristics faced an average of 8.5 entrants." Mean generic share of non-complex small molecules was 1.7 times larger than for complex small molecules, while the mean price discount from brand price was 1.6 times larger (price here reflecting manufacturer's revenues from sales to wholesalers and direct customers).

entrants consisting of the possibly cannibalized own brand, the successful first-to-file Paragraph IV challenger, and the authorized generic.[4]

In the case of a triopoly during 180-day exclusivity, at the retail level average prices for the ANDA and the AG are about 20 % less than the brand, although pharmacy acquisition prices (and hence, average revenues to the ANDA and AG manufacturers) are discounted even more, implying that during the 180-day exclusivity retail margins arc very substantial.[5] In spite of this modest retail price reduction during the 180-day exclusivity, early evidence suggested that within 4 weeks of LOE, the volume share of the combined successful ANDA challenger plus the AG was about 75 %.[6] Since the choice of who will be the AG is a decision made by the brand seeking to maximize its post-LOE profits (the AG could be an independent generic firm licensed to market the AG while paying the brand a royalty for the privilege, or a generic subsidiary of the brand), the more interesting combined share is that of the brand and the AG, particularly since according to the Federal Trade Commission (2011, p. 85), in recent years the royalty rate paid the brand by the AG has been in excess of 90 %. Evidence from the 2009–2013 time frame in the U.S. suggests that during the 180-day exclusivity, the brand volume share ranges from about 15 to almost 50 %, and the AG share from 20 to 30 %, with the combined brand plus AG share ranging between 50 and 65 %, while that for the successful first-to-file ANDA challenger is between 35 and 50 %.[7] To date, sample sizes in studies analyzing AG pricing patterns have been too small to detect whether pricing during the 180-day exclusivity differs depending on whether the AG is marketed by an independent generic or a subsidiary of the brand. The evidence does, however, suggest that the presence of an 180-day exclusivity period with restricted entry (either duopoly or triopoly) has no long-term effect on the extent of generic entry post-180-day exclusivity.[8] Notably, in recent years almost all brands at risk for initial LOE have faced patent challenges; these challenges have increasingly occurred at precisely 4 years following initial NDA approval which is the earliest time from initial NDA approval at which the brand's patent can be challenged. In most cases when there is a successful Paragraph IV challenger, the brand has responded with AG entry, although in some settlement situations generic entry has been delayed, or the brand agrees not to launch an AG.[9]

For many years conventional wisdom held that total molecule (brand plus all generic) utilization generally declined following LOE. This post-LOE decline has

[4] If multiple Paragraph IV filers submit their challenge on the same day, the various challengers share the 180-day exclusivity, resulting in a larger number of competitors. See Federal Trade Commission (2011, especially Chap. 7) for more details.

[5] See Federal Trade Commission (2002, 2011), and Aitken et al. (2013) for further details.

[6] Berndt et al. (2007). Also see Branstetter et al. (2011) for estimated effects of Paragraph IV entry on consumers' welfare in the U.S.

[7] Aitken et al. (2013).

[8] Berndt et al. (2007); Federal Trade Commission (2011); Aitken et al. (2013).

[9] Federal Trade Commission (2011, Chaps. 2 and 7, Aitken et al. (2013), Grabowski et al. (2014), and Drake et al. (2014).

been thought to reflect reduced marketing efforts by the brand as LOE approached and after it arrived, attempts by the brand to switch Rx-only to an over the counter version or to its next generation product in the same therapeutic class, as well as an absence of product differentiation marketing competition by generic manufacturers following LOE. However, in recent years a new phenomenon has emerged by which cross-molecule substitution from a patent protected brand to a generic version of another molecule results in total utilization of the off-patent generic molecule increasing following the brand's LOE. This occurred in 2006–2007 when Zocor (simvastatin) went off patent, inducing efforts by payers and their pharmaceutical benefit manager (PBM) agents to incent substitution toward simvastatin and away from the branded more costly Lipitor.[10] More recent data suggest the Zocor-simvastatin-Lipitor increase in post-LOE sales was not unique historically, and instead may become the norm as payers and their PBMs increasingly exercise their ability to effect cross-molecule substitution. Specifically, as reported in Aitken et al. (2013), among the top 50 prescribed molecules in 2013, for four of the six molecules experiencing initial LOE between 2009 and 2013, total utilization post-LOE increased, for one molecule it was relatively stable, and for only one molecule decrease in post-LOE utilization occurred, and that was only a very slight decrease.

In terms of number of generic entrants post-LOE, due to a combination of consolidation M&A activity among generic manufacturers and actual product exit, the total number of generic entrants in the US has tended to peak between 30 and 36 months following LOE.[11] In aggregate, over all products in 2009 the generic penetration rate (the proportion of all retail prescriptions dispensed as generics) was about 80 %, the unbranded generic revenue share was 10–15 % of total revenues, while branded products captured 75 % of revenues, with branded generics obtaining 10–15 % of total revenues.[12] By 2013 the generic penetration rate increased to 86 %, the unbranded generic revenue share increased to 17 %, branded products captured only 71 % of revenues, and branded generics obtained 12 % of total medicine spending.[13]

Stylized Facts Regarding European Small Molecule Generic Market Evolution

Although recent trends in generic efficiency improvements and average price reductions have accelerated, European small molecule generic efficiency rates have not been as high and average molecule prices have not fallen as much proportionately

[10] See Aitken et al. (2008) for details.

[11] Berndt and Aitken (2011); Reiffen and Ward (2007).

[12] IMS Institute for Healthcare Informatics (2011a, p. 21).

[13] IMS Institute for Healthcare Informatics (2014, p. 40).

as has occurred in the U.S. There are several reasons behind this historically less aggressive generic diffusion and pricing in European countries than in the U.S.

First, on-patent brand prices in Europe are generally lower than in the U.S. Even if a manufacturer launching a new drug launches it at parity pricing across the globe, in many European countries there are various forms of price controls, such as prohibitions on manufacturers raising prices more rapidly than some measure of overall national price inflation. This constrains European post-launch on-patent brand price growth, and so when patent expiration occurs, the brand's price relative to its price at time of initial launch is typically lower than in the U.S. This makes generics less of a "bargain" to European payers and consumers. It also implies that entry by generic manufacturers is not as profitable, thereby reducing incentives for, and mitigating the extent of, entry by generic manufacturers.

Second, and related to lower European brand prices than in the U.S., in most European countries the purchasing of medicines is centralized in national or regional governments, providing them with critical monopsonistic leverage in negotiating prices with manufacturers marketing patent-protected medicines. In some countries, such as the U.K. and Germany, in order to gain reimbursement from payers, manufacturers are required to provide data (via a national health technology assessment) showing their drug is at least as if not more cost-effective relative to existing treatments and their outcomes. Relative to this European standard, the U.S. market is much more fragmented, and up to this point in time has not demanded as much comparative cost-effectiveness data.

Third, in the U.S. the possibility of being the exclusive generic entrant for the first 180-days following loss of exclusivity acts as a powerful lure to generic firms, for during that 6 month time frame the successful generic challenger can monopolize substitution away from the off-patent brand, and by pricing just under the brand's umbrella price, the generic can capture a substantial temporary profit bonanza.[14] Comparable "Paragraph IV Challenge" provisions to the U.S. Hatch-Waxman legislation do not exist in most European countries.

Fourth, as has been pointed out convincingly by Danzon and Furukawa (2011), in the U.S. and in only several European markets, dispensing pharmacies face national healthcare reimbursement policies that direct whether brand-generic decision making is driven largely by pharmacies, by incentivizing patients with lower copayments for generic drugs, mandating generic for brand substitution, and inducing pharmacies to aggressively seek out the lowest-cost generic available among the generic manufacturers. By contrast to these pharmacy driven markets, Danzon and Furukawa describe physician-driven markets as ones in which physicians typically prescribe a specific off-patent molecule by brand name or the originator brand name, in which case generic suppliers are incentivized to compete on brand image rather than on price. Although there are differences among them, at the time Danzon and Furukawa (2011) were writing their article, they characterized the U.S., U.K.,

[14] For a discussion of the lucrative profitability of a successful Paragraph IV first to file challenger, even if the brand responds by launching an authorized generic, see, for example, Drug Channels (2011, 2012).

Netherlands and Canadian markets as pharmacy driven and Germany as becoming more so over time, whereas most other European markets (including France, Italy and Spain) were characterized by them as physician driven. Danzon–Furukawa document that pharmacy acquisition costs are generally lower in pharmacy driven markets, other things equal. Their empirical findings are largely corroborated in the eight European country study by Berndt and DuBois (2012).[15]

A related body of literature has examined the effective length of market exclusivity for small molecules across countries. Effective length of market exclusivity has been defined as the time period between regulatory approval of the drug and initial multi-source generic entry (entry may be delayed beyond patent expiration due to the presence of other market exclusivity provisions, such as the 6 month pediatric exclusivity extension in the US). Danzon and Furukawa (2011) report remarkable homogeneity in mean exclusivity length across the nine countries in their sample, close to 12 years. A similar length of time is reported by Hemphill and Sampat (2012) for the U.S., with "evergreening" attempts by brands to extend exclusivity through filing of additional patents offset roughly equally by "prospecting" patent challengers from generic firms. Hemphill and Sampat (2012) report relative stability with a mean of about 12 years for small molecules between 1991 and 2001. Slightly different findings have been reported by Grabowski and Kyle (2007), who suggest that for small molecules in the U.S., effective patent life has declined slightly over the last few decades.

Due to the very small number of biosimilar entries approved to date in the EU and US, there do not appear to be any studies yet comparing effective market exclusivity durations for biologics across countries.

With virtually no biosimilars in the US on which to provide pricing and entry empirical evidence, some insights might be gained by examining studies of specialty drugs as they lose exclusivity in the U.S. because like biologics many of the specialty drugs are injectable or infused and quite costly when under patent protection. To the extent biosimilar entrants will be therapeutic substitutes rather than be rated as fully interchangeable by the FDA,[16] one might plausibly view the amount of entry and the degree of price decrease from entry observed after LOE in specialty drugs in the U.S. as providing an upper bound to the extent of entry and magnitude of price effects likely to occur as biologics go off patent in the U.S.

[15] Berndt and DuBois (2012) amend the Danzon and Furukawa (2011) classification slightly, characterizing France as becoming pharmacy driven in 2006 (when pharmacies were first given strong financial incentives to substitute generics for brands) and Germany in 2007.

[16] According to the FDA (2014b, p. 15), biosimilarity means that "the biological product is highly similar to the reference product notwithstanding minor differences in clinically inactive components," and that "there are no clinically meaningful differences between the biological product and the reference product in terms of the safety, purity and potency of the product." Regarding interchangeability decisions, in the same FDA (2014b, pp. 5–6) document, draft guidance specified four possible process decision outcomes: definitely not interchangeable; might be interchangeable if enough clinical and analytical data to support it is developed; looks like it is interchangeable but needs more clinical and analytical data to support it; and biosimilar is a "fingerprint" of the innovator product, as demonstrated by analytical data, and no further clinical data is required.

The upper bound interpretation is supported by the fact that many generic specialty injectable or infused drugs are essentially small molecules dissolved in water. They do not have the manufacturing complexity issues that arise in creating biologics, although they do have some manufacturing complexity relating to ensuring that they are sterile and may require thermal and/or lighting-constrained distribution and storage.

In this context, it is useful to examine recent research by Conti and Berndt (2014) focusing on 41 cancer molecules (15 oral, 26 physician-administered injectable/infused) that faced initial LOE in the U.S. between 2001 and 2007. A number of the conclusions are particularly relevant. First, entry by generics following the brand's LOE was generally quite modest, and certainly much smaller than that typically observed for small molecule tablets and capsules: the mean number of ANDA sponsors entering a new molecule formulation after LOE ranged from 1.66 in 2003 to 4.9 in 2007, with what appears to be an upwards trend in entry count in 2006 and 2007 compared to previous years. Among several specialty drugs, exit by the branded manufacturer in the first few years following LOE was observed, as well as delayed and sequential ANDA entry into a given molecule undergoing LOE. Among another subset of drugs that were always generic between 2001 and 2007, the average number of manufacturers declined from 3.04 in 2001 to 2.3 in 2007, suggesting that generic manufacturers of cancer drugs may have been exiting from producing very old generic drugs and instead entering into segments experiencing initial LOE that offered potentially more profitable opportunities. Market size as measured by brand revenues for the molecule pre-LOE, the number of distinct indications for which the molecule was FDA approved or reimbursed by Medicare, and oral (as distinct from injected/infused) formulations increased the extent of entry. Although no monoclonal antibodies experienced initial LOE in this 2001–2007 sample time period, the results suggest that to the extent current on-patent monoclonal antibodies are large revenue products (such as Humira™ and Rituxan™), and are used to treat multiple indications, we can expect that as monoclonal antibody agents experience initial LOE in the US in the near future, they will attract a substantial number of biosimilar or biobetter entrants.

In this sample branded prices rise and generic prices fall in response to LOE and generic entry, with the brand price increases being considerably larger for injected/infused than for oral drugs, but decelerating as the number of generic manufacturers increased. While generic prices of oral formulations fall rapidly and steeply as the number of manufacturers increases, for injected/infused drugs generic prices fall more modestly as the number of manufacturers increased.

In terms of total generic plus brand volume following LOE, a common finding among the specialty cancer drugs was that total volume post-LOE was greater than that pre-LOE; this result may be unique to cancer drugs since a common phenomenon in oncology, but not widely observed elsewhere, is that newly approved drugs are combined with old off-patent oncologic agents in cocktail combination treatments. This positive volume impact is also larger for oral than for injected/infused cancer drugs.

Manufacturing Distinctions of Biopharmaceuticals

The manufacturing processes for biologic drugs are considerably more complex and costly than for traditional small molecules that are formulated as oral tablets/capsules. Classic small molecules pharmaceuticals such as aspirin are chemically synthesized and manufactured into tablets or capsules to be taken orally by patients. Biopharmaceuticals, or biologics, such as insulin and monoclonal antibodies, are large molecules usually produced by unicellar organisms (such as yeast or microbes) or by immortalized mammalian cells in large fermentation vessels. The large molecules are then purified from the cellular broth and formulated for administration to patients by injection or infusion to avoid their gastro-intestinal tract which would digest these biologic products. Because of their scientific and manufacturing differences, small molecule and biologic pharmaceuticals are subject to somewhat different regulatory rules and clinical handling which in turn may affect their development, distribution, reimbursement, competition, product life cycles and so economic incentives. Prior research suggests that the commercial experiences of biologics and small molecules differs significantly.[17]

Due to their size and propensity to fold into different conformations, biologics can be difficult to chemically characterize fully, as is typically done with small molecule pharmaceuticals. A famous set of rules for successful small molecule properties was developed by a Pfizer chemist and his colleagues after examining many successful, and failed, candidate drugs.[18] Now called the Lipinski rule of five, some of the rules state that successful drugs have a molecular weight less than 500, the number of hydrogen donating groups is less than five, and the total number of hydrogen atoms should be less than ten. Biologics violate all these rules. For example, the molecular weights of somatropin, erythropoietin alpha and filgrastim are 22,124, 30,400 and 18,800 respectively. Each is a folded protein consisting of a chain of amino acids. Depending on the product, the chain ranges from 165 to 191 amino acids long. Hence, since any single amino acid has at least two oxygen atoms and a nitrogen atom, each of these biologics clearly violates most of Lipinski's rules for a successful pharmaceutical.

Due to their size and complexity, fully characterizing a biologic through physiochemical means such as is used for pharmaceuticals or biological assays is not currently possible. Furthermore, the links among such features and medically important characteristics such as bioequivalence, interchangeability among products, immunogenicity, pharmacokinetics/dynamics, metabolism and even safety and efficacy are not yet well understood.[19]

The molecular size of biologics raises manufacturing and quality control issues that also confront biosimilar manufacturers. Biomanufacturing is complex, requiring isolation of the DNA or RNA to produce the protein, insertion of that DNA/RNA

[17] Trusheim et al. (2010).

[18] Leeson (2012).

[19] Schellekens (2005).

into a cell line and optimization of it for production, establishment of fermentation conditions, development of purification processes and the logistics of packaging, distribution and storage to preserve the protein. The result is that biopharmaceutical production costs are relatively high with cost of goods sold ranging from 15 to 30 % for today's products,[20] significantly higher than that for small molecule pharmaceuticals. It is unclear if biosimilars will achieve lower production costs or experience higher costs than the reference product(s). Some estimate that biosimilar manufacturers will be unable to exploit the economies of scale or match the accumulated expertise of established players, and so may incur substantially higher unit costs.[21] However, new bioproduction technologies may enable substantially lower unit production costs through lower capital costs and with higher product yields through modern cell lines, but whether they will still be designated as fully interchangeable products with the brand by regulators is uncertain.[22]

Given this regulatory uncertainty (on which, more below) and potentially greater average production costs, it may be more attractive economically for biologic manufacturers to pursue a "biobetter" product entry strategy with an NDA/BLA rather than an abbreviated biosimilar application, consciously differentiating their product from the brand that is now off-patent, rather than seeking biosimilar approval, even though with the latter the number of clinical investigations and costs are likely to be smaller. Other factors affecting the choice between a biosimilar or biobetter strategy include the ease and speed of patient recruitment, the speed with which manufacturing facilities are approved, the willingness of prescribers and payers to adopt biosimilars, concerns regarding immunogenicity, regulatory restrictions on marketing claims, and legal liability. Many of these factors are likely to vary across geographies. Hence with regulatory, commercial and scientific uncertainty all playing critical roles, it is plausible that the choice between biosimilar and biobetter strategies will differ in the US and EU, and depend on more than simply the degree to which EMA-FDA regulatory policies are harmonized or divergent.

With this as background, we now examine the accumulating evidence on biosimilar uptake trends in Europe since 2006.

European Evidence on Biosimilar Uptake

The Committee for Medicinal Products for Human Use (CHMP) of the European Medicines Agency (EMA) issued draft guidelines on similar biological medicinal products in November 2004, which were adopted by CHMP in September 2005 and came into effect on 30 October 2005.[23] According to the guidelines, products will be evaluated on a case-by-case basis but the general approval pathway will be abbre-

[20] Ziegler and Santagostino (2011).

[21] Kelley (2009).

[22] Morrow (2006).

[23] European Medicines Agency (2005).

viated relative to an entirely new biologics product application. This framework includes an overarching set of principles; general guidelines on quality, safety and efficacy; and guidelines specific to product classes. To date, the EMA has issued class-specific guidelines in seven classes,[24] and guidance is under development for several other major biologics product classes including recombinant follicle stimulation hormone, and recombinant interferon beta.[25] The EMA has also approved biosimilars in five product classes—somatropins, erythropoietins, granulocyte colony-stimulating factors (G-CSFs), tumor necrosis factor alpha (TNF-α, infliximab), and human follicle stimulating hormone (FSH, follitropin α).[26] An appendix to this paper provides further details on the reference products, biosimilars, and non-reference products in these five product classes.

Table 1 lists the EMA action history on biosimilars from the inception of the program up to April 18, 2014. Applications currently under review and those which were submitted and then withdrawn prior to EMA action (such as the Marvel applications for insulin biosimilars) are not included.

Sandoz's Omnitrope human growth hormone product was the first biosimilar approved in the EU (April 12, 2006) with Pfizer's Genotropin serving as the reference product.[27] As of June 2010, 14 biosimilars for somatropin, epoietin and filgrastim/lenograstim had been approved and were being marketed in the EU: seven for filgrastim (a granulocyte colony stimulating factor), two for somatropin (growth hormone), and five for short-acting epoietin.

The mid-2010 to 2013 lull having only limited EMA approval action activity was broken in the autumn of 2013 with CHMP recommendations for an eighth filgrastim biosimilar, for the first two biosimilars for a monoclonal antibody and for a fertility hormone. Remsima and Inflectra are both biosimilars for infliximab using Janssen's Remicade as the reference product. Monoclonal antibodies are perhaps the largest class of biologic products, both in terms of numbers of products as well as global revenues. From 2006 through mid-2010 the EMA had ruled on 15 biosimilar applications, approving all but one.

In the middle period (2010–2013) the EMA subsequently withdrew approval for two of the products and approved none. In addition it saw Marvel LifeSciences withdraw its applications for three insulin biosimilars.[28] The EMA noted that the company had stated that, "the decision to withdraw is in order to have sufficient time to repeat and submit bioequivalence T1D (type 1 diabetes) PK/PD

[24] The product-specific biosimilar guidelines include recombinant erythropoietins, low-molecular-weight heparins, recombinant interferon alpha, recombinant granulocyte-colony stimulating factor, somatropin, recombinant human insulin, and monoclonal antibodies. See European Medicines Agency (2006a, b, c, 2009a, b, 2010, 2012a, b, c).

[25] European Medicines Agency (2010, 2011a, b, c, 2012a, c (a finalized version of 2010))

[26] European Medicines Agency (2013a).

[27] European Medicines Agency (2013a, p. 4). BioPartners' Valtropin was approved by the EU on the same day as Sandoz' Omnitrope with Humatrope as the reference product, but Valtropin has not been marketed.

[28] European Medicines Agency (2012b).

Table 1 European biosimilar regulatory reviews and current marketing status[a]

EMA biosimilar rulings

Trade name	Active substance	Biosimilar sponsor	Reference product	Therapeutic area	Decision	Date
Omnitriop	Somatropin	Sandoz	Genotropin	Tumer Syndrome, Pituitary Dwarfism, Prader-Willi Syndrome	Approve	April 12, 2006
Valtropin	Somatropin	BioPartners	Humatrope	Tumer Syndrome, Pituitary Dwarfism	Approve	April 24, 2006
					Withdrawn	May 20, 2012
Alpheon	Interferon alpha-2a	BioPartners	Roferon-a		Reject	June 28, 2006
Abseamed	Epoetin alpha	Medice	Eprex	Chronic Kidney Failure, Anemia, Cancer	Approve	August 28, 2007
Binocrit	Epoetin alpha	Sandoz	Eprex	Chronic Kidney Failure, Anemia	Approve	August 28, 2007
Epoetin alfa Hexal	Epoetin alpha	Hexal	Eprex	Chronic Kidney Failure, Anemia, Cancer	Approve	August 28, 2007
Retracrit	Epoetin zeta	Hospira	Eprex	Anemia, Autologous Blood Transfusion, Cancer, Chronic Kidney Failure	Approve	December 18, 2007
Silapo	Epoetin zeta	STADA	Eprex	Anemia, Autologous Blood Transfusion, Cancer, Chronic Kidney Failure	Approve	December 18, 2007
Biograstim	Filgrastim	CT Arzneimitel	Neupogen	Cancer, Hematropoietic Stem Cell Transplantation, Neutropenia	Approve	September 15, 2008
Filgrastim ratiopharm	Filgrastim	Ratiopharm	Neupogen	Cancer, Hematropoietic Stem Cell Transplantation, Neutropenia	Approve	September 15, 2008
					Withdrawn	July 20, 2011
Ratiograstim	Filgrastim	Ratiopharm	Neupogen	Cancer, Hematropoietic Stem Cell Transplantation, Neutropenia	Approve	September 15, 2008
Tevagrastim	Filgrastim	Teva	Neupogen	Cancer, Hematropoietic Stem Cell Transplantation, Neutropenia	Approve	September 15, 2008
FilgrastimHexal	Filgrastim	Hexal	Neupogen	Cancer, Hematropoietic Stem Cell Transplantation, Neutropenia	Approve	February 6, 2009

(continued)

Table 1 (continued)

EMA biosimilar rulings

Trade name	Active substance	Biosimilar sponsor	Reference product	Therapeutic area	Decision	Date
FilgrastimZarzio	Filgrastim	Sandoz	Neupogen	Cancer, Hematropoietic Stem Cell Transplantation, Neutropenia	Approve	February 6, 2009
Nivestim	Filgrastim	Hospira	Neupogen	Cancer, Hematropoietic Stem Cell Transplantation, Neutropenia	Approve	June 8, 2010
Remsima	Infliximab	Celltrion	Remicade	Rheumatoid Arthritis, Crohn's Disease, Ulcerative Colitis, Ankylosing Spondylitis, Psoriatic Arthritis and Psoriasis	Approve with monitoring	September 9, 2013
Inflectra	Infliximab	Hospira	Remicade	Rheumatoid Arthritis, Crohn's Disease, Ulcerative Colitis, Ankylosing Spondylitis, Psoriatic Arthritis and Psoriasis	Approve with monitoring	September 9, 2013
Ovaleap	Follitropinalfa	Teva	Gonal-f	Anovulation	Approve with monitoring	September 27, 2013
Grastofil	Filgrastim	Apotex Europe	Neupogen	Neutropenia	Approve with monitoring	October 18, 2013

[a]European Medicines Agency (2013a, b, 2014)

(pharmacokinetic/pharmacodynamic) data on each clamp study in order to comply with the planned new insulin guideline…, at a validated CRO (contract research organization)."[29] This experience demonstrated the impact that regulatory guidance regarding the level of evidence required from sponsors to demonstrate biosimilarity can have on the number and timing of biosimilars. Compared to small molecule generics, the amount of scientific and clinical effort required for biosimilar determination by regulatory authorities such as the EMA is significantly greater.

Instead of documenting efficacy and safety of the follow-on biologic via clinical trials, for therapeutic substitutability a follow-on biologic must prove biosimilarity. To date the EMA has required at least one Phase II or Phase III clinical trial for a biosimilar to demonstrate similar safety and efficacy to its reference molecule. Notably, the EMA framework does not result in any findings of interchangeability, with questions of substitutability being left to the member state competence to regulate. Local substitution laws differ among EU member states, with some (e.g., Spain, Sweden) including explicit prohibition on automatic substitution for biologics (i.e., prohibiting mandatory pharmacy-level substitution).[30] This contrasts with US policy, by which the FDA approves applications as therapeutic substitutes or interchangeable therapies. Within the EU framework, the EMA also determines the extent to which biosimilarity of a biologic for treatment of one indication can be extrapolated to other indications for which the reference biologic had received EMA approval.[31]

Of note in the US, there has been considerable activity at the state legislature level, with legislation being introduced frequently following the principles advocated for by organizations such as the Biotechnology Industry Organization (BIO). Specifically, the principles BIO advocates that states should follow include: (1) substitution should occur only when the FDA has designated a biologic product as interchangeable; (2) the prescribing physician should be able to prevent substitution; (3) the prescribing physician should be notified of the substitution; (4) the patient, or the patient's authorized representative, should, at a minimum, be notified of the substitution; and (5) the pharmacist and the physician should keep records of the substitution.[32] Other countries have used a European-like approach, including Canada (where biosimilars are termed "subsequent entry biologics", or "SEBs") and Japan. Australia adopted the EU guidelines in August 2008.[33]

Because the phenomenon of biologic patent expiration is only a relatively recent development, the evidence on pricing of biologics to date is rather sparse. One publicly available peer-reviewed article is that by Calfee and DuPre (2006), who find that while the first generation of biologics (such as the "branded generic" insulins) were priced substantially lower in Europe than in the U.S., the later generation

---

[29] European Medicines Agency (2012b).

[30] Ehmann (2010).

[31] European Commission (2013, p. 27).

[32] Biotechnology Industry Association (2013). For additional discussion, see Karst (2013a), who also provides a state-by-state legislation scorecard.

[33] Grabowski et al. (2013, p. 3).

single source branded biologics were priced close to parity in Europe and the U.S. Whether prices of more recent cohorts of newly launched branded biologics are parity priced in Europe and the US is unknown.

However, a recent publication, by Grabowski et al. (2011, updated in 2013, p. 24) surveys various U.S. studies that have projected biosimilar price discounts relative to pre-entry U.S. brand prices. Grabowski et al. (2007a, b) projected a 10–30 % discount in year 1, the Congressional Budget Office (2008) a 20 % discount in year 1 increasing to 40 % by year 4, Steve Miller and Jonah Houts (2007) of Express Scripts 25 % in year 1, and Roland (Guy) King (2007) of Avalere Health a 20 % discount in year 1 increasing to 51 % in year 3. Notably, each of these projected discounts for biosimilars is considerably less than the discounts achieved in recent years by generic small molecules in the U.S.,[34] but is in the range of injected/infused cancer specialty drugs experiencing initial LOE in the U.S. in 2001–2007.[35]

Before presenting empirical evidence from biosimilars in the EU, we digress briefly to focus on some measurement issues. Because some medicines are used to treat diverse conditions having very different dosages across individuals (such as those based on weight or body mass index) and indications, measuring the volume of these multipurpose medicines is very challenging. The IMS Health Midas sales data in local currencies and extended units that we employ are derived from ex-manufacturer invoices; these data therefore reflect revenues received by manufacturers, they exclude wholesale and retail margins, and therefore do not reflect actual reimbursements by national health authorities or other payers to the retail sector. The local currency sales have been converted to US dollars at contemporaneous quarterly varying exchange rates, for all countries.

Some researchers transform these data into days of therapy utilizing the World Health Organization (WHO) Defined Daily Dosage (DDD) metric. The WHO Collaborating Center for Drug Statistics and Methodology defines the DDD as follows:

> "The DDD is the assumed average maintenance dose per day for a drug used for its main indication in adults…. It should be emphasized that the defined daily dose is a unit of measurement and does not necessarily reflect the recommended or Prescribed Daily Dose. Doses for individual patients and patient groups will often differ from the DDD and will necessarily have to be based on individual characteristics (e.g. age and weight) and pharmacokinetic considerations…. Drug consumption data presented in DDDs only give a rough estimate of consumption and not an exact picture of actual use. The DDD provide a fixed unit of measurement independent of price and dosage form (e.g., tablet strength) enabling the researcher to assess trends in drug consumption and to perform comparisons between population groups…. *The DDD is nearly always a* ***compromise*** *based on a review of the available information including doses used in various countries when the information is available. The DDD is sometimes a dose that is rarely if ever prescribed, because it is an average of two or more commonly used doses.*"[36]

[34] Aitken et al. (2008), Berndt and Aitken (2011) and Aitken et al. (2013).

[35] Conti and Berndt (2014).

[36] World Health Organization (2009, pp. 1–2); italics and bold in original text. For further details concerning DDD, see World Health Organization (2003 (Chap. 6), 2011); also see International Federation of Pharmaceutical Manufacturers and Associations (2006).

For our purposes, it is useful to note that the WHO DDD assigned to a drug is time invariant, and is identical across countries and dosage strengths. While it would be preferable to utilize the IMS Health daily average consumption (DACON) metric derived and updated from actual retail prescription data, as in Berndt and Aitken (2011), currently IMS Health DACON data are only available for some countries, and they are often not available for medicines dispensed outside the retail sector (thereby excluding biologics administered in hospitals or outpatient clinics).

Other possible volume measures include extended units, standard units, and eaches. *Extended units* are the number of tablets, capsules, milliliters, ounces, etc. of a product shipped in each unit. This number is calculated by multiplying the number of units by the product size. According to IMS, "...extended units are not meaningful above the package level, because a product may have different forms and strengths and therefore a different type of unit for each presentation."[37] *Standard units* represent the number of dose units sold for a particular product. Examples of standard units are the number of tablets sold, the number of 5-ml doses for liquid products sold, or the number of vials sold. According to IMS personnel, "Standard units enable you to compare sales volume data for products across different product forms and dosing regimens. For example, you can compare solid to liquid forms more precisely by equating the number of milliliters of a liquid preparation—such as 5 ml of liquid—to the standard solid dosage of one tablet. Standard units are defined for all product forms, allowing you to make accurate comparisons among several product forms."[38] It is our understanding that standard units replaced earlier measures based in part on eaches, and that standard unit measures for MIDAS and other IMS data bases such as the U.S. National Sales Perspective are not available before 2006. Regarding *eaches*, IMS personnel indicate that eaches represent "the number of single items (such as vials, syringes, bottles, or packet of pills) contained in a unit or shipping package and purchased by providers and pharmacies in a specific time period. An each is not a single pill or dosage of medicine (unless one package consists of a single dose). An each may be the same as a unit if the unit does not subdivide into packages. Eaches are usually used to look at injectable products. Eaches are most meaningful at the package level, since packages and their subunits may contain different quantities of strengths and volumes."[39]

As we shall see when discussing biosimilar utilization studies appearing in the existing literature, various researchers have differed in their choice of volume measure. DDD requires assumptions about actual clinical use, whereas standard units is more directly observable. Thus for gross market analyses, standard units may be a preferable measure to DDD. However, it is our understanding that standard units do not always account properly for unit size differences (e.g., a vial is a vial regardless of ml volume). In the results of our utilization research presented later in this section, we employ standard units; by contrast, Grabowski et al. (2011, 2013) employ DDDs

[37] From email correspondence with Terry McMonagle at the IMS Institute for Healthcare Informatics, September 4, 2013, 11:15 am.

[38] See footnote 37.

[39] See footnote 37.

as their unit of volume measure, while in Grabowski (2013) volume measures are entirely omitted and only dollar revenue share data are presented.

One other issue meriting discussion is that considerable diversity exists across molecules and countries in the setting in which biologics are dispensed such as retail pharmacies, physicians' offices or hospitals.[40] Not only might this affect which volume measure is most appropriate for a particular molecule, but because of the extensive tendering that occurs for hospitals in Europe, actual and average measured prices of biologic molecules could depend on the composition of dispensing sites. As best we understand it, to the extent the tendering process results in rebates paid by manufacturers to national or regional health authorities, these rebates are unlikely to be reflected in the invoice prices monitored by the IMS MIDAS data system.

We begin our overview of biosimilar diffusion in certain EU countries by discussing the Grabowski et al. (2013) results that utilized 2007–2009 DDD data for five large EU countries, and then update and compare our new analyses using more recent IMS Health MIDAS standard unit measures and a larger number of countries.

Table 2 above summarizes biosimilar DDD volume shares in five large European countries—France, Germany, Italy, Spain and the U.K.—for three molecules—somatropin, erythropoietin alpha, and granulocyte colony stimulating factor (G-CSF) from 2007 to 2009, as reported in Grabowski et al. (2013, Table 2, p. 6). The most striking finding is the absence of any pattern—the extent of biosimilar penetration varies substantially both across therapies within a country, and across countries for the same therapy. Through 2009, Germany exhibited the highest level of aggregate demand and market share for any biosimilar product (a 62 % market share for erythropoietin alpha in 2007). According to one analyst, Germany's influential Federal Healthcare Committee which has jurisdiction over which products and services are reimbursed, has in fact embraced biosimilars wholeheartedly, and reinforced its preference by implementing a reference pricing system. Germany also has placed specific targets or quotas for physicians and sickness funds for biosimilars that vary geographically. Finally, Germany has become the dominant source of biosimilar manufacturing in Europe.[41]

Relative to that in Germany, the uptake of biosimilars in other European countries through 2009 was much slower. Recall that while EMA approval is necessary for a biosimilar to be marketed in EU member countries, actual sales may be delayed since reimbursement must still be negotiated between manufacturers and regional/national government payers. This reimbursement approval delay may be partly responsible for later biosimilar entry dates in several European countries.

For erythropoietin alpha, in Germany the biosimilar products accounted for 62 % of total biosimilar and innovator erythropoietin products sold in 2009, within 2 years of its launch. However, the cross-country heterogeneity in biosimilar takeup is sub-

[40] See Walsh (2013) for examples and discussion.

[41] The analyst's comments are referenced in the note below Table 2 of this document; also see Walsh (2013).

Table 2 Initial biosimilar competition in selected EU countries: market share evidence biosimilar unit share of the molecular entity, 2007–2009

	France (%)	Germany (%)	Italy (%)	Spain (%)	U.K. (%)
Somatropin					
2007	2	3	6	1	0
2008	10	6	17	1	0
2009	16	8	27	5	1
Erythropoietin alpha					
2007	0	0	0	0	0
2008	0	35	0	0	0
2009	4	62	0	4	1
Granulocyte colony stimulating factor					
2007	–	–	–	–	–
2008	0	1	0	0	2
2009	7	17	N/A	9	21

Notes: Taken from Grabowski et al. (2013, p. 6). An endnote adds that "Data are based on IMS Midas data as reported in Rovira et al. (2011). Biosimilar share of unit sales are measured based on Defined Daily Dose. Biosimilar G-CSF was not launched until 2008, so biosimilar shares for 2007 are not reported in Table 3. For G-CSF in Italy in 2009 the biosimilar share is recorded as N/A to reflect insufficient data for calculating a biosimilar share—fewer than 5,000 DDDs were reported in the data for combined innovator and biosimilar unit sales in Italy that year."

stantial; for France and Spain the erythropoietin biosimilar share was only 4 %, and even less at 1 and 0 % for the U.K. and Italy, respectively.

Cross-country patterns are quite different for G-CSF (filgrastim). As seen in Table 2 above, biosimilar shares for G-CSF in 2009 ranged from 7 % in France to 21 % in the U.K., with Spain at 9 % and Germany at 17 % being in between. Biosimilar sales of G-CSF in 2009 failed to reach minimum reporting thresholds in Italy.

Finally, cross-country patterns for biosimilar somatropin differ from those both for erythropoietin alpha and G-CSF. As seen in the top panel of Table 2, in 2009 biosimilar somatropin volume shares were on average larger than for the other biosimilar molecules, with Italy at 27 % having the largest share, followed by France at 16 %, Germany 8 %, Spain 5 % and the U.K. 1 %.

European Relative Biosimilar Volume Shares, 2007–2012

To shed light on the importance of measurement issues concerning whether one measures volume based on DDD vs. standard units, in Table 3 we present standard unit based volume shares for the same five countries for years 2007–2012 as in Grabowski et al. (2013); note that the years 2007–2009 overlap in both Tables 2 and 3, facilitating a direct comparison. We also expand the set of countries to include several smaller ones in northern Europe and Scandinavia: Belgium, Finland,

Table 3 Biosimilar standard unit share of the molecular entity

	France (%)	Germany (%)	Italy (%)	Spain (%)	UK (%)	Belgium (%)	Finland (%)	Norway (%)	Sweden (%)
Somatropin									
2007	1	1	1	0	0	0	0	0	0
2008	5	2	2	0	0	0	0	0	0
2009	7	2	3	1	0	5	0	0	3
2010	9	3	4	1	1	6	0	0	7
2011	10	4	4	2	1	7	5	0	7
2012	11	5	4	3	1	7	5	0	8
Erythropoietin alpha and zeta									
2007	0	2	0	0	0	0	0	0	0
2008	0	41	0	0	0	0	0	0	1
2009	4	58	0	3	3	0	0	0	19
2010	9	64	5	19	7	0	100	0	49
2011	11	69	13	31	10	0	100	100	62
2012	16	68	21	39	7	0	100	100	70
G-CSF (fHgrastim)									
2007	0	0	0	0	0	0	0	0	0
2008	0	0	0	0	1	0	0	0	0
2009	6	8	1	9	20	0	0	25	5
2010	26	15	9	23	53	0	15	0	40
2011	35	23	36	33	71	0	33	0	70
2012	48	27	60	51	81	0	65	20	86

Norway and Sweden. We begin with a comparison limiting our attention to the five large EU countries, and then consider whether the smaller northern European and Scandinavian countries exhibit similar or divergent trends.

For the five large EU countries considered by Grabowski et al. (2013), a comparison of molecular shares for the 2007–2009 overlapping years in Tables 2 and 3 suggests a pattern in which biosimilar shares based on standard units tend generally to be somewhat smaller—in some cases considerably smaller—than those based on DDDs. For example, in 2009 for France, Germany, Italy, Spain and the U.K. the standard unit (DDD) shares for biosimilar somatropin are, respectively, 7 (16 %), 2 (8 %), 3 (27 %), 1 (5 %), and 0 % (1 %); for erythropoietin alpha and zeta the respective molecular shares in 2009 are 4 (4 %), 58 (62 %), 0 (0 %), 3 (4 %) and 3 % (1 %), while forfilgrastim(G-CSF) they are 6 (7 %), 8 (17 %), 1 (N/A), 9% (9 %), and 20 % (21 %).

Looking at years beyond 2009 for the five large EU countries in the first five columns of Table 3, we observe that each of the three products continued to follow a distinct adoption pattern. Across product classes, the most successful of these three biosimilar product classes is filgrastim (G-CSF), the least successful is somatropin, with erythropoietin alpha and zeta being in between. Filgrastim reaches 50 % market share or higher in most large EU countries by 2012 while somatropin exceeds 10 % share in only one (France) with erythropoietin achieving the most varied market shares, ranging from 7 % in the UK and 16 % in France to 39 % in Spain and 68 % in Germany, the highest single large EU country market share observed.

As with the DDD data, the standard unit data shows that adoption in the first years (2007–2009) was cautious across all five large EU countries and three products, with the exception of erythropoietin in Germany which achieved 50 % market share in only its third year, 2009. In addition to clinicians taking a cautious view to using biosimilars, this slow diffusion may reflect the range of dates for biosimilar regulatory and reimbursement decisions for the products among these five large EU countries.

By 2012, among these five large EU countries, biosimilar filgrastim has become the most widely successful biosimilar product to date, at least as measured in terms of standard unit market shares.[42] Perhaps surprisingly given its rapid initial use of the first biosimilar erythropoietin alpha, Germany has the lowest filgrastim biosimilar market share at 27 %. By 2012, the biosimilar filgrastim share in the U.K. has grown to an impressive 81 %, in Italy it increased from 36 % in 2011 to 60 % in 2012, and in both France and Spain it captured about half the filgrastim product market (48 and 51 % respectively)..

Biosimilar somatropin has continued its gradual but low penetration over time among these five large EU countries, with the greatest penetration being but 11 % for France in 2012, and with all other countries experiencing single digit market shares, usually below 5 %. Of the three products, somatropin is the only one in which France leads the five country usage.

Biosimilars to erythropoietin alpha (including the zeta forms that also used Eprex as a reference product, see Table 1) have maintained and leveled off their market share in Germany at 64–69 %, but the most rapid growth of this biosimilar molecule among the five largest EU countries has occurred in Spain (from 3 to 39 % between 2009 and 2012) and in Italy (from 0 to 21 % between 2009 and 2012). France has seen a steady increase in the biosimilar erythropoietin alpha share, but at 16 % in 2012 this share is still quite low. Finally, in the U.K. the biosimilar erythropoietin alpha share has an uneven trend, increasing from 3 to 10 % between 2009 and 2011, but then falling to 7 % in 2012.

Another common theme from Tables 2 and 3 is that there is significant heterogeneity across the largest EU countries in the penetration of a given biosimilar product, and across products in penetration by country. For example, at 81 % the U.K. has the greatest penetration of biosimilar filgrastim, Germany has the largest penetration of biosimilar erythropoietin alpha at 68 %, and France the highest penetration of biosimilar somatropin at 11 %. Heterogeneity in biosimilar uptake is the dominant theme.

---

[42] We note in passing that the 81 % biosimilar share reported for filgrastim in the UK in 2012 in Table 3 based on standard unit measures is very similar to the part-year 2013 83 % share for filgrastim in the UK reported in Walsh (2013, slide 8).

Germany adopted biosimilar erythropoietin alpha quickly and has achieved nearly double the relative usage in 2012 with 68 % biosimilar compared to the second place country Spain at 39 % and more than four times the 16 % standard unit market share in France. For somatropin, however, Germany's market share quickly fell behind that of France and in 2012 somatropin biosimilars had only achieved 5 % market share in Germany, half that in France. Moreover, with filgrastim, Germany has achieved the lowest biosimilar penetration of the five countries with only 27 % market share, merely about half of most other countries and only a third that of the leading country, U.K. with 81 %. Biosimilar filgrastim is also strong in Italy where its 60 % biosimilar penetration places second, the country's highest relative rank. Notably, these diffusion rates do not support earlier observations by Senior (2009), and Grabowski (2013) and Grabowski et al. (2011, 2013) based on data ending in 2009, suggesting a German exceptionalism due to its centralized and biosimilar encouraging Federal Healthcare Committee, its biosimilar reference pricing system, specific targets or quotas for physician and sickness funds for biosimilars, and its role as the main source of biosimilar manufacturing in Europe. Moreover, in 2012, Germany lagged in biosimilar use in two of the three product areas, with the U.K. and France leading in the other two.

Perhaps most surprisingly, Spain and Italy which have arguably suffered the most from the austerity of the Great Recession, do not lead in biosimilar use in any of the three products (see Table 3). This lower use of biosimilars in the most cost constrained countries may be changing, however. From 2010 to 2012, Italy's use of biosimilar filgrastim lept from last place at 9 % biosimilar penetration to a second place 60 % penetration in 2012. Similarly, Italy's acceptance of biosimilar erythropoietin moved from essentially none to 12 %, surpassing the penetration achieved in France and the U.K. We conclude, therefore, that among the five largest EU countries, the adoption of biosimilars remains dynamic with a stable equilibrium between biosimilar and branded drugs apparently not yet achieved.

It may be instructive, however, to expand the analysis to examining biosimilar uptake trends in smaller EU countries, such as Belgium and the Scandinavian countries of Finland, Norway, and Sweden; biosimilar uptake shares for these four countries are presented in the final four columns of Table 3. Perhaps the most striking result is that while Belgium has been very slow in converting to biosimilars, the Scandivanian countries initially delayed in biosimilar uptake, but since 2010 their penetration of biosimilars has been dramatically rapid and deep. For somatropin (the top panel in Table 3), biosimilar penetration has been modest—greater than that of France but less than that of the four other large EU countries. For erythropoietin alpha and zeta, however, by 2011 both Finland and Norway achieved 100 % biosimilar penetration, with Sweden in 2012 at 70 % being even more biosimilar-friendly than Germany at 68 %. In the case of filgrastim, while Norway at 20 % biosimilar volume share is lower than that of any of the large five EU countries, at 65 % Finland is second only to the U.K. among the large five EU countries, and Sweden tops them all with an 86 % biosimilar standard unit volume market share. In stark contrast to the Scandinavian countries, Belgium's adoption of biosimilars is strikingly small—only 7 % of somatropin in 2012, and no biosimilar adoption through 2012 for both erythropoietin alpha and zeta and filgrastim. It should be noted that volumes in these countries tend to be quite low and therefore substantial share changes may be driven by alterations in but a few contracts or by a few medical groups.

In summary, Germany experienced a rapid uptake of biosimilars, but the adoption rate has stabilized since about 2009. Scandinavian countries delayed their initial adoption, but their acceptance of biologics has been very rapid and deep since then. Belgium's transition to biosimilars has been among the slowest and shallowest among the nine EU countries examined here.

European Relative Biosimilar Revenue Shares, 2007–2012

It would normally be plausible to expect that the revenue weighted shares of biosimilars will be much lower than standard unit market shares, since it is usually assumed that biosimilars will be priced significantly lower than the reference product and other branded, first to market products in the same class. Table 4 shows the revenue market shares of biosimilars in the same format used for standard unit market shares in Table 3, not only for the three products in the five largest EU countries studied over the 2007–2009 time period by Grabowski et al. (2013), but also updated for 2010–2012 and expanded to four other northern European and Scandinavian countries. Again the analysis is based on aggregated quarterly IMS MIDAS data in which revenues were collected in local currencies and converted to US Dollars using the exchange rate in effect for that period. Data for the five large EU countries considered by Grabowski, Long and Mortimer are presented in the first five columns of Table 4, while that for the additional four countries we examine are in the final four columns.

Like the standard unit market shares in Table 3, even within the five largest EU countries the dollar shares in Table 4 exhibit a wide range of values among products and countries. However, some of the relative positions of the countries change. For instance, Germany biosimilar filgrastim use in 2012 came in lowest at 27 % among the five largest EU countries when using standard unit measures (Table 3) but biosimilar filgrastim's share doubles to 54 % in Germany when using dollar share (Table 4), placing it in a tie for second with Italy, but still trailing the UK's 81 % share. For somatropin, Spain's second to last place in standard units (3 %) rises to a 10 % share using dollar revenue share placing it in first place before France whose dollar share at 9 % is slightly lower than the 11 % standard unit share. The relative country rankings among the five largest EU countries for erythropoietin alpha/zeta remain unchanged.

Turning to standard unit (Table 3) vs. dollar (Table 4) shares for the three biosimilar molecules in the four additional countries (Belgium, Finland, Norway and Sweden), for somatropin we observe Belgium's dollar shares are slightly smaller than its standard unit shares (although both are small), but for Finland and Sweden the dollar shares are larger than the standard unit shares. Across all nine countries, Norway has the smallest and France the largest somatropin standard unit shares, and while Norway continues to have the smallest somatropin biosimilar dollar

Table 4 Biosimilar dollar share of the molecular entity

	France (%)	Germany (%)	Italy (%)	Spain (%)	UK (%)	Belgium (%)	Finland (%)	Norway (%)	Sweden (%)
Somatropin									
2007	1	1	1	0	0	0	0	0	0
2008	3	2	2	1	0	0	1	0	0
2009	5	2	4	2	1	3	3	0	4
2010	7	4	5	5	1	3	4	0	11
2011	8	5	6	7	1	4	5	0	11
2012	9	6	5	10	2	5	6	0	14
Erythropoietin alpha and zeta									
2007	0	1	0	0	0	0	0	0	0
2008	0	32	0	0	0	0	16	1	2
2009	3	50	0	2	4	0	55	3	21
2010	10	60	4	9	7	0	84	3	52
2011	12	69	9	15	9	0	97	54	65
2012	19	72	17	26	8	0	99	66	73
G-CSF (fHgrastim)									
2007	0	0	0	0	0	0	0	0	0
2008	0	0	0	0	1	0	0	0	0
2009	5	22	1	7	19	0	0	12	5
2010	20	35	7	17	54	0	11	5	29
2011	28	47	30	24	73	0	29	4	57
2012	43	54	54	48	81	1	54	9	80

shares (at zero), Sweden's dollar shares for somatropin biosimilars is larger than any of the other eight EU countries. Notice also that in Finland and Sweden, the switch from branded to biosimilar occasionally happens completely within a single year, e.g., Finland in 2010 and Norway in 2011 for erythropoietin alpha and zeta. A plausible hypothesis is that the volumes in these countries are quite small, so that simply switching a small number of contracts (perhaps just one?) results in dramatic product share changes.

Whether measured in terms of standard unit or dollar shares, diffusion of biosimilar versions of erythropoietin alpha and zeta, as well as filgrastim, in to Belgium is virtually nil. In sharp contrast, while the diffusion of biosimilar erythropoietin alpha and zeta among the five large EU countries was most rapid and deep for Germany (whether measured in shares of standard units or dollars), in both Norway and Sweden the extent of diffusion of this biosimilar molecule is close to that of Germany, and in the case of Finland the biosimilar share of erythropoietin alpha and zeta approaches 100 %.[43] Relatively slow and shallow adoption of biosimilar fil-

[43] The dollar and standard unit shares do not show complete concordance for Finland and Norway. The underlying IMS MIDAS data shows some minimal reference product dollar sales in spite of zero standard unit shipments. This could be due to reporting timing differences or other idiosyncratic causes affecting the very small values.

grastimis observed in Belgium and Norway, but as with erythropoietin alpha and zeta, adoption of biosimilar filgrastim by 2012 is very substantial in both Finland and Sweden. Although the relative standard unit/dollar share sizes differ among the four small EU countries for somatropin (standard unit shares larger than dollar shares for Belgium, smaller for Finland and Sweden, both zero for Norway), for erythropoietin alpha and zeta by 2012 standard unit (dollar) shares for Finland and Norway are at 100 % (99 and 66 %, respectively), for Sweden in all years the dollar shares of this biosimilar molecule are slightly greater than the standard unit shares. Finally, the same country patterns of share inequalities holds in the case of filgrastim: standard unit shares are greater than dollar shares for Finland and Norway, but the reverse inequality is observed in Sweden.

It is not clear what accounts for the differences between the standard unit and dollar shares. As reported above, while in most cases dollar shares are larger than shares in standard units, this is not always the case (e.g., somatropin in France, epoieten alpha and zeta in Spain, and filgrastim in Italy, Spain, Finland, Norway and Sweden). As discussed earlier, standard units may not be perfectly comparable if vial sizes or relative dosing forms (infusion, number injections from multi-dose vials, single use syringe injection) vary among countries and/or biologic products. Moreover, if rebates to government payers are not incorporated in the invoice data, nominal invoice prices could considerably overstate net of rebate payments; this might occur in the context of hospital tendering practices.[44] What is clear is that whether measured by standard unit or dollar shares, the diffusion of biosimilars among the four smaller EU countries is slowest in Belgium, followed by Norway, and generally most rapid in Finland and Sweden. Moreover, compared with Germany, the country adopting biosimilars most rapidly among the five largest EU economies, biosimilar adoption in Finland and Sweden is on a par if not more rapid and deep than in Germany.

European Relative Biosimilar/Brand Prices

Table 5 illustrates another surprising feature when comparing standard unit shares in Table 3 with dollar shares in Table 4: if for a given country/molecule/year the dollar share is greater than the standard unit share, then a logical inference is that the relative prices of the biosimilars appear to be *higher than* the reference and other branded products with which they compete. Table 5 shows the share changes between Tables 3 and 4—standard unit share minus dollar share. It shades in light grey those instances in which the standard unit share is smaller than the dollar share. This implies that the relative pricing of biosimilars might be higher, thus increasing

[44] We note, however, that in our nine country sample, for somatropin and filgrastim, only in Spain is dispensing limited to the hospital setting; in all other countries for these two molecules, dispensing occurs in both the hospital and retail setting. For erythropoietin alpha, in both Spain and Belgium dispensing occurs only in the hospital setting, and in all other countries dispensing takes place in both the hospital and retail settings.

Table 5 Biosimilar unit share minus dollar share of the molecular entity

	France (%)	Germany (%)	Italy (%)	Spain (%)	UK (%)	Belgium (%)	Finland (%)	Norway (%)	Sweden (%)
Somatropin									
2007	1	0	0	0	0	0	0	0	0
2008	2	0	0	0	0	0	−1	0	0
2009	2	0	−1	−1	0	2	−3	0	−2
2010	2	−1	−1	−4	0	3	−4	0	−4
2011	2	−1	−1	−5	0	3	0	0	−4
2012	2	−1	−1	−7	−1	2	−1	0	−5
Erythropoietin alpha and zeta									
2007	0	1	0	0	0	0	0	0	0
2008	0	10	0	0	0	0	−16	−1	−1
2009	0	7	0	0	−1	0	−55	−3	−2
2010	−1	4	2	10	0	0	16	−3	−3
2011	−1	0	3	17	1	0	3	46	−3
2012	−3	−4	4	13	−1	0	1	34	−3
G-CSF (filgrastim)									
2007	0	0	0	0	0	0	0	0	0
2008	0	0	0	0	0	0	0	0	0
2009	2	−14	0	2	1	0	0	13	0
2010	6	−21	2	7	0	0	4	−5	11
2011	7	−25	6	8	−2	0	4	−4	13
2012	5	−27	7	3	0	−1	11	11	6

Gray shaded means Unit Share is *less than* dollar share. This implies Biosimilars having *higher* prices

biosimilar dollar share. Alternatively, the analysis might be interpreted as indicating that standard units are imperfectly comparable across these products, or as noted earlier, the dollar shares exclude rebates from manufacturers to national/regional government authorities. Several patterns are worth noting, however. First, by country across all three products and 6 years, shaded cells occur for all nine countries, with half or more of the cells being shaded in Germany, the U.K., and Sweden. In France and Belgium, less than one third of the cells are shaded. Countries with between one third and one half of the cells being shaded include Italy, Spain, Finland, and Norway. Second, if one instead looks for molecule specific patterns, one finds that across the 6 years and nine countries, 33 of the 54 or 61 % of the cells are shaded for somatropin, for erythropoietin alpha and zeta, only 35 % of the cells are shaded, and for filgrastim the share of shaded cells is only 24 %. Hence, while the existence of biosimilar prices being apparently greater than for the branded competitors of the same molecule is widespread, it is particularly prevalent in Germany, the U.K. and Sweden, and for somatropin. Alternatively, this apparently counterintuitive finding concerning relative prices is observed least frequently in France, Finland, Norway and Sweden, and for filgrastim. It will be important for future research to focus on assisting researchers in understanding and interpreting these seemingly paradoxical pricing findings.

Several other characteristics of the EU biosimilar market merit attention, based on the IMS data. First, among the five largest EU countries, entry by biosimilar manufacturers is quite limited outside Germany. While Germany has multiple biosimilar entrants in all three products plus some cross-border shipments, Novartis/Sandoz is the only biosimilar entrant for two (somatropin and erythropoietin alpha) out of the three products in the other EU countries. With Germany again being the exception, since the early wave no new biosimilars have been entering these three product markets in the last few years. In most of the five largest EU countries the number of branded products outnumbers the number of biosimilar products by a factor of two or more. This market behavior contrasts with small molecule generics where multiple entrants over time are common.[45] This could reflect few other products losing their market exclusivity in recent years, but could also be indicative of the challenges of biosimilar entry, such as a preference for biobetters over biosimilars (e.g., long-acting Aranesp and Neulasta vs. short-acting biosimilars for erythropoietin/Eprex and filgrastim/Neupogen), particularly in countries such as Germany and Sweden.

The evidence from all nine EU countries demonstrates that at least for some products, biosimilars will be widely accepted by clinicians. The evidence further suggests that biosimilar usage in each country that chooses to allow them will be unique depending on the regulatory, reimbursement and clinical actions taken within each country. A somewhat surprising finding from this research is that at least up through 2012, in an ever more globalized therapeutic marketplace, biosimilar usage to date exhibits a distinctly local result for each product. Whether this is simply a differential initial market experience phenomenon with greater convergence and uniformity delayed but ultimately achieved, remains to be seen.

More generally, given differences in health care systems and cultures, biosimilar market development and share uptake may differ not only among the EU countries, but also in a systematic way between EU countries and the U.S. Although one could hypothesize that given the more litigious environment in the U.S., the FDA may decide to proceed more cautiously and require more clinical data than the EMA has in the past, the fact that the FDA approved Sandoz's enoxaparin sodium ANDA (referencing Sanofi's brand name Lovenox) without requiring any additional clinical evidence, whereas the EMA required additional clinical data to approve a biosimilar application for a low molecular weight heparin, seems to suggest that even such broad generalizations may not be valid.

Immunogenicity and INN Naming Implications for Biosimilar Adoption

Biological medicines have a higher risk than small molecule medicines of immunogenicity–being recognized by the body as "foreign" and inducing unwanted immune reactions.[46] As noted in European Commission (2013, pp. 32–33), immunogenicity

[45] Aitken et al. (2013), and Berndt and Dubois (2012).

[46] European Commission (2013, pp. 32–33).

is a significant safety element assessed by regulatory authorities considering approval of a new biologic, and is assessed in clinical trials by extensive testing and characterization of short and long term anti-product immune responses. Determining whether the original branded biologic and a potential biosimilar have similar short and long-term immune responses may take a considerable time and patient exposures. European regulators have often balanced this risk with granting patient access as early as legally possible by requiring extensive post-marketing risk management efforts. To the extent physicians, payers and patients are concerned about potential immunogenicity variations between the original branded biologic and the biosimilar, and potential emergence of immunogenicity after switching from the original branded biologic to the less costly biosimilar (and perhaps back), the pace at which the biosimilar is adopted may be diminished, limited perhaps only to new patients or to those not being satisfactorily treated by the original branded biologic and so unlikely to switch back to the original brand. Therefore, immunogenicity concerns may constrain biosimilar penetration more than for small molecule generics.

These immunogenicity concerns have created concerns whether approved biosimilars should have the same International Non-proprietary Name (INN) as the reference brand name biologic product or unique INNs recognizing the inevitably unique characteristics of each biologic product. For generic small molecules approved through an ANDA in the US or analogous application procedure in the EU, generics generally have identical INNs as their branded reference product. Although there are some exceptions (more on this below), for the most part brand-name drug makers and biotechnology manufacturers want biosimilars to have unique, non-proprietary or generic names to distinguish the medicines from the original biologics. From their perspective, distinct names would lessen confusion in the marketplace and via distinct product tracking and tracing through the product distribution channel process, would contribute to ensuring patient safety. For example, Geoff Eich, an Amgen spokesman, was quoted as stating "They should not all share the same name. I want to know which product was given the patient so I can work with the physician to understand what may have gone awry. We need distinguishable names because that's what tells us who to contact."[47]

Generic drug manufacturers, however, along with many pharmacies, health insurers, unions, pension plans and pharmaceutical benefit management organizations, disagree and believe that creating a new INN for biosimilars would, in fact, create confusion and inhibit adoption of lower priced medicines. They argue that pushing for distinct INNs is essentially a smokescreen and an attempt by branded biotechnology companies to blunt their revenue declines, especially in the lucrative U.S. market.[48] For example, Richard Davies, Hospira's chief commercial officer, has been quoted as saying, "Having the same name is clearly important for market uptake. We see the naming argument more around whether the products are different, but they're not.... Having the same name will help with market formation."[49]

[47] As quoted in FiercePharma (2014).

[48] Staton (2014).

[49] Staton (2014).

While much of this debate focuses on whether physicians would be more willing to substitute a biosimilar for the branded original if the two had identical INNs, and thereby contribute to lowering healthcare costs, there is also considerable disagreement about the extent to which distinct INNs would make it easier to identify, monitor and address safety issues. Monitoring via tracking and tracing biosimilars could be accomplished even when biosimilars had identical INNs, provided they had unique NDC codes, as well as information on lot number if these are also included in patient health records and prescription labels. Currently, while all small molecule generics have unique NDC codes they are not generally recorded and printed on each dispensed prescription. Alternatively, one could have a hybrid policy in which each biosimilar of a molecule had a common INN followed by a hyphen and the name or abbreviated name of the distributor.

The INN naming controversy has been particularly prominent recently in the U.S. For example, the Generic Pharmaceutical Association filed a citizen's petition with the FDA in September 2013 requesting that all biosimilars share the same INN as the original biologics. Brand name manufacturer Novartis, which has as a wholly owned subsidiary the generic manufacturer Sandoz, obviously has ambivalent concerns. Interestingly, in October 2013 the Novartis Group of companies filed a citizen's petition requesting that FDA "require that a biosimilar be identified by the same (INN)…as the reference product." Johnson and Johnson filed a petition arguing that biosimilar names should not be identical to the underlying biologic, and Amgen filed a massive 89-page document in December 2013 that detailed seven arguments supporting its case for distinguishable non-proprietary names.[50]

Amgen's position on INN naming conventions for a biosimilar is an interesting one, for it historically has been a major innovator biotechnology company, but currently is considering biosimilar entry. As Sanford Bernstein analyst Geoffrey Porges noted, "The company is clearly straddling two business opportunities that sometimes seem in conflict with each other—a defender of the innovative products and a participant in biosimilar products. That tension is going to continue to be difficult for them to manage".[51] Although it provided few details and specifics, Amgen announced in early 2013 it was planning to launch six biosimilars beginning in 2017—versions of four cancer drugs (Avastin™ Herceptin™, Rituxan™ and Erbitux™) and two rivals of Amgen's Enbrel franchise (Humira and Remicade). An Amgen spokesman hinted at the possibility of different launching strategies in the U.S. and emerging economies, thereby viewing biosimilars as more of an emerging market opportunity, stating "We feel that these medicines are very valuable and in many parts of the world patients have no access to them because they are expensive."[52] Bernstein analyst Porges expanded on this, saying, "They are a bellwether for the industry. The real focus is on getting into lower-priced markets and lower-priced products and driving business through cost efficiencies."[53]

---

[50] Karst (2013b),

[51] As quoted in Berkrot (2013).

[52] See footnote 51.

[53] See footnote 51.

An alternative interpretation of the Amgen strategy is that, given the substantial regulatory uncertainties in the U.S. and to a lesser extent in the EU, a risk diversification strategy focused on building low-cost, highly productive biologic manufacturing capacity might produce a higher return portfolio of biosimilar, innovator and biobetter products through differential pricing and distribution based on local market ability to pay and regulatory stringency. Might an attractive strategy be to gain approval for and market the same biologic formulation as a biobetter in the US (and perhaps Europe), but as a biosimilar in less wealthy and more price-sensitive regions of the globe?

Economic Incentives Facing Biosimilar and Biobetter Developers

The European experiences described above demonstrate that sufficient incentives already exist to induce some manufacturers to develop and introduce biosimilars. The experience also demonstrates, that with perhaps the exception of Germany, the number of biosimilar manufacturers for any specific biologic in a regulatory jurisdiction is generally less than five and sometimes limited to a single firm. For some products in some jurisdictions, the number of reference or other branded products significantly exceeds the number of introduced products. For instance, in France seven branded manufacturers have marketed somatropin since 2005 while only a single biosimilar somatropin, from Novartis, has been marketed since 2007. As noted in section "European Relative Biosimilar Volume Shares, 2007–2012" above, in some smaller EU countries such as Belgium and the Scandinavian countries of Finland, Norway, and Sweden the introduction of a single biosimilar under national contract has garnered significant share, presumably facilitated by national contracting. However, from the perspective of potential biosimilar manufacturers, the EU experience to date might suggest that the markets may support relatively few biosimilar manufacturers.

The reasons for the low numbers of biosimilar entrants cannot be inferred directly from the sales (standard unit and dollar share) evidence examined above. Some qualitative observations and resulting hypotheses can be made from the described development, regulatory and manufacturing processes as well as the immunogenicity phenomenon particularly relevant to biologics.

First, section "European Evidence on Biosimilar Uptake" above highlights that the EU biosimilar regulatory processes require substantial clinical development of at least one significantly sized trial. Recall that small molecule generics generally require no clinical trials conducted by the generic manufacturer, but only require very small bioequivalence studies based on cross-over designs. The US regulatory process for biosimilars also requires clinical trial evidence and moves further in distinguishing biosimilarity from interchangeability, with the latter requiring additional original clinical trial evidence from the biosimilar applicant. Qualitatively, a biosimilar manufacturer faces product development costs greater than those required of a small molecule generics manufacturer. The costs and risks, however, are qualitatively lower than those facing a novel pharmaceutical manufacturer. The biosimilar developer knows a priori that a reasonably replicated biologic will likely prove effective and incur similar safety risks as the reference product. These U.S. required clinical development investments entail multiple years, which from an economic valuation perspective delays and lowers, through discounting, the value of future profits. Compared to small molecule generics, it would appear that particularly U.S. biosimilar sponsors face greater development costs, time delays and some, albeit low, risk of clinical failure.

Second, section "European Relative Biosimilar Volume Shares, 2007–2012" above documents that biosimilar market share growth can reach levels at or above 80 %, (filgrastim in UK and Sweden, erythropoietin in the Scandinavian countries) but may also barely exceed single digit market shares even after multiple years of availability (somatropin in all examined countries). These represent relatively slow, low and variable clinical adoption rates compared to small molecule generics. It is not possible to determine from this data the extent to which physician and patient concerns about immunogenicity, interchangeability or simply true biosimilarity are responsible for the relatively slow EU market penetration by biosimilars. Other reasons such as branded product price reductions, continued sales and marketing efforts, and payer contracting delays may also reasonably play roles. Unlike the case for generic small molecules, national payers in EU countries have to this point not designated biosimilars as being fully interchangeable with their reference products. As a result, physicians may take a wait and see attitude as the biosimilar develops a track record, and thereby delay peak adoption of the biosimilar. Regardless of the detailed causes, it appears that biosimilar manufacturers face relatively low, variable and slow adoption which in turn should lower their financial expectations and enthusiasm for investment in biosimilars.

Third, as discussed in section "Manufacturing Distinctions of Biopharmaceuticals", biologic manufacturers traditionally face higher manufacturing costs than small molecules which could limit the downward pricing flexibility for biosimilar products. Section "Manufacturing Distinctions of Biopharmaceuticals" also notes that biologic manufacturing technologies have seen dramatic yield increases through improvements in production organisms and their growing conditions which lower costs per dose. In addition, new single-use manufacturing equipment is lowering the required capital costs and minimum efficient scale for biomanufacturing overall compared to the large, stainless steel tank production approaches of most biologic reference products. While each biologic product will face unique circumstances, it appears likely that biosimilar manufacturers, being free to adopt the newest technologies, should enjoy lower manufacturing costs than the reference products which should improve biosimilar expected financial returns.

Fourth, section "European Relative Biosimilar/Brand Prices" surprisingly indicated that biosimilar pricing may be close to or even higher than reference product pricing. The pattern was not consistent across countries or products and while the relative pricing may be similar, the absolute price level for the biosimilar products is likely lower than the branded pricing prior to biosimilar entry. Unlike with small molecule generics, at least some branded reference products attempt to defend their markets through continued marketing and importantly, price reductions. While it appears that biosimilars may obtain higher relative pricing compared to generic small molecules, it is not guaranteed. In summary, it is not qualitatively clear that the potentially favorable biosimilar pricing offsets the challenges above.

On net, it is not clear whether biosimilar manufacturers can contemplate adequate financial returns in either absolute terms or relative to small molecule generic manufacturing. Compared to novel biologics, some factors are favorable to biosimilars: development is shorter, less costly and more likely to succeed; manufacturing costs are likely lower. However, other factors are not favorable: adoption appears relatively slow with highly variable peak market shares with likely lower pricing (although exceptions have been observed), and mandatory substitution through interchangeability designation has not been implemented. But in comparison to small molecule generics, nearly all the factors are less favorable, with pricing perhaps being the exception. These economic incentives, between those of novel therapeutics and generic small molecule medicines may help explain the relatively few observed biosimilar entrants in the EU. Generic small molecule manufacturers contemplating biosimilar entry may find the incentives insufficient to build the required extensive clinical development skills, especially if they perceive the financial outcomes directionally lower, more uncertain, or both. Novel therapeutic sponsors may find that the mixed incentives do not overcome their cultural preferences and operational infrastructure tuned for high risk, high commercial intensity and high margin therapeutics

Novel therapeutic sponsors may find the incentives for biobetters superior to those for biosimilars. Drug developers create biobetters by incrementally improving the pharmacologic or convenience characteristics of the reference product. For instance, developing an extended release version of the biologic through pegylation, which involves direct structural change to the molecule or through a formulation approach, may constitute a biobetter option. Biobetters are considered new products by regulators, requiring safety testing and clinical trials. However, since the underlying molecular biology and clinical performance of the reference product are well established, the safety testing can sometimes be streamlined and the clinical trial outcomes more easily predicted. Compared to a completely novel therapeutic, the development timelines are likely to be shorter and the clinical outcomes relatively low risk. The resulting time and cost savings improve financial returns for the investment. Upon commercialization, biobetters may be differentiated from the reference product, unlike biosimilars. This differentiation may allow for some pricing advantages for the biobetter and nearly always can provide a value rationale to the physician and patient for switching to the biobetter. Biobetter developers may therefore expect greater revenues from stronger pricing, larger ultimate market share and faster attainment of those larger shares. This approach of shorter, less expensive and less risky development resulting in a differentiable product for a known market may provide attractive risk-adjusted expected returns and so earn a place in the portfolio of a novel pharmaceutical firm. A possibly attractive mixed global strategy alternative, discussed earlier, would gain approval for and market the same biologic formulation as a biobetter in the US (and perhaps Europe), but as a biosimilar in less wealthy and more price-sensitive regions of the globe, thereby taking advantage of newly available more efficient biologic manufacturing processes that reduce global production costs, but more importantly achieving differentiated product pricing with higher pricing and adoption rates in markets able to pay for the biobetter improvements while not forfeiting access to jurisdictions preferring only biosimilarity, with of course the usual *trans*-jurisdiction arbitrage shipment and public perception issues that such price discrimination strategies incur.

Some Final Observations on Other Issues Affecting Biosimilar Adoption

Notably, large traditional pharmaceutical companies as well as established innovator biotechnology companies have announced diverse strategies involving entry into the biosimilar/biobetter product spaces. For example, already in 2008 Merck announced a biosimilar development division Merck Bioventures with a commitment to invest $1.5 billion in which an initial target was a follow-on version of Amgen's erythropoietin-stimulating agent Aranesp™. Citing scientific complexities, Merck abandoned the Aranesp™ project in 2010, and instead focused on a follow-on to Amgen's multi-indication immunomodulator Enbrel™. A year later, in 2011 Amgen made a surprise announcement concerning a judicial decision that delayed considerably the patent expiration of Enbrel™, and shortly thereafter Merck Bioventures announced it was abandoning development of an Enbrel™ follow-on. Later in February 2013, Merck announced it was changing its strategy of developing follow-on biologics, and instead of going it alone Merck was reorganizing into a partnership with Samsung Bioepis, itself a joint venture formed by South Korean conglomerate Samsung and BiogenIdec; It also announced it was merging Merck Bioventures into a division of Merck Research Laboratories. In the Merck and Samsung Bioepis partnership, Merck ceded all development work to Samsung Bioepis, but gained exclusive global commercialization rights.[54]

The Samsung Bioepis joint venture was formed shortly before the Merck and Samsung Bioepis partnership announcement, and involved an agreement between Samsung Biologics, a newly minted development and manufacturing group, and BiogenIdec. Samsung officials initially described plans to develop a biosimilar to Rituxan™, a blockbuster treatment developed in partnership between BiogenIdec and Genentech, but at the December 2011 announcement of the formation of Samsung Bioepis, BiogenIdec officials went to some pains to make clear that Samsung Bioepis would not involve Rituxan™ or any of BiogenIdec's branded therapies. Within this joint venture, Samsung Biologics was responsible for sales and marketing, while BiogenIdec focused on manufacturing. According to BiogenIdec CEO George Stangos, "The manufacturing facilities have costs to run

[54] Carroll (2013a, b).

them, so the more products you run through them, the more efficient they are. To set ourselves up commercially could be a big distraction. I'd like a partner to take over that…This relationship will allow us to leverage our world-class protein engineering and biologics manufacturing capabilities, while maintaining focus on our mission of discovering, developing and delivering innovative therapies for patients worldwide"[55] A subsequent news story reported that two of the anti-inflammatory monoclonal antibodies targeted by the Samsung Bioepis joint venture were Remicade™ and Enbrel™.[56]

Although terms of the Samsung Bioepis joint venture did not allow it to pursue development of a biosimilar to Roche's Rituxan (marketed in the U.S. by BiogenIdec), in a separate alliance with the large contract research organization Quintiles, Samsung was able to attempt to develop a biosimilar to Rituxan™. In October 2012, however, a Korean newspaper reported that Samsung was halting development of its biosimilar to Rituxan™, because of "some internal reasons" speculated to involve difficulties in complying with recent regulatory guidance from the U.S. FDA. A Samsung spokesperson stated that Samsung was "speaking with FDA officials about clinical development requirements and a way forward for the SAIT101 program." Samsung's termination followed by about a month an earlier announcement by Israeli-based Teva that it was terminating its own late-stage development of a Rituxan™ biosimilar.[57] About 6 months later, in April 2013, Celltrion—another South Korean-based biotechnology company—and its partner Hospira announced the delay and potential termination of their late-stage development of a biosimilar version of Rituxan™. By mid-2013 both Teva and Samsung-Hospira had bailed out on their attempts to develop a biosimilar to Roche's Rituxan™, but observers have noted that other companies, such as Novarti's generic Sandoz unit and Boehringer Ingelheim, were continuing to pursue Rituxan™ biosimilars.[58] Although Samsung may have exited from development of a biosimilar to Roche's Rituxan™, Samsung has not gotten out of the biosimilar development business. In October 2013, Samsung and Roche announced they had clinched a "long-term strategic manufacturing agreement" whereby Samsung will manufacture an undisclosed number of Roche's cell-based products at its two manufacturing facilities in Incheon, South Korea.[59]

Another alliance announced with great fanfare in 2009 involved Israeli-based generic manufacturer Teva and the Swiss-based contract manufacturing firm Lonza Group, a Novartis subsidiary. Four years later, in July 2013, Teva and Lonza announced termination of the agreement to develop a series of biologics, including biosimilars. According to a Lonzo spokesman, "In our assessment those investments in biosimilars will require more capital than initially planned and will also take more time until they reach the market…This is why we intend in the future to

[55] As quoted in Carroll (2011).

[56] Carroll (2013c).

[57] McBride (2012).

[58] McBride (2013).

[59] Palmer (2013).

limit our role by focusing on our core expertise in the areas of contract manufacturing and cell line development." Rather than retreating from developing follow-on biologics, Teva vowed to continue developing them, stating that "Teva has a track record of success in the biologics arena and we plan to continue and build on that success."[60]

Interestingly, Teva has in fact been able to achieve some success with developing and gaining approval to market follow-on bioloigcs, not only in the EU but also in the US. However, success has been achieved not via the biosimilar pathway of the US's section 351(a) that allows applicants to use data from the innovative product to support their application, but instead via the recently enacted 351(k) provision by which Teva filed a traditional Biologics License Application (BLA). In October 2013 Teva received FDA approval to launch brand name Granix™ having the exact same recombinant active ingredient, filgrastic, as Amgen's Neupogen™; simultaneously, however, Teva withdrew its BLA for Lonquex, a long-acting pegylated version of filgrastim that Amgen markets under the US brand name of Neulasta™ and whose INN is pegfilgrastim. Neupogen™ is a once-daily medication, while Neulasta™ can be used less frequently on a once per chemotherapy dose regimen. Teva has another long-acting version of filgrastim, an albumin-fusion version, under BLA review at the FDA, which is called balugrastim. Because both the Granix™ and balugrastim applications are filed as BLAs, they are not subject to the biosimilar INN naming dispute, and therefore Teva has given them INNs of tbo-filgrastim and balugrastim. As noted earlier in this manuscript, Teva's Ratiograstim™, a filgrastim biosimilar, has been on the EU market since 2008 and its follitropin alfa biosimilar was just recently approved by CHMP in late 2013. In terms of the potential for Granix™ and balugrastim in the US market, analysts have noted that Amgen's Neupogen™ and Neulasta™ both are often on many pharmacy benefit plans' highest tier—number three or four. At a lower price, Granix™ and balugrastimc may stand a good chance of being listed on the second, "preferred", tier of those plans if they can overcome payer concerns of general overuse of the class in oncology, offering patients an opportunity to have a lower cost-sharing burden at a time when they will have many medical expenses.[61]

The diversity and volatility of strategies for follow-on biologics is striking. Presently there does not appear to be any observable convergence or dominant set of strategies. Although some firms exiting biosimilar alliances reference unexpected regulatory and capital cost developments, whether they are instead moving to a biobetter rather than biosimilar pattern is unclear. Predicting what will happen to biosimilars, biobetters, and other follow-on biologics in the US and EU is therefore speculative and highly likely to be inaccurate, particularly if a single theme is purported. It does appear likely that the extent of entry and decrease in price from branded innovator is likely to be much smaller than has been observed for generic small molecules in the US approved via ANDAs. Early indications are that the evolving market dynamics may be closer to what has been observed for injected

[60] As quoted in pmlive.com (2013).

[61] Gardner (2013).

specialty drugs that have been approved as AP-rated to the brand in the US. However, to date there has been little if any published research on the EU experience to confirm these preliminary observations and validate the analogy.

An alternative way to obtain insights on likely future follow-on biologic paths involves examining time paths of costs and revenues in simulation models comparing net present values of BLA/biobetter, biosimilar and entirely novel BLA paths for traditional molecules in selected therapeutic areas. Financial simulators driven by evidence based parameters can illustrate the range of incentives and challenges facing a drug developer considering an investment decision—whether that be in biosimilar, bio-better or a traditional novel therapeutic. Such evidence based simulations have been successfully employed to estimate the costs of drug development generally,[62] for therapeutic modalities,[63] for stratified medicine and companion diagnostic development,[64,65] and for adaptive licensing policy.[66] Beyond aiding individual firm investment choices, such simulations have informed public policy discussions regarding potential market inefficiencies in the drug development innovation chain, the effectiveness of the therapeutic innovation eco-system, and actions that might improve the level and productivity of therapeutic R&D investments. Similarly, further work on evidence based, financial simulators focused on biosimilar and biobetter market sub-segments could raise the level of public discussion regarding the future paths facing biologic therapeutics and point to the key externally observable evidence that might distinguish among competing perspectives and market dynamics. In addition since end of product life cycle revenues for a biologic or small molecule pharmaceutical often occur more than a quarter century after initial development, any differences among traditional novel compounds, novel biologics, biosiimilars and biobetters in the time at which significant events occur during the product life cycle—e.g., patent applications, regulatory filings, initiation of clinical trial phases and patient recruitment, FDA and EMA marketing approvals, launch and post-launch marketing efforts, initial loss of exclusivity (LOE), and extent and speed at which revenues are eroded post-LOE—can have significant impacts on net present value calculations. Note that the sensitivity of NPV calculations to differences among these molecule types in significant event timing will likely increase along with the choice of discount rate (that might also differ among traditional novel compounds, novel biologics, biosimilars and biobetters).

In summary, the biosimilar phenomenon is a relatively recent one involving complex options to develop, regulate, market and utilize biosimilars via an abbreviated BLA, biobetters via a BLA, or entirely novel but traditional small molecule medicines via NDAs. Understanding the historic and likely future evolution of these medicines will require addressing serious measurement issues involving volumes

---

[62] DiMaisi et al. (2003).

[63] DiMasi and Grabowski (2007).

[64] Trusheim et al. (2011).

[65] Trusheim and Berndt (2012).

[66] Baird et al. (2013).

and prices, and carefully analyzing descriptive market data trends for biologics and related specialty products in the EU, the US and other major global geographic regions, as well as how biosimilar adoption will alter incentives to develop biobetters or entirely new products.[67] But such understanding will also depend critically on simultaneously developing and implementing evidence-based financial, epidemiological, and clinical simulation models whose design, structure and underlying data are likely to differ substantially across therapeutic sub-segments.

Acknowledgements The research reported on here was funded by an educational grant from Pfizer Limited, Surrey, UK to Berndt Associates LLC. The funding source had no role in the design and conduct of the study; collection, management, analysis, or interpretation of the data; and while it provided comments on a draft version of this manuscript, it had no role in the preparation, review, submission, or approval of the manuscript for publication. The authors thank Kirsten Axelsen, Adam Heathfield, Jake Lebiecki and Danielle Rollman of Pfizer for comments on an earlier draft of this manuscript, and participants at the TIGER Forum 2014 at the Toulouse School of Economics in Toulouse, France, June 2, 2014. The statements, findings, conclusions, views, and opinions contained and expressed herein are those of the authors and are based in part on IMS MIDAS™ data obtained by Berndt Associates LLC under license from IMS Health (rights reserved), and are not necessarily those of IMS Health, its affiliates or subsidiaries, or the institutions with whom the authors are affiliated. Any errors or misstatements are our own.

Appendix

Biologic Molecules with Biosimilar Entry in Europe and Elsewhere

Short-Acting Epoietin Recombinant (Erythropoietin, Alpha, Beta, Theta, Zeta)

According to the US Food and Drug Administration, "Erythropoietin is a glycoprotein whose main function is to stimulate the proliferation and differentiation of erythroid precursors in the bone marrow. Erythropoietin is produced mainly in the kidneys, though several other tissues produce lesser amounts of the growth factor."[68] Approved by the FDA on June 1, 1989, Epogen/Procrit (epoetinalfa) was produced in Chinese Hamster Ovary cells that have modified through recombinant DNA technology to encode the gene for human erythropoietin, and was initially approved for the treatment of anemia in patients with chronic renal failure. Epogen/Procrit was subsequently approved for the treatment of anemia due to ziduvodine therapy in HIV-infected patients (1991) and for the treatment of anemia in patients with non-

[67] According to Brennan (2014), as of June 2014 24 countries have established biosimilar pathways or have approved follow-on biologics.

[68] Epoietinalfa, June 24, 2011 Division Director Summary Review, STN BL 103234/5166, p. 5 of 38. Available online at Drugs@FDA.

myeloid malignancies whose anemia is due to the effect of concomitantly administered chemotherapy (1993). Both these supplemental approvals were based on demonstration of a reduction in the proportion of patients receiving shaded blood cell (RBC) transfusions.

Within Europe, the epoietin alpha reference product is Johnson & Johnson's Erypo or Eprex. As of June 2011, there were five approved biosimilar products: Binocrit (Sandoz/Novartis), Epo A (Hexal/Novartis), Abseamed (Medici), Retacrit (Hospira) and Silapo (Stada), all approved between August 28, 2007 and December 18, 2007.[69] Using the IMS Health classification scheme, there are two non-referenced products in the epoietin biosimilar accessible market (defined as an original product, granted market exclusivity at the start of its commercial life in Europe, whose exclusivity is now expired, with the product never have been referenced, or may have been referenced but the referencing biosimilar has not yet launched): Roche's NeoRecormon, and Teva's Eporatio/Biopoin.[70] In the U.S., in addition to having been approved by the FDA for treating anemia in cancer patients on chemotherapy, anemia in chronic renal failure patients, and anemia in zidovudine-treated, HIV-infected patients, epoietin alfa is approved for the reduction of allogenic blood transfusion in surgery patients. As of 2011, Epogen/Procrit was available in 2000, 3000 and 4000 units/ml 1 ml single-dose vials for subcutaneous injection or intravenous solution administration, in 20000 units/ml 1 ml multidose vials, in 10000 units/ml 1 ml single-dose and 2 ml multidose vials, and in 40000 units/ml single-dose vials for subcutaneous injection or intravenous solution administration.

Another erythropoietin stimulating agent (ESA) approved in both the US and EU is Amgen's darbepoetin alfa (brand name Aranesp in the US). Darbepoietin is distinguished from epoietin agents primarily because of Aranesp's longer serum half-life, implying generally less frequent dosing than the epoietins.[71] Currently Aranesp is patent-protected in the US and EU, with its earliest reported year of key US patent expiry being 2024; for Epogen this US patent expiry date is 2013.[72]

Growth Hormone for Children Born Small for Gestational Age—SGA (Somatropin Molecule)

Of the approximately 2.5 % of children who are born small for gestational age (SGA), 10–15 % fail to "catch up" by age two. Children who do not catch up by age two, if left untreated, are destined in many cases to have compromised final height, relative to the norm for the population. A relative height measure is SDS—the number of standard deviations an individual at a particular age is away from the

[69] Grabowski et al. (2013, p. 4).

[70] IMS Health (2011), slides 10, 11 and 40.

[71] Drug Facts and Comparisons (2011), p. 154.

[72] Grabowski (2013, slide 5).

age-specific population mean. Though there are differences across and within countries on the measure of SDS triggering treatment, growth hormone supplementation in children born SGA can enhance growth velocity, height SDS, and predicted adult height. Aside from the known adverse effects of growth hormone therapy, of concern in treating these children is a risk of accelerating bone age beyond chronological age, with the possibility of precipitating precocious puberty and compromising final stature on that basis.[73]

On August 24, 1995, the FDA approved Pharmacia and Upjohn's NDA # 020280 application for Genotropin (somatropin recombinant) for children born SGA who fail to manifest catch-up growth by 2 years of age, caused by an inadequate secretion of endogenous growth hormone. Over the years the FDA has approved a number of supplemental indications (e.g., growth failure associated with chronic renal insufficiency, with Noonan syndrome, with Prader-Willi syndrome, with Turner syndrome, in adults with either adult- or childhood-onset growth hormone deficiency, and others) as well as several related somatropin products, such as Omnitrope (Sandoz), Serostim (Serono), Humatrope (Eli Lilly), Nutropin (Genentech), Salzen (Serono), Tev-Tropin (Gate), HumatroPen (Eli Lilly), Zorbtive (Serono), Norditropin (Novo Nordisk), Accretropin (Cangene), and Nutropin AQ, Nutropin AQ NuSpin 5, NuSpin 10 and NuSpin 20 (Genentech). Most of these formulations are subcutaneous injection, lyophilized power for solution, although some products, such as Norditropin, involve pen or two-chamber cartridge delivery systems, with a reconstitution device used to mix the diluent and powder.[74] Somatropin must not be injected intravenously. Genotropin lyophilized powder contains somatropin of rDNA origin, a polypeptide hormone. The amino acid sequence of the product is identical to that of human growth hormone of pituitary origin (somatropin). Genotropin is synthesized in a strain of *Escherichia coli* that has been modified by the addition of the gene for human growth hormone.[75]

According to IMS Health, as of June 2011 both Pfizer's (who acquired rights with the acquisition of Pharmacia and Upjohn) Genotropin and Eli Lilly's Humatrope were reference products, the two approved biosimilar products were Novartis Sandoz' Omnitrope and Somatropin (unknown manufacturing laboratory), whereas the non-referenced products included Sanofi Aventis' Maxomat, Nova Nordisk's Norditropin, Ipsen's Nutropinaq, Merck Serono's Saizen, and Ferring's Zomacton.[76]

[73] Genotropin, FDA Center for Drug Evaluation and Research, Application Number 20-280/S-031, Review—Administrative Documents, dated July 23, 2001. Available online at Drugs@FDA, Approval History, NDA 020280, 07/25/2001 031, p. 1 of 3, letter from David G. Orloff, M.D., Director, Division of Metabolic and Endocrine Drug Products to file NDA 20-280/S-031.

[74] "Somatropin", Drug Facts and Comparisons (2011), pp. 523–526.

[75] Genotropin Draft Package Insert, p. 3of 15, NDA 20-280/S-031, available online at Drugs@ FDA, Approval History, NDA 020280, 07/25/2001.

[76] IMS Health (2011), slide 41.

Granulocyte-Colony Stimulating Factor (G-CSF), Filgrastim and Lenograstim Molecules

On February 20, 1991, the US FDA approved Amgen's filgrastim (trade name Neupogen) to decrease the incidence of infection, as manifested in febrile neutropenia, in patients with non-myeloid malignancies receiving myelosuppressive anticancer drugs associated with a significant incidence of severe neutropenia with fever. On April 2, 1998, the FDA granted a supplemental NDA approval for acute myeloid leukemia (AML) adult patients receiving induction or consolidation chemotherapy, for reducing the time to neutrophil recovery and the duration of fever.[77] Other approved[78] uses of filgrastim include in patients with nonmyeloid malignancies undergoing myeloablative chemotherapy followed by bone marrow transplantation, and for the mobilization of hematopoietic progenitor cells into the peripheral blood for collection by leukapheresis.

Neutropenia is condition with an abnormally low number of neutrophils in the blood—the body's primary cellular defense system against bacteria and fungi. Neutrophils also help heal wounds and ingest foreign debris, such as embedded splinters. People who have severe neutropenia (fewer than 500 neutrophils per microliter of blood) can rapidly succumb to infection because their bodies lack the means to fight the invading organisms. Neutrophils mature in the bone marrow in about 2 weeks. After entering the blood stream, they circulate for about 6 h, searching for infective organisms and other intruders. When they find one, they migrate into the tissues, attach themselves to the intruders, and produce toxic substances that kill and digest the intruders. This reaction may damage healthy tissue in the area of the infection. The entire process produces an inflammatory response in the infected area, which appears on the body's surface as redness, swelling, and heat. Neutropenia has several causes. The number of neutrophils can decrease because bone marrow production isn't adequate or because large numbers of white blood cells are destroyed in the circulation. Aplastic anemia, and certain rare genetic diseases such as infantile genetic agranulocytosis and familial neutropenia cause decreases in the number of white blood cells. Certain drugs, especially chemotherapies used in cancer treatment, impair the bone marrow's ability to produce neutrophils. Growth factors that stimulate the production of white blood cells, particularly granulocyte colony-stimulating factor (G-CSF) and granulocyte-macrophage colony-stimulating factor (GM-CSF) can eliminate neutropenic episodes in cyclic neutropenia.[79]

In the U.S., Neupogen is still marketed exclusively by Amgen. Four dosage forms are approved—two in single use vials, and two as pre-filled injectable syringes.

[77] Neupogen, "Review and Summary Basis of Approval", Application Number 103353/000, Center for Drug Evaluation and Research, February 20, 1991. "Supplement 1036, Letter, April 2, 1998. Available online at Drugs@FDA.

[78] "Filgrastim", Drug Facts and Comparisons (2011), p. 163.

[79] "Neutropenia", in White Blood Cell Disorders, ch. 158, in Robert Berkow, Editor, *The Merck Manual of Medical Information*, (Home Edition), Whitehouse Station, NJ: Merck Research Laboratories, 1998, pp. 761–763.

Neupogen is produced by Escherichia coli (E coli) bacteria into which has been inserted the human granulocyte colony-stimulating factor gene. The protein has an amino acid sequence that is identical to the natural sequence predicted from human DNA analysis, except for the addition of an N-terminal methionine necessary for expression in E coli. Because Neupogen is produced in E coli, the product is non-glycosylated and thus differs from G-CSF isolated from a human cell. Neupogen is a sterile, clear, colorless preservative-free liquid for parenteral administration. In order to maintain clinical benefit, chronic daily administration is required.[80]

A second form of recombinant human granulocyte colony-stimulating factor is lenograstim (brand name Granocyte, Chugai Pharmaceuticals, marketed in the EU by Sanofi Aventis), a Chinese hamster ovary-derived G-CSF, indistinguishable from native G-CSF, and differing from filgrastim which is an Escherichica coli-derived G-CSF and is non-glycosylated, having an extra methionine group at the N-terminal end of the peptide chain.[81] Granocyte (lenograstim) is not available in the U.S., but is a non-referenced product in the EU, having brand names in addition to Sanofi Aventis' Granocyte, Euprotin (Almirall), Myelostim (Italfarmaco) and Roche's Neutrogin. As of June 2011, a number of biosimilars were approved in the EU using Amgen's filgrastim (Neupogen) as the reference product. These biosimilars include a Biograstim (CT Arzzneimittel), Novartis'/Sandoz filgrastim Zarzio, Teva's Tevagrastim, Ratiopharm's Ratiograstim, Hexai's Filgrastim Hexal, and Hospira's Nivestim, all approved between September 15, 2008 and June 8, 2010.[82]

Neulasta (pegfilgrastim) is a second-generation injectable granulocyte colony stimulating factor approved in both the US and EU. Pegfilgrastim is distinguished from filgrastim agents primarily because of pegfilgrastim's longer serum half-life, implying generally less frequent dosing than the filgrastims. Specifically, whereas filgrastim requires daily dosing, pegfilgrastim is administered only once per chemotherapy cycle.[83] Currently Neulasta is patent-protected in the US and EU. According to Grabowski (2013, Slide 5), while the earliest reported year of key US patent expiry is 2013 for Amgen's Neupogen, for its Neulasta it is 2015.

Infliximab (Remicade)

The EMA's Committee on Human Medicinal Products (CHMP) approved Hospira's biosimilar application for infliximab (injection) on September 10, 2013 with Centocor's (Johnson & Johnson) Remicade serving as the reference product. Infliximib is a monoclonal antibody, an immunomodulator distributed as a

[80] "Neupogen (Filgrastim) Drug Information: User Reviews, Drug Side Effects…", p. 1 of 3 in RxList…, last reviewed June 4, 2012. b Available online at http://www.rxlist.com/neupogen-drug.htm.

[81] Kim et al. (2003), p. 1 of 14.

[82] IMS Health (2011), slide 42. According to Grabowski et al. (2013, p. 4), Ratiopharm's Filgrastim ratiopharm was approved by the EU on September 15, 2008, but was withdrawn on July 20, 2011.

[83] "Pegfilgrastim", Drug Facts and Comparisons (2011), p. 166.

lyophilized powder for solution. In the EU it is approved for treating ankylosing spondylitis (a chronic inflammatory disease of unknown origin, first affecting the spine and adjacent structures and commonly progressing to eventual fusion of the involved joints[84]), psoriatic arthritis and psoriasis (see Table 1 in main text). In addition to these three indications, in the U.S. infliximab (Remicade) is approved by the FDA for treatment of Crohn disease, fistulizing Crohn disease, rheumatoic arthritis, and ulcerative colitis.[85] Infliximab is the first monoclonal antibody approved as a biosimilar in the EU. Like several other immunologic agents, infliximab has a risk of serious infections, since patients treated with infliximab are at an increased risk for developing serious infections that may lead to hospitalization or death. According to Drug Facts and Figures (2011, p. 2822), "Most patients who developed those infections were taking concomitant immunosuppressants such as methotrexate or corticosteroids."

Since EMA approval of biosimilar infliximab (referenced to Remicade) occurred just several months ago, data on its uptake within EU countries are not yet available. Indeed, it is likely that Hospira has not yet obtained reimbursement approval from any of the EMA member countries.

Follitropin Alfa (Gonal-F)

The EMA's Committee on Human Medicinal Products (CHMP) approved Teva's biosimilar application for follitropin alfa (injection) on September 27, 2013 with Merck Serono's Gonal-F serving as the reference product. Follitropin alfa is a human follicle stimulating hormone (FSH) distributed as a sterile, clear solution for subcutaneous injection. In the EU it is approved for treating an ovulation (the failure of the ovaries to release an egg during an ovulation cycle[86]), stimulation of multifollicular development in women undergoing superovulation for assisted reproductive technologies (ART) such as in vitro fertilization (IVF), gemete intra-fallopian transfer and zygote intra-fallopian transfer, and for ovaleap in association with a luteisising hormone (LH) preparation for women with severe LH and FSH deficiency. It is also approved in the EU for the stimulation of spermatogenesis in men who have congenital or acquired hypogonadotropic hypogonadism with concomitant human chorionic gonadotropin (hCG) therapy.[87] In the U.S. follitropin alfa FDA approved indications are limited to an ovulation and ART treatments.[88] Follitropin alfa is the first fertility biosimilar approved in the EU.

---

[84] "Ankylosing spondylitis" in Anderson et al. (1998), pp. 94–95.

[85] Drug Facts and Figures (2011), pp. 2822–2823.

[86] "Folliropin Alpha" Drug Facts and Comparisons (2009), p. 350.

[87] European Medicines Agency (2013c).

[88] Food and Drug Administration (2014a).

Since EMA approval of biosimilar follitropin alfa (referenced to Gonal-F) occurred just several months ago, data on its uptake within EU countries are not yet available. Indeed, it is likely that Teva has not yet obtained reimbursement approval from any of the EMA member countries.

References

Aitken ML, Berndt ER (2011) Medicare part D at age five: what has happened to seniors' prescription drug prices? Report by the IMS Institute for Healthcare Bioinformatics, July

Aitken ML, Berndt ER, Cutler DM (2008) Prescription drug spending trends in the U.S. Looking beyond the turning point. Health Aff 28(1):W151–W160

Aitken ML, Berndt ER, Bosworth M, Cockburn IM, Frank RG, Kleinrock M, Shapiro BT (2013) The regulation of prescription drug competition and market responses: patterns in prices and sales following loss of exclusivity. Working Paper 19487, October. National Bureau of Economic Research, Cambridge

Anderson KN, Anderson LE, Glanze WD (eds) (1998) Mosby's medical, nursing, & allied health dictionary, 5th edn. Mosby-Year Book, St. Louis

Baird Lynn G, Trusheim MR, Eichler H-G, Berndt ER, Hirsh G (2013) Comparison of stakeholder metrics for traditional and adpative development and licensing approaches to drug development. Ther Innov Regulat Sci 47(4):474–483

Berkrot B (2013) UPDATE 3—Amgen biosimilar push takes aim at blockbusters, 7 February. http://uk.reuters.com/assets/print?aid=UKL1NOB72N220130207. Accessed 8 Feb 2013

Berndt ER, Aitken ML (2011) Brand loyalty, generic entry and price competition in pharmaceuticals in the quarter century after the 1984 Waxman-Hatch legislation. Int J Econ Busi 18(2): 203–224

Berndt ER, Dubois P (2012) Cross-national trends in the daily cost of pharmaceutical treatments, 2004–2010. Discussion Paper No. 9140, September. Center for Economic Policy Research, London. www.cepr.org/poubs/dps/DP9140.asp

Berndt ER, Newhouse JP (2012) Pricing and reimbursement in us pharmaceutical markets. In: Danzon PM, Nicholson S (eds) The oxford handbook of the economics of the biopharmaceutical industry. Oxford University Press, New York, pp 201–265

Berndt ER, Mortimer R, Bhattacharjya A, Parece A, Tuttle E (2007) Authorized generic drugs, price competition, and consumers' welfare. Health Aff 26:790–799

Biotechnology Industry Organization (2013) BIO principles on patient safety in the substitution of biologic products, press release, 24 January. http://www.bio.org/advocacy/letters/bio-principles-patient-safety-substitution-biologic-prod

Branstetter LG, Chatterjee C, Higgins MJ (2011) Regulation and welfare: evidence from paragraph IV generic entry in the pharmaceutical industry, Working Paper 19487, October. National Bureau of Economic Research, Cambridge

Brennan Z (2014) Global biosimilar regulations diverging, experts say. BioPharma Reporter.com, 20 June. http://www.biopharma-reporter.com/content/view/print/933696

Calfee JE, DuPre E (2006) The emerging market dynamics of targeted therapeutics. Health Aff 25(5):1302–1308

Carroll J (2011) Biogen Idec signs up with Samsung on $300 million biosimilars venture. FierceBiotech, 6 December. http://www.fiercebiotech.com/story/biogen-idec-signs-samsung-300-million-biosimilars-v. Accessed 18 April 2014

Carroll J (2013a) Star-crossed Merck reorganizes troubled biosimilars effort around Samsung pact. FierceBiotech, 20 February. http://www.fiercebiotech.com/story/star-crossed-merck-reorganizes-troubled-biosimilars. Accessed 18 April 2014

Carroll J (2013b) Biosimilars face big delays as blockbuster knockoffs hit a roadblock. FierceBiotech, 27 February. http://www.fiercebiotech.com/story/biosimilars-face-big-delays-blockbuster-knockoffs-hit. Accessed 18 April 2014

Carroll J (2013c) Biogen Idec preps for a big leap into the anti-TNF blockbuster biosimilars market. FierceBiotech, Story 3, 17 December 2013. www.fiercebiotech.com

Congressional Budget Office (2008) Biologics price competition and innovation act of 2007, S.1695

Conti RM, Berndt ER (2014) Specialty drug prices and utilization after loss of U.S. Patent exclusivity, 2001–2007. Working Paper 20016, March. National Bureau of Economic Research, Cambridge

Cook A (1998) How increased competition from generic drugs has affected prices and returns in the pharmaceutical industry. The Congress of the United States, Congressional Budget Office, Washington DC. http://www.cbo.gov/

Danzon PM, Furukawa MF (2011) Cross-national evidence on generic pharmaceuticals: Pharmacy vs. physician-driven markets. Working Paper 17226, National Bureau of Economic Research, Cambridge

DiMaisi JA, Hansen RW, Grabowski HG (2003) The price of innovation: new estimates of drug development costs. J Health Econ 22:151–185

DiMasi JA, Grabowski HG (2007) The cost of biopharmaceutical R&D: is biotech different? Manage Decis Econ 28:469–479

Drake KM, Starr MA, McGuire TG (2014) Do 'reverse payment' settlements of brand-generic pharmaceutical patent disputes constitute an anticompetitive pay-for-delay. Paper presented at the ASHEcon Fifth Biennial Meetings, Los Angeles, 23 June. www.gma-us.com

Drug Channels (2011) Ranbaxy makes three: the battle for generic lipitor profits, 1 December. http://www.drugchannels.net/2011//12/ranbaxy-makes-three-battle-for-generic.html

Drug Channels (2012) Pfizer's lipitor strategy and the 2012 generic monster, 15 March. http://www.drugchannels.net/2012/03/pfizers-lipitor-strategy-and-2012.html

Drug Facts and Comparisons (2009) 2009 Edition. Wolters Kluwer Health, St. Louis MO

Drug Facts and Comparisons (2011) 2011 Edition. Wolters Kluwer Health, St. Louis MO

Ehmann F (2010) Biosimilars—regulation strategies and pathway in the EU (and U.S.). http://www.dvfa.de/files/die_dvfa/kommissionen/life_science/application/pdf/2_Falk_Ehmann_EMEA.pdf

Ellison SF, Cockburn IM, Griliches Z, Hausman JA (1997) Characteristics of demand for pharmaceutical products: an examination of four cephalosporins. RAND J Econ 28(3):426–446

European Commission (2013) What you need to know about Biosimilar Medicinal Products. Consensus Information Paper, Enterprise and Industry. http://ec.europa.eu/enterprise/sectors/healthcare/competitiveness/process_on_corporate_responsibility/platform_access/index_en.htm#h2-6

European Medicines Agency (2005) Guideline on similar biological medicinal products, Committee for medicinal products for human use (CHMP), 30 October. EMA/CHMP/437/04. http://www.emea.eu.int

European Medicines Agency (2006a) Annex to guideline on similar biological medicinal products containing biotechnology-derived proteins as active substance: non-clinical and clinical issues: guidance on similar medicinal products containing recombinant granulocyte-colony stimulating factor. EMEA Document Number CHMP/31329/2005, 22 February. http://www.emea.europa.eu/docs/en_GB/document_library/Scientific_guideline/2009/09/WC500003955.pdf

European Medicines Agency (2006b) Annex to guideline on similar biological medicinal products containing biotechnology-derived proteins as active substance: non-clinical and clinical issues: guidance on similar medicinal products containing recombinant human soluble insulin. EMEA Document Number CHMP/32775/05, 22 February. http://www.emea.europa/eu/docs/en_GB/document_library/Scientific_guidelines/2009/09/WD500003957.pdf

European Medicines Agency (2006c) Annex to guideline on similar biological medicinal products containing biotechnology-derived proteins as active substance: non-clinical and clinical issues: guidance on similar medicinal products containing recombinant somatropin. EMEA Document

Number CHMP/94528/2005, 22 February. http://www.emea.europa/eu/docs/en_GB/document_library/Scientific_guidelines/2009/09/WD500003956.pdf
European Medicines Agency (2009a) Guideline on clinical and non-clinical development of similar biological medicinal products containing low-molecular-weight Heparin. EMEA Document Number CHMP/BMWP/118264/07, 16 March. http://www.ema.europa.eu/docs/en_GB/document_library/Scientific_guideline/2009/09/WC500003927.pdf
European Medicines Agency (2009b) Non-clinical clinical development of similar biological medicinal products containing recombinant interferon alfa. EMEA Document Number CHMP/BMWP/102046/2006, 23 April. http://www.ema.europa.eu/docs/en_GB/document_library/Scientific_guideline/2009/09/WC500003930.pdf
European Medicines Agency (2010) Guideline on similar biological medicinal products containing monoclonal antibodies (Draft), 18 November. EMEA Document Number CHMP/BMWP/403543/2010. http://www.ema.europa.eu/docs/en_GB/document_library/Scientific_guideline/2010/11/WC50009936.pdf (Circulated November 2010 and comment period closed May 2011)
European Medicines Agency (2011a) Guideline on non-clinical and clinical development of similar biological medicinal products containing low-molecular-weight Heparin. EMEA Document Number CHMP/BMWP/118264/07, 16 March. http://www.ema.europa.eu/docs/en_GB/document_library/Scientific_guideline/2009/09/WC500003927.pdf
European Medicines Agency (2011b) Guideline on non-clinical and clinical development of similar biological medicinal products containing recombinant human follicle stimulation hormone (r-hFSH). EMEA Document Number CHMP/BMWP/671292/2010, 17 November. http://www.ema.europa.eu/docs/en_GB/document_library/Scientific_guideline/2011/11/WC500117986.pdf (Circulated November 2011 and comment period closed May 2012)
European Medicines Agency (2011c) Guideline on similar biological medicinal products containing interferon beta, 15 December. EMEA Document Number CHMP/BMWP/652000/2010. http://www.ema.europa.eu/docs/en_GB/document_library/Scientific_guideline/2012/01/WC500120652.pdf (Circulated December 2010 and comment period closed May 2012)
European Medicines Agency (2012a) Work plan for the biosimilar medicinal products. Working Party 2013, Committee for Medicinal Products for Human Use (CHMP), 13 December. EMA/CHMP/BMWP/610801/2012. http://www.ema.europa.eu/docs/en_GB/document_library/Work_programme/2009/11/WC500014447.pdf
European Medicines Agency (2012b) Marvel LifeSciences Ltd withdraws its marketing authorization applications for Solumarv, Isomarv and Combimarv, Press release, 27 November. EMA/747975/2012. http://www.ema.europa.eu/ema/index.jsp?curl=pages/news_and_events/news/2012/11/news_detail_001665.jsp&mid=WC0b01ac058004d5c1
European Medicines Agency (2012c) Guidelines on similar biological medicinal products containing monoclonal antibodies – non-clinical and clinical issues. Committee for Medicinal Products for Human Use (CHMP), 30 May. EMA/CHMP/BMWP/403543/2010. http://www.emea.eu.int
European Medicines Agency (2013a) European medicines agency recommends approval of first two monoclonal antibodies, Press release, 13 December. EMA/390722/2013. http://www.ema.europa.eu/docs/en_GB/document_library/Press_release/2013/06/WC500144941.pdf
European Medicines Agency (2013b) Gastofil summary of opinion, 25 July. EMA/CHMP/304525/2013. http://www.ema.europa.eu/docs/en_GB/document_library/Summary_of_opinion_-_Initial_authorisation/human/002150/WC500146600.pdf
European Medicines Agency (2013c) Ovaleap: EPAR-product information, 23 October, 2013. http://www.ema.europa.eu/ema/index.jsp?curl=pages/medicines/human/medicines/002608/human_med_001689.jsp&mid=WC0b01ac058001d124and, http://www.ema.europa.eu/docs/en_GB/document_library/EPAR_-_Product_Information/human/002608/WC500152906.pdf
European Medicines Agency (2014) European public assessment reports, Internet reporting system, 18 April, EMA/390722/2013. http://www.ema.europa.eu/ema/index.jsp?curl=pages%2Fmedicines%2Flanding%2Fepar_search.jsp&murl=menus%2Fmedicines%2Fmedicines.jsp&

mid=WC0b01ac058001d124&searchTab=searchByAuthType&alreadyLoaded=true&isNewQuery=true&status=Authorised&status=Withdrawn&status=Suspended&status=Refused&keyword=Enter+keywords&searchType=name&taxonomyPath=&treeNumber=&searchGenericType=biosimilars&genericsKeywordSearch=Submit

Federal Trade Commission (2002) Generic drug entry prior to patent expiration: an FTC study, July. http://www.ftc.gov/os/2002/07/genericdrugstudy.pdf

Federal Trade Commission (2011) Authorized generic drugs: short-term effects and long-term impact, August. http://www.ftc.gov/opa/2011/08/genericdrugs.shtm

FiercePharma.com (2014) As biosims win nods in Europe and India, debate over naming rights heats up. Story #4, January 29, 2014. www.fiercepharma.com

Food and Drug Administration (2014a) Approved labelling, Gonal-F, follitropinalfa. Available at http://www.accessdata.fda.gov/drugsatfda_docs/nda/2004/021684s000_Gonal-F_Rtt_Pen_Prntlbl.pdf

Food and Drug Administration (2014b) Draft guidance for industry: clinical pharmacology data to support a demonstration of biosimilarity to a reference product. Center for Drug Evaluation and Research, Center for Biologics Evaluation and Research. http://www.fda.gov/Drugs/GuidanceComplianceRegulatory Information/Guidances/default.htm

Gardner J (2013) Biosimilar battle emerges as Teva cues up Neupogen rival. 24 October. http://www.epvantage.com/Universal?View.aspx?type=Story&id=464236. Accessed 18 April 2014

Generic Pharmaceutical Association (2013) Generic industry by the numbers. Generic Pharmaceutical Association 2012 Annual Report. www.gphaonline.org

Grabowski HG (2013) How will US market for biosimilars evolve? Powerpoint presentation at Bates White LLC, Washington DC, 10 June

Grabowski HG, Kyle MK (2007) Generic competition and market exclusivity period in pharmaceuticals. Manage Decis Econ 28:491–502

Grabowski HG, Cockburn IM, Long G et al (2007a) The effect on federal spending of legislation creating a regulatory framework for follow-on biologics: key issues and assumptions. Unpublished white paper, August. http://public.econ.duke.edu/Papers//PDF/0907_H_Grabowski_I_Cockburn_G_Long_et_al_Effect_on_Federal_Spending_of_Follow_on_Biologics.pdf

Grabowski HG, Ridley DB, Schulman KA (2007b) Entry and competition in generic biologics. Manage Deci Econ 28:439–451

Grabowski HG, Long G, Mortimer R (2011b) Implementation of the biosimilar pathway: economic and policy issues. Seton Hall Law Rev 41(2):511–557

Grabowski HG, Long G, Mortimer R (2013) Biosimilars. In: Danzon P (ed) Encyclopedia of health economics. Elsevier, New York, www.analysisgroup.com

Grabowski HG, Guha R, Brain C, Taub A (2014) Pharmaceutical patent challenges and their implications for innovation and generic competition. Powerpoint presentation given at the ASHEcon Fifth Biennial Conference, Los Angeles, June 23. www.econ.duke.edu

Hemphill CS, Sampat BN (2012) Evergreening, patent challenges, and effective market life in pharmaceuticals. J Health Econ 31:327–339

Hurwitz MA, Caves RE (1988) Persuasion or information? Promotion and the shares of brand name and generic pharmaceuticals. J Law Econ 31:299–320

IMS Health (2011) Biosimilar accessible market: size and biosimilar penetration. Powerpoint presentation prepared for EFPIA-EGA-EuropaBio, December

IMS Institute for Healthcare Informatics (2011a) The use of medicines in the United States: review of 2010, April

IMS Institute for Healthcare Informatics (2011b) The global use of medicines: outlook through 2015, May

IMS Institute for Healthcare Informatics (2014) Medicine use and shifting costs of healthcare: a review of the use of medicines in the United States in 2013, April

International Federation of Pharmaceutical Manufacturers & Associations (2006) Measuring trends in pharmaceutical consumption: what do we want to measure and why? Powerpoint

presentation and CD-ROM of Workshop, International Conference Centre, Geneva, 17 March. www.ifpma.org
Karst KR (2013a) The biosimilars state legislation scorecard. FDA Law Blog, 5 September. www.hpm.com
Karst KR (2013b) Amgen advocates for distinguishable non-proprietary names in substantial comments to GPhA and Novartis biosimilar naming citizen petitions. FDA Law Blog, 22 December. www.hpm.com
Kelley B (2009) Industrialization of mAb production technology. MAbs 1(5):443–452
Kim H, Park SK, Suh O-K, Oh JM (2003) Comparison of lenograstim and filgrastim on haematological effects after autologous peripheral blood stem cell transplantation with high-dose chemotherapy. Curr Med Res Opin 19(8):14
King R (Guy) (2007) Modeling federal cost savings of follow-on biologics. Avalere Health. http://www.avalerehealth.net/research/docs/Follow_on_Biologic_Modeling_Framework.pdf
Kulkarni K, Foy H (2012) Pfizer scraps insulin deal with India's Biocon. Reuters Business & Financial News, Breaking US & International News, 13 March 2012. http://www.reuters.com/assets/print?aid=USBRE82C05920120313
Leeson P (2012) Drug discovery: chemical beauty contest. Nature 481:455–456. doi:10.1038/481455a
McBride R (2012) Report: Samsung stung with delay on Rituxan biosimilar. Fierce Biotech, 17 October. http://www.fiercebiotech.com/story/report-samsung-stung-delay-rituxan-biosiimilar. Accessed 20 April 2014
McBride R (2013) UPDATE: Roche sees another Rituxan biosimilar trial bite the dust. Fierce Biotech, 18 April. http://www.fiercebiotech.com/story/roches-sees-another-rituxan-biosimilar-bite-dust/. Accessed 20 April 2014
Miller S, Houts J (2007) Potential savings of biogenerics in the United States: executive summary. Express Scripts, February. http://www.express-scripts.com/research/studies/pharmacybenefitresearch/specialtypharmacyservices/docs/potentialSavingsBiogenericsUS.pdf
Morrow KJ (2006) Disposable bioreactors gaining favour: new components and systems improve process reliability and reduce cost. Gen Eng News 26(12):42–45
Olson LM, Wendling BW (2013) The effect of generic drug competition on generic drug prices during the Hatch-Waxman 180-day exclusivity period. Working Paper No. 317, April. Federal Trade Commission, Bureau of Economics, Washington DC. www.ftc.gov
Palmer E (2013) UPDATED: Samsung will make some Roche bioilogics. FiercePharma Manufacturing, 23 October. http://www.fiercepharmamanufacturing.com/story/samsoung-will-make-some-roche-biologi. Accessed 20 April 2014
Pmlive.com (2013) Lonza ends biosimilars deal with Teva. pmlive.com/pharma_news/lonza_ends_biosimilar_deal_with_teva_492305. Accessed 18 April 2014
Reiffen DE, Ward ME (2005) Generic drug industry dynamics. Rev Econ Stat 87(1):37–49
Reiffen DE, Ward ME (2007) 'Branded Generics' as a strategy to limit cannibalization of pharmaceutical markets. Manage Decis Econ 28:251–265
Rovira J, Espin J, Garcia L, de Labry AO (2011) The impact of biosimilars entry in EU markets. Andalusian School of Public Health. http://ec.europa.eu/enterprise/sectors/healthcare/files/docs/biosimilars_market_012011_en.pdf
Saha A, Grabowski HG, Birnbaum HM, Greenberg PE, Bizan O (2006) Generic competition in the U.S. pharmaceutical industry. Int J Econ Bus 13(1):15–38
Schellekens H (2005) Follow-on biologics: challenges of the 'next generation'. Nephrol Dial Transplant 20(Suppl 4):iv31–iv36. doi:10.1093/ndt/gfh1085
Schneider CK, Borg JJ, Falk E et al (2012) In support of the European Union biosimilar framework. Nat Biotechnol 30(8):745–748
Scientia Advisors (2010) ScientiaBiosimilar Webinar, 7 July. www.scientiaadv.com
Scott Morton F (1999) Entry decisions in the generic pharmaceutical industry. RAND J Econ 30:421–440
Scott Morton F (2000) Barriers to entry, brand advertising and generic entry in the U.S. pharmaceutical industry. Int J Indust Organ 18:1085–1104

Senior M (2009) European biosimilars' market performance mirrors U.S. legislative progress: slow but steady. BioPharma Today, 19 May. http://www.biopharmatoday.com/2009/05/european-biosimilars-market-performance-mirrors-us-legislative-progress-slow-but-steady-.html

Somers J (2010) Effects of using generic drugs on medicare's prescription drug spending. The Congress of the United States, Congressional Budget Office, Washington DC, September. http://www.cbo.gov/

Staton T (2014) PBMs join chorus of support for easy names on biosimilar drugs. Fierce Pharma, 3 July. fiercepharma.com

Trusheim MR, Berndt ER (2012) Economic challenges and possible policy actions to advance stratified medicine. Pers Med 9(4):413–427

Trusheim MR, Aitken ML, Berndt ER (2010) Characterizing markets for biopharmaceutical innovations: do biologics differ from small molecules? Forum for Health Economics & Policy, Manuscript 1200, June, p 48. http://www.bepress.com/fhep

Trusheim MR, Burgess B, Hu SX, Long T et al (2011) Quantifying factors for the success of stratified medicine. Nat Rev Drug Discov 10:817–833

Walsh K (2013) Biosimilars' utilization and the role payers do play in driving uptake in Europe: an industry perspective. Powerpoint presentation given at the Biosimilar Medicines 11th EGA International Symposium, April. http://www.sandoz-biosimilars.com/cs/www.sandoz-biosimilars.com-v3/assets/media/shared/documents/presentations

Wang X (2012) Understanding current trends and outcomes in generic drug patent litigation: an empirical investigation. Unpublished Masters' honors thesis, Stanford University: Public Policy Program, May

Wiggins SN, Maness R (2004) Price competition in pharmaceuticals: the case of anti-infectives. Econ Inq 42(2):247–263

World Health Organization (2003) Introduction to drug utilization research. Collaborating Center for Drug Statistics Methodology, Norwegian Institute of Public Health, Oslo, Norway. http://www.whocc.no/use_of_atc_ddd. Accessed 27 June 2011

World Health Organization (2009) DDD: definition and general considerations. Collaborating Center for Drug Statistics Methodology, Norwegian institute of Public Health, Oslo, Norway, last updated 17 December 2009. http://www.whocc.no/ddd/definition_and_general_considera/. Accessed 27 June 2011

World Health Organization (2011) Use of ATC/DDD. Collaborating Center for Drug Statistics Methodology, Norwegian Institute of Public Health, Oslo, Norway, last updated 15 February 2011. http://www.whocc.no/use_of_atc_ddd. Accessed 27 June 2011

Wosinska M, Huckman RS (2004) Generic dispensing and substitution in mail and retail pharmacies. Health Affairs Web Excl W4-409 to W4-416, posted 28 July 2004. www.healthaffairs.org (An abstract of the article was published in the hardcopy edition of Health Affairs 23(5):284, September/October 2004)

Ziegler M, Santagostino A (2011) Manufacturing drives value in biopharmaceuticals. Pharma Manufacturing. http://www.pharmamanufacturing.com/articles/2011/085.html

Building Biobetters: The Regulatory Landscape

Emanuela Lacana, Lynne Yao, Anne Pariser, Amy Rosenberg, and Janet Woodcock

Introduction

A biobetter is an improved or optimized version of an existing biological drug, or a new biologic carefully designed to maximize clinical performance, i.e., safety and efficacy. For many diseases, particularly serious or life-threatening diseases with unmet medical needs, biobetters are important in optimizing clinical outcome where no curative therapy currently exists. Several contributors in this book explored a variety of strategies to optimize product performance

E. Lacana
Office of Biotechnology Products, Center for Drug Evaluation and Research, Food and Drug Administration, 71/2338 10903 New Hampshire Ave, Silver Spring, MD 20993, USA

A. Rosenberg (✉)
Division of Biotechnology Review and Research III, Office of Biotechnology Products, Center for Drug Evaluation and Research, Food and Drug Administration, 71/2338 10903 New Hampshire Ave, Silver Spring, MD 20993, USA
e-mail: amy.rosenberg@fda.hhs.gov

L. Yao
Office of New Drugs, Center for Drug Evaluation and Research, Food and Drug Administration, Silver Spring, MD, USA

A. Pariser
Office of Translational Sciences, Center for Drug Evaluation and Research, Food and Drug Administration, Silver Spring, MD, USA

J. Woodcock
Center for Drug Evaluation and Research, Food and Drug Administration, Silver Spring, MD, USA

A. Rosenberg, B. Demeule (eds.), *Biobetters*, AAPS Advances in the Pharmaceutical Sciences Series 19, DOI 10.1007/978-1-4939-2543-8_16

through the following: enhanced delivery to target organs and tissues; improvement in the activity and longevity of the product, largely through improving in vivo stability; and reduced immunogenicity. This chapter describes the current regulatory landscape and focuses specifically on the regulatory pathways that strongly promote the development of biobetters.

In recent years, considerable effort, incentives, and legislative mandates have redefined the drug development environment for patients with serious or life-threatening disorders, with unmet medical needs and for children. The FDA understands that the development of drugs in an efficient and successful process is important for all patients, and especially important for patients who suffer from diseases for which there is no approved treatment or for whom existing treatments are inadequate. The FDA has carefully considered the development pathway for drugs to increase the efficiency and success of these programs.

In particular, for serious or life-threatening diseases, there is recognition that patients and physicians are generally willing to accept greater risks or side effects from products that treat life-threatening or severely-debilitating illnesses (US Food and Drug Administration 2012a). The broadest flexibility in applying statutory standards for approval will be exercised in these situations, without compromising safety and effectiveness, and the drugs will be evaluated in light of the severity of the disease. General considerations and procedures incorporated into the development of drugs in this environment include early consultation with regulators focusing on the risk/benefit analysis, consideration for early access to experimental agents outside of clinical trials, where appropriate (i.e., expanded access or "compassionate use" (US Food and Drug Administration 2012b)), safeguards for patient safety, and continued study and evaluation in post-marketing phases, where appropriate and as needed. Beyond general considerations, the FDA has at its disposal a variety of programs to facilitate and expedite drug development.

Expedited Programs

The ability to apply flexibility in drug development programs will heavily depend upon foundational and translational information obtained during early drug development phases, and on frequent and early communication with the FDA to incorporate flexibility and scientific judgment into clinical planning without compromising the ability to define the efficacy and safety of the intervention.

In the US, there are four "expedited programs" available to developers of drugs for serious and life-threatening diseases with unmet medical needs: Fast Track, Breakthrough Therapy, Accelerated Approval, and Priority Review (see Table 1) (US Depart HHS 2014). Table 1 provides an outline of the important features of each program.

Table 1 Expedited programs

Program	Qualifying criteria: serious condition and...	Features
Fast track	Nonclinical or clinical data demonstrate potential to meet an unmet medical need • Or, QIDP[a]	Actions to expedite development and review, e.g., meetings • Rolling review
Breakthrough therapy	Preliminary clinical evidence indicates drug may demonstrate substantial improvement on a clinical significant endpoint over available therapies	All Fast Track features Intensive guidance on efficient drug development • Organizational commitment
Accelerated approval	Provides meaningful advantage over available therapies • Demonstrates effect on surrogate endpoint reasonably likely to predict clinical benefit or on a clinical endpoint that can be measured earlier than IMM[b] that is reasonably likely to predict an effect on IMM or other clinical benefit (i.e., intermediate clinical benefit)	Approval based on a surrogate or intermediate clinical endpoint reasonably likely to predict clinical benefit
Priority review	Would provide a significant improvement in safety or effectiveness • Or, other qualifying programs[c]	Shorter review clock goal for marketing application review (6 vs. 10 month for a standard review)

[a]*QIDP* qualifying infectious disease product
[b]*IMM* irreversible morbidity or mortality
[c]Other qualifying programs, i.e., supplements that propose a labeling change pursuant to a report on a pediatric study under the Best Pharmaceuticals for Children Act (Federal Food, Drug, and Cosmetic Act 2012a), QIDP (Federal Food, Drug, and Cosmetic Act 2012b), or any application or supplement for a drug submitted with a priority review voucher (Federal Food, Drug, and Cosmetic Act 2012c, d)

Fast Track (US Depart HHS 2014; Federal Food, Drug, and Cosmetic Act 2012e)

The Fast Track program is intended to facilitate development and expedite review of products intended to treat a serious or life-threatening disease and that have the potential to address an unmet medical need. Fast Track provides the opportunity for frequent interactions with the FDA review team in terms of milestone meetings, such as pre-IND, end-of-phase-1 and end-of-phase-2 meetings. Under Fast Track, designated drugs may submit marketing applications for "rolling review", where the application may be submitted in modules, rather than as a complete application (Federal Food, Drug, and Cosmetic Act 2012e). Importantly, fast track designation can be based on clinical or non-clinical findings that indicate the drug's potential to meet an unmet medical need. Fast Track's benefits apply to drug developmental phases rather than to application review or to the post-marketing period and would be appropriate to consider for development of a biobetter product for a serious disease or unmet medical need, such as exists for many rare diseases.

Breakthrough Therapy Designation (Federal Food, Drug, and Cosmetic Act 2012f)

Breakthrough Therapy Designation is a relatively new program, available as of 2012 (US Depart HHS 2014; Federal Food, Drug, and Cosmetic Act 2012f). Products are granted Breakthrough designation when preliminary clinical evidence indicates that the drug may offer substantial improvement on a clinically significant endpoint over available therapies. Breakthrough Therapies are eligible for all of the Fast Track features (such as rolling review), and similar to the Fast Track Program, focus on early collaboration and communication between drug developers and FDA during the drug development phase. However, designation as a Breakthrough Therapy provides additional, intensive FDA guidance on efficient product development and includes an organizational commitment involving senior managers and experienced cross-disciplinary review and project management staff. Breakthrough also differs from Fast Track in that eligibility for the Breakthrough program requires preliminary *clinical* evidence indicating that the drug may demonstrate substantial improvement over existing therapies for a serious or life-threatening disease, while Fast Track eligibility requires *either preclinical* or clinical data. Despite these differences, both the Fast Track and Breakthrough Programs are intended to foster and encourage early engagement and ongoing communication between drug developers and FDA during the developmental phase, specifically to the earlier clinical development phases (ideally before end-of-phase-2). Thus, breakthrough would be appropriate to consider for development of a biobetter product for a serious disease for which preliminary clinical evidence indicates the drug may demonstrate substantial improvement over available therapies.

Priority Review (US Depart HHS 2014)

Products eligible to receive priority review pertain to the following categories: those for serious conditions that, if approved, would provide a significant improvement in safety or effectiveness over available therapies (US Depart HHS 2014); products that qualify under special programs (i.e., qualifying infectious disease products QIDP (Federal Food, Drug, and Cosmetic Act 2012b)), supplements that propose a labeling change pursuant to a report on a pediatric study under the Best Pharmaceuticals for Children Act (BPCA) (i.e., studies submitted to fulfill a Pediatric Written Request); or an application or supplement for a drug submitted with a priority review voucher (PRV) (Federal Food, Drug, and Cosmetic Act 2012c, d). Priority review provides a marketing application review time goal of 6 months, rather than the standard review goal of 10 months (for new molecular entities and original biologics, there is an additional 2-month review period added to the timeline for both priority and standard reviews (US Depart HHS 2014)). Products that are intended to offer a major advance in treatment (such as a biobetter) or provide treatment when no adequate therapy exists are eligible to receive a Priority Review.

Accelerated Approval (US Depart HHS 2014; Federal Food, Drug, and Cosmetic Act 2012g)

Accelerated approval is a path for marketing approval for products intended for a serious or life-threatening disease or condition that may offer a meaningful advantage over existing therapies. Accelerated approval may be granted: "upon a determination that the product has an effect on a surrogate endpoint that is reasonably likely to predict clinical benefit, or on a clinical endpoint that can be measured earlier than irreversible morbidity or mortality, that is reasonably likely to predict an effect on irreversible morbidity or mortality or other clinical benefit, taking into account the severity, rarity, or prevalence of the condition and the availability or lack of alternative treatments"(US Depart HHS 2014).

Importantly, for drugs granted accelerated approval, postmarketing confirmatory trials have been required to verify and describe the anticipated clinical benefit or effect on irreversible morbidity or mortality. This pathway has been used to approve biologics for rare diseases and may be appropriate as an approval pathway for biobetter agents.

Selected Incentive Programs

There are number of incentive programs available in the U.S. that are intended to encourage and facilitate the development of therapies for specific public health needs and special populations. Some of the longest-standing and most well-known of these programs are the incentives for rare diseases available under the Orphan Drug Act and for children under BPCA. Newer incentive programs include the Neglected Tropical Disease Priority Review Voucher Program and the Rare Pediatric Disease Priority Review Voucher Program (see Box 1).

Orphan Drug Act (1983)

The Orphan Drug Act (ODA), enacted in 1983, is one of the first incentive programs passed by Congress. The ODA provides incentives intended to make the development of products designated to treat rare diseases financially viable. A rare or orphan disease is defined as a disease or condition that affects less than 200,000 people in the U.S. The financial incentives include the following:

- May be eligible for 7 years of marketing exclusivity upon approval
- Waiver of the orphan drug's marketing application user fee (approximately $2 million in 2014)
- Tax credits of up to 50 % of clinical trial costs incurred for studying an orphan drug in a rare disease

Since the passage of the ODA, over 450 products designated to treat a rare disease have been approved.

Box 1: Selected Incentive Programs

1. The Orphan Drug Act (1983)
   The Orphan Drug Act (ODA) was enacted in the US in 1983, which defines orphan drugs as those drugs intended to treat, prevent or diagnose diseases or conditions affecting fewer than 200,000 persons in the US, and that have shown promise, based on supporting evidence, in the treatment of the disease or condition. If designated as an Orphan drug, the drug sponsor is eligible to receive incentives intended to make the development of drugs to treat small populations financially viable, including:
   - Tax credits—up to 50 % of clinical trial costs
   - Waiver of the user fee
   - 7 years of marketing exclusivity upon approval
2. Best Pharmaceuticals for Children Act (BPCA) (Federal Food, Drug, and Cosmetic Act 2012a; Best Pharmaceuticals for Children Act 2002)
   In 2002, the Best Pharmaceuticals for Children Act was passed. This legislation authorizes FDA to grant an additional 6 months of marketing exclusivity to sponsors who voluntarily complete pediatric clinical studies outlined in a Written Request (WR). A WR is a document issued by the FDA requesting submission of a certain study or studies to determine whether the use of a drug could provide a meaningful health benefit in the pediatric population.
3. Priority Review Voucher Programs (Federal Food, Drug, and Cosmetic Act 2012c, d)
   There are two PRV programs available for Neglected Tropical Diseases and Rare Pediatric Diseases. Briefly, a PRV is a voucher awarded to the developer of a qualifying drug upon approval of that drug, which may be applied to a future product to convert a standard review to a priority review (see expedited programs, Table 1). Alternatively, the voucher may be transferred or sold to another drug developer.

Best Pharmaceuticals for Children Act (Federal Food, Drug, and Cosmetic Act 2012a; Best Pharmaceuticals for Children Act 2002)

Many diseases that occur in adults also occur in children, and children are often treated with the same drugs as adults. However, prior to 1994, over 80 % of all drugs used in adults lacked specific information about use in pediatric populations. In order to advance the development of products used to treat pediatric populations, incentives for the development of products to treat pediatric populations were first incorporated into law in 1997. These incentives, first enacted under the Food and Drug Administration Modernization Act of 1997 (Food and Drug Administration Modernization Act 1997) ultimately formed the basis for The Best Pharmaceuticals for Children Act in 2002 (Best Pharmaceuticals for Children Act 2002). Under BPCA, FDA may grant drug manufacturers an additional 6 months of marketing exclusivity that may be added to any existing exclusivity or patent for completion and submission of studies that FDA considers a meaningful health benefit in children. These studies are established under a Written Request issued by the FDA. In addition to the incentives under BPCA, the Pediatric Research Equity Act (PREA), enacted in 2003, requires drug manufacturers, under certain circumstances, to perform pediatric studies. Since the passage of the incentive and requirement provisions under BPCA and PREA, over 500 labeling changes that provide additional effectiveness, safety, or dosing information for pediatric populations have been approved.

Neglected Tropical Disease and Rare Pediatric Disease Priority Review Voucher Programs (Federal Food, Drug, and Cosmetic Act 2012c, d)

The Neglected Tropical Disease (NTD) Priority Review Voucher (PRV) program, passed in 2007, encourages the development of new drug and biological products for prevention and treatment of certain tropical diseases. Through this program, FDA is authorized to award a voucher to the sponsor of a qualifying listed NTD at the time of approval of that drug. The voucher may be redeemed at a future date to convert a subsequent marketing application or supplement from a standard review (goal timeline of 10 months) to a priority review (goal timeline of 6 months), or the voucher may be transferred or sold to another drug sponsor. In 2012, additional legislation was passed by Congress to provide for a similar voucher program to provide additional incentives for rare pediatric diseases (RPD). Rare diseases, such as rare genetic diseases or childhood cancers, may qualify for this program, and could be applied to development of biobetter therapeutics, provided they meet qualifying criteria.

External Collaborations

FDA has recognized that efficient product development is often a collaborative effort among many stakeholders, including drug manufacturers, regulatory agencies, patients, academia, and advocacy groups. Therefore, it is important to maximize use of all opportunities for collaboration (such as meetings, workshops, etc.) in order to increase the efficiency and success of product development programs.

Clearly, collaboration with other federal agencies and global regulatory authorities can both help to encourage efficient product development by avoiding unintended duplication of efforts and potentially increasing resources to speed development efforts. Such collaborations have been successful in improving the number of successful product development programs (Davis 2011; Ramsey et al. 2011).

Conclusions

Despite the success of these scientific, regulatory and legislative initiatives to improve efficiency and success of biological product development, there remain considerable unmet medical needs for patients with serious diseases, and particularly for rare diseases. It is increasingly clear that close collaborations of all stakeholders including patients, industry, academia, regulators and other governmental agencies are critical to the success and efficiency of product development, especially for those products used to treat pediatric and rare diseases. Early planning and consultation with all stakeholders require availability of resources (e.g., time, experienced personnel, sustained funding and infrastructure). However, in many cases, important resources are not available during early phase product development. Thus, it seems prudent to develop incentive programs that aid in the earlier planning stages of product development in order to identify the best candidates that lead to approved treatments for patients in need. These incentive programs should target basic and translational science needs, and require planning and consultation with all stakeholders.

As described above, FDA can use a variety of programs to expedite and incentivize drug development. These regulatory programs are available to sponsors who make a conscious effort to develop a "better" product that provides for improved clinical outcome. Careful and thoughtful manufacturing design and a thorough understanding of the mechanism of action of a product can result in a drug that maximizes efficacy with a favorable safety profile. For example, if the mechanism of action of a product requires that the product gain entry into a specific organ or tissue to exert its pharmacological activity, a biobetter with an improved targeting mechanism may have a higher efficacy profile at lower doses than a currently marketed product.

There are several studies in the literature and chapters in this book that specifically investigate better targeting to affected tissues. For example, the deficiency in lysosomal acid alpha glucosidase (GAA) characteristic of Pompe disease, leads to accumulation of glycogen in the lysosome. Successful enzyme replacement therapy requires internalization of the therapeutic enzyme into target cells and delivery to the lysosome. Internalization and lysosomal delivery of GAA to the target organs occurs via the mannose-6P receptor, and binding to receptor requires the presence of phosphorylated mannose. Therefore, the carbohydrate side chains are critical for optimal delivery to the target organ and intracellular uptake, as well as to determine product half-life and route of elimination. The most affected organs in Pompe disease are the heart and skeletal muscle. Uptake in these organs, but especially skeletal muscle, is limited by numerous factors including low levels of the targeting moiety on the ERT, as well as low mannose 6-P receptor expression, the on the skeletal muscle. Moreover, the CNS is also affected in the most severe cases of PD, as well as in other lysosomal storage diseases and the ERTs to treat these conditions do not cross the blood brain barrier, thereby leaving this most vital of organs untreated. Thus, a multisystem approach is needed to establish the most optimal treatment approach including a "biobetter" GAA ERT which expresses higher levels of the targeting moiety or novel targeting moieties (e.g., IGF2 (Maga et al. 2013)) for skeletal muscle and the ability to cross the blood brain barrier to treat the CNS, as well as other agents that may increase receptor expression on skeletal muscle.

This type of approach has also been undertaken to better target the endothelial cells in multiple organs, as in the case of Fabry Disease. Hsu et al. (2011), developed a targeting approach in which nanocarriers were loaded with alfa galactosidase and antibody against ICAM 1, a surface molecule abundantly expressed in endothelial cells. The antibody/protein nanocarrier complex was efficiently internalized in model endothelial cells and localized to the endothelial surface of heart and kidney in mice. It is conceivable that a similar approach could be taken for other lysososmal diseases.

Thus, given the potential substantial improvements over the existing therapies, a biobetter may qualify for several of the expedited programs: (1) breakthrough designation, if the efficacy and safety profile is considered a significant improvement over existing therapies, (2) priority review, with a shorten review cycle to allow faster release to the market and (3) fast track to allow for rolling submission. Both the fast track and breakthrough designations provide for increased communication with the FDA, to facilitate efficient product development, and could facilitate the development of a truly biobetter product. Sponsors are encouraged to utilize other available incentive and expedited programs, such as the ODA and pediatric incentives, where applicable. Sponsors planning to develop biobetter products should communicate with Agency early and frequently in order to best take advantage of such expedited programs and develop a strategy that would allow for an efficient regulatory process.

References

Best Pharmaceuticals for Children Act (2002) Pub. L. 107–109 Reauthorized in 2007 by Pub. L. 110–85 (Food and Drug Administration Amendments Act), and permanently reauthorized in 2012 by Pub. L. 112–114 (Food and Drug Administration Safety and Innovation Act)

Davis PB (2011) Therapy for cystic fibrosis – the end of the beginning? N Engl J Med 365:1734–1735

Food and Drug Administration Modernization Act (1997) Pub. L. 105–115. 111 Stat. 2036. Section 111. Pediatric studies of drugs

Federal Food, Drug, and Cosmetic Act (2012a) 21 U.S.C. Section 355a: pediatric studies of drugs

Federal Food, Drug, and Cosmetic Act (2012b) 21 U.S.C. Section 355n-1: priority review for qualified infectious disease products

Federal Food, Drug, and Cosmetic Act (2012c) 21 U.S.C. Section 360n: priority review to encourage treatments for tropical diseases

Federal Food, Drug, and Cosmetic Act (2012d) 21 U.S.C. Section 360(ff): priority review to encourage treatments for rare pediatric diseases

Federal Food, Drug, and Cosmetic Act (2012e) 21 U.S.C. Section 356(b): designation of a drug as a fast track product

Federal Food, Drug, and Cosmetic Act (2012f) 21 U.S.C. Section 356(a): designation of a drug as a breakthrough therapy

Federal Food, Drug, and Cosmetic Act (2012g) 21 U.S.C. Section 356(c): accelerated approval of a drug for a serious or life-threatening disease or condition, including a fast track product

Federal Food, Drug, and Cosmetic Act (2012h) 21 U.S.C. Section 355c: research into pediatric uses for drugs and biological products

Hsu J, Serrano D, Bhowmick T et al (2011) Enhanced endothelial delivery and biochemical effects of α-galactosidase by ICAM-1 targeted nanocarriers for Fabry disease. J Control Release 149:323–331

Maga AJ, Zhou J, Kambampati R et al (2013) Glycosylation-independent lysosomal targeting of acid α-glucosidase enhances muscle glycogen clearance in Pompe mice. J Biol Chem 288:1428–1438

Orphan Drug Act (1983), Pub. L. 97–414. 96 Stat. 2049. Amended in 1984 by Pub. L. 98–551 to add a numeric prevalence threshold to the definition of rare diseases

Ramsey BW, Davies J, McElvaney NG et al (2011) A CFTR potentiator in patients with cystic fibrosis and the G551D mutation. N Engl J Med 365:1663–1672

US Depart HHS FDA (2014) Guidance for industry: expedited programs for serious conditions--drugs and biologics. Retrieved August 20, 2014 from http://www.fda.gov/downloads/Drugs/GuidanceComplianceRegulatoryInformation/Guidances/UCM358301.pdf

US Food and Drug Administration (2012a) Code of federal regulations title 21: drugs intended to treat life threatening and severely-debilitating illnesses. Pt. 312.80 (Subpart E). US Government Printing Office, Washington

US Food and Drug Administration (2012b) Code of federal regulations title 21: expanded access to investigational drugs for treatment use. Pt. 312.300 (Subpart I). US Government Printing Office. Washington

Index

A

Abbreviated New Drug Application (ANDA), 316–317, 321, 339, 340
Abciximab (RheoPro®), 227
Abicipar pegol, 242
Accelerated approval, 362, 363, 365
Acid alpha glucosidase (GAA), 368, 369
  ERT, 27
  glycosylated protein, 27
  MPR proteins, 25–26
  M6P tag, 25
  rhGAA efficacy (*see* Recombinant human GAA (rhGAA))
Acid maltase deficiency. *See* Pompe disease
Acute lymphoblastic leukemia (ALL), 244
Acute myeloid leukemia (AML), 244, 352
Acylation, 193
Adalimumab (Humira®), 222
Adnectin pharmacokinetic enhancer (AdPKE), 234
Ado-Trastuzumab Emtansine, 223
Albuferon, 278, 279
Albugranin, 278
Albumin-binding domain (ABD), 235, 237, 280–281
α1-antitrypsin (A1AT), 287
Alternative protein scaffolds, 223
Antibody-dependent cell-mediated cytotoxicity (ADCC), 205, 206, 208, 209, 214, 222, 275, 283
Antibody-dependent cellular phagocytosis (ADCP), 206
Antibody engineering approaches
  Fab region, 125–126
  Fc and hinge regions, 126
  modifications, 124–125
Antibody-like molecules
  advantages, 203
  bispecific targeting, 209–210
  clinically-tested bispecific format, 210–212
  constant region mediates, 206–207
  conventional monoclonal antibodies, 203
  engineering of, 204–205
  growth-factor induced oncogenic signaling
    EGFR blockers, 212–213
    ErbB2 blockers, 213
    HGFR/c-Met, 215
    MM-141 case study, 213–214
    VEGF blockers, 212
  immune effector function
    glycoengineering, 209
    isotype optimization, 207–208
    targeted mutagenesis, 208–209
  immunoglobulin G structure, 205–206
Anticalins
  β-barrel, 238
  BBP scaffold, 240
  Aβ peptide, 241
  disease-relevant protein antigens, 239
  ED-B positive cells, 240
  HBP, 238
  immunotherapy of cancer, 240
  Lcn1 and Lcn2, 239
  Lcn1/Tlc scaffold, 240
  lipocalins, 237
  PRS-050, 240
  T-cell response, 239
  VEGF-A blocks, 240
Anti-drug antibodies (ADA), 51–53, 63–66, 69, 72, 73, 234, 240, 252, 282, 284

A. Rosenberg, B. Demeule (eds.), *Biobetters*, AAPS Advances in the Pharmaceutical Sciences Series 19, DOI 10.1007/978-1-4939-2543-8

Antigen-binding fragments, 223
Antigen-specific VNAR fragments, 230
Anti-HER2 Affibodies, 236
AQB-101, 295–296
Aranesp™ (Amgen), 294, 296, 339, 345, 350
Aurograb, 226
Authorized generic (AG) manufacturers, 316–317

B
Balugrastim, 347
BBB. *See* Blood–brain barrier (BBB)
Bence Jones homo-dimers, 228
Bence-Jones proteins, 230
Best Pharmaceuticals for Children Act (BPCA), 364–367
Biologics License Application (BLA), 347, 348
Biopharmaceuticals, 322–323, 343–344
Biotechnology Industry Organization (BIO), 327
BiTE®, 243–244
Blinatumomab, 244, 245, 249
Blood–brain barrier (BBB)
  disruption, 45
  inborn errors of metabolism, 41, 42
  intra-thecal enzyme delivery, 42–44
  MPSI (*see* Mucopolysaccharidosis type I (MPSI))
  MPSII (*see* Mucopolysaccharidosis type II (MPSII))
  MTH technology
    CMT and RMT systems, 45
    insulin/Tf, 45
    MAb-enzyme fusion protein, 47–48, 59
    peptides, 45
    peptidomimetic monoclonal antibodies, 45
    TH fusion protein, 46–47
  nanoparticles, 45
Breakthrough therapy designation, 364

C
Campylobacter jejuni, 287
Caplacizumab, 229
Cation-dependent MPR (CD-MPR), 25–26
Cation-independent MPR (CI-MPR), 25–26
CD4 immunoadhesin, 271
Cerebrospinal fluid (CSF) compartment, 42–44
Certolizumab pegol, 228
Charge variants, 107, 111, 166
Cimzia, 228, 246, 283
Committee for Medicinal Products for Human Use (CHMP)
  EMA (*see* European Medicines Agency (EMA))
  follitropin alfa (injection), 354
  infliximab (remicade), 353–354
Complementarity-determining regions (CDRs), 222, 224, 227
Complement-dependent cytotoxicity (CDC), 207, 208, 214, 222, 275, 283
Critical product quality attribute (CQA), 138
C-terminal peptide (CTP), 296
Current good manufacturing practices (cGMP), 316

D
Daclizumab, 222
Daily average consumption (DACON), 329
Darleukin, 226
Deamidation
  charge variants, 107
  evidence, 110–111
  extrinsic factors, 109
  intrinsic factors, 108–109
  mechanism, 107–108
Defined Daily Dosage (DDD), 328–333
Dermatomyositis, 70
Designed ankyrin repeat protein (DARPin) scaffold, 232, 241–243, 253
Digoxigenin-binding Anticalin (DigiCal), 240
Diketopiperazine (DKP) formation, 115, 116
Disulfide scrambling/rearrangement
  cell-based binding assays and ELISA, 121
  dimers/oligomers, 121
  extrinsic factors, 120–121
  Fab-arm exchange, 122–123
  IgG2-A and IgG2-B isoforms, 122
  intrinsic factors, 120
  mechanism, 119
  thioether linkages, 121–122
  trisulfide variants, 121
Disulfide-stabilized Fv fragments (dsFv), 225
Domain antibodies (dAbs), 229
Dual-affinity re-targeting (DART®) protein, 244–245

E
Efungumab (Mycograb™), 226
Elastin-like polypeptides (ELPs), 293–294
Elonva®, 296
EMA. *See* European Medicines Agency (EMA)

Enbrel™, 345, 346
Endosomal recycling, 271
Enzyme replacement therapy (ERT)
  factors, 9, 13
  LSD
    Fabry disease, 13
    Gaucher disease, 13, 14
    IPD, 13–14, 16
    MPS, 13
    types, 9–12
  revolutionized treatment, 9–10
  and rhGAA efficacy (*see* Recombinant human GAA (rhGAA))
Epi-3D, 68
EpiSweep, 68
Epogen/Procrit, 349, 350
ErepoXen®, 287
ERT. *See* Enzyme replacement therapy (ERT)
Erythropoietin (EPO), 274–275
Erythropoietin stimulating agent (ESA), 350
EUROCALIN Consortium, 241
European Medicines Agency (EMA), 325–326
  BIO, 327
  DDD, 327–329
  erythropoietin alpha, 330, 331
  extended units, 329
  G-CSFs, 324, 331
  guidelines, 323, 324
  human follicle stimulating hormone (FSH), 324
  IMS Health DACON data, 329
  Inflectra, 324
  phase II or phase III clinical trial, 327
  Remsima, 324
  somatropins, 324, 330, 331
  standard units, 329
  tumor necrosis factor alpha (TNF-α), 324
Extra-domain B (ED-B), 226, 240

F
Fabry disease, 13, 16, 33, 369
Fast Track program, 363
Fc *N*-glycan hydrolysis
  evidence, 114
  extrinsic factors, 113–114
  intrinsic factors, 113
  mechanism, 111–112
Fibroblast growth factor 21 (FGF21), 234
Fibronectin type III domain (FN3), 233
Follitropin alfa (Gonal-F), 354
FynomAb® technology platform, 246
Fynomers®, 246

G
GAA. *See* Acid alpha glucosidase (GAA)
Gaucher disease, 3–7, 9, 10, 13
G-CSF. *See* Granulocyte colony-stimulating factor (G-CSF)
Genzyme Corporation, 7
Glycogen storage disorder type II (GSD II). *See* Pompe disease
GlycoPolymer technology, 295
Granix™, 347
Granulocyte colony-stimulating factor (G-CSF), 191–192, 284, 285, 289, 295, 324, 330–332, 351–353

H
Hatch-Waxman legislation, 316, 319
Heparosan, 287–289
Hepatocyte growth factor receptor (HGFR/c-Met), 211, 215, 235
HESylation®, 288, 297
Histamine-binding protein (HBP), 238
Human serum albumin (HSA)
  ABD, 280–281
  ABD034, 281
  acidic protein, 277
  advantage, 280
  AlbudAb™, 282
  Albugranin and Albuferon, 278
  Albuviritide, 279
  CD4, 278
  endogenous and exogenous ligands, 281
  FcRn interaction, 277
  immunogenicity, 279
  isoelectric point, 277
  Liraglutide, 280
  PC-DAC™, 279
  pharmaceutically relevant proteins, 278
  yeast-based expression systems, 278
Hunter's syndrome. *See* Mucopolysaccharidosis type II (MPSII)
Hydrolytic reactions
  deamidation
    charge variants, 107
    evidence, 110–111
    extrinsic factors, 109
    intrinsic factors, 108–109
    mechanism, 107–108
  disulfide rearrangement
    cell-based binding assays and ELISA, 121
    dimers/oligomers, 121
    extrinsic factors, 120–121
    Fab-arm exchange, 122–123

Hydrolytic reactions (*cont.*)
IgG2-A and IgG2-B isoforms, 122
intrinsic factors, 120
mechanism, 119
thioether linkages, 121–122
trisulfide variants, 121
Fc *N*-glycan hydrolysis
evidence, 114
extrinsic factors, 113–114
intrinsic factors, 113
mechanism, 111–112
mitigating effects
antibody engineering approaches, 124–126
antibody formulation, 124
peptide bond hydrolysis
β-elimination mediated hydrolysis, 115, 116
clipping, 115
direct hydrolysis, 115, 116
DKP formulation, 115, 116
evidence, 118
extrinsic factors, 117
intrinsic factors, 117
Hydroxyethyl starch (HES), 287–288, 298

I
Ig novel antigen receptors (IgNARs), 230
Immuna® screening platform, 225
Immune Thrombocytopenic Purpura (ITP), 70
Immunogenicity, 339–342
antigen specific immune tolerance induction, 69–70
deimmunization, 66–69
drug-induced immunosuppression, 66
Pompe disease, 65
in protein therapeutics, 63–64
Tregitopes
characteristics, 70
immunosuppressive effects, 70–71
IVIG, 70
mechanism of action, 71–72
in silico immunogenicity score, 70
Tregitope-mediated tolerance induction, 72
Infantile Pompe disease (IPD), 13–14, 16
Infliximab (Remicade), 353–354
International Non-proprietary Name (INN), 340–341

K
Kawasaki syndrome (KS), 70
Kineret®, 288

L
Linear solenoid, 241
Lipinski rule of five, 322
Lipocalins, 237
Loop-DARPins, 253
Loss of exclusivity (LOE), 316, 318, 321
Lysosomal storage disease (LSD)
Fabry disease, 13
Gaucher disease, 13, 14
IPD, 13–14, 16
MPS, 13
types, 9–12

M
MAbs. *See* Monoclonal antibodies (MAbs)
Macrophage-targeted glucocerebrosidase
accumulating material, 3, 4
carbohydrate unit, 5
catabolism, 3, 4
enzymatic modification, 5, 6
mannose-terminal glycoform, 5, 6
Merrimack's bispecific antibody MM-111, 213
Microbial transglutaminase, 288
Miller Fisher syndrome, 238
Mitigating effects
antibody engineering approaches
Fab region, 125–126
Fc and hinge regions, 126
modifications, 124–125
antibody formulation, 124
immunogenicity
antigen specific immune tolerance induction, 69–70
deimmunization, 66–69
drug-induced immunosuppression, 66
Tregitopes (*see* Tregitopes)
Molecular assessment (MA)
antibody discovery technology
advantages, 157–159
eukaryotic expression systems, 156
humanization, 156
phage display, 156
prokaryotic expression, 156
variant pool/library construction, 157
yeast and mammalian display, 156
CDR sequences, 169–170
de-risking process development
concentration-dependent precipitation, 163–164
phase separation, 162–163
physical instability, 160–161
site-specific chemical degradation, 160
viscosity, 162

experimental tools
accelerated temperature stability study, 165–167
material requirements, 164, 165
miscellaneous study, 169
oxidation hotspots, 167
solubility study, 168–169
viscosity, 167–168, 173–174
FIH trials, 154
improved stability, 171–173
vs. traditional approach, 154, 155
two different mAbs *vs.* different antigen targets, 170–171
two different mAbs *vs.* same antigen targets, 171
Molecular entity, Europe
DDD volume shares, 330, 331
dollar share, 337–339
erythropoietin alpha, 330, 331
revenue shares, 335–337
volume shares, 331–335, 343
Molecular Trojan Horse (MTH) technology
CMT and RMT systems, 45
insulin/Tf, 45
MAb-enzyme fusion protein, 47–48, 59
peptides, 45
peptidomimetic monoclonal antibodies, 45
TH fusion protein, 46–47
Monoclonal antibodies (mAbs)
Adalimumab, 222
biobetters, 223
bispecific constructs
BiTE®, 243–244
DART® protein, 244, 245
Fcab™s, 245
Fynomers®, 246
TandAb®s, 245
CDR-grafting, 222
CDR sequences, 222
enhancing immune effector functions, 222–223
Fab to domain antibody
Abciximab, 227
antigen-binding fragments, 223
antimicrobial therapy, 226
camelid VHH domains, 228–229
Caplacizumab, 229
CDRs, 224
Certolizumab pegol, 228
dAb, 230
disadvantages, 224
framework region(s), 224
functional antibody fragment, 227
Fv fragment, 225
"heavy-chain" antibodies, 228
hormones or cytokines insofar, 223
hypervariable region(s), 224
immunotoxin, 226
intact antigen-binding site, 224
Moxetumomab pasudotox, 226–227
Ozoralizumab, 229
PENTRA®body technology, 226
pharma and biotech companies, 228
Ranibizumab, 227
sdIF, 229, 230
single chain Fv fragment (scFv), 225, 226
SPECT, 227
hydrolytic reactions (*see* Hydrolytic reactions)
MA (*see* Molecular assessment (MA))
mouse hybridoma cell culture, 221
phage display techniques, 222
protein scaffolds (*see* Protein scaffolds)
structure
constant and variable domains, 84
Fab-CDRs, 83–85
FcR-IgGFc interactions, 86–87
hinge region, 87
"Y" shaped tetrameric IgG molecule, 82–83
Motavizumab, 276
Moxetumomab pasudotox, 226–227
MTH technology. *See* Molecular Trojan Horse (MTH) technology
Mucopolysaccharidosis (MPS), 13
Mucopolysaccharidosis type II (MPSII)
HIRMAb-IDS fusion protein
brain homogenate volume of distribution (VD), 55, 57
differential distribution, 56, 58
intracellular IDS enzyme activity, 55, 56
mouse, 59
organ uptake, 55–56, 58
time-response study, 55, 56
Mucopolysaccharidosis type I (MPSI)
cTfRMAb-IDUA fusion protein, 53–54
HIRMAb-IDUA fusion protein
humans and monkeys, 49–50
immune responses, 51–53
pharmacokinetics, 50–51
Muromonab-CD3/Orthoclone (OKT3®), 221
Muscle targeting. *See* Pompe disease
Myozymer, 30

N

Nanoparticle tracking analysis (NTA), 94
Neglected Tropical Disease (NTD), 367
Network Biology platform, 213
Neulasta, 353

Neutropenia, 351–352
Neutrophil gelatinase-associated lipocalin (NGAL), 238
New Drug Application (NDA), 316, 317
NexP™ technology (Alteogen), 295
Non-Ig scaffolds, 254

O
Oncofetal Fn, 240
Orphan Drug Act (ODA), 365, 366
Ozoralizumab, 229

P
Paclitaxel poliglumex (Opaxio™), 289
Particle-flow imaging (PFI), 93
PASylation®, 291–293
Pegasys®, 278
Pegdinetanib, 234
Pegylated immunonanoparticles, 45
Peptide bond hydrolysis
  β-elimination mediated hydrolysis, 115, 116
  clipping, 115
  direct hydrolysis, 115, 116
  DKP formulation, 115, 116
  evidence, 118
  extrinsic factors, 117
  intrinsic factors, 117
Pharmaceutical benefit manager (PBM), 318
Plasma half-life extension
  chemical coupling, 288–289
  endosomal recycling, 270, 271
  Fc fusion strategies, 272–273
    Aprolixr, 274
    APSCOVERYT technology, 275
    Arcalyst®, 274
    CD4 immunoadhesin, 271
    cytotoxic T lymphocyte-associated antigen 4, 274
    EloctateT, 274
    Eylear, 274
    FcRn, 271
    hybrid Fc (hyFc) technology, 274–275
    hybrid proteins via protein A affinity, 274
    LAPS-GLP/GCG, 275
    MetMAb, 274
    MIMETIBODYT platform, 277
    non-Ig molecules, 271
    peptibody platform, 276–277
    RSV, 276
    TNF receptor 1, 274
    UnibodyT format, 274
  genetic fusion
    PASylation®, 291–293
    XTEN (Amunix), 289–291
  glycosylated IgGs, 269, 271
  HSA (*see* Human serum albumin (HSA))
  immunoglobulins (Igs), 269
  PEGylation
    Adagen®, 282, 286
    clinical trials, 283
    Fleximer, 287
    KRYSTEXXA™, 284
    Lonquex™, 285
    Mircera, 284
    N-or O-glycosylation sites, 285
    Oncaspar, 286
    Pegasys, 283, 284
    PHF or Fleximer®, 286
    PLEGRIDY, 283
    PSA, 287
    ReCODE™ technology, 285
    single unpaired thiol side chain, 284
    TheraPEG™ conjugation procedure, 285
    TransCon, 285
  protein glycosylation, 294–296
  renal filtration, 270, 271
Polyethylene glycol (PEG) molecules, 193, 196, 240, 242, 275, 282–286, 291, 298
Poly-l-amino acids, 289
Polymyositis, 70
polysialic acid (PSA), 287
Pompe disease
  adult/late-onset form, 23–24
  clinical manifestations, 23
  disease process, 24
  GAA
    ERT, 27
    glycosylated protein, 27
    MPR proteins, 25–26
    M6P tag, 25
    rhGAA efficacy (*see* Recombinant human GAA (rhGAA))
  pathogenesis, 24
p53 protein, 186
Preformed Conjugate-Drug Affinity Complex technology (PC-DAC™), 279
Priority review voucher (PRV) program, 364, 367
Pronectins™, 234
Protein oxidation
  generation, 138–141
  His oxidation, 143
  hydroxyl radicals, 142
  intramolecular Diels-Alder reaction, 142
  Met-252 and Met-428 oxidation, 141

mitigation strategy, 145
pharmaceutical proteins, 144–145
Trp oxidation, 141–143
Tyr oxidation, 141–142
Protein scaffolds
Adnectins, 233–235
Affibodies, 235–237
Anticalins (*see* Anticalins)
clinical development, 246–251
DARPins, 241–243
definition, 230–231
research on, 231
single domain Ig fragments and advanced alternative, 232
sources, 231
Pseudomonas exotoxin A, 226, 227

R
Ranibizumab, 227
Rare pediatric diseases (RPD), 367
Reactive oxygen species (ROS), 187, 188
Recombinant human GAA (rhGAA)
dosage, 27
glycoengineering
enzymatic modification, 28–29
increase CI-MPR protein level, 32–33
nanocarriers, 33–34
peptide-based tag (IGF-2), 31–32
synthetic glycans conjucation, 26, 29–30
yeast strains plus glycosidase treatment, 30–31
large-scale production, 27
Regulatory landscape
expedited programs
Accelerated approval, 365
breakthrough therapy designation, 364
Fast Track program, 363
PRV, 364
external collaborations, 367–368
incentive programs
BPCA, 366–367
NTD, 367
ODA, 365, 366
PRV program, 367
RPD, 367
Remicade™, 346
Respiratory syncytial virus (RSV), 276
rhGAA. *See* Recombinant human GAA (rhGAA)
Rituxan™, 345–346
Rituximab, 222

S
Second/third generation antibodies, 223
Single chain Fv fragment (scFv), 125–126, 225–228, 240, 243, 244, 246, 252, 253, 280
Single-photon emission computed tomography (SPECT), 227
Small for gestational age (SGA), 350–351
Small modular immunopharmaceuticals (SMIPs), 253
Small molecules
European market, 318–321
U.S. market, 316–318

T
TandAb®s, 245
Tanzeum, 278
TheraPEG™ conjugation procedure, 285
Therapeutic proteins
acylation, 193
autophagy, 189
degradation/aggregation, 183–184
dysfunctional protein, 184
human biological environments, 185
infection, 185–186
nitric oxide, 186
oxidation, 187–189
p53, 186
ROS, 187, 188
fusion, 195–196
instability, 184
in vivo aggregation
antibody aggregates, 99, 101
biosimilar protein drug formulation, 102
electron microscopy, 94–96, 99–100
fibrils, 100, 101
globular structures, 100, 101
large aggregates formation, 99–101
light microscopy, 93, 95–96
materials, 93
NTA, 94
originator product, 102
particle size distributions, 95, 97–99
PFI, 93
liposomal encapsulation, 194–195
oxidation-resistant forms, 191–192
PEGylation and glycosylation, 192–194
physicochemical and biological considerations, 197
plasma lipoproteins, 99, 101
proteolytic degradation, 192
tissue distribution, 189
Trastuzumab, 93, 108, 110, 126, 184, 213, 222, 236

Tregitopes
characteristics, 70
immunosuppressive effects, 70–71
IVIG, 70
mechanism of action, 71–72
in silico immunogenicity score, 70
Tregitope-mediated tolerance induction, 72
Tresiba®, 280

V
Veltis® technology, 278

X
XTEN (Amunix), 289–291
Xtend Bevacizumab, 276
XYOTAX™, 289